Fungal Biology

Series Editors
Vijai Kumar Gupta, National University of Ireland, Galway, Ireland
Maria G. Tuohy, National University of Ireland, Galway, Ireland

About the Series

Fungal biology has an integral role to play in the development of the biotechnology and biomedical sectors. It has become a subject of increasing importance as new fungi and their associated biomolecules are identified. The interaction between fungi and their environment is central to many natural processes that occur in the biosphere. The hosts and habitats of these eukaryotic microorganisms are very diverse; fungi are present in every ecosystem on earth. The fungal kingdom is equally diverse consisting of seven different known phyla. Yet detailed knowledge is limited to relatively few species. The relationship between fungi and humans has been characterized by the juxtaposed viewpoints of fungi as infectious agents of much dread and their exploitation as highly versatile systems for a range of economically important biotechnological applications. Understanding the biology of different fungi in diverse ecosystems as well as their interactions with living and nonliving is essential to underpin effective and innovative technological developments. This series will provide a detailed compendium of methods and information used to investigate different aspects of mycology, including fungal biology and biochemistry, genetics, phylogenetics, genomics, proteomics, molecular enzymology, and biotechnological applications in a manner that reflects the many recent developments of relevance to researchers and scientists investigating the Kingdom Fungi. Rapid screening techniques based on screening specific regions in the DNA of fungi have been used in species comparison and identification, and are now being extended across fungal phyla. The majorities of fungi are multicellular eukaryotic systems and, therefore, may be excellent model systems by which to answer fundamental biological questions. A greater understanding of the cell biology of these versatile eukaryotes will underpin efforts to engineer certain fungal species to provide novel cell factories for production of proteins for pharmaceutical applications. Renewed interest in all aspects of the biology and biotechnology of fungi may also enable the development of "one pot" microbial cell factories to meet consumer energy needs in the 21st century. To realize this potential and to truly understand the diversity and biology of these eukaryotes, continued development of scientific tools and techniques is essential. As a professional reference, the *Fungal Biology Series* will be very helpful to all people who work with fungi and useful both to academic institutions and research teams, as well as to teachers, and graduate and postgraduate students with its information on the continuous developments in fungal biology with the publication of each volume.

More information about this series at http://www.springer.com/series/11224

Diane Purchase
Editor

Fungal Applications in Sustainable Environmental Biotechnology

Springer

Editor
Diane Purchase
Faculty of Science and Technology,
Department of Natural Sciences
Middlesex University
London
UK

ISSN 2198-7777 ISSN 2198-7785 (electronic)
Fungal Biology
ISBN 978-3-319-42850-5 ISBN 978-3-319-42852-9 (eBook)
DOI 10.1007/978-3-319-42852-9

Library of Congress Control Number: 2016945785

Printed on acid-free paper

This Springer imprint is published by Springer Nature
The registered company is Springer International Publishing AG Switzerland

Preface

> In the 21st century, I think the heroes will be the people who will improve the quality of life, fight poverty and introduce more sustainability.
>
> —Bertrand Piccard, solar airplane 'Solar Impulse' pilot

With finite resources and the pressure of the growing global population, we need a plan to stimulate action in areas of critical importance for humanity and the planet. In September 2015, the United Nations adopted an aspirational set of 17 Sustainable Development Goals with 169 targets to tackle these challenges; the 'Transforming Our World: the 2030 Agenda for Sustainable Development' (also known as the Sustainable Development Goals, SDG) was launched. As fungal biotechnologists, our research has direct applications that contribute towards many of these goals.

Fungi are an extremely diverse, versatile and robust group of organisms with complex interactions within the ecosystem. For instance, whilst they can cause devastating and costly damage to their hosts as pathogens, their host specificity can be harnessed to provide a sustainable solution for biological pest control. This book aims to provide a timely examination of fungal biotechnology and its potential applications. The collection presented herein includes outstanding research reports and reviews drawing on international expertise to elucidate key cross-disciplinary advances in environmental fungal biotechnology.

Without doubt fungi offer numerous applications in sustainable environmental biotechnology, however, many of the processes still have not found industrial applications or received the attention they deserve. It is clear that despite the advances, more research is required to realise the potential of sustainable fungal environmental biotechnology. I sincerely hope this book contributes to the body of knowledge of sustainable biotechnological applications of fungi and serves as a useful reference for any 'hero' who works with this fascinating group of organisms to improve the welfare of our planet and mankind.

London, UK Diane Purchase

Acknowledgements

This book would not be possible without the contributions of our group of experts who donated time from their incredibly busy schedules to compose its chapters. Their collective knowledge affords us a glimpse into their research, the incredible workings of fungi and how they can be applied to provide sustainable environmental solutions. To all of our experts: I really appreciate the discussions we have had and the opportunities to exchange knowledge. It has been a pleasure working with you all.

I would like to thank Eric Stannard and Brian Halm at Springer-Nature for all their professional advice and assistance.

My heartfelt appreciation to my family, who are always there for me.

Final and special thanks to Thomas W.K. Young Ph.D., my mentor, who first opened the door to the *Fifth Kingdom* all those years ago.

Contents

Contributors

Gregory Bonito Department of Plant, Soil and Microbial Sciences, Michigan State University, East Lansing, USA

Tomáš Cajthaml Institute for Environmental Studies, Faculty of Science, Charles University in Prague, Prague 2, Czech Republic; Institute of Microbiology, Academy of Sciences of the Czech Republic V.V.I., Prague 4, Czech Republic

Ernestina Castro-Longoria Department of Microbiology, Center for Scientific Research and Higher Education of Ensenada (CICESE), Ensenada, Baja California, Mexico

Wai Kit Chan Department of Natural Sciences, Faculty of Science and Technology, Middlesex University, London, UK

Bryan Cook School of Sciences, Royal Melbourne Institute of Technology, University, Bundoora, VIC, Australia

Stefano Covino Institute for Environmental Studies, Faculty of Science, Charles University in Prague, Prague 2, Czech Republic; Institute of Microbiology, Academy of Sciences of the Czech Republic V.V.I., Prague 4, Czech Republic

Fernanda Marisa da Cunha Embrapa Instrumentation, São Carlos, SP, Brazil

Ewere Job Egede Department of Natural Sciences, Faculty of Science and Technology, Middlesex University, London, UK

Alexander O. Emoghene Department of Microbiology, Faculty of Life Sciences, University of Benin, Benin City, Nigeria

Cristiane Sanchez Farinas Embrapa Instrumentation, São Carlos, SP, Brazil; Graduate Program of Chemical Engineering, Federal University of São Carlos, São Carlos, SP, Brazil

Anthony E. Futughe Department of Natural Sciences, School of Science and Technology, Middlesex University, London, UK

Geoffrey Michael Gadd Geomicrobiology Group, School of Life Sciences, University of Dundee, Dundee, Scotland, UK

Anastasia P. Galanopoulou Microbial Biotechnology Group (MBG), Department of Biology, National and Kapodistrian University of Athens, Attica, Athens, Greece

Hemda Garelick Department of Natural Sciences, Faculty of Science and Technology, Middlesex University, London, UK

Andrii P. Gryganskyi L F Lambert Spawn Co., Coatesville, PA, USA

Dimitris G. Hatzinikolaou Microbial Biotechnology Group (MBG), Department of Biology, National and Kapodistrian University of Athens, Attica, Athens, Greece

Huw Jones Department of Natural Sciences, Faculty of Science and Technology, Middlesex University, London, UK

Souvik Kusari Department of Chemistry and Chemical Biology, Institute of Environmental Research (INFU), TU Dortmund, Dortmund, Germany

Tao Lu Key Laboratory of Biomass Chemical Engineering of Ministry of Education, Department of Chemical and Biological Engineering, Zhejiang University, Hangzhou, China

Aidyn Mouradov School of Sciences, Royal Melbourne Institute of Technology, University, Bundoora, VIC, Australia

Diane Purchase Department of Natural Sciences, Faculty of Science and Technology, Middlesex University, London, UK

Susana Rodríguez-Couto Ceit-IK4, Unit of Environmental Engineering, San Sebastian, Spain; IKERBASQUE, Basque Foundation for Science, Bilbao, Spain

Adya P. Singh Scion (Formerly, Forest Research Institute), Te Papa Tipu Innovation Park, Rotorua, New Zealand

Tripti Singh Scion (Formerly, Forest Research Institute), Te Papa Tipu Innovation Park, Rotorua, New Zealand

Federica Spina Department of Life Sciences and Systems Biology, Mycotheca Universitatis Taurinensis (MUT), University of Turin, Turin, Italy

Michael Spiteller Department of Chemistry and Chemical Biology, Institute of Environmental Research (INFU), TU Dortmund, Dortmund, Germany

Tatiana Stella Institute for Environmental Studies, Faculty of Science, Charles University in Prague, Prague 2, Czech Republic; Institute of Microbiology, Academy of Sciences of the Czech Republic V.V.I., Prague 4, Czech Republic

Giovanna Cristina Varese Department of Life Sciences and Systems Biology, Mycotheca Universitatis Taurinensis (MUT), University of Turin, Turin, Italy

Wen-Xuan Wang Chair of Environmental Chemistry and Analytical Chemistry, Department of Chemistry and Chemical Biology, Institute of Environmental Research (INFU), TU Dortmund, Dortmund, Germany

Dirk Wildeboer Department of Natural Sciences, Faculty of Science and Technology, Middlesex University, London, UK

Christopher Wright Department of Plant, Soil and Microbial Sciences, Michigan State University, East Lansing, USA; Easygrow Mushrooms and Composting LLC, Farmington Hills, MI, USA

Shan-Jing Yao Key Laboratory of Biomass Chemical Engineering of Ministry of Education, Department of Chemical and Biological Engineering, Zhejiang University, Hangzhou, China

Qi-Lei Zhang Key Laboratory of Biomass Chemical Engineering of Ministry of Education, Department of Chemical and Biological Engineering, Zhejiang University, Hangzhou, China

Shi-Hong Zhang College of Plant Sciences, Jilin University, Changchun, China

Introduction

> Nature has been experimenting with fungi for a billion years, perfecting a lot of powerful survival tools. We can use these tools in fantastic ways – to revive damaged ecosystems, to help offset global warming, and even to prevent diseases.
>
> —Paul Stamets
> Recipient of the first "Gordon and Tina Wasson Award"
> from Mycological Society of America.

This book intends to showcase the latest research and development findings in fungal biotechnology. It examines current advances and future prospects and discusses both limitations and potential of fungal environmental biotechnology. The book has been designed such that readers can take each individual chapter as a standalone article or as part of a series on a particular theme related to a UN Sustainable Development Goal (SDG): food security, pollution reduction (waste and wastewater), biofuel production and fungal biosynthesis of new products (see Fig. 1). There are a number of inter-related topics between the chapters, readers can cross-reference and pursue in-depth knowledge of a topic by referring to the information suggested in the footnotes.

One area that deserves emphasis is endophytic fungi, for instance, arbuscular mycorrhizae. These symbiotic organisms have received a lot of attention and are crucial to ensuring soil fertility and plant health; they also play an important role in conferring tolerance to pollutants. Although this book has not dedicated a specific chapter to this amazing group of mycobionts, information on their contributions and functions can be found in many of the chapters presented therein.

The theme of Part I of this book is associated with the SDG Goal 2: *Zero Hunger*—the chapters in this section examine the role of fungi in sustainable food production, including resilient agricultural practices and applications to strengthen drought tolerance. Chapter 1 explores the use of fungi as a food source as well as their involvement in the production of compost essential for plant growth and fertility. It also discusses composting as a means of soil contaminant remediation. Chapter 2 presents research on the identification of the genes that improve abiotic

Fig. 1 A chart of the United Nations Sustainable Development Goals adopted by the General Assembly and promoted by the UN. The five areas explored by this book are outlined (*Source* http://www.globalgoals.org/; access 27 April 2016)

stress resistance in extremophilic fungi and their potential in protecting valuable crops. Chapter 3 provides an overview and the experimental data on using fungi as a sustainable alternative to agrochemicals to control plant pathogens.

The narrative for Part II is SDG Goal 6: *Clean Water and Sanitation*—the articles explore ways to improve water quality and substantially increase recycling and safe reuse of water. Chapter 4 provides a comprehensive review of biosorption mechanisms and biodegradation of fungi and their application to treat wastewater and sewage sludge. The unique ability of white-rot fungi to remove xenobiotics in different industrial wastewater is detailed in Chap. 5. Chapter 6 addresses the bioremediation of emerging micropollutants in wastewater from municipal treatment plants by fungal whole-cell systems or enzymes. Chapter 7 describes a partnership of microalga and fungi that shows additive and synergistic effects on wastewater treatment, with the additional benefit of converting biomass into value-added chemicals and biofuels.

Continuing on the pollution reduction theme, Part III of this book addresses SDG Goal 15: *Life on Land*—three chapters focus on mycoremediation (using fungi to remediate and restore the ecosystems) of different groups of pollutants. Chapter 8 critically reviews the use of fungi and fungal enzymes in the remediation of persistent organopollutants, while the emphasis of Chap. 9 is on the decomposition process of lignocellulosic materials by white- and brown-rot fungi. Chapter 10 examines fungal remediation of heavy metals and metalloids in contaminated land, presents a case study of an extremophilc fungus and investigates the application of metal-resistant fungi to metal nanoparticles production.

Part IV focuses on SDG Goal 7: *Affordable and Clean Energy*—the chapters detail advances in bioenergy research where biomass is transformed to biofuel. Chapter 11 describes the process of biorefinery and the role of fungi in biofuel production. Chapter 12 relates the progress made on the bioprocesses of fungal cellulase in the production of bioenergy.

Part V alludes to SDG Goal 12: *Responsible Consumption and Production*—the articles present the advances of fungi as candidates to contribute towards the efficient use of natural resources. Chapter 13 introduces fungi as a sustainable alternative to biosynthesis nanoparticles and Chap. 14 explores the role of endophytes in the production of secondary metabolites and other natural products. Chapter 15 critically reviews the importance of geomycology and evaluates the role of fungi in environmental biotechnology, including metal recovery, production of biominerals or metallic elements with catalytic properties.

Diane Purchase

Chapter 1
Fungi in Composting

Christopher Wright, Andrii P. Gryganskyi and Gregory Bonito

1.1 Introduction

Composting is a biological process by which microorganisms decompose organic matter through physical and chemical activities (Said-Pullicino et al. 2007; Bonito et al. 2010). When carried out properly, composted organic matter becomes stabilized, free of diseases and weed seeds; when amended to soils or brewed as compost teas and applied to aboveground plant biomass, it can benefit plant fertility and suppress soilborne and foliar diseases (Scheuerell and Mahaffee 2002; Malandraki et al. 2008; Hadar and Papadopoulou 2012). The process of composting can improve local and regional sustainability by diverting organic materials from waste streams while generating useful soil amendments for agricultural systems (Neher et al. 2013).

Microbiota such as fungi, bacteria (including actinomycetes) and archaea are critical in driving chemical transformations of organic matter during the composting process (Ryckeboer et al. 2003). These organisms are also likely to influence the effects of compost and compost tea on plants (Scheuerell and Mahaffee 2002; Hadar and Papadopoulou 2012). One fundamental challenge of compost research for over a century has been to characterize the microbial succession and genetic diversity of individual taxa, and to determine their relative contribution in transforming and stabilizing organic matter (Waksman et al. 1939; Miller 1992; Ryckeboer et al. 2003; Neher et al. 2013). Such rudimentary knowledge is critical for both verifying

C. Wright · G. Bonito (✉)
Department of Plant, Soil and Microbial Sciences, Michigan State University,
East Lansing 48824, USA
e-mail: bonito@msu.edu

C. Wright
Easygrow Mushrooms and Composting LLC, Farmington Hills, MI 48336, USA

A.P. Gryganskyi
L F Lambert Spawn Co., Coatesville, PA 19320, USA

D. Purchase (ed.), *Fungal Applications in Sustainable Environmental Biotechnology*, Fungal Biology, DOI 10.1007/978-3-319-42852-9_1

the removal of plant pathogens through the process and for identifying the active microbiota in compost that are beneficial to soil and plant health, mushroom cultivation, and bioremediation (Waksman and Nissen 1932; Anastasi et al. 2008; Malandraki et al. 2008; Viti et al. 2010). Research is needed to establish a basis for defining guidelines for desirable microbiological properties of compost (Neher et al. 2013).

In this chapter we discuss the role of fungi in composting. We first describe the general composting process, and research into municipal composting and vermicomposting. We then discuss the history and science of composting for cultivating mushrooms. Finally, we discuss the role of fungal composts in remediating pollutants in the environment.

1.1.1 The Composting Process

The general composting process has been characterized as having four distinct phases (Bonito et al. 2010). During the initial mesophilic phase, organic materials start to heat up through biological activity and can reach temperatures as high as 65 °C. The second distinct phase is a prolonged thermophilic phase in which temperatures surpass 65 °C for an extended period of time. The thermophilic phase is critical to eliminating pathogens and seeds from composting materials. During the third (middle mesophilic) phase, temperatures decrease to 50 °C. During the final curing or maturation phase organic matter and biological heat stabilize, resulting in a dark humified substrate that eventually reaches ambient temperatures.

Microbial succession is strongly influenced by the method of composting and the duration of its peak temperature, which differs by composting method (López-González et al. 2015). The three main types of commercial composting methods are windrow, aerated static pile (ASP), and vermicomposting (Neher et al. 2013). Windrow and ASP composting have an extended thermophilic phase that excludes most eukaryotic life. Thus, biological activity during thermophilic phases is exclusively prokaryotic. In contrast, vermicomposting lacks a thermophilic phase and therefore has a higher abundance and influence of fungi (Huang et al. 2013). In order for organic matter to compost well it must be physically broken into smaller particles, which increases surface area and allows the material to be physically accessible to microbial growth. Ground materials are conveyed into long rows called windrows or ASP for composting (Waksman et al. 1939; Ryckeboer et al. 2003; Neher et al. 2013).

Although biological activity drives the composting process, physical variables such as temperature and moisture content have critical influences on microbial activity and dynamics. Maintaining proper and even moisture content is important to the composting process; if too dry then biological activity slows or stops, if too wet then conditions go anaerobic, and methane and ammonia are generated. A moisture content between 50–75 % is recommended for proper composting (Guo et al. 2012). The chemical environment, particularly the availability of oxygen and nitrogen, also has a significant influence on microbial respiration and activity during composting (Guo et al. 2012; Nada 2015), and hence on compost thermo-regulation, resulting in feedback effects on microbial succession during composting. The composition of the starting organic substrates also has a strong influence on microbial community composition during composting (Waksman et al. 1939; Ryckeboer et al. 2003; Neher et al. 2013). Point controls for the composting process are based on adjusting the moisture content, oxygen availability, and carbon to nitrogen ratio of the starting substrates, as discussed below in regards to mushroom compost.

1.2 Fungi in Commercial Composting

Fungi play an important role in the decomposition and stabilization of organic substrates because they are able to metabolize cellulose, hemi-cellulose, lignin, and other complex carbon substrates (Floudas et al. 2012; De Gannes et al. 2013). Early studies to identify fungi in compost relied upon culture-dependent methods such as dilution plating and isolations using selective media (Waksman et al. 1939). This approach is valuable in obtaining living fungi that are directly involved in the composting process, but this approach is biased against slower growing fungi or those that are not readily culturable, and biased toward those that grow easily or quickly (Bonito et al. 2010). From pioneering culture-based studies of fungi in compost, and those that followed, it became clear that viable fungal propagules are present in all the major compost phases, but are most active and diverse during mesophilic phases. Some of the fungi most commonly isolated from compost include *Acremonium, Alternaria, Aspergillus, Chaetomium, Cladosporium, Emericella, Fusarium, Geotrichum, Mortierella, Mucor, Penicillium, Pseudallescheria, Scopulariopsis*, and *Trichoderma*; fungi that are able to grow at elevated thermophilic temperatures (>65 °C) and are isolated from thermophilic compost phases include *Absidia, Aspergillus, Chaetomium, Coprinus, Mucor, Paecilomyces, Penicillium, Rhizomucor, Scytalidium*, and *Thermomyces* (Waksman et al. 1939; Ghazifard et al. 2001; Vijay et al. 2002; Ryckeboer et al. 2003; Anastasi et al. 2005). Numerous review articles on the biology of composting have listed bacterial and fungal taxa that have been isolated from various composting organic materials and phases (De Bertoldi et al. 1983; Ryckeboer et al. 2003).

With the advent of PCR and molecular approaches based on DNA analyses, it became possible to assess fungal diversity during composting more completely (Marshall et al. 2003). Early molecular approaches were based on fingerprinting assessments that allowed for rough estimates of species diversity and communities (Cahyani et al. 2004; Halet et al. 2006). However, sequence-based approaches (particularly high-throughput amplicon sequencing) are now standard in the field as they allow fungi to be identified and phylogenetically characterized and offer improved resolution for determining microbial dynamics during composting (Bonito et al. 2010; Neher et al. 2013). From the few sequence-based studies on fungi in compost that have been published, there are some clear patterns that have emerged. It is clear that although the composting process is characterized by a succession of species, fungi belonging to Ascomycota are particularly abundant in all composting stages (Bonito et al. 2010; De Gannes et al. 2013; Neher et al. 2013). Similarly, it has been shown that both the composting process and starting substrates are important factors in defining the structure and succession of fungal communities during composting (De Gannes et al. 2013; Neher et al. 2013).

1.2.1 Factors Driving Fungal Diversity and Succession During Composting

Fungal diversity is high and dynamic during composting. Lower temperatures and water moisture during composting tends to favor fungi over bacteria (Mehta et al. 2014). The most comprehensive sequence-based studies find that fungal diversity in compost is in the order of 10s–100s of taxa, with a continual shift (succession) in fungal taxa over time (Bonito et al. 2010; De Gannes et al. 2013; Neher et al. 2013). While it may be possible to tabulate lists of fungi detected in compost, starting substrates, compost recipes, and composting methods that drive fungal community structure and succession during composting, the differences between studies and facilities indicate some functional redundancy. Further, a large number of the fungi detected in compost have yet to be properly classified at a species level. Still, many fungi in compost are detectable and classifiable. These include fungal pathogens of humans (e.g., *Candida tropicalis* and other spp., *Trichosporon asahii* and other spp., *Bipolaris spicifera*, *Scedosporium* spp., *Chaetomium funicola, Aspergillus fumigatus*) and plants (e.g., *Fusarium solani* complex and other spp., *Alternaria* spp., *Botryosphaeria dothidea, Nectria* spp.), which can be present and sometimes abundant at early phases during composting (Bonito et al. 2010; De Gannes et al. 2013). Accordingly, care should be taken by compost operators during preparation and composting to limit exposure and spread of airborne fungal spores and colonies.

The succession of fungi varies with the composition of substrates being composted and is driven by the availability of particular carbon and nutrient resources, and some general trends in fungal succession have been observed during composting. These include the turnover of species during thermophilic phases, with a dominance of Dothideomycetes, Eurotiales, and Sordariomycetes (De Gannes et al. 2013; Neher et al. 2013). Following the thermophilic phase fungal diversity increases and complex microbial-based food webs develop (Bonito et al. 2010; De Gannes et al. 2013; Neher et al. 2013). The inclusion of manure in compost also affects fungal communities that develop. For example, *Cladosporium, Chaetomium, Ascobolus*, and *Penicillium* are detectable in manure-based composts, while *Mucor* and *Mortierella* are detected in composts containing hay, and *Coprinus* and *Stropharia* in hardwood composts (Ryckeboer et al. 2003; Neher et al. 2013). Consequently, fungal community structure during composting may be manipulated through the selection and mix of starting substrates in order to tailor composts for particular applications.

1.3 Vermicomposting

Vermicomposting is the process of bio-oxidative composting and stabilization of organic matter utilizing earthworms in conjunction with fungi and other microorganisms living within the earthworms and substrate (Edwards and Bohlen 1996; Pižl and Nováková 2003). Although earthworms drive the process of vermicomposting by conditioning the substrate, altering biological activity of fungi and other microbiota, modifying the substrate through their digestive enzymes and increasing the surface area for microbiological activity, it is the microbes that carry out the biochemical degradation of organic matter (Aira et al. 2006). The most frequently employed earthworms in vermicomposting are *Eisenia fetida* and *Eisenia andrei*, commonly called "red wigglers," two similar epigeic worms that live in compost or litter rather than in mineral soil, confining their activities to the top 10–15 cm of their substrate. In contrast to traditional composting (also referred to as "thermophilic composting"), vermicomposting is an odorless, nonthermal (mesophilic), aerobic process.

In nature, earthworms (*Lumbricus terrestris*) ingest soil and organic matter such as animal droppings and plant litter (phytomass) and evacuate vermicast, more commonly known as worm castings, and slough off mucus which is utilized by microbiota (Brown and Doube 2004; Abbasi et al. 2015). Worldwide, earthworms and their associated microbes mineralize billions of tons of phytomass each year, helping to move and turnover soil, and transport and disperse microorganisms including fungi during the process (Brown and Doube 2004). The importance of earthworms in the decomposition of organic matter was first brought to the public's attention by Darwin (1892). However, it was not until the early twentieth century that commercial vermicomposting of organic agricultural wastes for large-scale use by farmers and growers was introduced (Barrett 1948; Oliver 1949). The United States domestic vermicomposting industry quickly grew, and commercial sales of both earthworms and composts expanded. Unfortunately, claims made by entrepreneurs in the industry were often exaggerated, and scientific scrutiny of these claims discredited early vermicomposting technology (Bouché 1987). However, recent investigations into the mechanics of vermicomposting have demonstrated the usefulness of employing vermicomposting in agricultural, municipal, and industrial settings (Edwards et al. 2010). As a result, vermicomposting has been gaining popularity and is increasingly being used in countries across the world including Italy (Beffa et al. 1998), the Philippines (Edwards et al. 2010), the Czech Republic (Pižl and Nováková 2003), India (Pramanik 2010), and Hong Kong (Edwards et al. 2010).

Vermicompost consists primarily of vermicast, a nutrient-rich organic substance suitable for application as a plant fertilizer, along with other organic matter in various stages of decomposition. During vermicomposting, organic matter is stabilized and pathogenic bacteria are reduced due to synergistic interactions among earthworms, fungi and other microorganisms (Edwards and Fletcher 1988). Vermicompost is rich in nutrients, beneficial soil microbes, and plant growth-promoting substances; the quality and fertility of vermicomposted organic matter are significantly increased over the initial substrate (Edwards and Bohlen 1996). Vermicomposting has been shown to have several advantages over thermal composting in terms of processing time, microbial richness, decreased phytotoxicity, particle size reduction, and an increase in available nutrients including NO_3^-, PO_4^{3-}, SO_4^{2-}, and K^+ (Singh and Suthar 2012). The fact that traditional composting is thermophilic while vermicomposting is mesophilic leads to differences in fungal community composition between thermal composting and vermicomposting (Beffa et al. 1998).

1.3.1 Fungi in Vermicomposting; Interactions Between Earthworms and Fungi

Fungi function in vermicomposting both as a source of enzymes to help facilitate decomposition as well as a preferred food source for the earthworms (Edwards and Fletcher 1988; Schönholzer et al. 1999; Pižl and Nováková 2003). Correspondingly, earthworms affect fungal populations through grazing and dispersal. Fungi have been found to be a favored food source for earthworms, which demonstrate preferential feeding of specific fungal species (Cooke 1983). Earthworms have been found to impact fungal community composition in vermicompost by influencing spore germination and creating microsites that may or may not favor the development of fungi (Brown 1995; Tiunov and Scheu 2000), although there is some contradictory evidence of this (Beffa et al. 1998). It has been demonstrated that *E. andrei* can survive on fungi as its sole food source (Bouwman 1998). Although earthworms consume fungi along with organic matter, fungal biomass and diversity in the finished vermicompost is generally equal to, or higher than, the fresh substrates (Pižl and Nováková 2003; Pramanik 2010). Fungal biomass in vermicompost is also higher when earthworms are present, and lower when earthworms leave the substrate (Aira et al. 2007).

The species of fungi active in vermicomposting can be variable by site, substrate utilized, and batch (Anastasi et al. 2004; Neher et al. 2013). One study using a culture-independent next generation sequencing approach found members of the *Arthrobotrys* (nemotode-trapper fungi), Pezizales, Microascaceae, *Zopfiella*, Agaricomycetes, and *Mortierella* to be highest in abundance in vermicompost (Neher et al. 2013). This is in contrast to the dominant fungi they found in windrow composting, which included unclassified Sordariomycetes, *Acremonium*, and an unclassified Basidiomycota group, and those found in aerated static pile composting which included an unclassified family of Pezizales, various unclassified Sordariomycetes and unclassified Sordariales.

Another study using traditional culture-based methods found *Aspergillus fumigatus, A. flavus, Geotrichum candidumi, Penicillium expansum, P. roquefortii, Fusarium ventricosum*, and *Rhizopus stolonifer* to be the dominant fungal taxa in vermicompost, with *A. fumigatus, G. candidum, Mucor circinelloides, Scopulariopsis brevicaulis, and P. expansum* dominating the intestines of the earthworms (Pižl and Nováková 2003). One study that incorporated animal waste as an initial substrate noted a dominance of keratinolytic fungi such as *Chrysosporium* spp. and *Scopulariopsis* spp. in the vermicompost (Anastasi et al. 2005). Several studies have noted a predominance of *Aspergillus* spp., taxa that are commonly found as saprotrophs of food waste (Pižl and Nováková 2003; Anastasi et al. 2004; Neher et al. 2013). Due to the great variability of fungi detected in different batches of vermicompost, close monitoring of the fungal community composition during incubation is necessary in order to facilitate the production of high-quality vermicompost (Peters et al. 2000). Functional characterization of the fungal community in vermicompost can also be employed to signify potential functional applications of different species of fungi in vermicomposting (Anastasi et al. 2005).

Vermicompost is generally found to harbor a greater diversity of fungi than the starting substrate. Vermicomposting also tends to result in a greater diversity of fungi than either windrow or aerated static pile composting, possibly due to the fact that the digestive tract of earthworms houses taxonomically distinct communities of fungi that add to the overall fungal diversity of vermicompost (Neher et al. 2013). However, many of the fungi found in vermiculture substrates and earthworm intestines are typical of those normally involved in soil organic matter decomposition.

Earthworm-processed substrates tend to have increased mycelial biomass in comparison to starting substrates. One study that employed vermicomposter bins comprised of vertically stacked layers, each layer representing a different phase of interaction between microbes and earthworms, found that those layers that housed the highest density of earthworms also had the highest fungal biomass (Aira et al. 2007). Fluctuations of water content and temperature have also been noted to affect the biomass of fungi in vermiculture systems (Pižl and Nováková 2003).

Earthworms have been found to activate fungal growth, triggering increased cellulose decomposition during vermicomposting. In a study by Aira et al. (2007) where vertically stacked vermireactors were utilized to vermicompost a pig manure slurry, β-glucosidase activity in vermireactors containing earthworms was increased 1.5 times that of control vermireactors that did not contain earthworms. β-glucosidase activity in the earthworm vermireactors was also increased 3–4 times over that of the initial starting slurry. Cellulase activity in most vertical layers of the vermireactors containing earthworms was also shown to be almost twice that of the control vermireactors without earthworms, affecting a rate of cellulose loss in vermireactors containing earthworms to almost twice that of the control vermireactors. Although cellulytic activity has been reported in species of earthworms, including *E. fetida* (Zhang et al. 2000), this activity is generally attributed to their fungi and other gut microbiota, which live in a mutualistic symbiosis with the earthworms (Lavelle and Spain 2001; Doube and Brown 1998).

Changes in fungal community composition during vermicomposting have also been noted. Huang et al. (2013) found that although Saccharomycetes, Lecanoromycetes, and Tremellomycetes dominated the starting materials of a vermicompost of vegetable wastes from Japanese markets, the finished vermicompost was dominated by Sordariomycetes, followed by Agaricomycetes, Pezizomycetes, Eurotiomycetes, Saccharomycetes, and Orbiliomyetes. They found that in contrast to thermal composting of the same material, which resulted in a narrowing of the fungal taxa found in the final compost, vermicomposting resulted in a more diverse fungal community in the final vermicompost. They also noted that the presence of several species of Sordariomycetes in the vermicompost may be an indicator of when vermicompost is mature, confirming results of Bonito et al. (2010) who found most sequences in mature compost belonged to the Sordariomycetes and Pezizomycetes.

Because fungi and other organisms play an important role in earthworm ecology, it has been suggested that manipulating the microbial community in vermiculture processing may be one way to increase the productivity of vermicomposting (Pižl and Nováková 2003). In a study by Pramanik (2010), inoculating the starting substrates of coconut coir and sugar bagasse with *Aspergillus niger* (a cellulytic fungus) and *Trichoderma viridae* (a ligninolytic fungus) significantly increased the nitrogen, phosphorous, and potassium content of the vermicompost as compared to an uninoculated control. Inoculating fresh substrates with fungi also decreased the time needed for both completing the vermicomposting process and stabilizing the total organic carbon content of the vermicompost. Pižl and Nováková (2003) found that pre-inoculating cattle manure substrate with fungi, including *A. flavus*, also resulted in an increase in the growth rate of *E. andrei.*

1.3.2 Vermicompost Employed as a Suppressive Toward Phytopathogens

Finished vermicompost has been shown to be suppressive toward plant pathogens such as *Fusarium oxysporum* (Szczech 1999), *Phytophthora nicotianae* var. *nicotianae* (Szczech et al. 1993), *Plasmodiophora brassicae* (Szczech 1999), and *Phytophthora infestans* (Kostecka et al. 1996). A vermicompost water extract was also shown to be effective in controlling *Blumeria graminis* f. sp. *horde*, a powdery mildew of barley. One study characterized the mechanism of phytopathogen suppression by comparing the effects of heat-sterilized versus unsterilized vermicompost on the suppression of *Fusarium oxysporum* f. sp *lycopersici* (Szczech 1999). They found that vermicompost lost its suppressive activity after heating, and that sterilized water extracts stimulated rather than hindered the growth of *F. oxysporum* in vitro, concluding that it was the biological and not chemical component of vermicompost that explained its suppressive action.

1.4 History and Science of Mushroom Composting

1.4.1 Composting for Mushroom Cultivation

The science of composting began with the cultivation of the button mushroom (*Agaricus bisporus*), the most cultivated mushroom species worldwide, and remains essential in button mushroom agriculture today (Lambert 1929; Waksman and Nissen 1932; Jurak et al. 2015). *Agaricus* species are unable to colonize raw plant substrates, but composting organic materials transforms them into a suitable substrate for mushroom cultivation.

The first records of mushroom growing appeared during the reign of Louis XIV (1638–1715) in France (Rettew and Thompson 1948). Gardeners, supplying their production to the royal court, found agarics growing on the beds of horse/donkey manure compost that had be prepared for growing melons and other fruits and vegetables. Selman Waksman was one of the first microbiologists to try to ascertain and control the composition of microbiota involved in the process in order to improve the button mushroom industry (Waksman 1932). The paddy straw mushroom (*Volvariella volvacea*) is another mushroom that is cultivated on composted rice straw and has been done so for centuries (Chang 2006), yet the exact records of time when this mushroom was first cultivated is unknown. Composting of raw plant materials to different degrees of substrate decomposition is also used for other mushroom species including *Morchella* spp., *Lepista* spp., and *Psilocybe* spp. (Guinberteau et al. 1991; Nicholas and Ogame 2006; Mortimer et al. 2012). Development of composting techniques is linked to technical progress of human civilizations and is related to the improvements in mechanization, electrification, and automation of agricultural production processes. The most advanced techniques today are used for cultivation of agaric mushrooms as described below.

Composts of various plant residues are used widely in the mushroom growing industry (Chang and Miles 1989; Rajapakse et al. 2010). Most mushroom growing technologies are based on previous processing of plant substrates with the two major goals: eliminating/inhibiting of harmful microbes and chemical degradation of the substrates to make them usable for following digestion by cultivated mushroom. The level of degradation depends on the lignocellulolytic capacity of enzymes of cultivated mushroom, which include wide range of laccases, peroxidases, xylanases, and ligninases (Lo et al. 2001). Many cultivated mushrooms (such as *Agaricus* spp.) are unable to degrade cellulose and lignin in plant cell walls (Morin et al. 2012). Usually substrate processes use high temperature alone or in combination with chemical changes of substrate composition caused by microorganisms, a process known as substrate fermentation or composting. The result of the composting process is a biologically active substrate free of pathogens and pests that provides sufficient nutrition, water, and air for mushroom growth.

1.4.2 Substrates for Composting

When producing compost for mushroom production it is important to understand that agarics do not grow on fresh manure, neither in nature nor in domestication. Manure must first be composted to produce mushrooms, and fresher manures require longer composting cycles. When selecting manure for mushroom composting there is a suite of characteristics to consider that may affect the resulting microbial community. First, it is best to source manure from healthy animals. Second, the presence of antibiotics in horse or chicken manure can persist during the composting and may affect the composting process and subsequent mushroom development (Waksman and Reneger 1934; Andaluri et al. 2012). Similarly, the presence of growth stimulants, hormones, and other chemicals (such as those used in racehorsing) could alter the microbiology of compost in deleterious ways.

All composts begin with parent materials or substrates. Substrates vary in their nutrient stoichiometry and produce different microbial communities for different applications. Perhaps the most productive substrate for growing agarics is horse manure mixed with straw. Its unique characteristics (N content, C:N ratio, porous texture beneficial in aeration, etc.) and availability in most countries make horse manure the preferable compost ingredient for most mushroom growing enterprises. It is possible to substitute horse manure with other animal manure (cows, pigs, sheep etc.) but the resulting compost is not as productive for mushroom growing. In many cases chicken manure (grown for meat, not for egg production) can serve as a suitable substitute for horse manure, but the high N content can be problematic (Noble and Gaze 1994).

Traditionally mushroom growers call horse manure compost natural and chicken manure compost synthetic (Gibbons et al. 1991). There are also many recipes combining both horse and chicken manure, which are called semisynthetic (Gbolagade et al. 2006). Some mushroom growers work only with organic sources of nitrogen, yet their composts are regarded as synthetic. No shortage of recipes for making composts exist and nearly each mushroom compost facility uses its own recipe depending on substrate availability, price, transportation costs, and other factors. Compost recipes (and especially composting techniques) change slightly between seasons given the nature of the substrates being used. While an appropriate ratio of manure to straw is important to achieve a desirable C:N ratio, the composting process itself is more important for successful commercial production than slight variations in the specific substrates being used (Noble et al. 2002). Proper compost can be characterized by many features, but only a handful are easily measurable. These include pH, N content, C/N ratio, electrical conductivity, temperature, and percent moisture (Day et al. 2001). Other qualitative or semiquantitative features can be taken into account based on the personal experience of the compost producer. These can be as important in controlling the composting process as are other physical and chemical properties.

Straw is a carbohydrate with a high C:N and is essential to the composting process (Hein 1930; Straatsma et al. 1995). Although horse manure contains many undigested plant stems, its texture is too fine for optimal compost structure, thus the addition and uniform mixing of straw is essential for successful composting. Given that straw has a C:N ratio of around 70, nitrogen has to be amended to straw via manure or other nitrogen sources in order to obtain an optimal C:N ratio (15–20) for proper composting (Noble et al. 2002). One of the goals of composting straw is to make the polysaccharides available for enzyme digestion by agarics. This is achieved after dissolving the outer wax layer of the straw surface, losing the tubular structure of the straw and breaking cellulose polymers into smaller pieces (Chen et al. 2000). Cutting or shredding the straw improves its digestibility and water uptake. Wheat or rye straw is most often used, but these can be partially or completely substituted by other cereal straws such as barley, oat and hay, or non-straw ligninocellulytic substances such as corncobs or cottonseed hulls (Noble et al. 2002). Aside from these main ingredients, several other components have traditionally been used in mushroom composting: feather, blood, and fishmeal as well as urine have been used to increase the N and protein content of the substrate; molasses, and sugar have been used to supply microbes with additional energy sources to accelerate their activity, stabilize their pH, and improve aeration (Noble et al. 2002). Chalk, gypsum and/or limestone have also been added to increase the buffering capacity and physical properties of compost, and to prevent greasiness and the inter-adhesion of compost particles, which reduces biological activity (Miller et al. 1990).

A succession of complex biochemical reactions occur during composting, which is essentially an aqueous matrix. Maintaining proper moisture content is therefore an essential component of the composting process. When water is limited during composting, biological activity is reduced; however, too much water limits aerobic activity and is therefore detrimental to the process (Richard et al. 2002; Said-Pullicino et al. 2007; Guo et al. 2012). During the first phase of composting water additions are particularly important. All components of the compost should be thoroughly mixed after each addition of water. The physical tubular structure of straw and wax layer on the cell surface are the main obstacles to water absorption. Composts can be moistened by spraying, preferably using warm or hot water rather than cold water, which can slow biological activity. Pressing the compost piles with tractors or other heavy equipment increases the ability to add water to compost. However, watering is still inefficient and creates large volumes of nutrient-rich aqueous waste. Modern composting facilities have collecting systems for the salvage water and its reuse in the composting cycle.

Finished composts have several characteristic features and their correct evaluation is based not only on its physical and chemical properties, but also on the personal experience of the manufacturer. Smells such as ammonia, which appears during fermentation and pasteurization, should disappear. Free ammonia is detrimental to mushroom mycelium, therefore preventing ammonia production during composting is important for mushroom crops (Wiegant et al. 1992). In preparing compost for mushroom production, one of the main criterion is to target a total nitrogen content of 1.5–2 % (mostly bound in proteins, not in ammonium salts or nitrites), as this correlates to the percentage found in harvested mushrooms (Royse and Chalupa 2009).

1.4.2.1 Mushroom Composting: Phase 1—Fermentation

Historically, compost for mushroom growing was prepared through a single fermentation phase known as Phase 1. The purpose of this phase was to properly mix all compost ingredients and raise them to a high temperature (>65 °C) through the metabolic activity of microbes inhabiting the manure and straw. Today, compost piles are usually placed outdoors on concrete pads, with proper aeration and access for mixing the compost. Perforated floors improve aeration in the pile and help to decrease the unpleasant odors caused by anoxia. Traditional phase 1 compost piles have approximate dimensions of 1.6 by 1.8 m. Their length depends upon the amount of compost being prepared. In winter months, compost piles are built up to 50 % higher and wider. Temperatures rise unevenly in compost windrows therefore regular turning or mixing is needed. A cross section of a fermenting compost windrow can be divided into three major parts: the outer non-fermented part ($\sim$20 cm deep), the internal anaerobic zone where anaerobic fermentation processes occur, and the main body where microbial fermentation occurs at temperatures close to 80 °C. To prevent the development of these zones compost is mixed at regular intervals of time (usually 2–3 days), watering at every turning. Traditionally, this process was done with hand forks and is still used today in countries with cheap labor. Post World War II manufacturers produced machines designed for high turning capacity that were adapted for compost mixing. The main purpose of turning the compost is to allow all parts of compost pile to go through the same fermentation process, which enables the even decomposition of the straw as well as making the nitrogen from manure more digestible for the mycelium. Today, due to the high level of mechanization, this phase continues for 11–14 days compared to 1–1.5 months at the beginning of twentieth century.

1.4.2.2 Mushroom Composting: Phase 2—Pasteurization

Wide adaptation of this phase over the past few decades has created a virtual revolution in mushroom compost making. The second phase essentially repeats the first one but under controlled conditions (Agrawal 2014). There are two main goals of this phase: (i) the elimination of all unwanted organisms (including molds, insect larvae and nematodes) and, (ii) the conditioning of the compost including ammonia removal to a level below 0.07–0.10 % (the level detectable by the human nose) which inhibits mycelial growth. This phase is done in specially built insulated chambers that are equipped with steam generators and ventilators for air exchange.

Capacities of several tons of starting material are typical of middle and large compost producers, which have several of such chambers, allowing them to build an efficient production schedule. In some farms these rooms serve a dual purpose as growing rooms. Because of the large volume of substrate to be composted, filling and emptying operations are mechanical. Compost made in pasteurization chambers (tunnels) can be placed in beds, trays or in bulk.

The pasteurization or thermophilic phase is usually 6–7 days in duration. First the temperature in the compost is held near 58 °C for pasteurization, and then at 48 °C for conditioning, with a final drop to 25 °C, which is the optimal for mycelial growth. During this time the temperature of the substrate is elevated due to the activity of thermophilic microorganisms, which are mostly bacteria. The main advantage of this phase is the elimination of pests, seeds, and disease organisms, which otherwise might survive and grow following the first phase of composting. Through pasteurization, the quality of the compost dramatically increased due to a more uniform physical environment. The quality and physical characteristics of compost can be slightly different on the top (too dry) and in the bottom (too wet) depending on how this step is carried out. Optimal moisture for mycelial growth is 64–67 %. This can be checked by a "fist test" where one takes a handful of compost with their bare hands and squeezes; well prepared compost will lack free water but will produce a characteristic hissing sound when squeezed and will not stain the palm of the hand. Once compost has reached maturity it is ready to be spawned with mushroom mycelium. Compost maturity is evaluated by several criteria: density, nutrient availability, as well as the microbial community and their metabolic products favorable for agarics growth. The combination of phases 1 and 2 for mushroom growing is called the conventional composting process (Miller 1992).

1.4.2.3 Mushroom Composting: Phase 3—Spawning, Colonization and Conditioning

This phase is an extension of the final portion of the phase 2 pasteurization stage: conditioning at 25 °C. After the temperature lowers in the compost it can then be spawned and colonized by the *Agaricus* mycelium. This phase may persist for weeks to months, but it usually takes two weeks for the spawn to colonize the compost. This phase has become more popular among mushroom growers because of several advantages it offers compared composts that have only undergone phase 1 and 2 composting: a higher yield per square foot, a shorter production cycle, and a lesser danger of contamination. However, processes occurring in phase 3 are associated with mycelial colonization of the compost rather than composting itself. Some mushroom growers employ a phase 4 composting stage wherein the compost is covered with casing soil that mushroom mycelium colonizes prior to delivery to a growing facility.

1.4.3 Indoor Composting

The compost preparation methods described above represent the basic compost processes used by the mushroom growing industry. However, adaptations are often made to this process. For example, indoor composting, introduced by Lambert in 1941, has been growing globally over last two decades. Major developments in indoor composting were achieved in the 1980s when it became widely adapted in the Netherlands, Italy, France, and other European countries. Complaints about unpleasant odors emanating from nearby composting facilities stimulated the adoption of indoor composting. Indoor composting has a very short phase 1 which is often accomplished outdoors. The substrates and amendments are well mixed and loaded into specially designed chambers where the mixture is processed. Heat is generated by microbes inhabiting the manure and straw and the developing compost is watered and aerated. One advantage of indoor composting is that watering is done only once at the beginning of the process. This is due to the high efficiency of blenders used to mix the starting materials, and results in a much shorter processing time. Other advantages of indoor composting are (i) the absence of foul odors, which is important in meeting regional and national air quality and pollution standards, and (ii) improved control and stability of the process compared to outdoor composting systems. Indoor composts also consistently give 5–10 % higher yields of fresh mushrooms than outdoor-processed composts. This also results in savings realized in the reduction of surface area for storage and compost processing areas, picking time, energy, and raw material use efficiency, along with improved control over all parameters of composting and sanitary conditions. However, there are some disadvantages to indoor composting including higher costs of running and maintaining equipment and facilities, and closer monitoring of compost temperature and the activity of beneficial microbes. Currently, indoor compost is mostly practiced in Europe but is gaining in popularity for American mushroom growers.

1.4.4 Biology of Mushroom Composting

The biological processes of composting are understudied and we still only have a rough estimate of the microbial diversity in composting. The most well-characterized fungi are *Mycothermus thermophilus* (previously also described as *Scytalidium*, *Humicola* and *Torula*) (Natvig et al. 2015). Other non-beneficial fungi that can tolerate the higher temperatures of composting and whose presence is suggested as the feature of poor quality compost belong to the genera *Aspergillus*, *Chaetomium*, *Rhizomucor*, and *Talaromyces*. Considerable efforts in characterizing fungi during mushroom composting were carried out by the Mushroom Experimental Station, Horst, (The Netherlands) and Penn State University, State College, Pennsylvania (U.S.A.) (Ivors et al. 2000).

1.4.5 Spent Mushroom Compost

Spent mushroom compost is not a waste material. In fact, it is well-decomposed plant biomass, enriched with proteins and microelements. It can serve as valuable organic fertilizer for agriculture. However, it can be overly saline, especially if a mushroom farm had a severe infection caused by fungal pathogens such as by *Mycogone*, *Verticillium*, *Trichoderma*, etc. This is because salt is regularly used in organic mushroom growing to control disease without the use of selective fungicides. Spent mushroom compost can also be used to remediate environmental pollutants, as discussed below.

1.5 Role of Thermophilic Microorganisms During Composting

Although this chapter is focused on fungi in compost we would be remiss if we did not mention the role of bacteria during composting, particularly those that are able to tolerate thermophilic conditions. Bacteria are responsible for much of the complex organic matter transformations that occur during the pasteurization phase of composting. A succession of microbial populations turnover during the composting process. These successions are associated with the heating and cooling of different phases of the composting process (Alfreider et al. 2002; Neher et al. 2013). Bacteroidetes appear to be particularly abundant following the thermophilic phase of windrow composts (Alfreider et al. 2002; Neher et al. 2013), while Firmicutes and gamma-Proteobacteria are at a higher relative abundance following the thermophilic phase of ASP (Alfreider et al. 2002; Neher et al. 2013) and precomposting phase for vermicomposts (Alfreider et al. 2002; Neher et al. 2013). Microbiota active in conditioning compost play a crucial role in compost applications (such as plant disease suppression) and are essential to producing quality composts suitable for agricultural and commercial applications.

1.6 Mycoremediation: Fungi, Compost, and Remediation of Pollutants and Polycyclic Aromatic Hydrocarbons[1]

With the advent of the industrial age, a demand for increasing amounts of fuel, pharmaceuticals, pesticides, and other industrial chemicals resulted in these substances being inadvertently released into the air, water, and soil (Alexander 1995).

[1]Additional information on fungal treatment of organic pollutants can be found in Chap. 8: *Mycoremediation of organic pollutants: principles, opportunities and pitfalls.*

Initial experiments to remediate polluted soils included physical and chemical processes such as excavation, solvent extraction, incineration, and soil washing. However, these processes are often found to be unsuitable as they are either prohibitively expensive or ineffective because contaminants are simply relocated rather than remediated. The use of microbes in remediating soils and water has been found to be an attractive alternative to these processes as microorganisms are uniquely adapted to physically and chemically interact with soil, compost and pollutants and can inexpensively mediate the transformation or mineralization of a wide range of environmental contaminants (Reddy 1995; Head 1998).

Many fungi are effective in remediating organo-xenobiotics because of the oxidative enzymes they genetically encode. These enzymes facilitate the breakdown of recalcitrant organic substrates such as lignocellulose during composting and are also effective at mineralizing (to CO_2) a wide variety of xenobiotic substances and environmental pollutants (Reddy 1995). This feature is attributable to the broad specificity of fungal ligninolytic enzymes, which are employed by fungi in nature to deconstruct the lignin component of lignocellulose (Sinsabaugh 2005). Lignin is a stereochemically complex, aromatic and heterogeneous biopolymer that comprises 20–30 % of the dry mass of woody plants (Boominathan and Reddy 1992). The heterogeneous chemical structure of lignin requires the action of nonspecific oxidative enzymes in order to completely deconstruct the complex polymer.

The fact that many environmental contaminants share structural similarity to lignin makes them ideal substrates for fungal ligninolytic enzymes. To that end, scientists have been investigating strategies of remediation with fungi, known as mycoremediation, ever since 1985 when the fungus *Phanerochaete chrysosporium* was found to be effective at degrading a variety of important environmental pollutants (Bumpus 1989). Since that time, a number of species of fungi including *Irpex lacteus, Pleurotus ostreatus, Trametes versicolor, Bjerkandera adusta, Lentinula edodes, Coriolopsis polysona, Cladosporium resinae,* and others have demonstrated the ability to degrade a variety of environmental pollutants (Bhatt et al. 2002; Singh 2006). Pollutants that have been successfully targeted for mycoremediation include polycyclic aromatic hydrocarbons (PAHs) (Bhatt et al. 2002; Singh 2006), polychlorinated biphenyls (PCBs) (Šašek et al. 1993), dioxanes and dioxins (Bhatt et al. 2002; Singh 2006), monoaromatics (benzene and toluene) (Pointing 2001), ammunitions (Kaplan 1992; Bennett 1994), and diesel fuel (Batelle 2000).

There are four main types of fungal ligninolytic enzymes: lignin peroxidases (LiPs), manganese-dependent peroxidases (MnPs), versatile peroxidases (VPs), and laccases (LACs). LiPs are glycosylated heme-containing proteins that house a tryptophan residue on the surface of the enzyme which facilitates the direct oxidation of phenolic and non-phenolic compounds via long-range, multistep, electron transfer. The products of this oxidation are intermediate radicals, which can then undergo nonenzymatic reactions including side-chain cleavage and demethylation (Dashtban et al. 2010). MnPs are also heme-containing proteins and possess two to three residues that bind and oxidize Mn^{2+} to Mn^{3+}. MnPs are believed to be one of the most important fungal enzymes in lignin deconstruction (Hofrichter 2002). They also likely play a major role in degrading xenobiotic compounds in the environment, and MnPs purified from *Lentinula edodes* have been demonstrated to degrade chlorophenols (Singh 2006). VPs are hybrid enzymes that possess both the tryptophan residue found in LiPs, as well as the Mn-binding sites found in MnPs. They may oxidize both phenolic and non-phenolic compounds, as well as Mn^{2+} (Heinfling et al. 1998). LACs are glycosylated oxidoreductases that couple the oxidation of phenolic and other aromatic and nonaromatic substrates with the reduction of molecular oxygen (O_2) to water (H_2O). The free-radical products of this oxidation are unstable, and may undergo further enzymatic or nonenzymatic reactions (Kunamneni et al. 2007). In this manner, the free-radical reactions of fungal ligninolytic enzymes often occur as chain reactions, with the formation of one lignin radical leading to the formation of several others, and the end result being the deconstruction or mineralization of the substrate (Barr and Aust 1994).

Fungi are uniquely equipped to seek out and metabolize pollutants in compost and soils. First, the filamentous nature of their growth generates enormous mechanical pressure that allows them to effectively invade and penetrate compost, soil and other substrates (Howard et al. 1991). Second, they are also one of the few groups of organisms that produce lignin-degrading enzymes (Reddy 1995). Third, they secrete their ligninolytic enzymes extracellularly (Sinsabaugh 2005). Secretion enables the powerful oxidative activity of fungal ligninolytic enzymes to permeate regions of compost and the soil matrix that would otherwise be inaccessible. Lastly, they are able to intercellularly translocate nutrients throughout their mycelial thallus, thus enabling them to grow through nutritionally deficient zones in order to seek out and metabolize both nutritional substrates and pollutants (Boswell et al. 2002).

Composts offer a rich source of residual nutrients, enzymes, xenobiotic-degrading fungi, and other microbes, and have been employed to treat soils contaminated with a variety of contaminants (Koster and Brons 1984; Anastasi et al. 2008).

Augmenting materials such as compost with fungi, a process known as mycoaugmention, has been demonstrated to increase the ability of composts to remediate soil pollutants via transformation, mineralization, or decreasing the bioavailability of contaminants (Lau et al. 2003; Zeng et al. 2007). The strategy for mycocomposting remediation is to mix contaminated soil with initial compost starting materials that have been inoculated with one or more ligninolytic fungi, with remediation occurring as the compost matures (Semple et al. 2001). Aerobic composting usually achieves a higher level of decomposition for most pollutants, so adding a bulking agent such as straw or bark chips increases the porosity and aerobicity of the compost and keeps the temperature below 45 °C (Semple et al. 2001). Another strategy is to mix mature compost in with the contaminated soil and incubate. Spent mushroom compost, a byproduct of the mushroom growing industry, has also been used for the mycoremediation of contaminated soils (Eggen 1999; Okparanma et al. 2013).

Several investigators have demonstrated the effectiveness of using mycoaugmented compost to bioremediate PAHs, a group of organic compounds of which 16 are listed by the USEPA as priority pollutants. PAHs contain two or more fused benzene rings and are often formed by the incomplete combustion of hydrocarbons such as coal and gasoline. PAHs are hydrophobic and tend to sorb to organic matter, which makes them less available for biodegradation and thus harder to remediate (Lau et al. 2003). However, the penetrability of extracellularly secreted fungal ligninolytic enzymes mitigates this difficulty, and the degradation of PAHs by microbes is the major process of removal from the environment (Kilbane 1998; Bumpus et al. 1985; Yuan et al. 2001). In a study by Okparanma et al. (2013), spent *Pleurotus ostreatus* substrate compost was tested in vitro for its ability to degrade PAHs in oil drilling cuttings, a muddy mixture of rocks and particulates that are released from holes bored for oil drillings. They found that after 56 days of composting, the overall degradation of all PAHs was between 80.25 and 92.38 % between replicate samples, including a degradation of 97.98 % for acenapthene and 100 % for flourene, phenanthrene, and anthracene, all of which are 3-ring PAHs. They noted that the degradation of different fractions of PAHs differed as a function of molar mass and ring group, as well as the amount of fungal compost applied, with smaller ring groups being degraded more completely than larger ring groups, and increased degradation with increasing amounts of fungal compost applied. In another study by Lau et al. (2003), spent mushroom compost of *Pleurotus pulmonarius* was used *in situ* to degrade a variety of PAHs in a sandy-loam soil. They realized complete degradation of 200 mg PAH kg^{-1} naphthalene, phenanthrene, benzo[a]pyrene, and benzo [g, h, i]perylene by 5 % (by volume) spent mushroom compost in their microcosms. They attributed the PAH removal to biotic activity, as sterilization of the soil completely eliminated any PAH degradation.

Eggen (1999) compared the ability of spent *Pleurotus ostreatus* mushroom compost (after fruiting body production) compared to fresh compost (before fruiting body production) to degrade PAHs in aged creosote-contaminated soil from an abandoned wood preservation site.

The difference between homogeneously mixing the compost with the soil versus layering the compost and soil was also investigated, as was the effect of adding fish oil to the mixture as a surfactant to increase the solubility of the hydrophobic PAHs. It was found that all treatments degraded over 73 % of all 3-ring PAHs, with the individual compounds acenaphthene, fluorene, and phenanthrene being degraded over 80 % in all treatments up to 99 % in the mixed treatment with spent mushroom compost and fish oil. The degradation of 4 ring PAHs was less consistent than the degradation of 3-ring PAHs, with the most effective treatment degrading 91 % of fluoroanthene, 83 % pyrene, 87 % benzo[a]anthracene, and 79 % chrysene. Spent mushroom compost had a tendency to be more effective at degrading 4–5 ring compounds, than fresh substrate, and mixing the soil and compost was more effective overall than layering the two.

Spent mushroom compost has also been demonstrated to effectively degrade pentachlorophenol (PCP), the most commonly used pesticide worldwide

(Hattemer-Frey and Travis 1989). Chiu et al. (1998) used spent *Pleurotus pulmonarius* mushroom compost to degrade PCP, finding a degradative capacity of 19 mg PCP g^{-1} in the spent mushroom compost 3 days post spiking. They found that both degradation and bioabsorption played a role in removing extractable PCP from the compost. In another study, Okeke et al. (1993) used spent sawdust compost from three strains of *Lentinula edodes* (shiitake mushroom) to mycoremediate soils contaminated with PCP, and found that the three strains were able to degrade the PCP by 60.5, 57.3 and 44.4 %, respectively, whereas an addition of H_2O_2 to the soil increased the amount of degradation to 69.6, 65.6, and 60.5 %, respectively, implicating peroxidase enzymes as the agents of degradation.

Ligninolytic fungi have also been isolated from compost and employed to detoxify pollutants in soil in situ. Anastasi et al. (2009) isolated three lignolytic basidiomycete (Basidiomycota) fungi, two strains whose ITS rDNA sequences were 100 % similar to *Trametes versicolor*, and another that was 99 % similar to two *Bjerkandera* spp. and *Lopharia spadicea*. They grew the isolates on straw and then mixed the straw in microcosms containing gamma-ray-sterilized soil spiked with 100 mg kg^{-1} pyrene, a highly toxic 4-ring PAH which is considered an indicator compound by the USEPA for monitoring PAH-contaminated soils (Saraswathy and Hallberg 2005). After 28 days, they found a 56 % decrease in the level of pyrene in the microcosms that contained the fungal isolates. Along with a decrease in pyrene, they also found an increase in the germination index of *Lepidium sativum* seeds from 74.1 to 154.3 %, indicating that phytotoxicity of the soil was reduced and growth was stimulated in comparison to control microcosms without pyrene or mycoaugmentation. Enzyme assays indicated that LAC activity was the highest of all ligninolytic enzymes during the first 20 days, the period when the majority of pyrene degradation occurred.

A major explosive ordnance used by the US military and in a widespread array of explosives and munitions is 2, 4, 6-trinitrotoluene (TNT) (Boileau et al. 2000). Over the last 100 years the commercial production and subsequent disposal of TNT has led to contamination of the soil and groundwater of disposal sites across the United States, with the Army estimating that over 1.2 million tons of soil have been contaminated with TNT across the country (Hampton and Sisk 1997). Composting was the first process to be tested for remediating contaminated military sites (Craig et al. 1995). However, both static and windrow composting were found to be ineffective in mineralizing TNT, with the end result being the binding of the majority of TNT to the soil organic matter with little mineralization (Pennington et al. 1995). However, several fungi, including *Phanerochaete chrysosporium*, have been demonstrated to affect significant mineralization of TNT in both aqueous and soil substrates, even when initial concentrations were up to 100 mg L^{-1} in an aqueous substrate, and 10 g L^{-1} in a soil substrate (Fernando et al. 1990; Fernando and Aust 1994). Other species of fungi found to be effective in mineralizing TNT include *Stropharia rugoso-annulata* and *Clitocybula dusenii* (Scheibner et al. 1997).

In addition to oxidizing organopollutants, mycoaugmented composting has been shown to be effective in the stabilization and reduction in the leachability and bioavailability of heavy metals. Although the exact mechanism is not known, it is speculated that sorption to polysaccharides, proteins, and other components of the outer wall of fungal hyphae play an important part, with many species of wood-inhabiting Basidiomycota showing promise for heavy metal biosorption (Baldrian 2003). Zeng et al. (2007), simulated the composting of a lead-contaminated soil inoculated with *Phanerochaete chrysosporium* by mixing soil, wheat straw, kitchen waste, and bran with a lead-nitrate solution containing 105 mg kg^{-1} dry weight Pb. It was found that after 80 days of composting, the exchangeable Pb content of the soil was reduced completely to 0 %. They concluded that augmenting the composting of lead-contaminated solid waste with ligninolytic fungi such as *P. chrysosporium* can reduce the bioavailablity of Pb and the hazards of Pb-contaminated soils.

Fungi and fungal composts offer an inexpensive and potentially effective means of remediating sites that have been contaminated by pollutants. As long as humans manufacture chemicals and inorganic substances that inevitably end up in our terrestrial and aquatic ecosystems, there will be a need for inexpensive, effective, and efficient methods to remediate contaminated sites. The deposition and persistence of pollutants in our environment is a significant health concern to all higher biota due to the health hazards they pose. Using agricultural by-products such as spent mushroom compost to remediate soils is an attractive means of putting them to good use rather than discarding them as waste. Further exploration of the potential of fungi to remediate contaminants is therefore warranted.

1.7 Conclusions

Fungi play an important role in composting processes, degrading recalcitrant compounds during composting, stabilizing organic matter while releasing nutrients and essential elements that are beneficial for plant growth and fertility. Fungi also enhance the activity of earthworms in vermicomposting both as a source of enzymes to help facilitate decomposition as well as a preferred food source for the earthworms. Fungal-mediated composts are agriculturally useful for amending soils and as a substrate for growing agaric mushrooms. Further, mycoaugmented composts are effective in remediating soils and have a great potential for remediating soils contaminated with a variety of organopollutants and xenobiotic compounds. Exploring the potential of fungal-augmented composts in remediating soils is an area that merits further investigation.

References

Abbasi SA, Nayeem-Shah M, Abbasi T (2015) Vermicomposting of phytomass: limitations of the past approaches and the emerging directions. J Clean Prod 93:103–114

Agrawal PK (2014) Microbial ecology of compost ecosystem: with special reference to mushroom compost. J Biol Sci Opin 2:45–50

Aira M, Monroy F, Domínguez J (2006) *Eisenia fetida* (Oligochaeta, Lumbricidae) activates fungal growth, triggering cellulose decomposition during vermicomposting. Microb Ecol 52:738–747

Aira M, Monroy F, Domínguez J (2007) *Eisenia fetida* (Oligochaeta: Lumbricidae) modifies the structure and physiological capabilities of microbial communities improving carbon mineralization during vermicomposting of pig manure. Microb Ecol 54:662–671

Alexander M (1995) How toxic are toxic chemicals in soil? Environ Sci Technol 29:2713–2717

Alfreider A, Peters S, Tebbe CC, Rangger A, Insam H (2002) Microbial community dynamics during composting of organic matter as determined by 16S ribosomal DNA analysis. Compost Sci Util 10:303–312

Anastasi A, Varese GC, Voyron S, Scannerini S, Marchisio VF (2004) Characterization of fungal biodiversity In compost and vermicompost. Compost Sci Util 12:185–191

Anastasi A, Varese GC, Marchisio VF (2005) Isolation and identification of fungal communities in compost and vermicompost. Mycologia 97:33–44

Anastasi A, Varese GC, Bosco F, Chimirri F, Marchisio VF (2008) Bioremediation potential of basidiomycetes isolated from compost. Bioresour Technol 99:6626–6630

Anastasi A, Coppola T, Prigione V, Varese GC (2009) Pyrene degradation and detoxification in soil by a consortium of basidiomycetes isolated from compost: role of laccases and peroxidases. J Hazard Mater 165:1229–1233

Andaluri G, Suri RPS, Kumar K (2012) Occurrence of estrogen hormones in biosolids, animal manure and mushroom compost. Environ Monit Assess 184:1197–1205

Baldrian P (2003) Interactions of heavy metals with white-rot fungi. Enzyme Microb Technol 32:78–91

Barr DP, Aust SD (1994) Effect of superoxide and superoxide dismutase on lignin peroxidase-catalyzed veratryl alcohol oxidation. Arch Biochem Biophys 311:378–382

Barrett TJ (1948) Harnessing the earthworm. A practical inquiry into soil-building, soil-conditioning, and plant nutrition through the action of earthworms, with instructions for intensive propagation and use of domesticated earthworms in biological soil-building. Q Rev Biol 23:361–361

Batelle CD (2000) Mushrooms: higher macrofungi to clean up the environment. Environmental Issues, Fall

Beffa T, Staib F, Lott Fischer J, Lyon PF, Gumowski P, Marfenina OE et al (1998) Mycological control and surveillance of biological waste and compost. Med Mycol 36(Suppl 1):137–145

Bennett JW (1994) Prospects for fungal bioremediation of TNT munition waste. Int Biodeterior Biodegradation 34:21–34

Bhatt M, Cajthaml T, Sasek V (2002) Mycoremediation of PAH-contaminated soil. Folia Microbiol 47:255–258

Boileau J, Fauquignon C, Hueber B. (2000). Explosives. In: *Ullmann's Encyclopedia of Industrial Chemistry*. Wiley-VCH Verlag GmbH & Co. KGaA

Bonito G, Isikhuemhen OS, Vilgalys R (2010) Identification of fungi associated with municipal compost using DNA-based techniques. Bioresour Technol 101:1021–1027

Boominathan K, Reddy CA (1992) Fungal degradation of lignin: biotechnological applications. In: Arora DK, Elander RP & Mukerji KG (eds), Handbook of applied mycology, vol 4. Fungal biotechnology. Marcel Dekker, Inc., New York, N.Y, p 763–822

Boswell GP, Jacobs H, Davidson FA, Gadd GM, Ritz K (2002) Functional consequences of nutrient translocation in mycelial fungi. J Theor Biol 217:459–477

Bouché MB (1987) Emergence and development of vermiculture and vermicomposting from a hobby to an industry, from marketing to a biotechnology from irrational to credible practices. In: *Selected Symposia and Monographs UZI*, pp. 519–531

Bouwman H (1998) Evaluation of a technique to obtain development-stage-synchronised earthworms (*Eisenia fetida*). Biol Fertil Soils 27:368–373

Brown GG (1995) How do earthworms affect microfloral and faunal community diversity? The significance and regulation of soil biodiversity. Springer, Netherlands, pp 247–269

Brown GC, Doube BM (2004) Functional interactions between earthworms, microorganisms, organic matter, and plants. Earthworm ecology. In: Ewards CA (ed) Earthworm ecology. CRC Press LLC, Boca Raton, FL, USA, pp 213–239

Bumpus JA (1989) Biodegradation of polycyclic hydrocarbons by *Phanerochaete chrysosporium*. Appl Environ Microbiol 55:154–158

Bumpus JA, Tien M, Wright D, Aust SD (1985) Oxidation of persistent environmental pollutants by a white rot fungus. Science 228:1434–1436

Cahyani VR, Matsuya K, Asakawa S, Kimura M (2004) Succession and phylogenetic profile of eukaryotic communities in the composting process of rice straw estimated by PCR-DGGE analysis. Biol Fertil Soils 40:334–344

Chang S-T (2006) Development of the culinary—medicinal mushrooms industry in China: Past, present, and future. Int J Med Mush. 8, doi:10.1615/IntJMedMushr.v8.i1.10

Chang S-T, Miles PG (1989) Edible mushrooms and their cultivation. CRC Press LLC, Boca Raton, FL, USA

Chen Y, Chefetz B, Rosario R, van Heemst JDH, Romaine CP, Hatcher PG (2000) Chemical nature and composition of compost during mushroom Growth. Compost Sci Util 8:347–359

Chiu SW, Chan YH, Law SC, Cheung KT, Moore D (1998) Cadmium and manganese in contrast to calcium reduce yield and nutritional values of the edible mushroom *Pleurotus pulmonarius*. Mycol Res 102:449–457

Cooke J (1983) The effects of fungi on food selection by *Lumbricus terrestris* L. Earthworm ecology. Springer, Netherlands, pp 365–373

Craig HD, Sisk WE, Nelson MD, Dana WH (1995) Bioremediation of explosives-contaminated soils: A status review. 10th annual conference on hazardous waste research. Manhattan, Kans, pp 168–179

Darwin C (1892) The formation of vegetable mould, through the action of worms, with observations on their habits. London: John Murray. 7th thousand. Corrected by Francis Darwin

Dashtban M, Schraft H, Syed TA, Qin W (2010) Fungal biodegradation and enzymatic modification of lignin. Int J Biochem Mol Biol 1:36–50

Day M, Shaw K, Stofella P, Kahn B (2001) Biological, chemical and physical processes of composting. Compost utilization in horticultural cropping systems, Lewis Publishers, Boca Raton, FL, USA, pp 17–50

De Bertoldi M, Vallini G, Pera A (1983) The biology of composting: a review. Waste Manag Res 1:157–176

De Gannes V, Eudoxie G, Hickey WJ (2013) Insights into fungal communities in composts revealed by 454-pyrosequencing: implications for human health and safety. Front Microbiol 4:164

Doube BM, Brown GG (1998) Life in a complex community: functional interactions between earthworms, organic matter, microorganisms, and plants. In: Edwards CA (ed) Earthworm Ecology. CRC Press, Boca Raton, FL, pp 179–211

Edwards CA, Bohlen PJ (1996) Biology and ecology of earthworms. Springer Science and Business Media

Edwards CA, Fletcher KE (1988) Interactions between earthworms and microorganisms in organic-matter breakdown. Agric Ecosyst Environ 24:235–247

Edwards CA, Arancon NQ, Sherman RL (2010) Vermiculture technology: earthworms, organic wastes, and environmental management. CRC press

Eggen T (1999) Application of fungal substrate from commercial mushroom production—*Pleurotus ostrearus*—for bioremediation of creosote contaminated soil. Int Biodeterior Biodegradation 44:117–126
Fernando T, Aust SD (1994) Biodegradation of toxic chemicals by white rot fungi. In: Chaudhry GR (ed) Biological degradation and bioremediation of toxic chemicals London. Chapman and Hall, London, pp 386–402
Fernando T, Bumpus JA, Aust SD (1990) Biodegradation of TNT (2,4,6-trinitrotoluene) by *Phanerochaete chrysosporium*. Appl Environ Microbiol 56:1666–1671
Floudas D, Binder M, Riley R, Barry K, Blanchette RA, Henrissat B et al (2012) The paleozoic origin of enzymatic lignin decomposition reconstructed from 31 fungal genomes. Science 336:1715–1719
Gbolagade JS, Fasidi IO, Ajayi EJ, Sobowale AA (2006) Effect of physico-chemical factors and semi-synthetic media on vegetative growth of *Lentinus subnudus* (Berk.), an edible mushroom from Nigeria. Food Chem 99:742–747
Ghazifard A, Kasra-Kermanshahi R, Far ZE (2001) Identification of thermophilic and mesophilic bacteria and fungi in Esfahan (Iran) municipal solid waste compost. Waste Manag Res 19:257–261
Gibbons WR, Maher AA, Todd RL (1991) Button mushroom production in synthetic compost derived from agricultural wastes. Bioresour Technol 38:65–77
Guinberteau J, Olivier JM, Tanne MN (1991) Improvement of *Lepista* species cultivation, technical factors and selection of strains. Mushroom Sci 2:615–621
Guo R, Li G, Jiang T, Schuchardt F, Chen T, Zhao Y, Shen Y (2012) Effect of aeration rate, C/N ratio and moisture content on the stability and maturity of compost. Bioresour Technol 112:171–178
Hadar Y, Papadopoulou KK (2012) Suppressive composts: microbial ecology links between abiotic environments and healthy plants. Annu Rev Phytopathol 50(50):133–153
Halet D, Boon N, Verstraete W (2006) Community dynamics of methanotrophic bacteria during composting of organic matter. J Biosci Bioeng 101:297–302
Hampton ML, Sisk WE (1997) Environmental stability of windrow composting of explosives-contaminated soils. In: Tedder DW (ed) Emerging technologies in hazardous waste management IX, division of industrial and engineering chemistry. American Society of Chemistry, Washington DC, pp 252–257
Hattemer-Frey HA, Travis CC (1989) Pentachlorophenol: environmental partitioning and human exposure. Arch Environ Contam Toxicol 18:482–489
Head IM (1998) Bioremediation: towards a credible technology. Microbiology 144:599–608
Heinfling A, Ruiz-Dueñas FJ, Martínez MJ, Bergbauer M, Szewzyk U, Martínez AT (1998) A study on reducing substrates of manganese-oxidizing peroxidases from *Pleurotus eryngii* and *Bjerkandera adusta*. FEBS Lett 428:141–146
Hein I (1930) Straw compost for mushroom culture. Mycologia 22:39–43
Hofrichter M (2002) Review: lignin conversion by manganese peroxidase (MnP). Enzyme Microb Technol 30:454–466
Howard RJ, Ferrari MA, Roach DH, Money NP (1991) Penetration of hard substrates by a fungus employing enormous turgor pressures. Proc Natl Acad Sci USA 88:11281–11284
Huang K, Li F, Wei Y, Chen X, Fu X (2013) Changes of bacterial and fungal community compositions during vermicomposting of vegetable wastes by *Eisenia foetida*. Bioresour Technol 150:235–241
Ivors KL, Collopy PD, Beyer DM, Kang S (2000) Identification of bacteria in mushroom compost using ribosomal RNA sequence. Compost Sci Util 8:247–253
Jurak E, Punt AM, Arts W, Kabel MA, Gruppen H (2015) Fate of carbohydrates and lignin during composting and mycelium growth of *Agaricus bisporus* on wheat straw based compost. PLoS ONE 10:e0138909
Kaplan DL (1992) Biological degradation of explosives and chemical agents. Curr Opin Biotechnol 3:253–260

Kilbane JJ II (1998) Extractability and subsequent biodegradation of PAHs from contaminated soil. Water Air Soil Pollut Focus 104:285–304

Kostecka J, Blazej JB, Kolodziej M (1996) Investigations on application of vermicompost in potatoes farming in second year of experiment. Zeszyty Naukowe Akademii Rolniczej W Krakowie 310:69–77

Koster IW, Brons HJ (1984) Respirometric testing method for biodegradability of xenobiotics using compost. J Environ Sci Health B 19:785–792

Kunamneni A, Ballesteros A, Plou FJ, Alcalde M (2007) Fungal laccase—a versatile enzyme for biotechnological applications. Commun Curr Res Educ Top Trends Appl Microbiol 1:233–245

Lambert EB (1929) The production of normal sporophores in monosporous cultures of *Agaricus campestris*. Mycologia 21:333–335

Lambert EB (1941) Studies on the preparation of mushroom compost. J Agric Res 415–422

Lau KL, Tsang YY, Chiu SW (2003) Use of spent mushroom compost to bioremediate PAH-contaminated samples. Chemosphere 52:1539–1546

Lavelle P, Spain AV (2001) Soil ecology. Kluwer Scientific, Amsterdam

López-González JA, Vargas-García MDC, López MJ, Suárez-Estrella F, Jurado MDM, Moreno J (2015) Biodiversity and succession of mycobiota associated to agricultural lignocellulosic waste-based composting. Bioresour Technol 187:305–313

Lo SC, Ho YS, Buswell JA (2001) Effect of phenolic monomers on the production of laccases by the edible mushroom pleurotus sajor-caju, and partial characterization of a major laccase component. Mycologia 93:413–421

Malandraki I, Tjamos SE, Pantelides IS, Paplomatas EJ (2008) Thermal inactivation of compost suppressiveness implicates possible biological factors in disease management. Biol Control 44:180–187

Marshall MN, Cocolin L, Mills DA, VanderGheynst JS (2003) Evaluation of PCR primers for denaturing gradient gel electrophoresis analysis of fungal communities in compost. J Appl Microbiol 95:934–948

Mehta CM, Palni U, Franke-Whittle IH, Sharma AK (2014) Compost: its role, mechanism and impact on reducing soil-borne plant diseases. Waste Manag 34:607–622

Miller FC (1992) Composting as a process based on the control of ecologically selective factors. In: Metting FB Jr (ed) Soil microbial ecology: applications in agricultural and environmental management. Marcel Dekker Inc, New York, pp 515–544

Miller FC, Harper ER, Macauley BJ, Gulliver A (1990) Composting based on moderately thermophilic and aerobic conditions for the production of commercial mushroom growing compost. Aust J Exp Agric 30:287–296

Morin E, Kohler A, Baker AR, Foulongne-Oriol M, Lombard V, Nagy LG et al (2012) Genome sequence of the button mushroom *Agaricus bisporus* reveals mechanisms governing adaptation to a humic-rich ecological niche. Proc Natl Acad Sci USA 109:17501–17506

Mortimer PE, Karunarathna SC, Li QH, Gui H, Yang XQ, Yang XF et al (2012) Prized edible Asian mushrooms: ecology, conservation and sustainability. Fungal Divers 56:31–47

Nada WM (2015) Stability and maturity of maize stalks compost as affected by aeration rate, C/N ratio and moisture content. J Soil Sci Plant Nutr 15(3):751–764

Natvig DO, Taylor JW, Tsang A, Hutchinson MI, Powell AJ (2015) *Mycothermus thermophilus* gen. et comb. nov., a new home for the itinerant thermophile *Scytalidium thermophilum* (*Torula thermophila*). Mycologia 107:319–327

Neher DA, Weicht TR, Bates ST, Leff JW, Fierer N (2013) Changes in bacterial and fungal communities across compost recipes, preparation methods, and composting times. PLoS ONE 8:e79512

Nicholas LG, Ogame K (2006) Psilocybin mushroom handbook: easy indoor and outdoor cultivation. Ed Rosenthal

Noble R, Gaze RH (1994) Controlled environment composting for mushroom cultivation: substrates based on wheat and barley straw and deep litter poultry manure. J Agric Sci 123:71–79

Noble R, Hobbs PJ, Mead A, Dobrovin-Pennington A (2002) Influence of straw types and nitrogen sources on mushroom composting emissions and compost productivity. J Ind Microbiol Biotechnol 29:99–110

Okeke BC, Smith JE, Paterson A, Watson-Craik IA (1993) Aerobic metabolism of pentachlorophenol by spent sawdust culture of "Shiitake" mushroom (*Lentinus edodes*) in soil. Biotechnol Lett 15:1077–1080

Okparanma RN, Ayotamuno JM, Davis DD, Allagoa M (2013) Mycoremediation of polycyclic aromatic hydrocarbons (PAH)-contaminated oil-based drill-cuttings. Afr J Biotechnol 10:5149–5156

Oliver GS (1949) Our friend, the earthworm. Organic Gardening

Pennington JC, Hayes CA, Myers KF, Ochman M, Gunnison D, Felt DR et al (1995) Fate of 2, 4, 6-trinitrotoluene in a simulated compost system. Chemosphere 30:429–438

Peters S, Koschinsky S, Schwieger F, Tebbe CC (2000) Succession of microbial communities during hot composting as detected by PCR-single-strand-conformation polymorphism-based genetic profiles of small-subunit rRNA genes. Appl Environ Microbiol 66:930–936

Pižl V, Nováková A (2003) Interactions between microfungi and *Eisenia andrei* (Oligochaeta) during cattle manure vermicomposting: the 7th international symposium on earthworm ecology. Pedobiologia 47:895–899

Pointing SB (2001) Feasibility of bioremediation by white-rot fungi. Appl Microbiol Biotechnol 57:20–33

Pramanik P (2010) Changes in microbial properties and nutrient dynamics in bagasse and coir during vermicomposting: quantification of fungal biomass through ergosterol estimation in vermicompost. Waste Manag 30:787–791

Rajapakse JC, Rubasingha P, Dissanayake NN (2010) The potential of using cost-effective compost mixtures for oyster mushroom (*Pleurotus* spp) cultivation in Sri Lanka. Trop Agric Res Ext 10, doi:10.4038/tare.v10i0.1868

Reddy CA (1995) The potential for white-rot fungi in the treatment of pollutants. Curr Opin Biotechnol 6:320–328

Rettew FG, Thompson GR (1948) Manual of mushroom culture. Mushroom Supply Company

Richard TL, Hamelers (bert) HVM, Veeken A, Silva T (2002) Moisture relationships in composting processes. Compost Sci Util 10:286–302

Royse DJ, Chalupa W (2009) Effects of spawn, supplement and phase II compost additions and time of re-casing second break compost on mushroom (*Agaricus bisporus*) yield and biological efficiency. Bioresour Technol 100:5277–5282

Ryckeboer J, Mergaert J, Vaes K, Klammer S, De Clercq D, Coosemans J et al (2003) A survey of bacteria and fungi occurring during composting and self-heating processes. Ann Microbiol 53:349–410

Said-Pullicino D, Erriquens FG, Gigliotti G (2007) Changes in the chemical characteristics of water-extractable organic matter during composting and their influence on compost stability and maturity. Bioresour Technol 98:1822–1831

Saraswathy A, Hallberg R (2005) Mycelial pellet formation by *Penicillium ochrochloron* species due to exposure to pyrene. Microbiol Res 160:375–383

Šašek V, Volfová O, Erbanová P, Vyas BRM, Matucha M (1993) Degradation of PCBs by white rot fungi, methylotrophic and hydrocarbon utilizing yeasts and bacteria. Biotechnol Lett 15:521–526

Scheibner K, Hofrichter M, Herre A, Michels J, Fritsche W (1997) Screening for fungi intensively mineralizing 2, 4, 6-trinitrotoluene. Appl Microbiol Biotechnol 47:452–457

Scheuerell S, Mahaffee W (2002) Compost tea: principles and prospects for plant disease control. Compost Sci Util 10:313–338

Schönholzer F, Hahn D, Zeyer J (1999) Origins and fate of fungi and bacteria in the gut of *Lumbricus terrestris* L. studied by image analysis. FEMS Microbiol Ecol 28:235–248

Semple KT, Reid BJ, Fermor TR (2001) Impact of composting strategies on the treatment of soils contaminated with organic pollutants. Environ Pollut 112:269–283

Singh D, Suthar S (2012) Vermicomposting of herbal pharmaceutical industry waste: earthworm growth, plant-available nutrient and microbial quality of end materials. Bioresour Technol 112:179–185

Singh H (2006) Fungal degradation of polychlorinated biphenyls and dioxins. In: Mycoremediation. John Wiley & Sons, Inc. pp 149–180

Sinsabaugh RL (2005) Fungal enzymes at the community scale. Mycology Ser 23:349

Straatsma G, Olijnsma TW, Gerrits JPG, Griensven LJLDV, Samson RA, Camp HJMOD (1995) Bioconversion of cereal straw into mushroom compost. Can J Bot 73:1019–1024

Szczech MM (1999) Suppressiveness of vermicompost against *Fusarium* wilt of tomato. J Phytopathol 147:155–161

Szczech M, Rondomański W, Brzeski MW, Smolińska U, Kotowski JF (1993) Suppressive effect of a commercial earthworm compost on some root infecting pathogens of cabbage and tomato. Biol Agric Hortic 10:47–52

Tiunov AV, Scheu S (2000) Microfungal communities in soil, litter and casts of *Lumbricus terrestris* L. (Lumbricidae): a laboratory experiment. Appl Soil Ecol 14:17–26

Vijay B, Sharma SR, Lakhanpal TN (2002) Role of thermophilic fungi in compost production for *Agaricus bisporus*. J Mycol Plant Pathol 32:204–210

Viti C, Tatti E, Decorosi F, Lista E, Rea E, Tullio M et al (2010) Compost effect on plant growth-promoting rhizobacteria and mycorrhizal fungi population in maize cultivations. Compost Sci Util 18:273–281

Waksman SA (1932) Mushroom nutrition: a group of problems in microbiology. J Bacteriol 23:81

Waksman SA, Nissen W (1932) On the nutrition of the cultivated mushroom, *Agaricus campestris*, and the chemical changes brought about by this organism in the manure compost. Am J Bot 19:514–537

Waksman SA, Reneger CA (1934) Artificial manure for mushroom production. Mycologia 26:38–45

Waksman SA, Umbreit WW, Cordon TC (1939) Thermophilic actinomycetes and fungi in soils and in composts. Soil Sci 47:37–61

Wiegant WM, Wery J, Buitenhuis ET, de Bont JA (1992) Growth-promoting effect of thermophilic fungi on the mycelium of the edible mushroom *Agaricus bisporus*. Appl Environ Microbiol 58:2654–2659

Yuan SY, Chang JS, Yen JH, Chang BV (2001) Biodegradation of phenanthrene in river sediment. Chemosphere 43:273–278

Zeng G, Huang D, Huang G, Hu T, Jiang X, Feng C et al (2007) Composting of lead-contaminated solid waste with inocula of white-rot fungus. Bioresour Technol 98:320–326

Zhang B-G, Li G-T, Shen T-S, Wang J-K, Sun Z (2000) Changes in microbial biomass C, N, and P and enzyme activities in soil incubated with the earthworms *Metaphire guillelmi* or *Eisenia fetida*. Soil Biol Biochem 32:2055–2062

Chapter 2
The Genetic Basis of Abiotic Stress Resistance in Extremophilic Fungi: The Genes Cloning and Application

Shi-Hong Zhang

2.1 Introduction

The major lineage of fungi was believed to have first arisen about 1000 million or so years ago, which was followed by land plants in approximately 700 million years ago (Heckman et al. 2001). From the biological and environmental evolution perspective, fungi are one of the earliest eukaryotes to colonize the ancient earth (Gray and Shear 1992; Horodyski and Knauth 1994). Considering the harsh physical environments on the ancient earth, to ensure the chances of survival, fungi need to be more tolerant or resistant to adverse environmental factors than the latter appeared plants or animals. Indeed, within the last few decades a number of fungal species (halophile, xerophiles, or thermophile) that can live in a variety of extreme environments have been isolated. For example, *Eurotium herbariorum* that can survive in 340 g/L total dissolved salts was isolated from the Dead Sea (Kis-Papo et al. 2001; Yan et al. 2005); *Sarcinomyces petricola* strain A95 (a representative strain of rock-inhabiting fungi) was isolated from a marble rock surface near the Philopappos monument on Musaios Hill in Athens (Gorbushina 2007); a thermophilic fungus *Thermomyces lanuginosus* is able to survive at 62 °C, the highest growth temperature recorded so far, was isolated from horse dung (Prasad and Maheshwari 1978); and in our laboratory, the fungal strain *Chaetomium thermophile* isolated from rotting corn stalk was observed to be capable of growing at 65 °C (Song and Zhang, unpublished data). In contrast, with the exception of plant symbionts harboring mycorrhizal and/or endophytic fungi, it is hard to find plants that are able to survive under any one of the above extreme conditions.

Generally speaking, plants are relatively sensitive to high levels of abiotic stresses such as drought, high salt, extreme low/high temperatures, therefore, they are less able to survive and thrive in extreme environments (Alpert 2000). However,

S.-H. Zhang (✉)
College of Plant Sciences, Jilin University, Changchun 130062, China
e-mail: zhang_sh@jlu.edu.cn

D. Purchase (ed.), *Fungal Applications in Sustainable Environmental Biotechnology*, Fungal Biology, DOI 10.1007/978-3-319-42852-9_2

when an association between a plant and a mycorrhizal or endophytic fungus is established in an extreme habitat, the plant together with the habitat-adapted fungal endophytes, are able to survive or even thrive in that environment (Rodriguez et al. 2008). Without the fungal endophytes, the same plant is unable to endure habitat-imposed abiotic and biotic stresses (Redman et al. 2011). Therefore, this chapter will address the importance of endophytic fungi to emphasize their potential application in plant stress tolerance biotechnology.

Though great achievements have been made in recent years in understanding the abiotic stress responses and molecular mechanisms in plants (Bhatnagar-Mathur et al. 2008; Gupta and Huang 2014; Zhan et al. 2015), crop-breeding practice of higher resistant varieties remains unsatisfactory due to the lack of abiotic stress resistance germplasms. Wild germplasms are of interests, as wild crops may have retained genetic information before the domestication and artificial selection of modern plants (Lam et al. 2010). Wild germplasms screening, however, is a daunting and time-consuming task; and most likely we will never find certain wild germplasms again because of their extinction due to modern agricultural practices and environmental changes. Considering the survival abilities in extreme conditions, extremophilic fungi may provide special or different resistance mechanism compared to plants. Why are endophyte-free plants so sensitive to extreme adversities? And why do fungal endophytes facilitate and ensure plants to thrive? Both are complicated questions but the genetic advantage retained in fungal endophytes might provide information to enlighten and enable us to use the extremophilic fungi efficiently. The genetic basis of abiotic stress resistance in fungi, particularly in extremophilic fungi, is a unique genetic resource to improve abiotic stress resistance of crops. Here, we highlight the abiotic stress resistance mechanisms and resistant genes in extremophilic fungi. In addition, application strategies for anti-abiostress genetic engineering are also discussed.

2.2 The Responsive Pathways to Abiotic Stress in Extremophilic Fungi[1]

Fungi like other eukaryotic organisms (such as plants) depend on signal-receiving and transmitting systems to respond to, survive and thrive under the imposed adverse conditions. In eukaryote microorganisms, the yeast *S. cerevisiae* is known to have only moderate levels of tolerance to salt, drought, extreme temperature and other stressors (Prista et al. 2002, 2005; Serrano and Gaxiola 1994). Thus, a wild type *S. cerevisiae* is not the best model organism, neither for salt tolerance nor for sensitivity to salt. However, *S. cerevisiae* was an excellent tool for genetic

[1]For additional information on stress responses in acidophilic fungi against heavy metals and metalloids, please see Chap. 9—*Mycoremediation of heavy metal/metalloid-contaminated soil: current understanding and future prospects.*

manipulation (fast growth rate and easy transformation) and has been widely applied in the field of resistance research. The high-osmolarity glycerol (HOG1) pathway which is an essential stress-signaling module has been extensively studied in fungi (Fig. 2.1): from the yeast *S. cerevisiae* to the filamentous fungus *Thichoderma harzianum* (Brewster et al. 1993; Delgado-Jarana et al. 2006). The HOG pathway is responsible for survival of the fungus in periods of high-osmolarity or oxidative stress (San José et al. 1996; Ikner and Shiozaki 2005; Enjalbert et al. 2006; Alonso-Monge et al. 2006). The activation of the HOG pathway leads to an increase in intracellular glycerol which provides protection against osmotic stress (Saito and Posas 2012). The activation of ScHog1 in *S. cerevisiae* can lead to downregulation of ergosterol biosynthesis and adaption to osmotic stress through changes in membrane fluidity (Montañés et al. 2011). In *Candida albicans*, mitochondrial function appears to be required for the activation of CaHog1 in response to oxidative stress (Thomas et al. 2013).

The HOG pathway has high sequence identity across many fungal species (Hayes et al. 2014). HOG1 is also found in extremophilic fungi. Nevo's group testified that *Eurotium herbariorum* HOG1 is highly similar to the homologs from non-extreme fungi such as *Aspergillus nidulans*, *S. cerevisiae*, *Schizosaccharomyces pombe* (Yan et al. 2005). However, it appears that HOG1 is found exclusively in fungi and no homologue gene has been detected in plants.

The HOG1 regulation system in yeast involves two pathways, the low osmolarity SHO1 pathway and high-osmolarity SLN1 pathway. The difference between the two pathways lays at the SHO1 and SLN1 sensors, but both pathways ultimately leads to glycerol biosynthesis and the glycerol concentration for osmotic balance (Brewster et al. 1993). Besides salt, HOG1 also responds to a variety of other stressors (Delgado-Jarana et al. 2006), suggesting cross-talking feature in the HOG pathways. Another stress response is the cell wall integrity (CWI) pathway

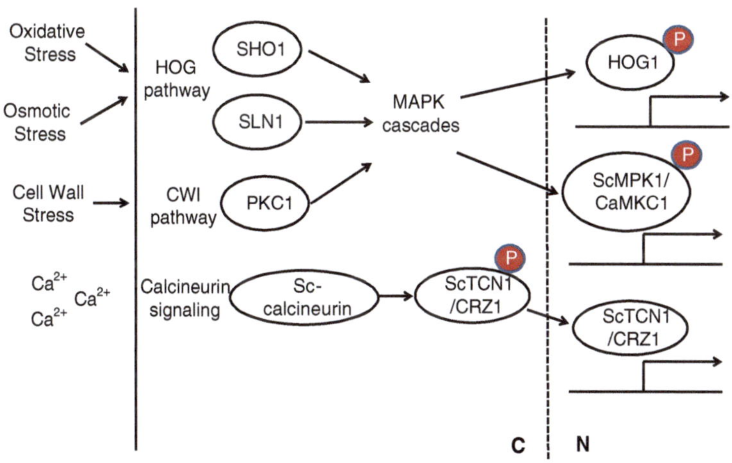

Fig. 2.1 Stress signaling pathways in fungi. *C* cytoplasm; *N* nucleus

(Fig. 2.1). The CWI pathway is involved in sensing cell wall stresses (Levin 2011). CWI is activated in response to abiotic stresses, such as osmotic pressure, cell wall damage, alteration of growth temperature, and is responsible for an osmotically stable cell wall as well as cell integrity during cell growth (Navarro-García et al. 1995, 2005; Ikner and Shiozaki 2005; Levin 2011). The function of the CW1 pathway has been described in detail for *S. cerevisiae*; it also operates in other fungal species, such as *C. albicans, Aspergillus fumigatus, Cryptococcus neoformans, Botrytis cinerea, Fusarium graminearum* and *Magnaporthe grisea* (Navarro-García et al. 1995; Xu et al. 1998; Hou et al. 2002; Kraus et al. 2003; Rui and Hahn 2007; Valiante et al. 2008). However, there is a dearth of information about this pathway in extremophilic fungi.

An additional stress responsive system required for salt stress tolerance in yeast is calcineurin; this protein phosphatase complex is dependent on calcium ion and calmodulin (Nakamura et al. 1993; Mendoza et al. 1994). Calcineurin is required for the genes transcription of sodium and calcium ion ATPases and a cell wall β-1,3 glucan synthase through regulating CRZ1/TCN1, the downstream zinc-finger transcription factor (Matheos et al. 1997; Stathopoulos-Gerontides et al. 1999). The salt-responsive calcineurin-CRZ1 pathway is also involved in yeast stress responses (Juvvadi et al. 2014). When CRZ1 was overexpressed in the industrial baker's yeast HS13 strain, tolerance to both salt and freezing was increased (Panadero et al. 2007).

The calcineurin pathway (Fig. 2.1), unlike HOG1, is highly conserved in eukaryotes from yeast to animals, which can be searched in the public nucleic acid sequence repository (http://www.ncbi.nlm.nih.gov/genbank). In plants, the physiology functions of calcineurin have been clarified (Luan et al. 1993; Allen and Sanders 1995), the osmotic stress resistance is associated with the increased expression of calcineurin pathway genes. In fungi, the calcineurin homologs PsCNA1/PsCNB1 from the wheat rust disease fungus *Puccinia striiformis* have been studied. Results indicated that the calcineurin signaling pathway participates in stripe rust morphogenetic differentiation, especially the formation of haustoria during the early stage of infection and during the production of urediospores (Zhang et al. 2012). Interestingly, calcineurin may be a multifunctional enzyme, because it was required not only for drug tolerance but also hyphal growth and virulence in *Candida tropicalis* (Chen et al. 2014). In contrast, relatively little is known about the calcineurin pathway in extremophilic fungal.

2.3 Osmoregulation in Extremophilic Fungi

To remain viable, fungi under extreme conditions must regulate and keep essential cellular processes. The fluidity and components of the plasma membrane play important roles in maintaining the cell membrane physiological functions and the adaption to extreme conditions (Turk et al. 2007; Zhang et al. 2015). Plasma membrane fluidity has been regarded as a typical indicator of fitness for survival in extreme environments (Turk et al. 2007). Unsaturated fatty acids are the key compounds in plasma membrane and cellular unsaturated fatty acids constitutions are directly controlled by fatty acid desaturases. In *Pichia pastoris* GS115, cellular fatty acids compositions were changed with the increased or decreased expression of desaturases; in addition, deletions of fatty acid desaturases give rise to increased resistance to adverse environmental stress (Zhang et al. 2015). In Turk's study (Turk et al. 2007), all the tested fungi showed increased plasma membrane fluidity in response to increased salt concentrations. However, when salinity exceeded their optimal range, the extremophilic fungi (*Hortaea werneckii*, *Crytococcus liquefaciens*) showed decreased plasma membrane fluidity, reflecting the limitation of cell membrane remodeling and suggesting extremophilic fungi could have different resistant mechanism.

When responding to osmotic stress, eukaryotic microorganism cells accumulate some metabolites inside the cells to equilibrate the cytoplasm osmotically with the outside of the cells (Brown 1978). Many polyols have been reported to contribute to fungal survival at high-salt concentrations or drought conditions. And among these compatible solutes, glycerol and trehalose have been extensively studied. Glycerol is the major product when extremophilic fungi, such as *Aspergillus glaucus*, grow on glucose-contained medium with high concentrations of NaCl (Liu et al. 2015). In the process of glycerol biosynthesis, a number of key enzymes determine the production of intracellular glycerol and therefore impact on osmotic stress tolerance. Glycerol 3-phosphate dehydrogenase encoded by the gene of GPD1 in *S. cerevisiae* is important for yeast survival under osmotic stress (Albertyn et al. 1994). The yeast glycerol 3-phosphatases gpp1p and gpp2p are also essential for glycerol biosynthesis, but their roles in the cellular responses to osmotic, anaerobic, and oxidative stress are different (Pahlman et al. 2001).

To conquer high-osmotic stress by biosynthesis of glycerol is inefficient and uneconomical; the active retention and uptake of glycerol become necessary when fungi are at high density. Aquaglyceroporins [AQGPs; (GlpFs in yeast)] transport glycerol along with water and other uncharged solutes are involved in osmoregulation in myriad species. The two genes encoding AQGPs in the yeast genome, Fps1 (Oliveira et al. 2003; Tamás et al. 1999) and Yfl054 (Hohmann et al. 2000; Oliveira et al. 2003), are functional glycerol facilitators. Fps1 plays a key role in yeast osmoregulation by regulating intracellular glycerol levels during changes in external osmolarity (Luyten et al. 1995; Hohmann et al. 2007; Ahmadpour et al. 2014), whereas the cellular function of Yfl054 remains uncertain (Oliveira et al. 2003). Recently, the AQGPs of arbuscular mycorrhizal fungi have received a lot of attention. The aquaglyceroporin GintAQPF2 from *Glomus intraradices*, a member of the γ subgroup (Xu et al. 2013), showed high activity when exposed to polyethylene glycol and high capacity to transport water, which is crucial for transformed yeast cells to survive osmotic stress (Li et al. 2013). Similarly, two AQGPs (Lacbi1:317173 and 391485) in the ectomycorrhizal basidiomycete *Laccaria bicolor*, belonging to the α subgroup (Xu et al. 2013), enabled much higher water permeability than the orthodox Aquaporins (AQP) Lacbi1:392091 (Dietz et al. 2011). Given the relationship between mycorrhizal fungus and its host plant, AQGPs' function may partially explain the higher stress tolerance in endophyte plant than in endophyte-free plant. In the halophilic fungus

Aspergillus glaucus, the aquaglyceroporin gene *AgGlpF* has been demonstrated to be a water/glycerol channel (Liu et al. 2015). Interestingly, *AgGlpF* functions not only in *S. cerevisiae* and *Neurospora crassa* but also in model plants such as *Arabidopsis thaliana*. When *AgGlpF* was expressed in *A. thaliana*, the transgenic lines survived under high osmotic pressure and particularly under drought stress (Liu et al. 2015).

Another metabolite associated with osmoregulation is trehalose, the highly stable disaccharide is commonly found in nature. Trehalose has multiple functions (Elbein et al. 2003) and is well known for osmoprotection where correlations between accumulation of trehalose and high resistance to various stresses have been observed (Crowe et al. 1992). However, an unbiased study carried out by Petitjean et al. (2015) casted doubt on this long-held belief that trehalose is an osmoprotectant. By combining the use of mutant strains expressing catalytically inactive variants of Tps1, MAL^+ yeast strains were able to accumulate trehalose from an exogenous supply, the authors found that the stress-protecting role of trehalose in the yeast was largely overestimated: trehalose actually was unable to protect yeast cells from dying; on the contrary, it is the Tps1 protein, the key enzyme for synthesis of trehalose, that played essential roles for yeast survival in response to temperature, oxidative and desiccation stresses (Petitjean et al. 2015). Though we do not know the concrete molecular mechanism of Tps1 on playing a secondary function, the phenomenon may indicate osmoregulation is not limited to polyols only. To uncover osmoregulation mechanisms, more polyols and moonlighting proteins must be further investigated.

2.4 Abiotic Stress Resistance Genes in Extremophilic Fungi

Like *S. cerevisiae*, the yeast *Debaryomyces hansenii* that is usually found in salty environments has been extensively investigated in recent years. As a salt-loving fungus, it is able to accumulate high concentrations of sodium without suffering from any adverse effect. It also grows well under additional stress factors such as high temperature and extreme pH in the presence of 0.25 M NaCl (Almagro et al. 2000). Through screening of *S. cerevisiae* transformants that contain the genomic library prepared from *D. hansenii* (Prista et al. 2002), a series of genes associated with salt tolerance were indentified and characterized. For example, the *DhGZF3* gene, which encodes a GATA transcription factor homologues to Dal80 and Gzf3 in *S. cerevisiae*, has been functionally analyzed in *D. hansenii*, but the gene was verified to be a negative transcription factor when it was expressed in *S. cerevisiae* (García-Salcedo et al. 2006). Using the cDNA library from the stress-tolerant basidiomycetes yeast *Rhodotorula mucilaginosa*, more than 100 *S. cerevisiae* transformants that are tolerant to concentrations of various osmolites have been screened by Gostinčar and Turk (2012). Among the sequenced clones, 12 genes mediated increased stress tolerance in the *R. mucilaginosa* transformants. Recently, from the *D. hansenii* genome database, Pereira et al. (2014) analyzed nine candidates of polyol/H(+) symporters by heterologous expression in *S. cerevisiae*. Five distinct polyol/H(+) symporters were confirmed, among which two symporters were tested to be specific for uncommon substrates as galactitol and D-(+)-chiro-inositol.

Interestingly, the stress tolerance genes in extremophilic fungi are scarcely reported and their functions need more research (Table 2.1). These genes could be of significant importance in transgenic biotechnology. Above all, the abiotic stress resistance genes isolated from extremophilic fungi appear to be more resistant than the homologs from non-extremophiles. *EhHOG*, as mentioned above, is the *E. herbariorum* MAPK kinase gene similar to the HOG1 homologs from *A. nidulans*, *S. cerevisiae*, *Schizosaccharomyces pombe*, and most other fungi; but HOG1 mutant complemented with EhHOG outperformed the wild type under high salt and freezing–thawing conditions (Yan et al. 2005), indicating the higher genetic fitness of EhHOG in comparison with the corresponding HOG from *S. cerevisiae*. Some genes isolated from the halophilic fungus *A. glaucus* were also found to be more resistant to osmotic stress than the common fungi such as *S. cerevisiae* and *Magnaporthe oryzae*. A yeast expression library containing full-length cDNAs of *A. glaucus* was constructed and used to screen salt resistance transformants in our laboratory. The ribosomal protein L44 (RPL44), one of the proteins of the large ribosomal subunit 60S, was obtained according to its association with salt resistance. In comparison with the homologous sequence from *M. oryzae*, MoRPL44 in a yeast expression system, the yeast cells with overexpressed *AgRPL44* were more resistant to salt, drought and heavy metals than yeast cells expressing *MoRPL44* at a similar level of stress. In addition, when *AgRPL44* was introduced into *M. oryzae*,

Table 2.1 Abiotic stress resistance genes in extremophilic fungi

Extremophilic fungi	Protein/function	Resistance gene	Effect on abiotic stress	Reference (s)
Eurotium herbariorum	High-osmolarity glycerol	*EhHOG*	Tolerance to high salt and freezing–thawing	Yan et al. (2005)
Aspergillus glaucus	Ribosomal protein L44	*AgRPL44*	Resistant to salt, drought, heavy metals	Liu et al. (2014)
A. glaucus	Ribosomal protein subunit	*AgRPS3aE*	Resistant to salt, sorbitol	Liang et al. (2015)
A. glaucus	Aquaglyceroporins	*AgglpF*	Tolerance to salt, sorbitol, $CuSO_4$	Liu et al. (2015)
A. glaucus	Chitinase	To be identified	Enhanced salt and drought	Zhang and Liu, unpublished
A. glaucus	Cellulose	To be identified	Enhanced salt and drought	Li et al., unpublished
A. glaucus	Glucanase	To be identified	Enhanced salt and drought	Zhang and Liu, unpublished

the transformants also displayed significantly enhanced tolerance to salt and drought, indicating the unique osmosis resistance ability from the halophilic fungus. Similar results were obtained in the studies of another ribosomal protein subunit of AgRPS3aE (Liang et al. 2015), a aquaglyceroporins of AgglpF (Liu et al. 2015), a 60S protease subunit and 14 other unknown or predicted genes including the cell wall degrading enzymes such as chitinase, cellulase and glucanase (Zhang and Liu, unpublished). The common features of all these genes are highly conserved, at least not specific to extremophilic fungi, but they obviously support transgenic cells or organisms surviving under stress conditions, suggesting special mechanisms to be uncovered in future and potential values for genetic engineering.

2.5 Genetic Application Strategies

Foreign gene transfer from unrelated organisms frequently happened in the course of plant evolution; and indeed, plants can benefit from the presence of foreign genes, such as CtHSR1 from the halophytic yeast *Candida tropicalis* transferring the ability to adapt to adverse environments (Martínez et al. 2015). Up to now, many genes from diverse fungi have been successfully transferred into plants. *T. harzianum* is commonly used as a mycoparasite fungus for agriculture biological control. On the other hand, *T. harzianum* has moderate levels of tolerance to stressors; it is regarded as an active agent with abilities to induce resistance to abiotic stress in plants and to promote plant growth (Dana et al. 2006; Shoresh et al. 2010). *T. harzianum* provides an excellent genetic pool for cloning multiresistance genes. For example, ThHog1 (Delgado-Jarana et al. 2006), HSP70 (Montero-Barrientos et al. 2008, 2010) and Thkel1 (Hermosa et al. 2011) have been successively characterized to be the genes responsible for resistance to salt or other stressors. Interestingly, some genes generally associated with cell wall degradation were shown to be associated with stress tolerance when they were transferred into plants (Nicolás et al. 2014). This is not dissimilar to the ribosomal protein subunits RPL44 and RPS3aE described above. Considering the moderate levels of tolerance in *T. harzianum*, the homologous genes from extremophilic fungi could be even more resistant. Therefore, it is important and necessary to identify and characterize more genes related to stress resistance regardless of their origin and novelty.

On the other hand, the potential applications of these fungal resistant genes are strengthened according to their functions in transgenic plants. Three major steps are required to obtain a transgenic plant line. Briefly: (i) the identification of a function determined gene; (ii) transfer of the gene into the target plants and (iii) the selection of a tolerance improved line. How to achieve an economical, efficient, genetically stable and biological safety transgenic plant would require further research and optimisation. Crops are often exposed to multiple stresses. One gene with multiple actions such as *Trichoderma* HSP70 is no doubt efficient and economical. Transgenic *Arabidopsis* containing HSP70 showed an enhanced tolerance to oxidative, osmotic and salt stresses (Montero-Barrientos et al. 2010). The highly conserved ribosomal protein subunits like RPL44 and RPS3aE are also promising candidates for creating tolerance enhanced crops (Liu et al. 2014; Liang et al. 2015). These genes are generally in the downstream of resistant pathway and likely to have direct contribution to stress tolerance. Therefore, other physiological traits in transgenic plants may not be seriously affected, even if all these genes are overexpressed.

As to other genes, the expression levels and patterns of transferred sequence may have many unpredicted impacts on the growth and development of transgenic plants: low levels of expression may have no anticipated function; however, highly levels of expression probably affect the bioassay or productivity of the plant. Thus, timely expression of transferred genes is in need. Inducible promoters are available in stress resistance genetic engineering. In plants, there exist many stress responsive genes, particularly in sensitive plants (Dey et al. 2015). The resistant gene or genes from an extremophilic fungus driven by a plant inducible promoter constitutes the so called "two-component sensor systems" (de Wit 1992). This strategy will solve the problem of excess cellular materials and energy (ATP) consumption.

2.6 Concluding Remarks

The three most pervasive environmental problems afflicting crop growth and development throughout the world are saline soils, drought and temperature extremes. Due to the lack of abiotic stress resistance in germplasms, conventional anti-stress breeding in plants is largely not successful. Fortunately, recent researches that focus on molecular mechanisms and genes cloning in extremophilic fungi offer hope in solving these problems. Based on the two significant osmotic-resistance pathways in yeast systems, a series of homolog resistance genes in extremophilic fungi have been characterized. And indeed, these genes were confirmed to be more resistant than their homologs in non-extremophiles, suggesting certain specific genetic base and novel mechanism exist in extremophilic fungi. More promising resistant genes will be exploited with the publication of the whole genome sequences of extremophilic fungi. Therefore, a reliable filamentous genetic model of exrememophilic fungi must be established as soon as possible, for to date, yeast is the only system that has been studied extensively.

T. harzianum cell wall degrading enzymes have shown to be effective candidates for promoting tolerance to various stresses; and several genes have been used to create multi-resistant plants (Nicolás et al. 2014). These homologous genes in extremophilic fungi should be identified as well, because of their special genetic background. Ribosomal proteins AgRPL44 and AgRPS3aE, which supported plants to survive under high osmosis conditions, are highly conserved in organisms, suggesting these genes can be prospective candidates for creating genetic modified crops without consideration of their biosafety.

As described above, salt tolerance genes are extensively studied in halophilic fungi. However, only a few thermophilic fungi have been examined: *T. lanuginosus*, *Myceliophthora thermophila* and *Thielavia terrestris*; and, these studies focused mainly on the biomass-degrading application and regulation mechanisms (Singh et al. 2003; Berka et al. 2011; McHunu et al. 2013). Genes contributing to temperature extremes should be uncovered in these thermophiles. Once identified, the high or low temperature resistance genes, with halotolerant genes, could be used to generate multiple resistant plants. In our laboratory, we are cloning and transforming the abiotic stress resistance genes observed from extremophilic fungi to plants (Liu et al. 2014, 2015; Liang et al. 2015). Additional genes and their functions will be elucidated soon.

Acknowledgments This work was partially supported by a grant of National Natural Science Foundation of China (31171794) and a project of Ministry of Agriculture of China (2011ZX08009-001). The author is very grateful to Dr. Diane Purchase who reviewed individual sentences and gave tremendous help to the preparation of this chapter. The author also likes to express his thanks to Dr. Liu Shou-An who kindly helped to revise the manuscript of this chapter.

References

Ahmadpour D, Geijer C, Tamás MJ, Lindkvist-Petersson K, Hohmann S (2014) Yeast reveals unexpected roles and regulatory features of aquaporins and aquaglyceroporins. Biochim Biophys Acta 1840:1482–1491

Albertyn J, Hohmann S, Thevelein JM, Prior BA (1994) GPD1, which encodes glycerol 3-phosphate dehydrogenase, is essential for growth under osmotic stress in *Saccharomyces cerevisiae*, and its expression is regulated by the high osmolarity glycerol response pathway. Mol Cell Biol 14:4135–4144

Allen GJ, Sanders D (1995) Calcineurin, a type 2B protein phosphatase, modulates the calcium-permeable slow vacuolar ion channel of stomatal guard cells. Plant Cell 7:1473–1483

Almagro A, Prista C, Castro S, Quintas C, Madeira-Lopes A, Ramos J et al (2000) Effects of salts on *Debaryomyces hansenii* and *Saccharomyces cerevisiae* under stress conditions. Int J Food Microbiol 56:191–197

Alonso-Monge R, Navarro-García F, Román E, Negredo AI, Eisman B, Nombela C et al (2006) The Hog1 mitogen-activated protein kinase is essential in the oxidative stress response and chlamydospore formation in *Candida albicans*. Eukaryot Cell 2(2):351–361

Alpert P (2000) The discovery, scope, and puzzle of desiccation tolerance in plants. Plant Ecol 151:5–17

Berka RM, Grigoriev IV, Otillar R, Salamov A, Grimwood J, Reid I et al (2011) Comparative genomic analysis of the thermophilic biomass-degrading fungi *Myceliophthora thermophila* and *Thielavia terrestris*. Nat Biotechnol 29(10):922–927. doi:10.1038/nbt.1976

Bhatnagar-Mathur P, Vadez V, Sharma KK (2008) Transgenic approaches for abiotic stress tolerance in plants: retrospect and prospects. Plant Cell Rep 27(3):411–424

Brewster JL, de Valoir T, Dwyer ND, Winter E, Gustin MC (1993) An osmosensing signal transduction pathway in yeast. Science 259:1760–1763

Brown AD (1978) Compatible solutes and extreme water stress in eukaryotic micro-organisms. Adv Microb Physiol 17:181–242

Chen YL, Yu SJ, Huang HY, Chang YL, Lehman VN, Silao FG et al (2014) Calcineurin controls hyphal growth, virulence, and drug tolerance of Candida tropicalis. Eukaryot Cell 13(7):844–854. doi:10.1128/EC.00302-13

Crowe JH, Hoekstra FA, Crowe LM (1992) Anhydrobiosis. Annu Rev Physiol 54:579–599

Dana MM, Pintor-Toro JA, Cubero B (2006) Transgenic tobacco plants overexpressing chitinases of fungal origin show enhanced resistance to biotic and abiotic stress agents. Plant Physiol 142:722–730

de Wit PJ (1992) Molecular characterization of gene-for-gene systems in plant-fungus interactions and the application of avirulence genes in control of plant pathogens. Annu Rev Phytopathol 30:391–418

Delgado-Jarana J, Sousa S, González F, Rey M, Llobell A (2006) ThHog1 controls the hyperosmotic stress response in Trichoderma harzianum. Microbiology 152(Pt 6):1687–1700

Dey N, Sarkar S, Acharya S, Maiti IB (2015) Synthetic promoters in planta. Planta. doi:10.1007/s00425-015-2377-2

Dietz S, von Bülow J, Beitz E, Nehls U (2011) The aquaporin gene family of the ectomycorrhizal fungus *Laccaria bicolor*: lessons for symbiotic functions. New Phytol 190:927–940. doi:10.1111/j.1469-8137.2011.03651.x

Elbein AD, Pan YT, Pastuszak I, Carroll D (2003) New insights on trehalose: a multifunctional molecule. Glycobiology 13:17R–27R

Enjalbert B, Smith DA, Cornell MJ, Alam I, Nicholls S, Brown AJP et al (2006) Role of Hog-1 stress-activated protein kinase in the global transcriptional response to stress in the fungal pathogen *Candida albicans*. Mol Biol Cell 17:1018–1032

García-Salcedo R, Casamayor A, Ruiz A, González A, Prista C, Loureiro-Dias MC et al (2006) Heterologous expression implicates a GATA factor in regulation of nitrogen metabolic genes and ion homeostasis in the halotolerant yeast *Debaryomyces hansenii*. Eukaryot Cell 5 (8):1388–1398

Gorbushina AA (2007) Life on the rocks. Environ Microbiol 9(7):1613–1631

Gostinčar C, Turk M (2012) Extremotolerant fungi as genetic resources for biotechnology. Bioengineered 3(5):293–297. doi:10.4161/bioe.20713

Gray J, Shear W (1992) Early life on land. Am Sci 80:444–456

Gupta B, Huang B (2014) Mechanism of salinity tolerance in plants: physiological, biochemical, and molecular characterization. Int J Genomics 701596. doi:10.1155/2014/701596

Hayes BME, Anderson M, Traven A, van der Weerden NL, Bleackley MR (2014) Activation of stress signalling pathways enhances tolerance of fungi to chemical fungicides and antifungal proteins. Cell Mol Life Sci 71:2651–2666

Heckman DS, Geiser DM, Eidell BR, Stauffer RL, Kardos NL, Hedges SB (2001) Molecular evidence for the early colonization of land by fungi and plants. Science 293(5532):1129–1133

Hermosa R, Botella L, Keck E, Jiménez JA, Montero-Barrientos M, Arbona V et al (2011) The overexpression in *Arabidopsis sthaliana* of a *Trichoderma harzianum* gene that modulates glucosidase activity, and enhances tolerance to salt and osmotic stresses. J Plant Physiol 168:1295–1302

Hohmann I, Bill RM, Kayingo I, Prior BA (2000) Microbial MIP channels. Trends Microbiol 8:33–38. doi:10.1016/S0966-842X(99)01645-5

Hohmann S, Krantz M, Nordlander B (2007) Yeast osmoregulation. Methods Enzymol 428:29–45. doi:10.1016/S0076-6879(07)28002-4

Horodyski RJ, Knauth LP (1994) Life on land in the precambrian. Science 263(5146):494–498

Hou Z, Xue C, Peng Y, Katan T, Kistler HC, Xu J-R (2002) A mitogen-activated protein kinase gene (MGv1) in *Fusarium graminearum* is required for female fertility, heterokaryon formation, and plant infection. Mol Plant Microbe Interact 15(11):1119–1127

Ikner A, Shiozaki K (2005) Yeast signalling pathways in the oxidative stress response. Mutat Res 569(1–2):13–27

Juvvadi PR, Lamoth F, Steinbach WJ (2014) Calcineurin as a multifunctional regulator: unraveling novel functions in fungal stress responses, hyphal growth, drug resistance, and pathogenesis. Fungal Biol Rev 28(2–3):56–69

Kis-Papo T, Grishkan I, Oren A, Wasser SP, Nevo E (2001) Spatiotemporal diversity of filamentous fungi in the hypersaline Dead Sea. Mycol Res 105:749–756

Kraus PR, Fox DS, Cox GM, Heitman J (2003) The *Cryptococcus neoformans* MAP kinase Mpk1 regulates cell integrity in response to antifungal drugs and loss of calcineurin function. Mol Microbiol 48(5):1377–1387

Lam HM, Xu X, Liu X, Chen W, Yang G, Wong FL et al (2010) Resequencing of 31 wild and cultivated soybean genomes identifies patterns of genetic diversity and selection. Nat Genet 42 (12):1053–1059. doi:10.1038/ng.715

Levin DE (2011) Regulation of cell wall biogenesis in *Saccharomyces cerevisiae*: the cell wall integrity signaling pathway. Genetics 189(4):1145–1175

Li T, Hu YJ, Hao ZP, Li H, Chen BD (2013) Aquaporin genes GintAQPF1and GintAQPF2 from *Glomus intraradices* contribute to plant drought tolerance. Plant Signal Behav 8:e24030. doi:10.4161/psb.24030

Liang X, Liu Y, Xie L, Liu X, Wei Y, Zhou X et al (2015) A ribosomal protein AgRPS3aE from halophilic *Aspergillus glaucus* confers salt tolerance in heterologous organisms. Int J Mol Sci 16(2):3058–3070. doi:10.3390/ijms16023058

Liu XD, Xie L, Wei Y, Zhou XY, Jia B, Liu J, Zhang SH (2014) Abiotic stress resistance, a novel moonlighting function of ribosomal protein RPL44 in the halophilic fungus *Aspergillus glaucus*. Appl Environ Microbiol 80(14):4294–4300. doi:10.1128/AEM.00292-14

Liu XD, Wei Y, Zhou XY, Pei X, Zhang SH (2015) *Aspergillus glaucus* aquaglyceroporin gene *glpF* confers high osmosis tolerance in heterologous organisms. Appl Environ Microbiol 81 (19):6926–6937. doi:10.1128/AEM.02127-15

Luan S, Li W, Rusnak F, Assmann SM, Schreiber SL (1993) Immunosuppressants implicate protein phosphatase regulation of K + channels in guard cells. Proc Natl Acad Sci USA 90:2202–2206

Luyten K, Albertyn J, Skibbe WF, Prior BA, Ramos J, Thevelein JM et al (1995) Fps1, a yeast member of the MIP family of channel proteins, is a facilitator for glycerol uptake and efflux and is inactive under osmotic stress. EMBO J 14:1360–1371

Martínez F, Arif A, Nebauer SG, Bueso E, Ali R, Montesinos C et al (2015) A fungal transcription factor gene is expressed in plants from its own promoter and improves drought tolerance. Planta 242(1):39–52. doi:10.1007/s00425-015-2285-5

Matheos DP, Kingsbury TJ, Ahsan US, Cunningham KW (1997) Tcn1p/Crz1p, a calcineurin-dependent transcription factor that differentially regulates gene expression in *Saccharomyces cerevisiae*. Genes Dev 11:3445–3458

McHunu NP, Permaul K, Abdul Rahman AY, Saito JA, Singh S, Alam M (2013) Xylanase superproducer: genome sequence of a Ccmpost-loving thermophilic fungus, *Thermomyces lanuginosus* strain SSBP. Genome Announc 1(3) e00388–13. doi:10.1128/genomeA.00388-13

Mendoza I, Rubio F, Rodriguez-Navarro A, Pardo JM (1994) The protein phosphatase calcineurin is essential for NaCl tolerance of *Saccharomyces cerevisiae*. J Biol Chem 269:8792–8796

Montañés FM, Pascual-Ahuir A, Proft M (2011) Repression of ergosterol biosynthesis is essential for stress resistance and is mediated by the Hog1 MAP kinase and the Mot3 and Rox1 transcription factors. Mol Microbiol 79(4):1008–1023

Montero-Barrientos M, Hermosa MR, Cardoza RE, Gutiérrez S, Nicolás C, Monte E (2010) Transgenic expression of the *Trichoderma harzianum* HSP70 gene increases *Arabidopsis* resistance to heat and other abiotic stresses. J Plant Physiol 167:659–665

Montero-Barrientos M, Hermosa R, Nicolás C, Cardoza RE, Gutiérrez S, Monte E (2008) Overexpression of a *Trichoderma* HSP70 gene increases fungalresistance to heat and other abiotic stresses. Fungal Genet Biol 45:1506–1513

Nakamura TY, Liu Y, Hirata D, Namba H, Harada S, Hirokawa T et al (1993) Protein phosphatase type 2B (calcineurin)-mediated, FK506-sensitive regulation of intracellular ions in yeast is an important determinant for adaptation to high salt stress conditions. EMBO J 12:4063–4071

Navarro-García F, Sánchez M, Pla J, Nombela C (1995) Functional characterization of the MKC1 gene of *Candida albicans*, which encodes a mitogen-activated protein kinase homolog related to cell integrity. Mol Cell Biol 15(4):2197–2206

Navarro-García F, Eisman B, Fiuza SM, Nombela C, Pla J (2005) The MAP kinase Mkc1p is activated under different stress conditions in *Candida albicans*. Microbiology 151(8):2737–2749

Nicolás C, Hermosa R, Rubio B, Mukherjee PK, Monte E (2014) *Trichoderma* genes in plants for stress tolerance- status and prospects. Plant Sci 228:71–78

Oliveira R, Lages F, Silva-Graça M, Lucas C (2003) Fps1p channel is the mediator of the major part of glycerol passive diffusion in *Saccharomyces cerevisiae*: artefacts and re-definitions. Biochim Biophys Acta 1613:57–71. doi:10.1016/S0005-2736(03)00138-X

Pahlman AK, Granath K, Ansell R, Hohmann S, Adler L (2001) The yeast glycerol 3-phosphatases gpp1p and gpp2p are required for glycerol biosynthesis and differentially involved in the cellular responses to osmotic, anaerobic, and oxidative stress. J Biol Chem 276:3555–3563

Panadero J, Hernández-López MJ, Prieto JA, Randez-Gil F (2007) Overexpression of the calcineurin target CRZ1 provides freeze tolerance and enhances the fermentative capacity of baker's yeast. Appl Environ Microbiol 73(15):4824–4831

Pereira I, Madeira A, Prista C, Loureiro-Dias MC, Leandro MJ (2014) Characterization of new polyol/H+ symporters in *Debaryomyces hansenii*. PLoS ONE 9(2):e88180. doi:10.1371/journal.pone.0088180

Petitjean M, Teste MA, François JM, Parrou JL (2015) Yeast tolerance to various stresses relies on the Trehalose-6P Synthase (Tps1) protein. Not on Trehalose J Biol Chem 290(26):16177–16190. doi:10.1074/jbc.M115.653899

Prasad AR, Maheshwari R (1978) Purification and properties of trehalase from the thermophilic fungus Humicola lanuginosa. Biochim Biophys Acta 525(1):162–170

Prista C, Soeiro A, Vesely P, Almagro A, Ramos J, Loureiro-Dias MC (2002) Genes from *Debaryomyces hansenii* increase salt tolerance in *Saccharomyces cerevisiae* W303. FEMS Yeast Res 2:151–157

Prista C, Loureiro-Dias MC, Montiel V, García R, Ramos J (2005) Mechanisms underlying the halotolerant way of *Debaryomyces hansenii*. FEMS Yeast Res 5(8):693–701

Redman RS, Kim YO, Woodward CJDA, Greer C, Espino L, Doty SL, Rodriguez RJ (2011) Increased fitness of rice plants to abiotic stress via habitat adapted symbiosis: a strategy for mitigating impacts of climate change. PLoS ONE 6(7):e14823. doi:10.1371/journal.pone.0014823

Rodriguez RJ, Henson J, Van Volkenburgh E, Hoy M, Wright L, Beckwith F et al (2008) Stress tolerance in plants via habitat-adapted symbiosis. ISME J 2(4):404–416. doi:10.1038/ismej.2007.106

Rui O, Hahn M (2007) The Slt2-type MAP kinase Bmp3 of *Botrytis cinerea* is required for normal saprotrophic growth, conidiation, plant surface sensing and host tissue colonization. Mol Plant Pathol. 8(2):173–184

Saito H, Posas F (2012) Response to hyperosmotic stress. Genetics 192(2):289–318

San José C, Monge RA, Pérez-Díaz R, Pla J, Nombela C (1996) The mitogen-activated protein kinase homolog HOG1 gene controls glycerol accumulation in the pathogenic fungus *Candida albicans*. J Bacteriol 178(19):5850–5852

Serrano R, Gaxiola R (1994) Microbial models and salt stress tolerance in plants. Crit Rev Plant Sci 13:121–138

Shoresh M, Harman GE, Mastouri F (2010) Induced systemic resistance and plant responses to fungal biocontrol agents. Annu Rev Phytopathol 48:21–43

Singh S, Madlala AM, Prior BA (2003) *Thermomyces lanuginosus*: properties of strains and their hemicellulases. FEMS Microbiol Rev 27(1):3–16

Stathopoulos-Gerontides A, Guo JJ, Cyert MS (1999) Yeast calcineurin regulates nuclear localization of the Crz1p transcription factor through dephosphorylation. Genes Dev 13:798–803

Tamás MJ, Luyten FC, Sutherland A, Hernandez J, Albertyn J, Valadi H et al (1999) Fps1p controls the accumulation and release of the compatible solute glycerol in yeast osmoregulation. Mol Microbiol 31:1087–1104. doi:10.1046/j.1365-2958.1999.01248.x

Thomas E, Roman E, Claypool S, Manzoor N, Pla J, Panwar SL (2013) Mitochondria influences CDR1 efflux pump activity, Hog1-mediated oxidative stress pathway, iron homeostasis and ergosterol levels in *Candida albicans*. Antimicrob Agents Chemother 57(11):5580–5599

Turk M, Abramović Z, Plemenitas A, Gunde-Cimerman N (2007) Salt stress and plasma-membrane fluidity in selected extremophilic yeasts and yeast-like fungi. FEMS Yeast Res 7(4):550–557

Valiante V, Heinekamp T, Jain R, Härtl A, Brakhage AA (2008) The mitogen-activated protein kinase MpkA of *Aspergillus fumigatus* regulates cell wall signaling and oxidative stress response. Fungal Genet Biol 45(5):618–627, 52

Xu J-R, Staiger CJ, Hamer JE (1998) Inactivation of the mitogen-activated protein kinase Mps1 from the rice blast fungus prevents penetration of host cells but allows activation of plant defense responses. Proc Natl Acad Sci USA 95(21):12713–12718

Xu H, Cooke JEK, Zwiazek JJ (2013) Phylogenetic analysis of fungal aquaporins provide insight into their possible role in water transport of mycorrhizal associations. Botany 91:495–504. doi:10.1139/cjb-2013-0041

Yan J, Song WN, Nevo E (2005) A MAPK gene from Dead Sea fungus confers stress tolerance to lithium salt and freezing-thawing: prospects for saline agriculture. Proc Natl Acad Sci USA 102(52):18992–18997

Zhan X, Zhu JK, Lang Z (2015) Increasing freezing tolerance: kinase regulation of ICE1. Dev Cell 32(3):257–258. doi:10.1016/j.devcel.2015.01.004

Zhang H, Guo J, Voegele RT, Zhang J, Duan Y, Luo H et al (2012) Functional characterization of calcineurin homologs PsCNA1/PsCNB1 in *Puccinia striiformis* f. sp. *tritici* using a host-induced RNAi system. PLoS ONE 7(11):e49262. doi:10.1371/journal.pone.0049262

Zhang M, Liu Z, Yu Q, Mao J, Zhang B, Xing L et al (2015) Deletion of genes encoding fatty acid desaturases leads to alterations in stress sensitivity in *Pichia pastoris*. FEMS Yeast Res 15 (4):fov020. doi:10.1093/femsyr/fov020

Chapter 3
Fungi as an Alternative to Agrochemicals to Control Plant Diseases

Alexander O. Emoghene and Anthony E. Futughe

3.1 Introduction

Majority of the populations living in developing countries are actively engaged in agriculture with a good percentage being small scale farmers, however, the turn out of their farm produce are low owing to crops crippling diseases. In Nigeria, smallholder farmers produce crops such as cocoa, cereals, potato, tomato, vegetable, yam, cassava, plantain, banana, orange, which are the raw materials for local industries and also contribute to the nation's economic development as foreign exchange earners (Oloruntoba 1989). Plant diseases account for considerable losses in crop production and storage. Currently, growers, particularly in developing countries like Nigeria still rely heavily on agrochemicals to prevent and/or control these crops threatening diseases. Despite the high effectiveness and ease of utilization, these agrochemicals can result in environmental contamination and pesticide residue presence on food, contributing to additional social and economic problems. Varieties of causal agents such as fungi, bacteria, viruses, nematodes amongst others have been implicated in plant diseases with an enormous reduction in crop yields globally. In most developing countries, Nigeria inclusive, crop losses are usually higher than their developed counterparts (FAO 2004).

A.O. Emoghene (✉)
Department of Microbiology, Faculty of Life Sciences, University of Benin, Benin City, Nigeria
e-mail: emoghene@uniben.edu; emogheney2k4@yahoo.co.uk

A.E. Futughe
Department of Natural Sciences, School of Science and Technology, Middlesex University, London, UK
e-mail: a.futughe@mdx.ac.uk; futugheanthony44@gmail.com

D. Purchase (ed.), *Fungal Applications in Sustainable Environmental Biotechnology*, Fungal Biology, DOI 10.1007/978-3-319-42852-9_3

Diseases caused by Oomycetes fungi of the order perenosporales present major problems world-wide. Important foliar diseases include late blight on potatoes, blue mould of tobacco, grape downy mildew, plus a wide range of other foliar blights and downy mildew on cereals, fruits, and vegetables (Coffey et al. 1984). In addition, *Phytophthora* and *Pythium* species are responsible for many pre- and post-harvest problems of fruits and vegetables including late blight of potato tubers (Barak et al. 1984), brown rot of citrus (Cohen 1981), and black pod of cocoa (McGregor 1984). Bacterial wilt of potato, tomato, eggplant, tobacco, groundnut and banana is caused by *Pseudomonas solanacerum* (Wheele 1969).

A serious shoot disease of *Amaranthus* spp. causing a blight of the young shoot which can result in a total crop failure, and associated with *Choanephora cucurbitarum* has been reported in Benin City, Nigeria (Ikediugwu 1981; Ikediugwu et al. 1994; Emoghene and Okigbo 2001). Evidence on the disease of crops such as fruits and vegetables and their known control measures are well documented in literature. A good example is strains of *Rhizoctonia* causing damping-off on a wide range of cultivated plants. These include cereals, potato, root and fodder crops, legumes, vegetables and ornamentals (Moore 1959). Sclerotia of *Rhizoctonia solani* are frequently found on potato tubers. *Botrytis cinerea* often cause damping-off of lettuce in association with *Pythium* and *Rhizoctonia. Uromyces appendiculatus* attack bean, *Puccinia asparagi* on asparagus, *Puccinia alli* on onion, leek and garlic, *Puccinia methae* on peppermint (Wheele 1969). A number of control measures which have been adopted include: (i) inspection and quarantine procedures, (ii) cultural methods and (iii) fungicide applications. Organomercurials such as methylmercury dicyandiamide (Panogen) and the compound tetramethythiuram disulphide (Thiram) are amongst those chemicals which have been commonly used. Similarly, copper fungicides including Bordeaux and Burgundy mixtures and some of the dithiocarbamates such as maneb and organic tin compounds have been applied to manage fungal plant pathogens. Other important classes of systemic fungicides such as carbamates, cymoxamil, acylanilides and alky phosphate have also been used in the control of crop diseases (Cohen and Coffey 1986).

The above control methods are effective but have their disadvantages. Fungicides may not be the most desirable means of disease control for several important reasons. Fungicides are heavily regulated and vary from country to country in their use and registration (Jones 1985). In addition, they are expensive, can cause environmental pollution, and may induce pathogenic resistance. They can also cause stunting and chlorosis of young seedlings (Jones 1985). Cultural methods can injure plants, are labour intensive, and are less attractive to commercial growers (Rytter et al. 1989).

The use of microorganisms to control crop pests and diseases is an exciting and rapidly advancing branch of biotechnology. Novel methods have been established by different researchers to control plant pests and plant diseases. For instance, Emoghene and Futughe (2011) reported a more sustainable control measure using soil solarization to control *Amaranthus viridis* shoot disease caused by *Choanephora cucurbitarum*. Biological control, a term first coined by Smith (1919) to denote insect pest control by the use of natural enemies, is another sustainable example. Biological control when effective is usually more enduring than any other control methods as reported by Baker and Cook (1982). Successful applications of biological control with the use of microorganisms against plant pathogens began with the control of crown gall with *Agrobacterium radiobacter* K84 (Kerr 1980), and seedling blight caused by *Pythium* and *Rhizoctonia* with *Trichoderma harizanum* (Harman and Bjorkman 1998). Ikediugwu et al. (1994) reported the biological control of the shoot disease of *Amaranthus viridis* caused by *Choanephora cucurbitarum* with *Bacillus subtilis*.

3.2 Agrochemicals

Agrochemicals including pesticides and fertilizers are considered the result of modern technology that depends on inorganic processes. Pesticide according to FAO (1989) is any substance or mixture of substances intended for preventing, destroying, or controlling any pest including vectors of human or animal diseases, unwanted species of plants or animals causing harm during, or otherwise interfering with, the production, processing, storage, or marketing of food, agricultural commodities, wood and wood products, or animal feedstuffs, or which may be administered to animals for the control of insects, arachnids or other pests in or on their bodies. Chemicals employed such as growth regulators, defoliants, desiccants, fruit thinning agents, or agents for preventing the premature fall of fruits, and substances applied to crops either before or after harvest to prevent deterioration during storage or transport are also included in the term. However, it excludes such chemicals used as fertilizers, plant and animal nutrients, food additives and animal drugs. The term pesticide is also defined by FAO in collaboration with UNEP (1990) as chemicals designed to combat the attacks of various pests and vectors on agricultural crops, domestic animals and human beings. The above definitions suggest that, pesticides are toxic chemical agents (basically organic compounds) that are intentionally introduced to attack crop pests and disease vectors in the environment. Pesticides are chemically synthesized compounds, devise or organisms that are applied routinely in agriculture in order to mitigate, destroy, attack or repel pests, pathogens and parasites. They include both organic and inorganic moieties and may be classified into different groups depending on their chemical compositions. Examples of these agrochemicals include organochlorines, organophosphates, carbamates, formamidines, thiocyanates, organotins, denitrophenols, synthetic pyrethroids and antibiotics (Bohmont 1990). Upon application, the fate of these agrochemicals in the soil and the transport processes that take place are dependent on: (i) the cumulative effects of the physicochemical characteristics such as adsorptivity, solubility, volatility and degradation rate; (ii) the soil's characteristic; (iii) application methods and (iv) the site condition (Jeong and Forster 2003).

3.3 Effect of Agrochemicals Usage

Over 15,000 metric tons of agrochemicals are applied in Nigeria annually, comprising about 135 pesticide chemicals marketed locally under 200 different produce brands and formulation, thereby making Nigeria one of the largest agrochemicals users in sub-Sahara African (Osibanjo and Adeyeye 1995). According to Kamrin (1997) the benefits of agrochemicals cannot be overemphasized, however, their uses are a source of environmental, human and other animal concerns. It has been estimated that over 98 % of sprayed insecticides and 95 % herbicides get to a destination other than their intended target, in addition to non-target species, air, water and soil (Miller 2004). Runoff of agrochemicals into aquatic environment is one of the causes of water pollution, while they can be air-borne and drifted to other fields, grazing areas, human settlements and undeveloped areas which can potentially affect other species. Repeated application can cause persistent resistance and sources of soil contamination. Agrochemicals poisoning incidence may occur as a result of misuse, storage near consumable food stuff or farm produce and the use of agrochemical containers for domestic purposes, such as the case of Iraq 1970 as reported by WHO (1990). People exposed to agrochemicals either accidentally or occupationally include: manufacturers, vendors/seller, mixers, transporters, loaders, operators of application equipment, growers, pickers and clean-up workers, and consumers of farm produce with pesticide residues. It has been estimated by WHO and UNEP that about 3 million workers in agriculture from developing countries suffer severe poisoning from agrochemicals each year with about 18,000 deaths (Miller 2004). As many as 25 million workers in developing countries may be affected with mild pesticide poisoning yearly (Jeyaratnam 1990; WHO 2006). Just recently, according to the Punch newspaper (2015) the deaths of 18 people in south-western Nigeria were attributed to strange disease probably associated with agrochemical poisoning.

3.4 Mechanisms of Fungi-Based Biocontrol of Plant Pathogens

Plants and fungi have different interactions resulting in different mechanisms of action. The most common mechanisms for fungi-based biocontrol of plant pathogens are: (i) parasitism, (ii) mutualism, (iii) predation, (iv) competition, (v) induced resistance and (vi) the production of antimicrobial substances. In order to interact, fungi must have some form of direct or indirect contact with the plant and/or plant's pathogen and often, several mechanisms act together to give the most effective biocontrol. Direct fungal-based biocontrol result from physical contact and a high-level of specificity for the plant's pathogen. In hyperparasitism, the plant's pathogen is directly attacked by a selective fungi-based biocontrol agent that destroys it or its propagules. Several fungal hypoparasites have been implicated in addition to those attacking the sclerotia e.g. *Coniothyrium minitans*, others attacking pathogenic fungal hyphae such as *Pythium oligandrum*. However, cases abound where a single fungal pathogen can be attacked by multiple hyperparasites. A good example is powdery mildew pathogens which are susceptible to different hyperparasites such as *Acremonium alternatum*, *Acrodontium crateriforme*, *Ampelomyces quisqualis*, *Cladosporium oxysporum* and *Gliocladium virens* (Milgroom and Cortesi 2004). Fungi predation, unlike hyperparasitism, is a more general, non-specific and less predictable levels of plant disease biocontrol. Some Fungi such as *Trichoderma* sp. exhibit predatory behaviour under nutrient-limited (e.g. cellulose) conditions by synthesizing a range of enzymes e.g. chitinase that are directed against pathogenic fungi cell walls like *Rhizoctonia solani* (Benhamou and Chet 1997). Genes encoding for cell wall degrading enzyme (CWDES) such as chitinolytic, glucanolytic, and proteolytic enzymes have been isolated and applied to enhance fungi-based biocontrol capabilities of *Trichoderma* strains (Elad et al. 1982; Chet et al. 1993; Lorito et al. 1993).

Indirect fungi-based biocontrol, in contrast, results from activities that do not involve targeting a plant's pathogen by a biocontrol active fungus. Reports have demonstrated that some lytic enzyme activity may induce indirect efficacy against plant pathogens e.g. oligosaccharides from fungal cell walls can stimulate plant host defences (Howell et al. 1988). According to Van Loon et al. (1998) and Ryu et al. (2004), substantial number of fungi products such as transglutaminase, elicitins and a-glucan in Oomycetes; chitin and ergosterol in all fungi; and xylanase in *Trichoderma* have elicited plant host defences. Stimulation and improvement of plant host defence mode of action by non-pathogenic fungi such as mycorrhizae is the most indirect form of 'antagonism' (Kloepper et al. 1980; Maurhofer et al. 1994; Lafontaine and Benhamo 1996).

Mycorrhizae are formed due to a mutualistic symbiosis between plants and fungi. A resting spore germinates upon perception of exudates from root of host plant resulting to an induced hyphal branching which heightened the tendencies of a direct symbiotic contact as illustrated in Fig. 3.1. This interaction enables ubiquitous root colonists assisting plants to take up nutrients especially phosphorus and micronutrients. Arbuscular mycorrhizal fungi also known as vesicular arbuscular mycorrhizal fungi start to form by continuous dichotomous branching of fungal hyphae about two days after its root penetration inside the cortical cell of the host plant. It is believed that arbuscules is the site of communication between the host plant and the fungus (Biermann and Linderman 1983). Arbuscular mycorrhizal fungi can prevent root infection during colonization by reducing the access sites and stimulating plant host defence. Linderman (1994) reported that arbuscular mycorrhizal fungi reduced root-knot nematode incidence. There are also various mechanisms allowing arbuscular mycorrhizal fungi to increase host plant's stress tolerance. One of such mechanisms includes the intricate network of fungal hyphae around the roots which prevent pathogen infection. Catska (1994) inoculated apple-tree seedlings with arbuscular mycorrhizal fungi *Glomus fasciculatum* and *Glomus macrocarpum* and observed suppressed apple replant disease caused by phytotoxic myxomycetes. Arbuscular mycorrhizal fungi also protect the host plant

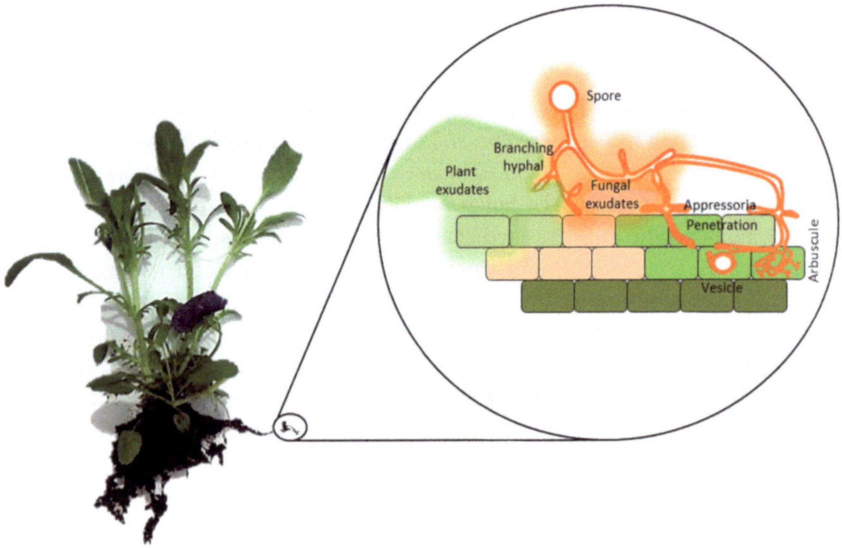

Fig. 3.1 Fungi-plant symbiotic relationship-mycorrhizea

against root-infecting pathogenic bacteria as reported by Garcia-Garrido and Ocampo (1989) where the damage on tomato caused by *Pseudomonas syringae* was reduced significantly as a result of mycorrhizal colonization of the tomato plant. The mechanisms include physical protection, chemical interactions and indirect effects (Fitter and Garbaye 1994). Enhanced nutrition to plant; morphological changes in the root by increased lignification; changes in the chemical composition of plant tissues such as antifungal chitinase, isoflavonoids are other mechanisms employed by arbuscular mycorrhizal fungi to indirectly suppress host plant pathogens (Morris and Ward 1992). Alleviation of abiotic stress and changes in the microbial content in the mycorrhizophere are also implicated mechanisms as reported by Linderman (1994).

Proliferation of ectomycorrhizae outside the root surface as against arbuscular mycorrhizal fungi, form a sheath around the root by the combination of mass of root and hyphae known as mantle. Multiple mechanisms in addition to antibiosis, fungistatic substances produce by plant roots in response to mycorrhizal infection and a physical barrier of the fungal mantle around the plant by ectomycorrhizal fungi give disease protection to the host plant (Duchesne 1994). According to Ross and Marx (1972) ectomycorrhizal fungi such as *Paxillus involutus* controlled effectively root rot caused by *Fusarium oxysporum* and *Fusarium moniliforme* in red pine. Inoculation of sand pine with *Pisolithus tinctorius*, another ectomycorrhizal fungus, controlled disease caused by *Phytophthora cinnamomi*. Literatures abound demonstrating post-harvest disease control by applying antagonistic microbes especially fungi (Table 3.1).

Table 3.1 Successful biocontrol of post-harvest diseases

Crop	Disease	Antagonist	Mechanism	Reference
Birch	Decay	*Trichoderma* sp.	Competition for nutrient and space	Shields and Atwells (1963)
Lemon	Green mold	*Trichoderma* sp.	Competition for nutrient and space	de Matose (1983)
Citrus	Green mold and Blue mold	*Trichoderma* sp. *and Bacillus* sp.	Competition for nutrient and space and antibiosis	Futughe (2007)
	Green mold	*Candida famata*	Induction of resistance	Arras (1996)
	Green mold	*Debaryomyces hansenii*	Competition for nutrient and space	Taqarort et al. (2008)
	Blue mold	*Cryptococcus laurentii*	Competition for nutrient and space	Liu et al. (2010)
	Green mold	*Wickerhamomyces anomalus*	Antibiosis	Platania et al. (2012)
Pine	Penicillium rot	*Trichoderma* sp.	Competition for nutrient and space	Lindgren and Harvey (1952)
Pineapple	Decay	Attenuated strains of *Penicillium* sp.	Competition for nutrient and space	Lim and Rorhbach (1980)
Potato	Soft rot	*Pseudomonas putida*	Antibiosis	Colyer and Mount (1984)
Stone fruit	Brown rot	*Bacillus subtilis*	Antibiosis	Pusey and Wilson (1984)
Straw berry	Botrytis rot	*Trichoderma* sp.	Competition for nutrient and space	Tronsmo and Dennis (1983)

3.5 Examples of Fungal-Based Biocontrol of Plant Pathogens

It is clear that there are a number of advantages in using fungal-based biocontrol against plant pathogens, including:

(i) Prevents environmental pollution of soil, air and water.
(ii) Maintains healthy biological control balance by avoiding adverse effects on beneficial organisms.
(iii) Less expensive than agrochemicals and devoid of resistance problems.
(iv) Fungi-based biocontrols are self-maintaining in simple application while agrochemicals need repeated applications.
(v) Very effective for soil-borne pathogens where agrochemical approach is not feasible.
(vi) Eco-friendly, durable and long-lasting.

(vii) Very high control potential by integrating fungicide resistant antagonists.
(viii) Helps in inducing system resistance among the crop species e.g. *Trichoderma* sp. resistant to fungicide such as Benomyl and Metalaxyl among others.

We present two examples to illustrate the potential of fungal-based biocontrol.

3.5.1 Against **Fusarium** *Wilt of Tomato* **(Licopersicon Esculentum** *Mill*) *by* **Trichoderma** *Species*

Tomato (*Licopersicon esculentum* Mill) is a very important fruit vegetable that is used extensively for salad, soups and stews. Industrially, ripe tomato fruits are processed into puree, sauce and juice (Purseglore 1977). Many countries around the world have large scale production of tomato with the United States, Italy, Spain and Bulgaria as the leading producers (Simons and Sobulo 1975; Purseglore 1977). Tomato has been cultivated almost all over Nigeria for decades with the most predominant area being the North and South Western regions (Erinle 1979; Denton and Swarup 1983). Crop crippling diseases are serious limitations to tomato production (Wheele 1969; Simons and Sobulo 1975; Erinle 1979; Adelana and Simons 1980; Denton and Swarup 1983). Bacterial and fungal wilts are the most commonly known field diseases of tomato. *Ralstania* (*Pseudomonas*) *solanacearum* and *Fusarium oxysporum* f. sp *lycopersici* are the most devastating in many growing belts of the world (Wheele 1969; Walker 1971; Prior et al. 1990), Nigeria inclusive (Erinle 1977; Osuinde and Ikediugwu 1995). *F. oxysporum* f. sp *lycopersici* and *Ralstania* (*P*) *solanacearum* which are causative agents of tomato wilt disease are soil inhibiting microorganisms and survive saprophytically in soil (Walker 1957, 1971; Park 1959). *Fusarium* wilt of tomato caused by *F. oxysporum* f. sp *lycopersici* is a serious economic problem in Southern Western Nigeria (Erinle 1977). Tomato wilts, like most soil-borne diseases of plant have proved extremely difficult to control by the application of agrochemicals which are expensive and hazardous to man and the environment. Currently, research into alternative sustainable control measures to agrochemicals is getting global attention. Efforts have been made in some parts of the world towards genetic (by using resistant cultivars) and biological control (biotechnology). However, the use of resistant cultivars has been complicated by the occurrence of more than one species of some wilt pathogens (Walker 1957), resulting in costly loss of resistance in the field, thereby, making biological control mostly favourable as it has attracted a growing market base with more diversified biotechnological products (Ardakani et al. 2009). A common form of biological control such as the use of fungi encourages the growth of microorganisms (e.g. fungi) antagonistic to the pathogen in the environment of the crop plant to the detriment of the pathogen (Alexander 1977; Baker and Cook 1982).

Trichoderma species, a fungus, has been used as an alternative to agrochemicals to control *Fusarium* wilt disease of tomato. Its potential was previously found to be antagonistic to *F. oxysporium* f. sp. *Lycopersici* in vitro. Seedlings of tomato inoculated with the pathogen (*F. oxysporium* f. sp. *Lycopersici*) alone revealed mild wilt symptoms by the following day and by the fourth (4th) day; plant sagged and wilted completely (Table 3.2). In contrast, fungi control of the plant disease was observed with the *Trichoderma* spp., depending on the concentration of spores and method of application (whether root-dip or direct soil inoculation), on whether the pathogen was applied simultaneously with antagonist, and on how long spores of the pathogen was allowed to grow ahead of the spores of antagonist (Osuinde et al. 2002). When the pathogen and antagonist were applied simultaneously, the result depending on the spores concentration and method of application: 10^3 spores/mL delayed symptom expression only for one day (Table 3.2). Mild wilt symptoms which affected 40 % and 80 % plants in root-dip and direct soil inoculation methods respectively was observed from day 2 to day 4. However, when 10^6 spores/mL of antagonist was applied by root-dip method, there was no wilt development at all as plants were healthy throughout the study period. But in the direct soil inoculation method, mild wilt was observed in 60 % of the seedlings by the 2nd and 3rd day. When the spores of the pathogen were allowed to grow one day (24 h) ahead of spores of antagonist, the result also depended on the concentration of the spores and method of application. When 10^3 spores/mL of antagonist was used, mild wilt was observed the following day in 80 and 100 % of seedling up to the 3rd day in root-dip and direct soil inoculation methods respectively. Nevertheless, when 10^6 spores/mL of antagonist was applied there was no wilt symptoms in plants in the root-dip method, while 40 % of plants developed mild wilt by 2nd day up to 4th day in the direct soil inoculation methods (Tables 3.2 and 3.3) (Osuinde et al. 2002). There was no effect on progress of wilt upon germination of spores of the pathogen after 2 day (48 h) ahead of spores of antagonist irrespective of the spore concentrations of antagonist and application methods. All the plant (100 %) were completely wilted by the following day and died two days later. All the plants, however, fair better and look healthier compared to plants treated with pathogen and antagonist. When antagonist alone was applied, there were no wilt symptoms whatsoever (Osuinde et al. 2002).

It was also observed that roots of tomato seedlings treated with antagonist and pathogen showed root rot (necrosis) depending on the concentration of the antagonist, method of application, and how long the pathogen was allowed to germinate ahead of antagonist. Roots of plants inoculated with antagonist by root-dip method had lower level of necrosis than those inoculated by direct soil inoculation method. All the roots of the plant (100 %) treated with pathogen alone had severe necrosis. In contrast, roots of plants inoculated with antagonist alone had no necrosis at all, rather, were better than those inoculated with antagonist and pathogen (Table 3.4) (Osuinde et al. 2002).

Table 3.2 Wilt disease development in tomato plants with time after inoculation with *Trichoderma* (antagonist) in *F. oxysporum* f. sp. *Lycopersici* (pathogen) infested soil

Wilt development with time (days)	1	2	3	4	5	6	7
Treatments—root dip method							
1a. Pathogen alone (control)	+	+	+	++	++	++	++
2a. Antagonist alone (control)	−	−	−	−	−	−	−
3a. Pathogen + 10^3 spores/mL antagonist simultaneously	−	+	+	+	−	−	−
4a. Pathogen + 10^6 spores/mL antagonist simultaneously	−	−	−	−	−	−	−
5a. Pathogen incubated 24 h + 10^3 spores/mL antagonist	+	+	+	−	−	−	−
6a. Pathogen incubated 24 h + 10^6 spores/mL antagonist	−	−	−	−	−	−	−
7a. Pathogen incubated 48 h + 10^3 spores/mL antagonist	++	D	D	D	D	D	D
8a. Pathogen incubated 48 h + 10^6 spores/mL antagonist	++	D	D	D	D	D	D
Treatment—direct soil inoculation							
1b. Pathogen alone (control)	+	+	+	++	++	++	++
2b. Antagonist alone (control)	−	−	−	−	−	−	−
3b. Pathogen + 10^3 spores/mL antagonist simultaneously	−	+	+	+	−	−	−
4b. Pathogen + 10^6 spores/mL antagonist simultaneously	−	+	+	−	−	−	−
5b. Pathogen incubated 24 h + 10^3 spores/mL antagonist	+	+	+	−	−	−	−
6b. Pathogen incubated 24 h + 10^6 spores/mL antagonist	−	+	+	+	−	−	−
7b. Pathogen incubated 48 h + 10^3 spores/mL antagonist	++	D	D	D	D	D	D
8b. Pathogen incubated 48 h + 10^6 spores/mL antagonist	++	D	D	D	D	D	D

Source Osuinde et al. (2002)
+ = Partial (Mild) wilt
++ = Complete wilt
− = No wilt
D = Plant death

The antagonist and pathogen were re-isolated from root segments of tomato plants after 7 days growth in the greenhouse study. The frequency of re-isolation of the antagonist and the pathogen differ greatly. The frequency of re-isolation of *Trichoderma* from treated plants was 60–100 % while that of *F. oxysporum* f. sp. *Lycopersici* was 30–50 % in root dip; 40–80 and 50–60 % in direct soil inoculation method respectively. Re-isolation of *Trichoderma* spp. and *F. oxysporum* f. sp. *Lycopersici* in the control plants was 100 % in the separate treatment. Colonies of antagonist (*Trichoderma* spp.) was far more numerous than the pathogen (*F. oxysporum* f. sp. *Lycopersici*) in the root washes (Osuinde et al. 2002) and this agrees with several reports that the high competitive ability, antibiosis and mycoporasitism of *Trichoderma* spp. made them persist on the rhizoplane (root-surface) and rhizosphere of plants and thus out-number other soil microorganisms especially the plant pathogens (Harman et al. 1980; Chet and Henis 1987; Sivan et al. 1987)

Table 3.3 Tomato plants (%) affected by wilt disease with time after inoculation with *Trichoderma* species in *F. oxysporum* f. sp. *Lycopersici* infested soil

Wilt development with time (days)							
	1	2	3	4	5	6	7
Treatments—root dip method							
1a. Pathogen alone (control)	100	100	100	100	100	100	100
2a. Antagonist alone (control)	0	0	0	0	0	0	0
3a. Pathogen + 10^3 spores/mL antagonist simultaneously	0	40	40	40	0	0	0
4a. Pathogen + 10^6 spores/mL antagonist simultaneously	0	0	0	0	0	0	0
5a. Pathogen incubated 24 h + 10^3 spores/mL antagonist	80	80	80	0	0	0	0
6a. Pathogen incubated 24 h + 10^6 spores/mL antagonist	0	0	0	0	0	0	0
7a. Pathogen incubated 48 h + 10^3 spores/mL antagonist	100	100	100	100	100	100	100
8a. Pathogen incubated 48 h + 10^6 spores/mL antagonist	100	100	100	100	100	100	100
Treatment—direct soil inoculation							
1b. Pathogen alone (control)	100	100	100	100	100	100	100
2b. Antagonist alone (control)	0	0	0	0	0	0	0
3b. Pathogen + 10^3 spores/mL antagonist simultaneously	0	80	80	80	0	0	0
4b. Pathogen + 10^6 spores/mL antagonist simultaneously	0	60	60	0	0	0	0
5b. Pathogen incubated 24 h + 10^3 spores/mL antagonist	100	100	100	0	0	0	0
6b. Pathogen incubated 24 h + 10^6 spores/mL antagonist	0	40	40	40	0	0	0
7b. Pathogen incubated 48 h + 10^3 spores/mL antagonist	100	100	100	100	100	100	100
8b. Pathogen incubated 48 h + 10^6 spores/mL antagonist	100	100	100	100	100	100	100

Source Osuinde et al. (2002)

3.5.2 *Against Post-harvest Blue Mould of Oranges* (Citrus Sinensis) *by Screened Microbial Antagonist*

Orange (*Citrus sinensis*) ranks among one of the most important fruits produced in Nigeria. The principal orange-producing region in Nigeria is the southern part of the country, from where it is exported to various markets all over Nigeria and even abroad. Orange accounts for over 90 % of total fruit production in the region (Lateef et al. 2004), however, post-harvest losses associated with fungal diseases

Table 3.4 Tomato plant (%) affected by wilt disease with time after inoculation with *Trichoderma* species in *F. oxysporum* f. sp. *Lycopersici* infested soil

(%) Plants wilt root necrosis and severity of root necrosis					
	4	3	2	1	0
Treatments—root dip method					
1a. Pathogen alone (control)	100	–	–	–	–
2a. Antagonist alone (control)	–	–	–	–	100
3a. Pathogen + 10^3 spores/mL antagonist simultaneously	–	–	40	60	–
4a. Pathogen + 10^6 spores/mL antagonist simultaneously	–	–	20	80	–
5a. Pathogen incubated 24 h + 10^3 spores/mL antagonist	–	–	40	60	–
6a. Pathogen incubated 24 h + 10^6 spores/mL antagonist	–	–	20	80	–
7a. Pathogen incubated 48 h + 10^3 spores/mL antagonist	100	–	–	–	–
8a. Pathogen incubated 48 h + 10^6 spores/mL antagonist	100	–	–	–	–
Treatment—direct soil inoculation					
1b. Pathogen alone (control)	100	–	–	–	–
2b. Antagonist alone (control)	–	–	–	–	100
3b. Pathogen + 10^3 spores/mL antagonist simultaneously	–	20	40	40	–
4b. Pathogen + 10^6 spores/mL antagonist simultaneously	–	–	60	40	–
5b. Pathogen incubated 24 h + 10^3 spores/mL antagonist	–	–	80	20	–
6b. Pathogen incubated 24 h + 10^6 spores/mL antagonist	–	–	80	20	–
7b. Pathogen incubated 48 h + 10^3 spores/mL antagonist	100	–	–	–	–
8b. Pathogen incubated 48 h + 10^6 spores/mL antagonist	100	–	–	–	–

Source Osuinde et al. (2002)
4 = Very severe necrosis all secondary root dead
3 = Considerable root necrosis, with little root regrowth above dead region
2 = Moderate root necrosis
1 = Very slight necrosis limited mainly to tips of a few secondary root
0 = Root system well developed and no visible lesions

are a major limiting factor of its shelf-life. Post-harvest blue and green moulds caused by *Penicillium italicum* and *Penicillium digitatum* respectively are among the most economically important post-harvest diseases of citrus globally. At below 10 °C, *P. italicum* grows faster than *P. digitatum* as a result; blue mould incidence becomes more important when citrus fruits are kept under cold storage over a long period of time (Palou et al. 2001). Agrochemicals such as imazalil, sodium ortho-phenyl phenate, or thiabendazole have been commonly used to control these diseases (Yildiz et al. 2005; Torres et al. 2007). These synthetic fungicides have been applied for many years with few or limited success owing to resistance development by the fungal pathogens (Holmes and Eckert 1999; Zamani et al. 2006). Moreover, the accumulation of these hazardous agrochemicals in the environment has generated public concern about their impact on human health, thus, creating an opportunity for sustainable alternative methods to control post-harvest diseases without harming either man or his environment. Biological control such as the used of antagonist fungi has been proposed as an alternative to agrochemical and considerable success has been recorded by utilizing antagonistic microorganisms for post-harvest disease control (Wilson et al. 1993). The use of fungi as an alternative to synthetic fungicides has other benefits such as reducing environmental pollution, effectively controlling post-harvest diseases and producing high quality and safe food (He et al. 2003).

Screening for potential antagonistic microorganisms to *P. italicum* from the phylloplane and soil in the orchard was carried out to investigate their efficacy in controlling post-harvest blue mould of orange fruit under in vitro and in vivo conditions. Three fungi genera, *Trichoderma*, *Aspergillus*, *Penicillium*; two yeasts of the genus *Saccharomyces* and a bacterium, *Pseudomonas* were isolated from the phylloplane of leaves, healthy orange fruits and from the orchard soil. The result varied from treatments when pathogen, *P. italicum* was allowed 24–48 h growth ahead of each antagonist depending on its exhibited mechanism against the pathogen. *Trchoderma* sp., a fast grower and good competitor of nutrient showed the best level of antagonism than the others as it stands out to be the best fungal-base biocontrol agent (Tables 3.5 and 3.6) (Emoghene et al. 2011). As can be observed in Table 3.5 antagonism of the pathogen by each of the antagonist varied. Biocontrol of the pathogen placed after 24 h, was significantly higher than that of 48 h. *Penicillium* sp. showed a gradual and steady control rate of the *P. italicum* than *Trichoderma* sp. which demonstrated a better control outcome in a markedly sharp increase in its antagonism (Emoghene et al. 2011). From Table 3.6, it was deduced that the later the antagonists were introduced after *P. italicum*, the better the antagonism. When the pathogen had growth prior to the inoculation of *Trichoderma* sp., inhibition of mycelial growth by mycoparasitism, hyphae interference or antifungal (antibiosis) production was highest than that of *Aspergillus niger* and *Penicillium* sp. (Emoghene et al. 2011).

The bacteria or yeast antagonists seeded on potato dextrose agar (PDA) plate with mycelial extension growth of *P. italicum*, inhibited growth at varying degrees within 6 days of measurement depending on inoculation time. It was observed that mycelial extension growth was best inhibited by *Pseudomonas* sp. followed by yeast when *P. italicum* was inoculated after one day. However, the reverse was the case after two days, with the mycelial extension growth best inhibited by Saccharomyces sp. followed by *Pseudomonas* sp. (Emoghene et al. 2011).

Biocontrol of *P. italicum*, the causative agent of post-harvest rot of orange fruit by *Trichoderma* sp. *Penicillium* sp., *Aspergillus niger*, *Pseudomonas* and *Saccharomyces* sp. showed different levels of antagonistic control efficacy. *Trichoderma* sp. showed a superior biocontrol efficacy and its antagonistic effect on different pathogens is well documented (Grondona et al. 1997; Kucuk and Kivanc 2005; Shaigan et al. 2008). *Trichoderma* sp. grows tropically toward hyphae of other fungi, coil around them in a lectin-mediated reaction and degrade cells of the target fungi. This mycoparasitism process limits growth and activity of most plant pathogenic fungi (Carsolio et al. 1999; Shaigan et al. 2008). *Trichoderma* spp. grows more rapidly than *P. italicum* in mixed culture and this gives it an

Table 3.5 Effects of inoculation time of *Penicillium italicum* on orange fruit rot controlled by fungal antagonist

Mycelial inhibition with time (mm/days)								
Antagonists	Inoculation time (hour)	0	1	2	3	4	5	6
Control	24	0.00 ± 0.0	15.5 ± 1.5	20.7 ± 7.7	29.0 ± 18.7	37.1 ± 16.6	40.5 ± 16.8	43.9 ± 16.1
Penicillium sp.	48	0.00 ± 0.0	0.00 ± 0.0	15.5 ± 1.7	19.7 ± 7.6	27.2 ± 18.2	35.3 ± 14.5	37.5 ± 17.7
		0.00 ± 0.0	0.00 ± 0.0	0.00 ± 0.0	14.3 ± 0.0	24.2 ± 1.1	31.9 ± 1.1	32.6 ± 10.0
Control	24	0.00 ± 0.0	0.00 ± 0.0	0.00 ± 0.0	12.0 ± 0.7	13.3 ± 1.4	16.3 ± 0.6	24.6 ± 1.8
Aspergillus	48	0.00 ± 0.0	0.00 ± 0.0	0.00 ± 0.0	0.00 ± 0.0	0.00 ± 0.0	7.2 ± 10.1	18.8 ± 8.8
niger		0.00 ± 0.0	0.00 ± 0.0	0.00 ± 0.0	0.00 ± 0.0	0.00 ± 0.0	16.7 ± 0.0	14.3 ± 0.0
Control	24	0.00 ± 0.0	0.00 ± 0.0	13.4 ± 1.6	27.3 ± 18.3	42.0 ± 11.5	48.3 ± 10.6	54.5 ± 8.4
Trichoderma	48	0.00 ± 0.0	0.00 ± 0.0	0.00 ± 0.0	12.5 ± 17.7	27.2 ± 18.2	41.7 ± 11.8	46.9 ± 9.8
sp.		0.00 ± 0.0	0.00 ± 0.0	0.00 ± 0.0	8.4 ± 11.8	8.4 ± 11.8	22.7 ± 8.4	32.5 ± 10.6

Source Emoghene et al. (2011)

Table 3.6 Effects of inoculation time of fungal antagonists on orange fruit rot control

Mycelial inhibition with time (mm/days)								
Antagonists	Inoculation time (hour)	0	1	2	3	4	5	6
Control	24	0.00 ± 0.0	15.5 ± 1.5	20.7 ± 7.7	29.0 ± 18.7	37.1 ± 16.6	40.5 ± 16.8	43.9 ± 16.1
Penicillium sp.	48	0.00 ± 0.0	5.6 ± 7.8	5.6 ± 7.8	5.6 ± 7.8	9.6 ± 0.6	5.6 ± 5.0	24.1 ± 1.3
		0.00 ± 0.0	15.0 ± 7.1	13.8 ± 5.4	12.9 ± 5.4	24.1 ± 1.3	27.5 ± 6.2	30.2 ± 2.3
Control	24	0.00 ± 0.0	0.00 ± 0.0	0.00 ± 0.0	12.0 ± 0.7	13.3 ± 1.4	16.3 ± 0.6	24.6 ± 1.8
Aspergillus	48	0.00 ± 0.0	0.00 ± 0.0	21.2 ± 17.1	21.8 ± 19.1	26.7 ± 26.0	26.7 ± 26.0	33.8 ± 19.5
niger		0.00 ± 0.0	5.0 ± 7.1	14.9 ± 11.6	21.7 ± 7.1	26.2 ± 7.2	26.2 ± 7.2	30.2 ± 7.3
Control	24	0.00 ± 0.0	0.00 ± 0.0	13.4 ± 1.6	27.3 ± 18.3	42.0 ± 11.5	48.3 ± 10.6	54.5 ± 8.4
Trichoderma	48	0.00 ± 0.0	0.00 ± 0.0	24.3 ± 6.1	24.3 ± 6.1	22.5 ± 3.5	27.2 ± 3.1	35.1 ± 2.5
sp.		0.00 ± 0.0	10.8 ± 2.4	24.5 ± 3.2	24.8 ± 3.6	32.4 ± 8.6	38.4 ± 11.8	52.3 ± 3.2

Source Emoghene et al. (2011)

important advantage in the competition for space and nutrients with plant pathogenic fungi (Barbosa et al. 2001). It can be deduced above that, introduction of the pathogen after the inoculation of the antagonist resulted to a better antagonism. This could give the antagonists enough time to grow, reproduce and sporulate using the available nutrient in a competitive manner, in addition to secreting enough antagonistic substances which affect the establishment of the pathogen. Therefore, colonization of the host could be prevented by early application of fungi antagonist to prevent infection and plant diseases, even though there was no significant difference in the time of application of either the pathogen or antagonist ($p > 0.05$) (Emoghene et al. 2011).

3.6 Conclusion

Control of fungal pathogen of plant diseases is based on the application of agronomic practices and pesticides; however, widespread use of agrochemicals inundates the agroecosystems with hazardous substances that impact the balance of the natural food chain. Coupled with the selection of resistant and more virulent plant pathogens resulting to escalation in the quantity of pesticides used. Researches are ongoing to develop new, alternative and sustainable methods to integrate or substitute the application of agrochemicals in an attempt to reduce ecological impact and financial cost of plant disease control. Antagonistic microorganisms especially fungi have been investigated in depth and considered as an attractive alternative to agrochemicals in the control of plant diseases. Fungi-based biofungicides have yielded successful and consistent results as depicted above; however, its application has been delayed owing to the poor relative understanding of the plant-microbe and microbe-microbe interactions in the antagonistic processes amongst others. Diverse microorganisms may have been used for biocontrol of plant diseases, but the most widely applied and researched are on isolates of genera of *Trichoderma*, *Bacillus* and *Pseudomonas* with *Trichoderma* being the most studied fungal biocontrol agent. The mechanism of action of *Trichoderma* spp. as effective biofungicides is well documented. Fundamental discoveries show that *Trichoderma* and other mycoparasites have developed a vast array of molecular technique to enhance their parasitic behaviour. It is agreed that *Trichoderma* produces different types of lytic enzymes that target the cell wall of fungi resulting to their death. Since fungal-biocontrol of plant pathogens are very diverse with different plant hosts, it is therefore very imperative to look for new and novel biocontrol fungi with different mechanisms. The greatest hope for fungi as alternative to agrochemicals lies in understanding its mechanism(s) of action as biofungicides and the pathogenesis of the pathogens. It is anticipated that this knowledge will open up new possibilities and innovative approaches for controlling plant diseases as agrochemicals usage is on the increase, particularly in developing countries like Nigeria and is no longer sustainable owing to adverse environmental effect and loss of human life.

References

Adelana BO, Simons JH (1980) An evaluation of some tomato varieties for yield and disease tolerance in South Western Nigeria. Nig J Agric Sci 2:29–34

Alexander M (1977) Introduction to soil microbiology, 2nd edn. Wiley, New York, p 467

Ardakani S, Heydari A, Khorasani N, Arjmandi R, Ehteshami M (2009) Preparation of new biofungicides using antagonistic bacteria and mineral compounds for controlling cotton seedling damping-off disease. J Plant Prot Res 49:49–55

Arras G (1996) Mode of action of an isolate of Candida famata in biological control of *Penicillium digitatum* in orange fruits. Postharvest Biol Technol 8:191–198

Baker KF, Cook RJ (1982) Biological control of plant pathogens. Freeman and Company, San Francisco, p 433

Barbosa MA, Rehn KG, Menezes M, Mariano RR (2001) Antagonism of *Trichoderma* species on *Cladosporium herbarum* and their enzymatic characterization. Braz J Microbiol 32:98–104

Barak E, Edgington LV, Ripley BD (1984) Bioactivity of the fungicide metalaxyl in potato tubers against some species of *Phytophthora*, *Fusarium*, and *Alternaria* related to polyphenoloxidase activity. Can J Plant Pathol 6:304–308

Benhamou N, Chet I (1997) Cellular and molecular mechanisms involved in the intersection between *Trichoderma harzianum* and *Pythium ultimum*. Appl Environ Microbiol 63:2095–2099

Biermann B, Linderman RG (1983) Use of vesicular-arbuscular mycorrhizal roots, intraradical vesicles and extraradical vesicles as inoculum. New Phytol 95:97–105

Bohmont BL (1990) The standard pesticide user's guide. Prentice Hall, Upper Saddle River, NJ

Carsolio C, Benhamou N, Haran S, Cortes C, Gutierrez A, Chet A, Herrera-Estrella A (1999) Role of the *Trichoderma harzianum* endochitinase gene, ech42, in mycoparasitism. Appl Environ Microbiol 65:929–935

Catska V (1994) Interrelationship between vesicular-arbuscular mycorrhiza and rhizosphere microflora in apple replant disease. Biol Plant 36:99–104

Chet I, Barak ZI, Oppenheim A (1993) Genetic engineering of microorganisms for improved biocontrol activity. In: Chet I (ed) Biotechnological prospect of plant disease control. Wiley-Liss, New York, pp 211–235

Chet I, Henis Y (1987) *Trichoderma* as a biocontrol agent against soil borne root pathogen. In: Chet I (ed) Innovative approaches to plant disease control. Wiley, New York, p 137

Coffey MD, Ohr HD, Guillemet FB, Cambell SD (1984) Chemical control of *Phytophthora cinnamomi* on avocado root stocks. Plant Dis 68:956–958

Cohen E (1981) Metalaxy for post-harvest control of brown rot of citrus fruit. Plant Dis 65:672–675

Cohen Y, Coffey MD (1986) Systemic fungicides and the control of oomycetes. Ann Rev Phytopathol 24:311–338

Colyer PD, Mount MS (1984) Bacterization of potatoes with *Pseudomonas putida* and its influence on postharvest soft rot diseases. Plant Dis 68:703–706

de Matose AP (1983) Chemical and biological factors influencing the infection of lemons by *Geotrichum candidum* and *Penicillium digitatum*. Ph.D dissertation. University of Carolina, Riverside

Denton L, Swarup V (1983) Tomato cultivation and its potential in Nigeria. Acta Hort 123:257–271

Duchesne LC (1994) Role of ectomycorrhizal fungi in biocontrol. In: Pfleger FL, Linderman RG (eds) Mycorrhizae and plant health. APS Press, St. Paul, MN, pp 27–45

Elad Y, Chet I, Henis Y (1982) Degradation of plant pathogenic fungi by *Trichoderma harzianum*. Can J Microbiol 28:719–725

Emoghene AO, Okigbo RN (2001) Phylloplane microbiota of *Amaranthus hybridus* and their effect on shoot disease caused by *Choanephora cucurbitarum*. Trop Agric (Trinidad) 79(2):90–94

Emoghene AO, Futughe AE (2011) Effect of soil solarization on *Amaranthus viridis* shoot disease caused by *Choanephora cucurbitarum*. Nig Ann Nat Sci 11(1):33–40
Emoghene AO, Adesina IA, Eyong MM (2011) Biocontrol of postharvest blue mold of orange fruit (*Citrus sinensis*) by screened microbial antagonists. Biol Environ Sci J Trop 8(1):123–128
Erinle ID (1977) Vascular and other wilt diseases of dry season tomatoes in the Nigerian Savanna, Samaru. Agric Newslett 19:35–41
Erinle ID (1979) Tomato diseases in Northern States of Nigeria. extension bulletin 31, agric. Extension and research Liason services, A. B. U., Zaria, Nigeria, 37 p
FAO (1989) International *Code* of conduct on the distribution and use of pesticides, Rome, Italy
FAO (2004) FAOSTAT: FAO statistical data bases. Rome, Italy. (http://faostat.fao.org)
Fitter AH, Garbaye J (1994) Interactions between mycorrhizal fungi and other soil microorganisms. Plant Soil 159:123–132
Futughe AE (2007) Biological control of postharvest diseases of orange (*Citrus aurantium*) with antagonists. B.Sc. dissertation. University of Benin, Benin City, Nigeria
Garcia-Garrido JM, Ocampo JA (1989) Effect of VA mycorrhizal infection of tomato on damage caused by *Pseudomonas syringae*. Soil Biol Biochem 21:165–167
Grondona I, Hermosa R, Tejada M, Gomis MD, Mateos PF, Bridge PD, Monte E, Garcia-Acha I (1997) Physiological and biochemical characterization of *Trichoderma harzianum,* a biological control agent against soilborne fungal plant pathogens. Appl Environ Microbiol 63:3189–3198
Harman GE, Chet I, Baker R (1980) *Trichoderma hamatum* effect on seed and seedling disease induced in radish and pea by *Pythium* species or *Rhizoctonia solani*. Phytopathology 73:1043–1046
Harman GE, Bjorkman T (1998) Potential and existing uses of *Trichoderma* and *Gliocladium* for plant disease control and growth enhancement. In: Harmon GE, Kubicek CP (eds) *Trichoderma and Gliocladium*: enzymes, biological control and commercial applications, vol 2. Taylor and Francis, London, UK, pp 229–265
He D, Zheng X, Yin Y, Sun P, Zhang H (2003) Yeast application for controlling apple postharvest diseases associated with *Penicillium expansum*. Bot Bull Acad Sin 44:211–216
Holmes GJ, Eckert JW (1999) Sensitivity of *Penicillium digitatum* and *P. italicum* to postharvest citrus fungicides in California. Phytopathology 89:716–721
Howell CR, Beier RC, Stipanovic RD (1988) Production of ammonia by *Enterobacter cloacae* and its possible role in the biological control of *Pythium* pre-emergence damping-off by the bacterium. Phytopathology 78:1075–1078
Ikediugwu FEO (1981) A shoot disease of *Amaranthus* spp. in Nigeria associated with *Choanephora cucurbitarum*. J Hort Sci 56:289–293
Ikediugwu FEO, Emoghene AO, Ajiodo PO (1994) Biological control of the shoot disease of *Amaranthus hybridus* caused by *Choanephora cucurbitarum* with *Bacillus subtilis*. J Hort Sci 69:351–356
Jeong H, Forster L (2003) Empirical investigation of agricultural externalities: effects of pesticide use and tillage system on surface water. Department of Agricultural, Environmental and Development Economics, The Ohio State University, Working Paper: AEDE-WP-0034–03, p 31
Jeyaratnam J (1990) Acute pesticide poisoning: a major global health problem. World Health Stat Q 43:139–144
Jones, R. K. (1985). Fungicides for bedding plants. Bedding plant Inc. *News.* **4**: 37–44
Kamrin MA (1997) Pesticide profiles: toxicity, environmental impact, and fate. Lewis Publisher, Boca Raton, New York, p 685
Kerr A (1980) Biological control of crown gall through production of agrocin 84. Plant Dis 64:25–30
Kloepper JW, Leong J, Teintze M, Schroth MN (1980) *Pseudomonas* siderophores: a mechanism explaining disease suppression in soils. Curr Microbiol 4:317–320
Kucuk C, Kivanc M (2005) Effect of the formulation on the viability of biocontrol agent, *Trichoderma harzianum* conidia. Afri J Biotechnol 4(5):483–486

Lafontaine PJ, Benhamon N (1996) Chitosan treatment: an emerging strategy for enhancing resistance of greenhouse tomato to infection by *Fusarium oxysporum* f.sp. *radicilycopersici*. Biocontrol Sci Technol 6:111–124

Lateef A, Oloke JK, Gueguim-kana EB (2004) Antimicrobial resistance of bacterial strains isolated from orange juice products. Afri J Biotechnol 3:334–338

Lim TK, Rorhbach KG (1980) Role of *Penicillium funiculosum* strains in the development of pineapple fruit disease. Phytopathology 70:663–665

Linderman RG (1994) Role of AM fungi in biocontrol. In: Pfleger FL, Linderman RG (eds) Mycorrhizae and plant health. APS Press, St. Paul, MN, pp 1–25

Lindgren RM, Harvey GM (1952) Decay control and increased permeability in Southern pine sprayed with fluoride solution. Proc For Prod Soc 5:250–256

Liu X, Fang W, Liu L, Yu T, Lou B, Zheng X (2010) Biological control of postharvest sour rot of citrus by two antagonistic yeasts. Lett Appl Microbiol 51:30–35

Lorito M, Harman GE, Hayes CK, Broadway RM, Woo SL, Di Piettro A (1993) Chitinolytic enzymes produced by *Trichoderma harziamum*. II. Antifungal activity of purified endochitinasc and chitobiosidase. Plytopathology 83:302–307

Maurhofer M, Hase C, Meuwly P, Metraux JP, Defago G (1994) Induction of systemic resistance to tobacco necrosis virus by the root-colonizing *Pseudomonas fluorescens* strain CHA0: influence of the *gacA* gene and of pyoverdine production. Phytopathology 84:139–146

McGregor AJ (1984) Comparison of cuprous oxide and metalaxyl with mixtures of these fungicides for the control of *Phytophthora* pod rot of cocoa. Plant Pathol 33:81–87

Milgroom MG, Cortesi P (2004) Biological control of chestnut blight with hypovirulence: a critical analysis. Ann Rev Phytopathol 42:311–338

Miller GT (2004) Sustaining the earth: an integrated approach, 6th edn. Thompson Learning, Inc., Pacific Grove, California, pp 211–216

Moore WC (1959) British parasitic fungi. Cambridge University Press, London, p 430

Morris PF, Ward EWR (1992) Chemoattraction of zoospores of the plant soybean pathogen, *Phytophthora sojae*, by isoflavones. Physiol Mol Plant Pathol 40:17–22

Oloruntoba BS (1989) Foreword, In: Progress in tree crop research in Nigeria, 2nd edn. Cocoa Research Institute Ibadan, Nigeria, pp I–ii

Osibanjo O, Adeyeye A (1995) Organochlorine pesticide residue in Nigeria market. Bull Environ Toxicol 54:460–465

Osuinde MI, Ikediugwu FEO (1995) *Fusarium oxysporum* and *Pseudomonas solanacearum* cause wilt disease of tomato (*Lycopersicon esculentum* Mill) in Southern Western Nigeria. Nig J Plant Prot 16

Osuinde MI, Aluya EI, Emoghene AO (2002) Control of Fusarium Wilt Tomato (*Licopersicon esculentum* Mill*)* by *Trichoderma* species. Acta Phytopathogica et Entomoligica Hungarica 37 (1–2):47–55

Palou L, Smilanick JL, Usall J, Vinas I (2001) Control of postharvest blue and green molds of oranges by hot water, sodium carbonate, and sodium bicarbonate. Plant Dis 85:371–376

Park D (1959) Some aspect of the biuology of *Fusarium oxysporum* Schl. Soil Ann Bot (NS) 23:35–49

Platania C, Restuccia C, Muccilli S, Cirvilleri G (2012) Efficacy of killer yeasts in the biological control of *Penicillium digitatum* on Tarocco orange fruits (*Citrus sinensis*). Food Microbiol 30:219–225

Prior P, Stera H, Cadet E (1990) Aggressiveness of strains of *Pseudomonas solanacearum* from the French West Indics (Martinique and Guadeloupe) on tomato. Plant Dis 74:962–965

Punch (2015) Pesticide responsible for Ondo death, says WHO. [online]. 20th April. Available from: http://www.punchng.com/news/pesticide-responsible-for-ondo-deaths-says-who/. Accessed 20th Aug 2015

Purseglore JC (1977) Tropical crops dicotyledons, vols 1 and 2. Longman, London, 536 p

Pusey PL, Wilson CL (1984) Postharvest biological control of stone fruit brown rot by *Bacillus subtilis*. Plant Dis 68:753–756

Ross EW, Marx DM (1972) Susceptibility sand pine to *Phytophthora cinnamomi*. Phytopathology 62:1197–1200

Rytter JL, Lukezic FL, Craig R, Moorman GW (1989) Biological control of Geranium rust by *Bacillus subtilis*. Phytopathology 79:367–370

Ryu CM, Farag MA, Hu CH, Reddy MS, Kloepper JW, Pare PW (2004) Bacterial volatiles induce systemic resistance in *Arabidopsis*. Plant Physiol 134:1017–1026

Shaigan S, Seraji A, Moghaddam SAM (2008) Identification and investigation on antagonistic effect of *Trichoderma* spp. on teas seedlings white foot and root rot (*Sclerotium rolfsii* Sacc.) in vitro condition. Pakis J Biol Sci 11:2346–2350

Shields JK, Atwell EA (1963) Effect of a mold. *Trichoderma viridis,* on decay of birch by four storage-rot fungi. For Prod J 13:262–265

Simons JH, Sobulo RA (1975) Methods for higher tomato yields in Western State of Nigeria. Publicity and Information Section, M. A. N. R., Ibadan, Nigeria

Sivan A, Ucko O, Chet I (1987) Biological control of Fusarium crown rot of tomato by *Trichoderma hazianum* under field conditions. Plant Dis 71:587–592

Smith HS (1919) On some phases of insect control by the biological method. J Econ Entomol 17:347–356

Taqarort N, Echairi A, Chaussod R, Nouaim R, Boubaker H, Benaoumar AA, Boudyach E (2008) Screening and identification of epiphytic yeasts with potential for biological control of green mold of citrus fruits. World J Microbiol Biotechnol 24:3031–3038

Torres R, Nunes C, Garcia JS, Abadias M, Vinas I, Manso T, Olmo M, Usall J (2007) Application of *Pantoea agglomerans* CPA-2 in combination with heated sodium bicarbonate solution to control the major postharvest diseases affecting citrus fruit at several Mediterranean locations. Eur J Plant Pathol 118:73–83

Tronsmo A, Dennis C (1983) The use of *Trichoderma* species to control strawberry fruit rots. Neth J Plant Pathol 83:449–455

UNEP (1990) Public health impact of pesticides use in agriculture. Geneva, Switzerland

Van Loon LC, Bakker PAHM, Pieterse CMJ (1998) Systemic resistance induced by rhizosphere bacteria. Ann Rev Phytopathol 36:453–483

Walker JC (1957) Plant pathology, 2nd edn. McGraw-Hill, New York 589 p

Walker JC (1971) *Fusarium* wilt in tomatoes. Am Phytopathol Soc Monogr 6:56

Wheele BEJ (1969) An introduction to plant diseases. Wiley, Chichester, p 374

WHO (1990) WHO recommended classification of pesticide by hazard and guidelines to classification 1990–1991. Unpublished WHO document WHO/PCS/90.1. Available on request from: division of environment health, WHO, 1211 Geneva 27, Switzerland

WHO (2006) WHO gives indoor use of DDT a clean bill of health for controlling Malaria. Division of environmental health, WHO, 1211 Geneva 27, Switzerland

Wilson CL, Wisniewski ME, Droby S, Chalutz E (1993) A selection strategy for microbial antagonists to control postharvest disease of fruit and vegetables. Sci Hortic 53:183–189

Yildiz F, Kinay P, Yildiz M, Sen F, Karacali I (2005) Effect of preharvest applications of $CaCl_2$, 2, 4-D and benomyl and postharvest hot water, yeast and fungicide treatments on development of decay on *Satsuma mandarins*. J Phytopathol 153:94–98

Zamani M, Tehrani AS, Ahmadzadeh M, Abadi AA (2006) Effect of fluorescent pseudomonads and *Trichoderma* sp. and their combination with two chemicals on *Penicillium digitatum* caused agent of citrus green mold. Commun Agric Appl Biol Sci 71:1301–1310

Part II
Providing Sustainable and Environmentally Friendly Solutions to Treat Wastewater

Chapter 4
Application of Biosorption and Biodegradation Functions of Fungi in Wastewater and Sludge Treatment

Tao Lu, Qi-Lei Zhang and Shan-Jing Yao

4.1 Introduction

Sewage treatment has always been one of the core problems of environmental protection as wastewaters contain a variety of harmful substances such as heavy metals, dyes and phenolics (Rangabhashiyam et al. 2014). Disposing the sludge generated from sewage treatment plant in a safe way is also challenging. The surface and ground waters in many parts of the world have been subject to pollution due to the emission of industrial wastewater and cannot be used as drinking water (Rangabhashiyam et al. 2014). For the basic needs of life, there is the increasing need of pollution control and water quality protection. Various water treatment methods using physical and chemical techniques have been applied. The main methods include: (i) filtration (Zouboulis et al. 2002), (ii) ion exchange (Kabsch-Korbutowicz and Krupinska 2008), (iii) solvent extraction (Lin and Juang 2002), (iv) advanced oxidation processes (Esplugas et al. 2002), (v) activated carbon adsorption (Kurniawan et al. 2006). But most of technologies mentioned above are limited due to their high cost. Biological treatment is a relatively economical when comparing to the traditional physical and chemical processes (Crini 2006). Biological technologies such as biological adsorption and microbial degradation are commonly applied to the treatment of industrial effluents because many microorganisms such as bacteria and fungi are able to concentrate selected substances and degrade different contaminants (Fu and Viraraghavan 2001a; McMullan et al. 2001; Volesky 2007; Chen et al. 2014a; Wang et al. 2015). Utilizing fungi for biological treatment have been extensively studied due to their vast amounts of biomass generated from fermentation industries (Zhou and Kiff 1991). On the fungal cell wall there exist many functional groups such as carboxyl,

T. Lu · Q.-L. Zhang · S.-J. Yao (✉)
Key Laboratory of Biomass Chemical Engineering of Ministry of Education, Department of Chemical and Biological Engineering, Zhejiang University, Hangzhou 310027, China
e-mail: yaosj@zju.edu.cn

D. Purchase (ed.), *Fungal Applications in Sustainable Environmental Biotechnology*, Fungal Biology, DOI 10.1007/978-3-319-42852-9_4

hydroxyl, amino, sulfonate, and phosphonate, which bring about the excellent adsorption property in fungi (Cabib et al. 1988; Bowman and Free 2006). Fungi also possess multiple mechanisms for degradation of organic and inorganic pollutants (Awasthi et al. 2014; Mishra and Malik 2014b). Thus, fungi are playing increasingly important roles in wastewater treatment. The following notable features of fungi made them excellent candidates for treatment processes: (i) high adsorption capacity, (ii) easy solid–liquid separation, (iii) good adverse resistance, and (iv) broad degradation ability. The main goal of this review is to provide up-to-date information pertaining to the application of biological adsorption and biodegradation functions of fungi in sewage treatment.

4.2 Biosorption Function

Many literatures have reported the fungal application for wastewater treatment in recent years. Filamentous fungi are promising materials to replace or supplement traditional treatment processes (Sharma et al. 2011). Many genera of fungi have been researched both in living or inactivated form (Srinivasan and Viraraghavan 2010). For living cells, the mechanism involves biosorption and biodegradation because fungi can produce laccase or other enzymes to mineralize organic pollutants (Raghukumar et al. 1996). For dead cells, the mechanism is biosorption without active metabolic transport process (Volesky 2007).

4.2.1 Performance of Wastewater Treatment for Dyes, Heavy Metals and Phenolic Compounds

The ingredient of wastewater is usually very complex, containing all kinds of pollutants. This paper mainly focuses on fungal biosorption effect on organic dyes, heavy metals and phenolic compounds.

(1) **Dyes**

Dye wastewater is one of the most difficult industrial wastewaters to treat. Particularly from the textile industry, more than 1.5×10^8 m^3 of colored effluents are discharged annually (Ip et al. 2010). These dyes may be mutagenic or carcinogenic in human beings and could lead to dysfunctions of the liver, kidneys, central nervous and reproductive system (Dincer et al. 2007; Shen et al. 2009). It is also recognized that color is the first obvious sign of sewage. The presence of trace amounts of dyes in water can be easily visible and undesirable (Banat et al. 1996; Robinson et al. 2001). Activated carbon is an effective adsorbent which was widely used for dye removal in sewage treatment with several advantages such as the large surface area and high adsorption capacity. But the usage of activated carbon would be limited in many cases because of its disposal problem and the high running cost (Xiong et al. 2010). The application of fungi for the removal of dyes is an attractive alternative to the colored sewage treatment (Solis et al. 2012). Fungal species such as *Pencillium oxalicum, Aspergillus niger, Trametes versicolor, Rhizopus stolonifer, Rhizopus oryzae.* have been widely reported as biosorbents for the removal of dye from aqueous solution (Abd El-Rahim et al. 2003; Bayramoğlu and Arica 2007; Binupriya et al. 2007; Srinivasan and Viraraghavan 2010; Solis et al. 2012; Akar et al. 2013; Rangabhashiyam et al. 2014). The representative examples of dye decolorization by fungi are tabulated in Table 4.1. Since color is the first obvious sign of the presence of sewage, the decolorization rate, i.e., the removal percent becomes one of the most important indices for colored wastewater treatment. From the table it is obvious that a wide variety of fungi are capable of decolorizing all kinds of dyes. With regard to a certain type of dye, different species of fungi tend to have different adsorption effectiveness. For example, the removal efficiency of Reactive Black 5 can exceed 99 % after 48 h treatment by using *Penicillium geastrivorus* (Yang et al. 2003), but the removal rate drops to 88 % when the adsorbent is change into *Aspergillus niger* over 60 h (Taskin and Erdal 2010), which suggests that dye adsorption in fungi is species specific. Meanwhile, some special types of fungi are found to have high biosorption effectiveness on multiple dyes that make them potential candidates to treat colored wastewater generated from industry contain a variety of dyes. For instance, mycelial pellets formed of marine-derived *Penicillium janthinellum* P1 have a broad spectrum of adsorption capacity (Fig. 4.1): its decolouration efficiency of Congo Red, Naphthol Green B, Eriochrome Black T, Amino Black 10B could exceed 99 %, and the removal rate of Neutral Magenta, Methyl Red, Acid Fuchsin, Crystal Violet, and Brilliant Green could reach up to 94.4, 82.1, 63.5, 56.9, and 63.0 %, respectively (Wang et al. 2015).

(2) **Heavy metals**

In addition to dyes, the existence of heavy metals in wastewater presents further complexity in processing. Industries that discharge heavy metals sewage includes: electroplating, battery manufacturing, mining engineering, printing, photography industry (Kadirvelu et al. 2001). Heavy metals in the environment may accumulate in the food chain and eventually cause great harm to human health (Kurniawan et al. 2006; Dal Bosco et al. 2006; Fu and Wang 2011). Although many biological materials can adsorb heavy metals, only those with sufficiently high metal-binding capacity and selectivity for heavy metals are appropriate for use in the biosorption process (Sag 2001). Fungal cell wall surface contains different functional groups, many of which are found to play vital roles in metal chelation (Rangabhashiyam et al. 2014). Fungi have a large capacity for heavy metal sorption from aqueous solutions and in certain circumstance even outperformed activated carbon (Rangabhashiyam et al. 2014). Fungal species such as *Aspergillus niger*, *Pleurotus ostreatus*, *Phanerochaete chrysosporium*, *Rhizopus arrhizus*, *Trametes versicolor*, and *Fusarium* sp. have been extensively researched in the removal of heavy metal ions as they are abundantly available and low cost in mass production. Biosorption capacities of heavy metal ions on various fungal species are compared in Table 4.2.

Table 4.1 Results of dye decolorization by fungi

Fungi	Dye	Percent removal (%)	Initial dye concentration (mg/L)	Time of contact	References
Rhizopus oryzae	Rhodamine B	90	100	5 h	Das et al. (2006)
Pencillium oxalicum	Reactive Blue 19	91	100	80 min	Zhang et al. (2003)
Aspergillus niger 31	Polar Red	94	300	8 d	Abd El-Rahim et al. (2003)
	Direct Blue 1	63.2	800	6 h	Bayramoğlu and Arica (2007)
Aspergillus niger	Acid Blue 29	80	50	30 h	Fu and Viraraghavan (2001b)
Penicillium geastrivorus	Reactive Black 5	>99	100	48 h	Yang et al. (2003)
Aspergillus niger	Reactive Black-5	88	100	60 h	Taskin and Erdal (2010)
Penicillium chrysogenum MT-6	Reactive Black-5	89	300	100 h	Erdal and Taskin (2010)
Trametes pubescens	Congo Red	98	100	60 min	Si et al. (2015)
Phanerochaete chrysosporium	Amido black 10B	98	1000	3 day	Senthilkumar et al. (2014)
Aspergillus lentulus	Acid Blue 120	90	100	12 h	Kaushik et al. (2014)
Thamnidium elegans	Reactive Red 198	98	100	75 min	Akar et al. (2013)
Penicillium janthinellum P1	Congo Red	>99	150	24 h	Wang et al. (2015)
	Naphthol Green B	>99	150	24 h	
	Eriochrome Black T	>99	150	24 h	
	Amino Black 10B	>99	150	24 h	
	Neutral Magenta	94.4	150	24 h	
	Methyl Red	82.1	150	24 h	
	Acid Fuchsin	63.5	150	24 h	
	Crystal Violet	56.9	150	24 h	
	Brilliant Green	63.0	150	24 h	

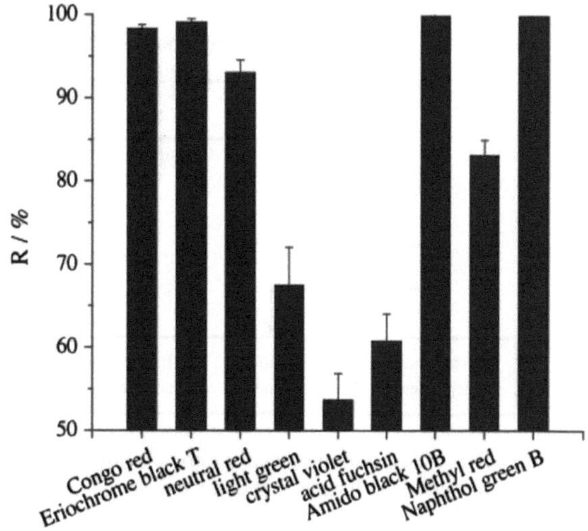

Fig. 4.1 The decolorization rate of *Penicillium janthinellum* P1 in treating nine different dyes (Wang et al. 2015)

The maximum hexavalent chromium adsorption capacities are ranging from 10.75 to 117.33 mg/g by different species of fungi, which shows the specific differences.

(3) **Phenolic compounds**

Phenol can cause harm to human health even in minute quantity (Senturk et al. 2009; Hank et al. 2014), thus US Environmental Protection Agency have taken stringent measures to lower phenol content in the wastewater to <1 mg/L (Banat et al. 2000). Many other phenolic compounds also have different degrees of toxicity, which are contained in sewages originated from petrochemical, phenol producing, coal conversion and other chemical processes (Hamdaoui and Naffrechoux 2007a, b). In recent years, many studies have focused on fungi that are able to biosorb phenols and chlorophenols. Table 4.3 shows the data on the biosorption capacities of phenol and phenolic compounds by various fungi.

4.2.2 Mechanisms

Due to the diversity of the fungi and the complexity of contaminants in wastewater,the mechanism of fungal biosorption is often difficult to characterize, except perhaps in the simplest laboratory systems where a variety of mechanisms may be operative under given conditions (Gadd 2009). There are variety of ways for the pollutant to be captured by fungal cell, thus biosorption mechanisms could be multiple and in many cases they are still not very well understood (Sag 2001). The biosorption mechanisms are classified to different types on the basis of cell metabolism status or pollutants sorption location (Fig. 4.2).

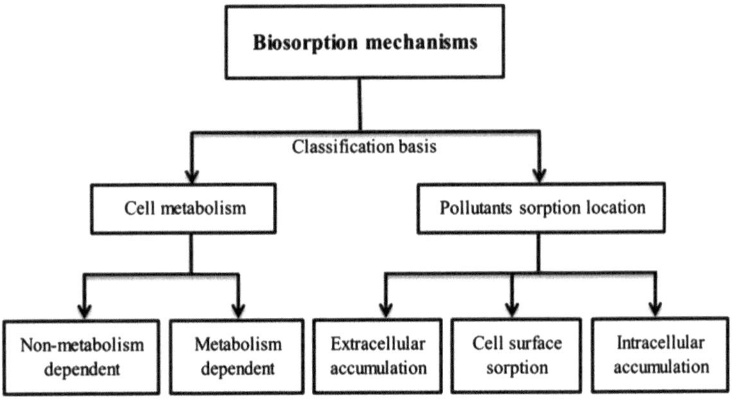

Fig. 4.2 The biosorption mechanisms are classified to different types on the basis of cell metabolism status or pollutants sorption location

Table 4.2 Biosorption capacities of heavy metals by fungi

Fungi	Heavy metal	Adsorption capacities (mg/g biomass)	References
Ganoderma lucidum	Zr(IV)	142.5	Hanif et al. (2015)
Coriolus versicolor	Zr(IV)	110.75	Amin et al. (2013)
Penicillium citrinum	U(VI)	127.3	Pang et al. (2011)
Rhizopus arrihizus	U(VI)	112.2	Wang et al. (2010)
Yarrowia lipolytica	Ni(II)	95.33	Shinde et al. (2012)
Pleurotus ostreatus	Cr(VI)	10.75	Javaid et al. (2011)
Aspergillus niger	Cr(VI)	117.33	Khambhaty et al. (2009)
Coriolus versicolor	Cr(VI)	62.89	Sanghi et al. (2009)
Rhizopus arrhizus	Cr(VI)	78	Aksu and Balibek (2007)
Saccharomyces cerevisiae	Cr(VI)	32.6	Ozer and Ozer (2003)
Surfactant-modified yeast	Cr(VI)	94.34	Bingol et al. (2004)
Trametes versicolor	Cu(II)	140.9	Subbaiah et al. (2011a)
Aspergillus niger	Cu(II)	20.91	Iskandar et al. (2011)
Penicillium simplicissimum	Cu(II)	16.18	Iskandar et al. (2011)
Trichoderma asperellum	Cu(II)	12.81	Iskandar et al. (2011)
Mucor rouxii	Zn(II)	53.85	Yan and Viraraghavan (2003)

(continued)

Table 4.2 (continued)

Fungi	Heavy metal	Adsorption capacities (mg/g biomass)	References
Trametes versicolor	Pb(II)	208.3	Subbaiah et al. (2011b)
	Cd(II)	166.6	
Amanita rubescens	Pb(II)	38.4	Sari and Tuzen (2009)
	Cd(II)	27.3	
Clitopilus scyphoides	Cd(II)	200	Moussous et al. (2012)
Auricularia polytricha	Cd(II)	63.3	Huang et al. (2012)
	Cu(II)	73.7	
	Pb(II)	221	

According to the correlation with the cell metabolism, fungal biosorption mechanisms can be divided into two types: (i) Non-metabolism dependent (passive uptake)—involving ion exchange, precipitation, complexation and physical adsorption (Veglio and Beolchini 1997), and (ii) Metabolism dependent (active uptake)—comprising an energy-driven process (Gadd 2009). Biosorption by dead fungal cells is a passive process which based on the interaction between the cell biomass and adsorbate. Dead cells capture pollutants through chemical functional groups on the cell wall which takes up most of the cellular dry weight. Passive uptake could also be present when the cell is metabolically active, however, it may be suppressed by cellular protective mechanisms against the toxic pollutants, e.g., active metal exclusion processes (Volesky 2007). Thus the dead fungal biomass holds promising biosorption capacity towards the toxic pollutants such as heavy metals and phenolic compounds (Kumar et al. 2008; Rao and Viraraghavan 2002). When living cells are used, the biosorption mechanisms become much more complicated. A lot of mechanisms may exist simultaneously since many cell metabolisms may be involved in biosorption. In addition, a variety of reaction, such as (i) adsorption, (ii) ion exchange, (iii) complexation, and (iv) precipitation may be affected by the change of the microenvironment around the cells, which can be altered by fungal cellular metabolism; for instance: nutrient uptake, respiration and metabolite release (Gadd and White 1993).

According to the location where the pollutants were captured and concentrated, fungal biosorption may also be classified as: (i) extracellular accumulation or precipitation, (ii) cell surface sorption (e.g., ion exchange, complexation, physical adsorption, precipitation), and (iii) intracellular accumulation (e.g., transport across cell membrane) (Muraleedharan et al. 1991). The chemical constitution and structural organization of the fungal cell wall are very complicated and all kinds of pollutant can either be bound in its surface or be deposited within its structure before they entry into the cytoplasm where they could be detained by other compounds or organelles. Fungal cell wall consists mainly of polysaccharides, proteins and lipids, offering diverse function groups, such as carboxyl (–COOH), phosphate (PO_4^{3-}), hydroxyl (–OH), amino (–NH_2), thiol (–SH) (Crini 2006) that are able to interact with adsorbed contaminants in different degrees. Many species of fungi have microfibrillar layer structures inside the

Table 4.3 Biosorption capacities of phenolic compounds by fungi

Fungi	Phenolic compounds	Adsorption capacities (mg/g biomass)	References
Aspergillus niger	Phenol	0.5	Rao and Viraraghavan (2002)
Emericella nidulans	2,4-dichlorophenol	9.1	Benoit et al. (1998)
	p-chlorophenol	3.0	
Rhizopus arrhizus	Pentachlorophenol	14.9	Bell and Tsezos (1987)
Trametes versicolor	Phenol	50	Kumar et al. (2009)
	o-chlorophenol	86	
	p-chlorophenol	112	
Schizophyllum commune	Phenol	120	Kumar and Min (2011)
	o-chlorophenol	178	
	p-chlorophenol	244	
Pleurotus sajor-caju	Phenol	89	Denizli et al. (2005)
	o-chlorophenol	159	
	p-chlorophenol	188	
	2,4,6-trichlorophenol	372	
Phanerochaete chrysosporium	Phenol	115	Denizli et al. (2004)
	o-chlorophenol	190	
	p-chlorophenol	228	
	2,4,6-trichlorophenol	421	

cell wall, which are composed of chitin or cellulose chains. The chitosan plays very important roles in fungal biosorption. Heavy metal ions could bind to the amine sites of chitin (R_2–NH) and chitosan (R–NH_2). Meanwhile, these amine sites also appear to be the major reactive groups for dyes, since intermolecular interactions of the dye molecules are most probable in chitosan–dye systems (Crini and Badot 2008). The Mucorales family (e.g., *Rhizopus arrhizus*), have outstanding biosorbent performances, which may be attributed to the high chitin content in their cell walls.

The diversity of chemical structures encountered in organic pollutants meant that their molecular size, charge, solubility, hydrophobicity, and reactivity, all affect the wastewater composition, choice of biosorbent and their biosorption efficiency. Pragmatically, of course, it may not be necessary to understand what mechanism is operative if the prime research goal is to identify an efficient biosorbent system (Gadd 2009).

4.2.3 Factors Influencing Biosorption Capacities

A variety of factors can affect biosorption. The type and nature of the fungal biomass or derived product can be very important. The properties of the biomass can be influenced by the age and growth condition of the fungal cells, which lead to changes in cell size, cell wall components, extracellular secretion, and other metabolic activities. Changes of cellular properties can also be achieved by physical and chemical pretreatment. The environment of biosorption and the use of bioreactors would dramatically influence the reactions between the biomass and contaminants thereby alter the biosorption efficiency.

(1) **Pretreatment for fungal biosorbents: physical or chemical treatment**

Physical processing of fungal biomass usually includes autoclaving, drying and crushing. Many researches in the biosorption of organic compounds, heavy metals or other toxicity pollutants with fungal biomass showed that enhanced sorption capacity was obtained using dead fungal biomass rather than the living cells (Rao and Viraraghavan 2002; Kumar et al. 2008). That may be because the living fungi can prevent toxic substances from entering into cells by cellular protective mechanisms. Physical treatments such as boiling, drying, lyophilisation or autoclaving will kill the fungal cells and improve the efficiency of sorption accordingly. Autoclaving could also rupture the fungal structure and expose the potential binding sites for certain adsorbate (Fu and Viraraghavan 2000). Drying can bring convenience to storage and transportation of fungal biomass and crushing will enhance the surface area so as to improve the adsorption rate.

Chemical treatments such as alkali treatment can improve biosorption capacity in some circumstances: chitin deacetylation resulting in the formation of chitosan-glucan complexes with higher metal affinities compare to the control group (Wang and Chen 2006). Acid pretreatment could change the negatively charged surface of fungal biomass to positively charged and thus increasing the attraction between fungal biomass and anionic dyes (Fu and Viraraghavan 2001b). In practical application, the specific pretreatment methods are determined by the types of adsorbates and fungal species. For instance, acid and alkali treatment decrease the adsorption capacity of Congo red onto fungal biomass of *Trametes versicolor*, while autoclaving could improve the removal percentage (Fig. 4.3).

(2) **The environment of biosorption: pH, temperature and salinity**

The solution pH value determines the surface electrical charge of fungal biomass and the ionic forms of contaminants. Therefore, solution pH affects both adsorbate chemistry and the fungal biomass binding sites. Heavy metal biosorption is strongly pH-dependent in almost all systems examined. Competition between cations and protons for binding sites leads to inferior biosorption efficiency of metals like Cu, Cd, Ni, Co, and Zn at low pH values (Gadd and White 1985; Shroff and Vaidya 2011). Conversely, anionic metal species like TcO_4^-, $PtCl_4^{3-}$, CrO_4^{2-}, SeO_4^{2-}, $Au(CN)_2^-$ may have a higher absorption rate at lower pH values. There also exists competition between cations, which can depress the biosorption of the metal. For dye biosorption, the decolorization rate usually has higher value when pH is lower (Fig. 4.3). For instance, the biosorption of Reactive Red 120 dye on the fungal biomass *Lentinus sajor-caju* increased as the pH was decreased, and similarly, maximum removal of reactive dye Remazol Black-B was found in the range of pH 1–2 and dropped sharply at higher values (O'Mahony et al. 2002; Arica and Bayramoğlu 2007).

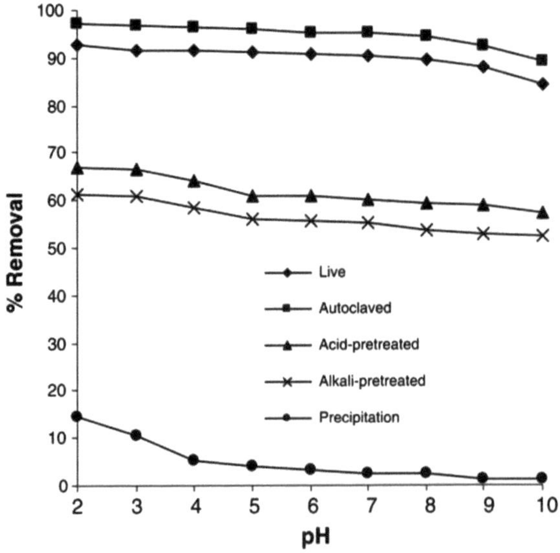

Fig. 4.3 Effect of pH on biosorption of Congo red fungal biomass of *Trametes versicolor* pretreated by different methods (Binupriya et al. 2008)

The effect on biosorption of temperature does not appear to be as strong as pH values. In some cases, higher temperature would enhance the biosorption efficiency due to the increase of the surface activity of fungal biomass and the kinetic energy of dye (Bakshi et al. 2006; Kaushik and Malik 2009). But at relatively high temperature, e.g., above 40 °C, the biosorption capacity often decrease in many cases (Iqbal and Saeed 2007; Erden et al. 2011), possibly due to the deactivation of the cell surface and destruction of some binding sites. Low temperature could restrain living cell metabolism systems and most of auxiliary processes which aid biosorption resulting in the decrease of biosorption efficiency (Gadd 2009).

Certain types of industrial wastewaters contain high salt concentration which may influence the biosorption processes. For example, the addition of 50 g/L salt resulting in a 28.8 % reduction in the biosorption capacity of Yellow RL dye of the *Rhizopus arrhizus* biomass (Aksu and Balibek 2010). However, some marine-derived fungi, like *Penicillium janthinellum* ZJU-BS-P1, have a strong tolerance with high salt concentrations (Fig. 4.4).

In addition to the above factors, the ionic strength, initial pollutant concentration and sorbent dosage would also affect the biosorption efficiency to a certain extent (Zhou and Banks 1993; Asgher et al. 2008; Khelifi et al. 2009; Levin et al. 2010).

(3) Biomass immobilization

The immobilization of fungal biomass may enhance biosorption capacity due to: (i) improved mechanical strength, (ii) increased porosity characteristics, (iii) less clogging, (iv) ease for regeneration and (v) multiple biosorbent recycle (Aksu and Gonen 2004; Aksu 2005; Solis et al. 2012). A number of materials have been successfully applied to immobilize fungal biomass. For example, nylon sponges, polyurethane foam, Luffa sponges, polystyrene foam, Ca-alginate beads, lignite

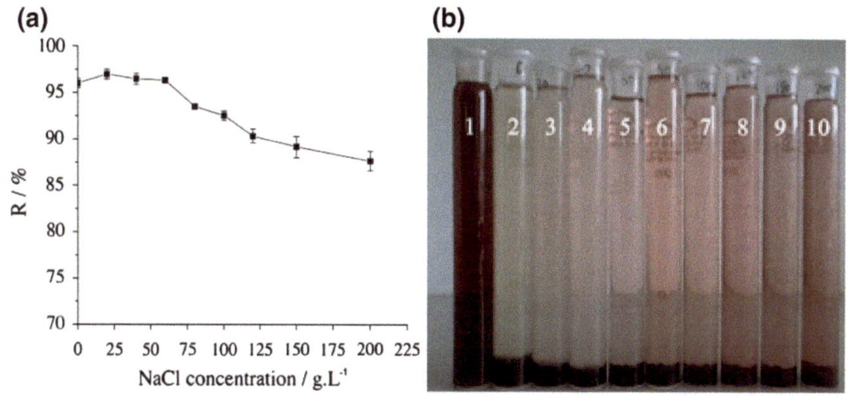

Fig. 4.4 **a** Effects of NaCl concentration on the biosorption of Congo red by marine-derived *Penicillium janthinellum*. **b** Congo red solutions after 24 h treatment (lane 1: control solution, lane 2–10: NaCl concentration 0, 20, 40, 60, 80, 100, 120, 150 and 200 g L^{-1}) (Wang et al. 2015)

granules, $ZrOCl_2$-activated pumice (Karimi et al. 2006; Maurya et al. 2006; Iqbal and Saeed 2007; Bohmer et al. 2010; Enayatzamir et al. 2010; Grinhut et al. 2011). Kocaoba and Arisoy (2011) observed that the biosorption capacity of *Pleurotus ostreatus* immobilized on Amberlite XAD-4 to remove Cr(III), Cd(II) and Cu(II) remained stable even after 10 cycles of sorption and desorption.

4.3 Biodegradation Function

Biodegradation is a kind of wastewater treatment that tackles the pollutants more thoroughly than biosorption and has a wide range of applications. In contrast to biosorption that can use dead fungal cells (and sometimes their effect is even better than living cells), biodegradation must be performed by living cells since the degradation process is controlled by enzymes secreted by fungi. Meanwhile, the wastewater sludge processing is one of the important applications of fungal degradation.

4.3.1 Biodegradable Application in Sludge Treatment

(1) **Roles of fungal laccases**[1]

Among the different types of enzymes produced by fungi, laccase is one of the best researched. It is a multicopper oxidase glycoprotein that is well known to be

[1]Additional information on the role of fungal laccase in wastewater treatment can be found in Chap. 5—*Potential of white-rot fungi to treat xenobiotic-containing waster* and Chap. 6—*Fungal bioremediation of emerging micropollutants in municipal wastewaters.*

ubiquitous in all kinds of living organisms. Fungal laccases play important roles in catalyzing the oxidation of a wide range of environmental pollutants, such as lignin, dyes and phenolic compounds (Aust and Benson 1993; Xu 1996; Giardina et al. 2010). Laccases consist of a sequence of polypeptide of about 500 amino acid residues and are linked to saccharides. Most fungal laccases are extracellular secretions and their molecular weights range is about 60–70 kDa (Baldrian 2006; Giardina et al. 2010). Laccases are nonspecific enzymes to their substrates, thus they are able to catalyze the oxidation of many organic contaminants including phenols (Reiss et al. 2011). They have also been used for the decolorization and detoxification of effluents from textile and paper making industries (Harms et al. 2011). As shown in Table 4.4, laccases has been used for the removal of many emerging organic contaminants from wastewater treatment plant effluents (Gasser et al. 2014a). For instance, laccases from *T. versicolor* could remove natural and synthetic estrogens, including estrone and 17β-estradiol, estriol from municipal wastewater (Auriol et al. 2007, 2008). In another batch study, *T. versicolor* was found to be able to remove 95–100 % of oxybenzone (Garcia et al. 2011). *Funalia trogii* ATCC 200800 had a strong ability to mineralize synthetic dyes by producing the laccase or MnP (Raghukumar et al. 1996). Laccase from *Trametes* sp. had been applied to biodegrade phenolic endocrine-disrupting chemicals including bisphenol A, nonylphenol, octylphenol, and ethynylestradiol (Tanaka et al. 2001).

(2) **Sludge treatment**

Traditional wastewater treatment generates plenty of sewage sludge which must be disposed of to maintain environmental protection (Zaidi 2008). Sludges contain

Table 4.4 Pollutants removal by laccases in various effluents

Sources of fungal laccase	Pollutants investigated	Percent removal (%)	References
Trametes versicolor	E1, E2, E3, EE2	100	Auriol et al. (2007)
Trametes versicolor	E1, E2, E3, EE2	97	Auriol et al. (2008)
Thielavia genus	BPA	98	Hommes et al. (2012)
Coriolopsis polyzona	BPA	93	Hommes et al. (2012)
Myceliophthora thermophila	E1	98	Lloret et al. (2013b)
	E2	≥97	
	EE2	≥99	
Coriolopsis polyzona	BPA	90	Demarche et al. (2012)
Coriolopsis gallica	BPA	≥85	Nair et al. (2013)
	diclofenac	30	
	EE2	≥85	
Thielavia genu	BPA	~66	Gasser et al. (2014b)

E1 estrone; *E2* 17 β–estradiol; *E3* estriol; *EE2* 17 α–ethinylestradiol; *BPA* bisphenol A more than 90 % of water along with organic solids that are problematic during transportation and treatment process (More et al. 2010). Therefore, the recovery and disposal of sludge is also an important issue in sewage treatment (Martins et al. 2004). The wastewater sludge contains a variety of microorganisms and organic matter. Fungi are saprophytic organisms and their nutrient requirement can be accomplished by the degradation of sludge. (Osiewacz 2002; Fakhru'l-Razi and Molla 2007). Filamentous fungi have great potential for sludge treatment and their functions include: (i) organic solids reduction, (ii) bioflocculation, (iii) pathogens removal, (iv) dewaterability, and (v) detoxification; a detailed account was given by More et al. (2010). Fungi have some advantages over bacteria in sludge treatment because of their strong capability to degrade more complex and variety of substrates (Khursheed and Kazmi 2011). Various fungi have been used for sludge treatment. For example: *Aspergillus niger* (Mannan et al. 2005), *Phanerochaete chrysosporium* (Molla et al. 2001), *Penicillium expansum* (Subramanian et al. 2008), *Trichoderma* sp. (Verma et al. 2005).

(3) **Extensive applications**

Residues of pesticides in the wastewater are harmful to the environment. For example, triclosan, which is a powerful bacteriostat, has been used extensively in soaps, shampoos, toothpastes, and disinfectants. It has long half-life and may potentially cause long term health risks in human body. Triclosan biodegradation yield can reach 71.91 % at about 7.5 mg/L initial concentration in semi-synthetic medium by using *Aspergillus versicolor* (Tastan and Donmez 2015). Coking wastewater contains various phenolic compounds and many other contaminants that are refractory, toxic and carcinogenic. When treated with *Phanerochaete chrysosporium*, a white-rot fungus, the removal rates of phenolic compounds and COD (Chemical Oxygen Demand) can achieve 84 and 80 %, respectively in 3 days (Lu et al. 2009). It is also found that laccase from a *Trametes* species has ability of degradation on polyunsaturated fatty acids and conjugated resin acids (Zhang et al. 2005). Thus, selected fungi and enzymes are also used for pitch removal (Singh and Singh 2014).

4.3.2 Factors Influencing Fungal Biodegradation

Various physicochemical operational parameters, including temperature, pH, nutrition, redox mediator, and the type of bioreactor, can influence the efficiency of fungal biodegradation.

(1) **Temperature, pH and nutrition**

Temperature is an important factor for all processes associated with fungal vitality. The maximum rate of biodegradation is generally related to the optimum growth temperature for each fungal species. It is worth mentioning that the oxidations catalyzed by laccases usually occur at ambient temperature (20–40 °C), thus laccases have become green and environmentally friendly for the elimination of pollutants (Wells et al. 2006). The pH has a great effect on the efficiency of biodegradation by influencing the enzyme activity. Fungi and yeast often show better biodegradation activities at acidic or neutral pH than bacteria (Khan et al. 2013). The effects of pH are also related to the transport of pollutant molecule across the cell membrane, which is considered as the rate limiting step for the biological catalysis (Kodam et al. 2005). For example, the removal rates of phenolic compounds and COD from coking wastewater by immobilized fungus *Phanerochaete chrysosporium* was significantly affected by pH and temperature (Fig. 4.5).

Enzymes production is dependent on the supply of nutrients to a certain degree. Certain types of wastewater, such as printing and dyeing effluents, lack nutrition, and require additional nutrients to improve the degradation efficiency of the fungi. Many studies on the azo dye decolorization were performed in the presence of additional carbon and nitrogen sources. The addition of glucose has frequently been demonstrated to improve the efficacy of azo dye degradation (Khan et al. 2013). But carbon sources seemed to be less effective than nitrogen sources (such as peptone, urea and yeast extract) in promoting biodegradation, probably due to the fact that nitrogen is the necessary building block of protein synthesis. Nitrogen sources can also regenerate NADH, which acts as an electron donor for the reduction of contaminants by microorganisms (Chang et al. 2000).

(2) **Mediator**

Laccases and other oxidases are able to oxidize small chemical compounds leading to radical formation (Canas and Camarero 2010). These radicals can act as redox mediators oxidizing compounds that might otherwise not be oxidized, thus broadening the substrate range. Redox mediators can also enhance many reductive processes under anaerobic conditions (Kodam et al. 2005). For instance, complete diclofenac removal could be achieved at pH 4 after 4, 2, and 0.5 h using no mediator, syringaldehyde, and 1-hydroxybenzotriazole, respectively (Lloret et al. 2013a).

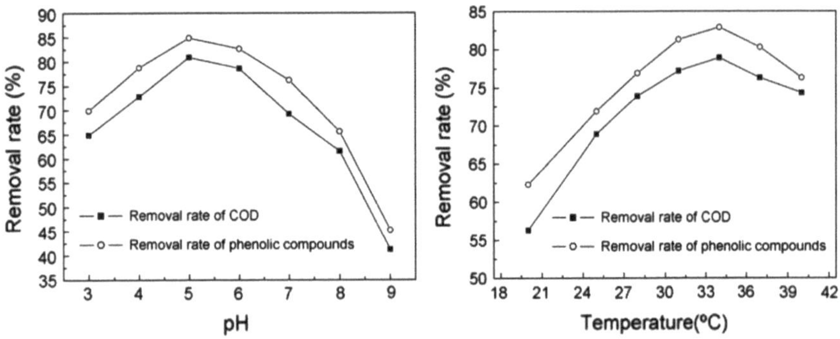

Fig. 4.5 Effect of pH on phenolic compounds removal rate (temperature 30 °C) and effect of temperature on phenolic compounds removal rate (pH 6.0) from coking wastewater by immobilized fungus *Phanerochaete chrysosporium*. Initial concentration of phenolic compounds and COD 313.5 and 3420 mg/L, respectively (Lu et al. 2009)

(3) **Use of bioreactors**

Bioreactor is an important tool for biodegradation. It can control the interaction pattern between fungal cells and pollutants in wastewater, thus influence the biodegradation efficiency greatly. Rotating drum, packed bed, fluidized bed, immobilized, and membrane bioreactors have been used as bioreactors. For instance, a membrane bioreactor using *Trametes versicolor* combined with reverse osmosis was effective for decolorization of dye wastewater (Kim et al. 2004). A wood-rotting fungal strain F29 decolorized 95–99 % Orange II in a continuous packed bed and fluidized bed bioreactor systems (Zhang et al. 1999). Immobilized bioreactors have been found to exhibit good biological activities and abilities for longtime operation (Srinivasan and Viraraghavan 2010).

4.4 Mix Fungi and Cooperation

Most studies demonstrate the effectiveness of a certain fungal strain to remove a particular contaminant; however, such specificity may limit the range of pollutants that can be treated by the fungus. Moreover, there are wide variations in the pollutant uptake capacity among different fungal strains. As industrial effluents contain various organic and inorganic contaminants, to develop a biological system capable of remediating all kinds of wastewaters, diverse types of microbial strains should be used in the form of a consortium (Mishra and Malik 2014a).

4.4.1 Cooperation Between Fungi

Different fungal strains usually possess different capacity to adsorb a specific pollutant. Therefore, the combination of fungi that exhibit different biosorption functions would enhance the total contaminants removal rate. For instance, a tripartite fungal consortium was studied for the abilities to remove metals (Cr^{6+} and Cu^{2+}) and dyes (Acid Blue 161 and Pigment Orange 34) from mixed waste streams (Mishra and Malik 2014b). The consortium consisted of *Aspergillus lentulus, Aspergillus terreus* and *Rhizopus oryzae* was significantly more effective than individual in removing the metals and dyes (Fig. 4.6).

The degradation function of fungi is very powerful, but there are large differences between the species. Therefore, the utilization of mix fungi has huge potential to improve the degradation efficiency. For instance, a mixed filamentous fungi culture (*Aspergillus niger* and *Penicillium corylophilum*) was used in a sewage sludge bioremediation study, and the highest removal of turbidity, total suspended solid and COD were achieved at 99, 98, and 93 %, respectively, by day 10 compared to the control (Rahman et al. 2014).

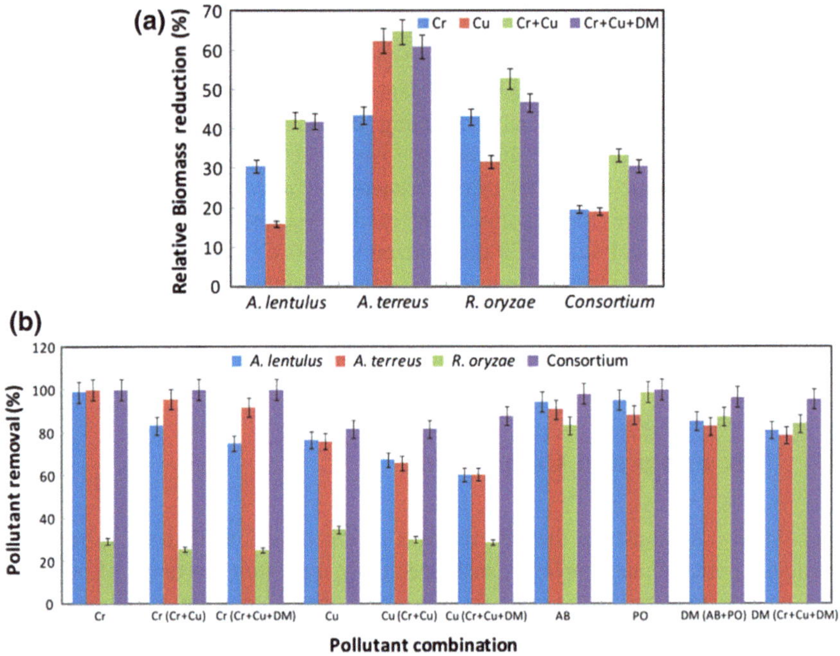

Fig. 4.6 **a** Performance of fungal consortium versus individual strain in terms of **a** relative biomass reduction and **b** pollutant removal from mixed pollutant. *AB* Acid Blue 161, *PO* Pigment Orange 34, *DM* dye mixture (Mishra and Malik 2014b)

During the liquid culture process, filamentous fungal could form mycelial pellets. As self-immobilized and bioactive particles, these pellets show advantages over mycelium for some industrial applications, such as: (i) strong surviving ability, (ii) fast settlement rates, (iii) easy solid–liquid separation and (iv) good reusability. Mycelia pellet can be used as a biological carrier for whole-cell immobilization due to its stable structural characteristics. As shown in Fig. 4.7, an innovative two-species whole-cell immobilization system was achieved simply by inoculating the *Pestalotiopsis* sp. J63 conidia into culture medium containing *Penicillium janthinellum* P1 pre-grown mycelia pellets and the resulting co-immobilization system was used for the treatment of paper mill effluent. Numerous insoluble fine fibers in the sewage were successfully and rapidly biodegraded and removed using this novel co-immobilization system (Chen et al. 2014a, b).

4.4.2 Cooperation Between Fungus and Bacterium

Bacteria have strong degradation ability on certain contaminants and they can be fixed by fungal mycelial pellets as biological carrier so as to form an immobilization

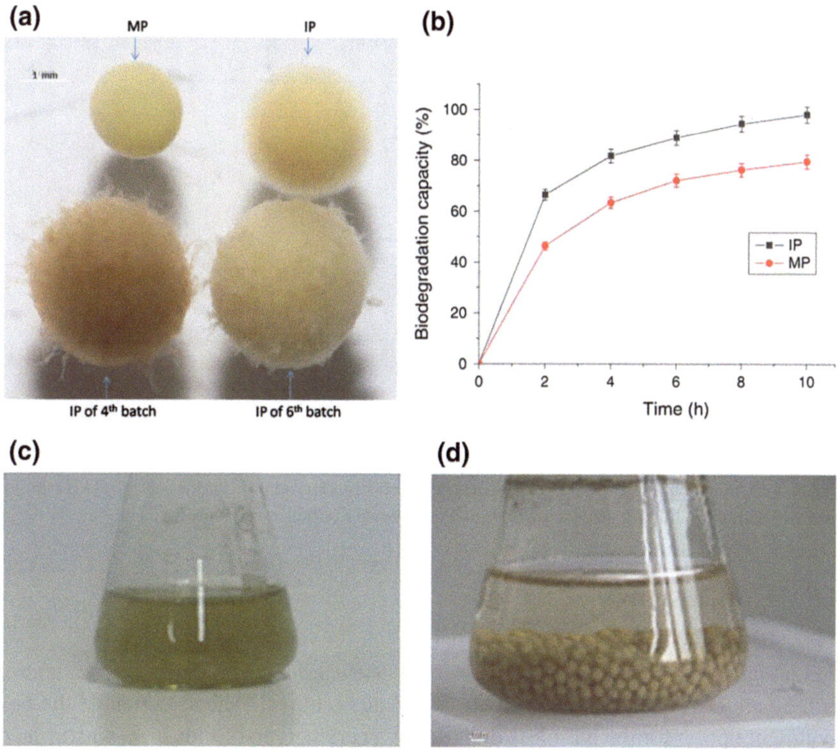

Fig. 4.7 The treatment of paper mill effluent by using a novel two-species whole-cell immobilization system. The immobilized pellets (IP) was made by inoculating the marine-derived fungus *Pestalotiopsis* sp. J63 spores into culture medium containing another fungus *Penicillium janthinellum* P1 pre-grown mycelia pellets (MP) for 2 days. **a** The different pellet sizes. **b** Biodegradation capacity of immobilized pellets and mycelia pellets in the process of wastewater treatment. **c** Diluted wastewater. **d** The effect of wastewater treatment using immobilized pellets. Modified and cited from Chen et al. (2014a)

system of fungi—bacteria that possesses multiple functions including biosorption and biodegradation. For instance, mycelial pellet of *Aspergillus niger* Y3 was used to immobilize the aniline degradation bacterium, *Acinetobacter calcoaceticus* JH-9 and other COD rapid removal bacteria. The combined mycelial pellets were applied in the SBR and the biological removal efficiency was about 0.9 mg aniline (L/d) (Zhang et al. 2011). A new azo dyes-decolorizing fungal strain *Penicillium* sp. QQ was used to immobilize *Sphingomonas xenophaga* QYY which has good azoreductase activity, the co-cultures were found to perform better than individual strains (Gou et al. 2009). Biosorption and direct biodegradation of polycyclic aromatic hydrocarbons (PAHs) in soil can be stimulated by *P. chrysosporium* and promoted synergistically by wild microorganisms (Chen and Ding 2012); the schematic diagram was shown in Fig. 4.8.

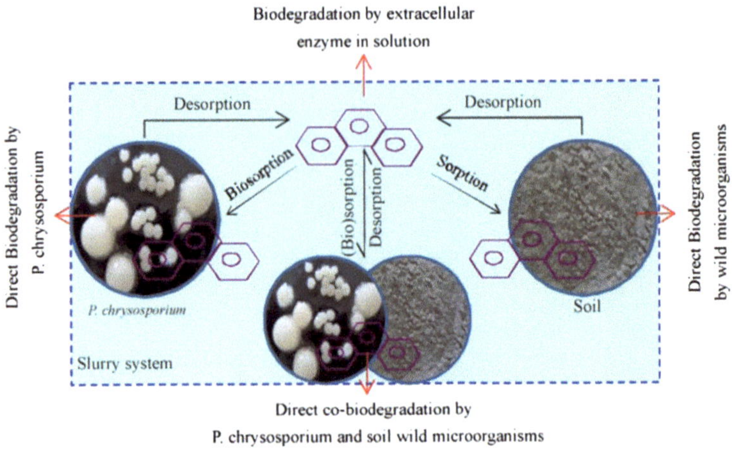

Fig. 4.8 The schematic diagram of (bio)sorption and biodegradation processes of PAHs in soil slurry systems containing *Phanerochaete chrysosporium* (Chen and Ding 2012)

4.5 Conclusions

Fungal biosorption and biodegradation of wastewater have received much attention as they are cost-effective methods for pollutants removal. The selection of the best treatment option for the harmless disposal of a certain type of industrial wastewaters is a difficult task because of their complex composition. The best way is often a combination of two or more species, and the choice of such consortium depends on the effluent composition, cost, toxicity of the degradation products and future use of the treated water (Solis et al. 2012). Most of researches on fungi in treatment of sewage and wastewater have been performed on a laboratory scale. Therefore, extensive laboratory works followed by series of pilot scale studies are essential for future industrial process applications.

Acknowledgments This work was financially supported by the National Natural Science Foundation of China, and Scientific Technology Program of Zhoushan City (2013C51021).

References

Abd El-Rahim WM, Moawad H, Khalafallah M (2003) Microflora involved in textile dye waste removal. J Basic Microb 43(3):167–174. doi:10.1002/jobm.200390019

Akar T, Arslan S, Akar ST (2013) Utilization of *Thamnidium elegans* fungal culture in environmental cleanup: a reactive dye biosorption study. Ecol Eng 58:363–370. doi:10.1016/j.ecoleng.2013.06.026

Aksu Z (2005) Application of biosorption for the removal of organic pollutants: a review. Process Biochem 40(3–4):997–1026. doi:10.1016/j.procbio.2004.04.008

Aksu Z, Balibek E (2007) Chromium(VI) biosorption by dried *Rhizopus arrhizus*: Effect of salt (NaCl) concentration on equilibrium and kinetic parameters. J Hazard Mater 145(1–2):210–220. doi:10.1016/j.jhazmat.2006.11.011

Aksu Z, Balibek E (2010) Effect of salinity on metal-complex dye biosorption by *Rhizopus arrhizus*. J Environ Manage 91(7):1546–1555. doi:10.1016/j.jenvman.2010.02.026

Aksu Z, Gonen F (2004) Biosorption of phenol by immobilized activated sludge in a continuous packed bed: prediction of breakthrough curves. Process Biochem 39(5):599–613. doi:10.1016/S0032-9592(03)00132-8

Amin M, Bhatti HN, Sadaf S (2013) Bioremediation of zirconium from aqueous solution by *Coriolus versicolor*: process optimization. J Chem Soc Pakistan 35(3):692–698

Arica MY, Bayramoğlu G (2007) Biosorption of Reactive Red-120 dye from aqueous solution by native and modified fungus biomass preparations of *Lentinus sajor-caju*. J Hazard Mater 149 (2):499–507. doi:10.1016/j.jhazmat.2007.04.021

Asgher M, Kausar S, Bhatti HN, Shah SAH, Ali M (2008) Optimization of medium for decolorization of Solar golden yellow R direct textile dye by *Schizophyllum commune* IBL-06. Int Biodeter Biodegr 61(2):189–193. doi:10.1016/j.ibiod.2007.07.009

Auriol M, Filali-Meknassi Y, Adams CD, Tyagi RD, Noguerol TN, Pina B (2008) Removal of estrogenic activity of natural and synthetic hormones from a municipal wastewater: efficiency of horseradish peroxidase and laccase from *Trametes versicolor*. Chemosphere 70(3):445–452. doi:10.1016/j.chemosphere.2007.06.064

Auriol M, Filali-Meknassi Y, Tyagi RD, Adams CD (2007) Laccase-catalyzed conversion of natural and synthetic hormones from a municipal wastewater. Water Res 41(15):3281–3288. doi:10.1016/j.watres.2007.05.008

Aust SD, Benson JT (1993) The fungus among us—use of white-rot fungi to biodegrade environmental-pollutants. Environ Health Persp 101(3):232–233. doi:10.2307/3431547

Awasthi MK, Pandey AK, Khan J, Bundela PS, Wong JWC, Selvam A (2014) Evaluation of thermophilic fungal consortium for organic municipal solid waste composting. Bioresour Technol 168:214–221. doi:10.1016/j.biortech.2014.01.048

Bakshi DK, Saha S, Sindhu I, Sharma P (2006) Use of *Phanerochaete chrysosporium* biomass for the removal of textile dyes from a synthetic effluent. World J Microbiol Biotechnol 22(8):835–839. doi:10.1007/s11274-005-9111-3

Baldrian P (2006) Fungal laccases—occurrence and properties. FEMS Microbiol Rev 30(2):215–242. doi:10.1111/j.1574-4976.2005.00010.x

Banat FA, Al-Bashir B, Al-Asheh S, Hayajneh O (2000) Adsorption of phenol by bentonite. Environ Pollut 107(3):391–398. doi:10.1016/S0269-7491(99)00173-6

Banat IM, Nigam P, Singh D, Marchant R (1996) Microbial decolorization of textile-dye-containing effluents: a review. Bioresource Technol 58(3):217–227. doi:10.1016/S0960-8524(96)00113-7

Bayramoğlu G, Arica MY (2007) Biosorption of benzidine based textile dyes "Direct Blue 1 and Direct Red 128" using native and heat-treated biomass of *Trametes versicolor*. J Hazard Mater 143(1–2):135–143. doi:10.1016/j.jhazmat.2006.09.002

Bell JP, Tsezos M (1987) Removal of hazardous organic pollutants by adsorption on microbial biomass. Water Sci Technol 19(3–4):409–416

Benoit P, Barriuso E, Calvet R (1998) Biosorption characterization of herbicides, 2,4-D and atrazine, and two chlorophenols on fungal mycelium. Chemosphere 37(7):1271–1282. doi:10.1016/S0045-6535(98)00125-8

Bingol A, Ucun H, Bayhan YK, Karagunduz A, Cakici A, Keskinler B (2004) Removal of chromate anions from aqueous stream by a cationic surfactant-modified yeast. Bioresour Technol 94(3):245–249. doi:10.1016/j.biotech.2004.01.018

Binupriya AR, Sathishkumar M, Kavitha D, Swaminathan K, Yun SE, Mun SP (2007) Experimental and isothermal studies on sorption of Congo red by modified mycelial biomass of wood-rotting fungus. Clean-Soil Air Water 35(2):143–150. doi:10.1002/clen.200700025

Binupriya AR, Sathishkumar M, Swaminathan K, Ku CS, Yun SE (2008) Comparative studies on removal of Congo red by native and modified mycelial pellets of *Trametes versicolor* in various reactor modes. Bioresour Technol 99(5):1080–1088

Bohmer U, Kirsten C, Bley T, Noack M (2010) White-rot fungi combined with lignite granules and lignitic xylite to decolorize textile industry wastewater. Eng Life Sci 10(1):26–34. doi:10.1002/elsc.200900024

Bowman SM, Free SJ (2006) The structure and synthesis of the fungal cell wall. BioEssays 28 (8):799–808. doi:10.1002/Bies.20441

Cabib E, Bowers B, Sburlati A, Silverman SJ (1988) Fungal cell-wall synthesis—the construction of a biological structure. Microbiol Sci 5(12):370–375

Canas AI, Camarero S (2010) Laccases and their natural mediators: biotechnological tools for sustainable eco-friendly processes. Biotechnol Adv 28(6):694–705. doi:10.1016/j.biotechadv.2010.05.002

Chang JS, Kuo TS, Chao YP, Ho JY, Lin PJ (2000) Azo dye decolorization with a mutant *Escherichia coli* strain. Biotechnol Lett 22(9):807–812. doi:10.1023/A:1005624707777

Chen BL, Ding J (2012) Biosorption and biodegradation of phenanthrene and pyrene in sterilized and unsterilized soil slurry systems stimulated by *Phanerochaete chrysosporium*. J Hazard Mater 229:159–169. doi:10.1016/j.jhazmat.2012.05.090

Chen HY, Guan YX, Yao SJ (2014a) A novel two-species whole-cell immobilization system composed of marine-derived fungi and its application in wastewater treatment. J Chem Technol Biotechnol 89(11):1733–1740. doi:10.1002/Jctb.4253

Chen HY, Wang MX, Shen YB, Yao SJ (2014b) Optimization of two-species whole-cell immobilization system constructed with marine-derived fungi and its biological degradation ability. Chin J Chem Eng 22(2):187–192. doi:10.1016/S1004-9541(14)60024-0

Crini G (2006) Non-conventional low-cost adsorbents for dye removal: a review. Bioresour Technol 97(9):1061–1085. doi:10.1016/j.biortech.2005.05.001

Crini G, Badot PM (2008) Application of chitosan, a natural aminopolysaccharide, for dye removal from aqueous solutions by adsorption processes using batch studies: A review of recent literature. Prog Polym Sci 33(4):399–447. doi:10.1016/j.progpolymsci.2007.11.001

Dal Bosco SM, Jimenez RS, Vignado C, Fontana J, Geraldo B, Figueiredo FCA et al (2006) Removal of Mn(II) and Cd(II) from wastewaters by natural and modified clays. Adsorption 12 (2):133–146. doi:10.1007/s10450-006-0375-1

Das SK, Bhowal J, Das AR, Guha AK (2006) Adsorption behavior of rhodamine B on *Rhizopus oryzae* biomass. Langmuir 22(17):7265–7272. doi:10.1021/La0526378

Demarche P, Junghanns C, Mazy N, Agathos SN (2012) Design-of-experiment strategy for the formulation of laccase biocatalysts and their application to degrade bisphenol A. New Biotechnol 30(1):96–103. doi:10.1016/j.nbt.2012.05.023

Denizli A, Cihangir N, Rad AY, Taner M, Alsancak G (2004) Removal of chlorophenols from synthetic solutions using *Phanerochaete chrysosporium*. Process Biochem 39(12):2025–2030. doi:10.1016/j.procbio.2003.10.003

Denizli A, Cihangir N, Tuzmen N, Alsancak G (2005) Removal of chlorophenols from aquatic systems using the dried and dead fungus *Pleurotus sajor caju*. Bioresour Technol 96(1):59–62. doi:10.1016/j.biortech.2003.11.029

Dincer AR, Gunes Y, Karakaya N, Gunes E (2007) Comparison of activated carbon and bottom ash removal of reactive dye from aqueous solution. Bioresour Technol 98(4):834–839

Enayatzamir K, Alikhani HA, Yakhchali B, Tabandeh F, Rodriguez-Couto S (2010) Decolouration of azo dyes by P*hanerochaete chrysosporium* immobilised into alginate beads. Environ Sci Pollut Res 17(1):145–153. doi:10.1007/s11356-009-0109-5

Erdal S, Taskin M (2010) Uptake of textile dye Reactive Black-5 by *Penicillium chrysogenum* MT-6 isolated from cement-contaminated soil. Afr J Microbiol Res 4(8):618–625

Erden E, Kaymaz Y, Pazarlioglu NK (2011) Biosorption kinetics of a direct azo dye Sirius Blue K-CFN by *Trametes versicolor*. Electron J Biotechn 14(2) doi:ARTN 8. doi:10.2225/vol14-issue2-fulltext-8

Esplugas S, Gimenez J, Contreras S, Pascual E, Rodriguez M (2002) Comparison of different advanced oxidation processes for phenol degradation. Water Res 36(4):1034–1042. doi:10.1016/S0043-1354(01)00301-3

Fakhru'l-Razi A, Molla AH (2007) Enhancement of bioseparation and dewaterability of domestic wastewater sludge by fungal treated dewatered sludge. J Hazard Mater 147(1–2):350–356. doi:10.1016/j.jhazmat.2007.01.060

Fu FL, Wang Q (2011) Removal of heavy metal ions from wastewaters: a review. J Environ Manage 92(3):407–418. doi:10.1016/j.jenvman.2010.11.011

Fu YZ, Viraraghavan T (2000) Removal of a dye from an aqueous solution by the fungus *Aspergillus niger*. Water Qual Res J Can 35(1):95–111

Fu YZ, Viraraghavan T (2001a) Fungal decolorization of dye wastewaters: a review. Bioresour Technol 79(3):251–262. doi:10.1016/S0960-8524(01)00028-1

Fu YZ, Viraraghavan T (2001b) Removal of CI Acid Blue 29 from an aqueous solution by *Aspergillus niger*. Aatcc Rev 1(1):36–40

Gadd GM (2009) Biosorption: critical review of scientific rationale, environmental importance and significance for pollution treatment. J Chem Technol Biotechnol 84(1):13–28. doi:10.1002/Jctb.1999

Gadd GM, White C (1985) Copper Uptake by *Penicillium-Ochro-Chloron*—influence of pH on toxicity and demonstration of energy-dependent copper influx using protoplasts. J Gen Microbiol 131:1875–1879 (Aug)

Gadd GM, White C (1993) Microbial treatment of metal pollution—a working biotechnology. Trends Biotechnol 11(8):353–359. doi:10.1016/0167-7799(93)90158-6

Garcia HA, Hoffman CM, Kinney KA, Lawler DF (2011) Laccase-catalyzed oxidation of oxybenzone in municipal wastewater primary effluent. Water Res 45(5):1921–1932. doi:10.1016/j.watres.2010.12.027

Gasser CA, Ammann EM, Shahgaldian P, Corvini PFX (2014a) Laccases to take on the challenge of emerging organic contaminants in wastewater. Appl Microbiol Biotechnol 98(24):9931–9952. doi:10.1007/s00253-014-6177-6

Gasser CA, Yu L, Svojitka J, Wintgens T, Ammann EM, Shahgaldian P et al (2014b) Advanced enzymatic elimination of phenolic contaminants in wastewater: a nano approach at field scale. Appl Microbiol Biotechnol 98(7):3305–3316. doi:10.1007/s00253-013-5414-8

Giardina P, Faraco V, Pezzella C, Piscitelli A, Vanhulle S, Sannia G (2010) Laccases: a never-ending story. Cell Mol Life Sci 67(3):369–385. doi:10.1007/s00018-009-0169-1

Gou M, Qu YY, Zhou JT, Ma F, Tan L (2009) Azo dye decolorization by a new fungal isolate, *Penicillium* sp QQ and fungal-bacterial cocultures. J Hazard Mater 170(1):314–319. doi:10.1016/j.jhazmat.2009.04.094

Grinhut T, Salame TM, Chen YN, Hadar Y (2011) Involvement of ligninolytic enzymes and Fenton-like reaction in humic acid degradation by *Trametes* sp. Appl Microbiol Biotechnol 91(4):1131–1140. doi:10.1007/s00253-011-3300-9

Hamdaoui O, Naffrechoux E (2007a) Modeling of adsorption isotherms of phenol and chlorophenols onto granular activated carbon—Part I. Two-parameter models and equations allowing determination of thermodynamic parameters. J Hazard Mater 147(1–2):381–394. doi:10.1016/j.jhazmat.2007.01.021

Hamdaoui O, Naffrechoux E (2007b) Modeling of adsorption isotherms of phenol and chlorophenols onto granular activated carbon—Part II. Models with more than two parameters. J Hazard Mater 147(1–2):401–411. doi:10.1016/j.jhazmat.2007.01.023

Hanif A, Bhatti HN, Hanif MA (2015) Removal of zirconium from aqueous solution by *Ganoderma lucidum*: biosorption and bioremediation studies. Desalin Water Treat 53(1):195–205. doi:10.1080/19443994.2013.837005

Hank D, Azi Z, Hocine SA, Chaalal O, Hellal A (2014) Optimization of phenol adsorption onto bentonite by factorial design methodology. J Ind Eng Chem 20(4):2256–2263. doi:10.1016/j.jiec.2013.09.058

Harms H, Schlosser D, Wick LY (2011) Untapped potential: exploiting fungi in bioremediation of hazardous chemicals. Nat Rev Microbiol 9(3):177–192. doi:10.1038/Nrmicro2519

Hommes G, Gasser CA, Howald CBC, Goers R, Schlosser D, Shahgaldian P, Corvini PFX (2012) Production of a robust nanobiocatalyst for municipal wastewater treatment. Bioresour Technol 115:8–15. doi:10.1016/j.biortech.2011.11.129

Huang HW, Cao LX, Wan YX, Zhang RD, Wang WF (2012) Biosorption behavior and mechanism of heavy metals by the fruiting body of jelly fungus (*Auricularia polytricha*) from aqueous solutions. Appl Microbiol Biotechnol 96(3):829–840. doi:10.1007/s00253-011-3846-6

Ip AWM, Barford JP, Mckay G (2010) A comparative study on the kinetics and mechanisms of removal of Reactive Black 5 by adsorption onto activated carbons and bone char. Chem Eng J 157(2–3):434–442. doi:10.1016/j.cej.2009.12.003

Iqbal M, Saeed A (2007) Biosorption of reactive dye by loofa sponge-immobilized fungal biomass of *Phanerochaete chrysosporium*. Process Biochem 42(7):1160–1164. doi:10.1016/j.procbio.2007.05.014

Iskandar NL, Zainudin NAIM, Tan SG (2011) Tolerance and biosorption of copper (Cu) and lead (Pb) by filamentous fungi isolated from a freshwater ecosystem. J Environ Sci China 23(5):824–830. doi:10.1016/S1001-0742(10)60475-5

Javaid A, Bajwa R, Shafique U, Anwar J (2011) Removal of heavy metals by adsorption on *Pleurotus ostreatus*. Biomass Bioenergy 35(5):1675–1682. doi:10.1016/j.biombioe.2010.12.035

Kabsch-Korbutowicz M, Krupinska B (2008) Removal of natural organic matter from water by using ion-exchange resins. Przem Chem 87(5):473–475

Kadirvelu K, Thamaraiselvi K, Namasivayam C (2001) Removal of heavy metals from industrial wastewaters by adsorption onto activated carbon prepared from an agricultural solid waste. Bioresour Technol 76(1):63–65. doi:10.1016/S0960-8524(00)00072-9

Karimi A, Vahabzadeh F, Bonakdarpour B (2006) Use of *Phanerochaete chrysosporium* immobilized on Kissiris for synthetic dye decolourization: involvement of manganese peroxidase. World J Microbiol Biotechnol 22(12):1251–1257. doi:10.1007/s11274-006-9169-6

Kaushik P, Malik A (2009) Fungal dye decolourization: recent advances and future potential. Environ Int 35(1):127–141. doi:10.1016/j.envint.2008.05.010

Kaushik P, Mishra A, Malik A, Pant KK (2014) Biosorption of textile dye by *Aspergillus lentulus* pellets: process optimization and cyclic removal in aerated Bioreactor. Water Air Soil Pollut 225(6). doi:Artn 1978 doi:10.1007/S11270-014-1978-X

Khambhaty Y, Mody K, Basha S, Jha B (2009) Kinetics, equilibrium and thermodynamic studies on biosorption of hexavalent chromium by dead fungal biomass of marine *Aspergillus niger*. Chem Eng J 145(3):489–495. doi:10.1016/j.cej.2008.05.002

Khan R, Bhawana P, Fulekar MH (2013) Microbial decolorization and degradation of synthetic dyes: a review. Rev Environ Sci Biotechnol 12(1):75–97. doi:10.1007/s11157-012-9287-6

Khelifi E, Ayed L, Bouallagui H, Touhami Y, Hamdi M (2009) Effect of nitrogen and carbon sources on Indigo and Congo red decolourization by *Aspergillus alliaceus* strain 121C. J Hazard Mater 163(2–3):1056–1062. doi:10.1016/j.jhazmat.2008.07.060

Khursheed A, Kazmi AA (2011) Retrospective of ecological approaches to excess sludge reduction. Water Res 45(15):4287–4310. doi:10.1016/j.watres.2011.05.018

Kim TH, Lee Y, Yang J, Lee B, Park C, Kim S (2004) Decolorization of dye solutions by a membrane bioreactor (MBR) using white-rot fungi. Desalination 168:287–293. doi:10.1016/j.desal.2004.07.011

Kocaoba S, Arisoy M (2011) The use of a white rot fungi (Pleurotus ostreatus) immobilized on Amberlite XAD-4 as a new biosorbent in trace metal determination. Bioresour Technol 102(17):8035–8039. doi:10.1016/j.biortech.2011.05.004

Kodam KM, Soojhawon I, Lokhande PD, Gawai KR (2005) Microbial decolorization of reactive azo dyes under aerobic conditions. World J Microbiol Biotechnol 21(3):367–370. doi:10.1007/s11274-004-5957-z

Kumar NS, Boddu VM, Krishnaiah A (2009) Biosorption of phenolic compounds by *Trametes versicolor* polyporus fungus. Adsorpt Sci Technol 27(1):31–46. doi:10.1260/026361709788921597

Kumar NS, Min K (2011) Phenolic compounds biosorption onto *Schizophyllum commune* fungus: FTIR analysis, kinetics and adsorption isotherms modeling. Chem Eng J 168(2):562–571. doi:10.1016/j.cej.2011.01.023

Kumar R, Bishnoi NR, Garima Bishnoi K (2008) Biosorption of chromium(VI) from aqueous solution and electroplating wastewater using fungal biomass. Chem Eng J 135(3):202–208. doi:10.1016/j.cej.2007.03.004

Kurniawan TA, Chan GYS, Lo WH, Babel S (2006) Comparisons of low-cost adsorbents for treating wastewaters laden with heavy metals. Sci Total Environ 366(2–3):409–426. doi:10.1016/j.scitotenv.2005.10.001

Levin L, Melignani E, Ramos AM (2010) Effect of nitrogen sources and vitamins on ligninolytic enzyme production by some white-rot fungi. Dye decolorization by selected culture filtrates. Bioresour Technol 101(12):4554–4563. doi:10.1016/j.biortech.2010.01.102

Lin SH, Juang RS (2002) Removal of free and chelated Cu(II) ions from water by a nondispersive solvent extraction process. Water Res 36(14):3611–3619. doi:Pii S0043-1354(02)00074-X. doi:10.1016/S0043-1354(02)00074-X

Lloret L, Eibes G, Moreira MT, Feijoo G, Lema JM (2013a) On the use of a high-redox potential laccase as an alternative for the transformation of non-steroidal anti-inflammatory drugs (NSAIDs). J Mol Catal B Enzym 97:233–242. doi:10.1016/j.molcatb.2013.08.021

Lloret L, Eibes G, Moreira MT, Feijoo G, Lema JM (2013b) Removal of estrogenic compounds from filtered secondary wastewater effluent in a continuous enzymatic membrane reactor. Identification of biotransformation products. Environ Sci Technol 47(9):4536–4543. doi:10.1021/es304783k

Lu Y, Yan LH, Wang Y, Zhou SF, Fu JJ, Zhang JF (2009) Biodegradation of phenolic compounds from coking wastewater by immobilized white rot fungus *Phanerochaete chrysosporium*. J Hazard Mater 165(1–3):1091–1097. doi:10.1016/j.jhazmat.2008.10.091

Mannan S, Fakhru'l-Razi A, Alam MZ (2005) Use of fungi to improve bioconversion of activated sludge. Water Res 39(13):2935–2943. doi:10.1016/j.watres.2005.04.074

Martins AMP, Pagilla K, Heijnen JJ, van Loosdrecht MCM (2004) Filamentous bulking sludge—a critical review. Water Res 38(4):793–817. doi:10.1016/j.watres.2003.11.005

Maurya NS, Mittal AK, Cornel P, Rother E (2006) Biosorption of dyes using dead macro fungi: Effect of dye structure, ionic strength and pH. Bioresour Technol 97(3):512–521. doi:10.1016/j.biortech.2005.02.045

McMullan G, Meehan C, Conneely A, Kirby N, Robinson T, Nigam P et al (2001) Microbial decolourisation and degradation of textile dyes. Appl Microbiol Biotechnol 56(1–2):81–87

Mishra A, Malik A (2014a) Metal and dye removal using fungal consortium from mixed waste stream: optimization and validation. Ecol Eng 69:226–231. doi:10.1016/j.ecoleng.2014.04.007

Mishra A, Malik A (2014b) Novel fungal consortium for bioremediation of metals and dyes from mixed waste stream. Bioresour Technol 171:217–226. doi:10.1016/j.biortech.2014.08.047

Molla AH, Fakhru'l-Razi A, Abd-Aziz S, Hanafi MM, Alam MZ (2001) In-vitro compatibility evaluation of fungal mixed culture for bioconversion of domestic wastewater sludge. World J Microbiol Biotechnol 17(9):849–856. doi:10.1023/A:1013844306960

More TT, Yan S, Tyagi RD, Surampalli RY (2010) Potential use of filamentous fungi for wastewater sludge treatment. Bioresour Technol 101(20):7691–7700. doi:10.1016/j.biortech.2010.05.033

Moussous S, Selatnia A, Merati A, Junter GA (2012) Batch cadmium(II) biosorption by an industrial residue of macrofungal biomass (*Clitopilus scyphoides*). Chem Eng J 197:261–271. doi:10.1016/j.cej.2012.04.106

Muraleedharan TR, Iyengar L, Venkobachar C (1991) Biosorption—an attractive alternative for metal removal and recovery. Curr Sci India 61(6):379–385

Nair RR, Demarche P, Agathos SN (2013) Formulation and characterization of an immobilized laccase biocatalyst and its application to eliminate organic micropollutants in wastewater. New Biotechnol 30(6):814–823. doi:10.1016/j.nbt.2012.12.004
O'Mahony T, Guibal E, Tobin JM (2002) Reactive dye biosorption by *Rhizopus arrhizus* biomass. Enzyme Microb Tech 31(4):456–463. doi:Pii S0141-0229(02)00110-2; doi:10.1016/S0141-0229(02)00110-2
Osiewacz HD (2002) Genes, mitochondria and aging in filamentous fungi. Ageing Res Rev 1 (3):425–442. doi:Pii S1568-1637(02)00010-7; doi:10.1016/S1568-1637(02)00010-7
Ozer A, Ozer D (2003) Comparative study of the biosorption of Pb(II), Ni(II) and Cr(VI) ions onto *S. cerevisiae*: determination of biosorption heats. J Hazard Mater 100(1–3):219–229. doi:10.1016/S0304-3894(03)00109-2
Pang C, Liu YH, Cao XH, Li M, Huang GL, Hua R et al (2011) Biosorption of uranium(VI) from aqueous solution by dead fungal biomass of *Penicillium citrinum*. Chem Eng J 170(1):1–6. doi:10.1016/j.cej.2010.10.068
Raghukumar C, Chandramohan D, Michel FC, Reddy CA (1996) Degradation of lignin and decolorization of paper mill bleach plant effluent (BPE) by marine fungi. Biotechnol Lett 18 (1):105–106. doi:10.1007/Bf00137820
Rahman RA, Molla A, Fakhru'l-Razi A (2014) Assessment of sewage sludge bioremediation at different hydraulic retention times using mixed fungal inoculation by liquid-state bioconversion. Environ Sci Pollut Res 21(2):1178–1187. doi:10.1007/s11356-013-1974-5
Rangabhashiyam S, Suganya E, Selvaraju N, Varghese LA (2014) Significance of exploiting non-living biomaterials for the biosorption of wastewater pollutants. World J Microbiol Biotechnol 30(6):1669–1689. doi:10.1007/s11274-014-1599-y
Rao JR, Viraraghavan T (2002) Biosorption of phenol from an aqueous solution by *Aspergillus niger* biomass. Bioresource Technol 85(2):165–171. doi:Pii S0960-8524(02)00079-2; doi:10.1016/S0960-8524(02)00079-2
Reiss R, Ihssen J, Thony-Meyer L (2011) *Bacillus pumilus* laccase: a heat stable enzyme with a wide substrate spectrum. Bmc Biotechnol 11. doi:Artn 9; doi:10.1186/1472-6750-11-9
Robinson T, McMullan G, Marchant R, Nigam P (2001) Remediation of dyes in textile effluent: a critical review on current treatment technologies with a proposed alternative. Bioresour Technol 77(3):247–255. doi:10.1016/S0960-8524(00)00080-8
Sag Y (2001) Biosorption of heavy metals by fungal biomass and modeling of fungal biosorption: a review. Sep Purif Method 30(1):1–48. doi:10.1081/Spm-100102984
Sanghi R, Sankararamakrishnan N, Dave BC (2009) Fungal bioremediation of chromates: conformational changes of biomass during sequestration, binding, and reduction of hexavalent chromium ions. J Hazard Mater 169(1–3):1074–1080. doi:10.1016/j.jhazmat.2009.04.056
Sari A, Tuzen M (2009) Kinetic and equilibrium studies of biosorption of Pb(II) and Cd(II) from aqueous solution by macrofungus (*Amanita rubescens*) biomass. J Hazard Mater 164(2–3):1004–1011. doi:10.1016/j.jhazmat.2008.09.002
Senthilkumar S, Perumalsamy M, Prabhu HJ (2014) Decolourization potential of white-rot fungus *Phanerochaete chrysosporium* on synthetic dye bath effluent containing Amido black 10B. J Saudi Chem Soc 18(6):845–853. doi:10.1016/j.jscs.2011.10.010
Senturk HB, Ozdes D, Gundogdu A, Duran C, Soylak M (2009) Removal of phenol from aqueous solutions by adsorption onto organomodified Tirebolu bentonite: equilibrium, kinetic and thermodynamic study. J Hazard Mater 172(1):353–362. doi:10.1016/j.jhazmat.2009.07.019
Sharma P, Kaur H, Sharma M, Sahore V (2011) A review on applicability of naturally available adsorbents for the removal of hazardous dyes from aqueous waste. Environ Monit Assess 183 (1–4):151–195. doi:10.1007/s10661-011-1914-0
Shen DZ, Fan JX, Zhou WZ, Gao BY, Yue QY, Kang Q (2009) Adsorption kinetics and isotherm of anionic dyes onto organo-bentonite from single and multisolute systems. J Hazard Mater 172(1):99–107. doi:10.1016/j.jhazmat.2009.06.139
Shinde NR, Bankar AV, Kumar AR, Zinjarde SS (2012) Removal of Ni (II) ions from aqueous solutions by biosorption onto two strains of *Yarrowia lipolytica*. J Environ Manage 102:115–124. doi:10.1016/j.jenvman.2012.02.026

Shroff KA, Vaidya VK (2011) Kinetics and equilibrium studies on biosorption of nickel from aqueous solution by dead fungal biomass of *Mucor hiemalis*. Chem Eng J 171(3):1234–1245. doi:10.1016/j.cej.2011.05.034

Si J, Yuan TQ, Cui BK (2015) Exploring strategies for adsorption of azo dye Congo Red using free and immobilized biomasses of *Trametes pubescens*. Ann Microbiol 65(1):411–421. doi:10.1007/s13213-014-0874-3

Singh AP, Singh T (2014) Biotechnological applications of wood-rotting fungi: a review. Biomass Bioenerg 62:198–206. doi:10.1016/j.biombioe.2013.12.013

Solis M, Solis A, Perez HI, Manjarrez N, Flores M (2012) Microbial decolouration of azo dyes: a review. Process Biochem 47(12):1723–1748. doi:10.1016/j.procbio.2012.08.014

Srinivasan A, Viraraghavan T (2010) Decolorization of dye wastewaters by biosorbents: a review. J Environ Manage 91(10):1915–1929. doi:10.1016/j.jenvman.2010.05.003

Subbaiah MV, Vijaya Y, Reddy AS, Yuvaraja G, Krishnaiah A (2011a) Equilibrium, kinetic and thermodynamic studies on the biosorption of Cu(II) onto *Trametes versicolor* biomass. Desalination 276(1–3):310–316. doi:10.1016/j.desal.2011.03.067

Subbaiah MV, Yuvaraja G, Vijaya Y, Krishnaiah A (2011b) Equilibrium, kinetic and thermodynamic studies on biosorption of Pb(II) and Cd(II) from aqueous solution by fungus (*Trametes versicolor*) biomass. J Taiwan Inst Chem Eng 42(6):965–971. doi:10.1016/j.jtice.2011.04.007

Subramanian SB, Yan S, Tyagi RD, Surampalli RY (2008) A new, pellet-forming fungal strain: its isolation, molecular identification, and performance for simultaneous sludge-solids reduction, flocculation, and dewatering. Water Environ Res 80(9):840–852. doi:10.2175/106143008x304703

Tanaka T, Tonosaki T, Nose M, Tomidokoro N, Kadomura N, Fujii T et al (2001) Treatment of model soils contaminated with phenolic endocrine-disrupting chemicals with laccase from Trametes sp in a rotating reactor. J Biosci Bioeng 92(4):312–316. doi:10.1263/Jbb.92.312

Taskin M, Erdal S (2010) Reactive dye bioaccumulation by fungus *Aspergillus niger* isolated from the effluent of sugar fabric-contaminated soil. Toxicol Ind Health 26(4):239–247. doi:10.1177/0748233710364967

Tastan BE, Donmez G (2015) Biodegradation of pesticide triclosan by *A. versicolor* in simulated wastewater and semi-synthetic media. Pestic Biochem Phys 118:33–37. doi:10.1016/j.pestbp.2014.11.002

Veglio F, Beolchini F (1997) Removal of metals by biosorption: a review. Hydrometallurgy 44(3):301–316. doi:10.1016/S0304-386x(96)00059-X

Verma M, Brar SK, Tyagi RD, Valero JR, Surampalli RY (2005) Wastewater sludge as a potential raw material for antagonistic fungus (*Trichoderma* sp.): role of pre-treatment and solids concentration. Water Res 39(15):3587–3596. doi:10.1016/j.watres.2005.07.001

Volesky B (2007) Biosorption and me. Water Res 41(18):4017–4029. doi:10.1016/j.watres.2007.05.062

Wang JL, Chen C (2006) Biosorption of heavy metals by *Saccharomyces cerevisiae*: a review. Biotechnol Adv 24(5):427–451. doi:10.1016/j.biotechadv.2006.03.001

Wang JS, Hu XJ, Wang J, Bao ZL, Xie SB, Yang JH (2010) The tolerance of *Rhizopus arrihizus* to U(VI) and biosorption behavior of U(VI) onto *R. arrihizus*. Biochem Eng J 51(1–2):19–23. doi:10.1016/j.bej.2010.04.010

Wang MX, Zhang QL, Yao SJ (2015) A novel biosorbent formed of marine-derived *Penicillium janthinellum* mycelial pellets for removing dyes from dye-containing wastewater. Chem Eng J 259:837–844. doi:10.1016/j.cej.2014.08.003

Wells A, Teria M, Eve T (2006) Green oxidations with laccase-mediator systems. Biochem Soc Trans 34:304–308

Xiong XJ, Meng XJ, Zheng TL (2010) Biosorption of CI Direct Blue 199 from aqueous solution by nonviable *Aspergillus niger*. J Hazard Mater 175(1–3):241–246. doi:10.1016/j.jhazmat.2009.09.155

Xu F (1996) Oxidation of phenols, anilines, and benzenethiols by fungal laccases: Correlation between activity and redox potentials as well as halide inhibition. Biochemistry-Us 35 (23):7608–7614. doi:10.1021/Bi952971a

Yan GY, Viraraghavan T (2003) Heavy-metal removal from aqueous solution by fungus *Mucor rouxii*. Water Res 37(18):4486–4496. doi:10.1016/S0043-1354(03)00409-3

Yang QX, Yang M, Pritsch K, Yediler A, Hagn A, Schloter M et al (2003) Decolorization of synthetic dyes and production of manganese-dependent peroxidase by new fungal isolates. Biotechnol Lett 25(9):709–713. doi:10.1023/A:1023454513952

Zaidi MK (2008) Wastewater reuse risk assessment, decision making, and environmental security: a technical report. NATO Science for Peace Security Programme, pp 445–450

Zhang FM, Knapp JS, Tapley KN (1999) Development of bioreactor systems for decolorization of Orange II using white rot fungus. Enzyme Microb Technol 24(1–2):48–53. doi:10.1016/S0141-0229(98)00090-8

Zhang S, Li A, Cui D, Yang JX, Ma F (2011) Performance of enhanced biological SBR process for aniline treatment by mycelial pellet as biomass carrier. Bioresour Technol 102(6):4360–4365. doi:10.1016/j.biortech.2010.12.079

Zhang SJ, Yang M, Yang QX, Zhang Y, Xin BP, Pan F (2003) Biosorption of reactive dyes by the mycelium pellets of a new isolate of *Penicillium oxalicum*. Biotechnol Lett 25(17):1479–1482. doi:10.1023/A:1025036407588

Zhang X, Renaud S, Paice M (2005) The potential of laccase to remove extractives present in pulp and white water from TMP newsprint mills. J Pulp Pap Sci 31(4):175–180

Zhou JL, Banks CJ (1993) Mechanism of humic-acid color removal from natural-waters by fungal biomass biosorption. Chemosphere 27(4):607–620. doi:10.1016/0045-6535(93)90096-N

Zhou JL, Kiff RJ (1991) The uptake of copper from aqueous-solution by immobilized fungal biomass. J Chem Technol Biotechnol 52(3):317–330

Zouboulis AI, Lazaridis NK, Grohmann A (2002) Toxic metals removal from waste waters by upflow filtration with floating filter medium. I. The case of zinc. Separ. Sci Technol 37(2):403–416. doi:10.1081/Ss-120000795

Chapter 5
Potential of White-Rot Fungi to Treat Xenobiotic-Containing Wastewater

Susana Rodríguez-Couto

5.1 Introduction

Industrial effluents from different industries contain a high load of pollutants, which could cause detrimental effects to the ecosystem if they are released without pre-treatment. Most of these compounds are xenobiotics i.e., strange to the biosphere, and are resistant to the biodegradation by the indigenous micro-organisms. In addition, most of them are harmful to living beings including humans. Therefore, they have to be removed before being released into the environment. However, the physico, chemical and physico-chemical in-use techniques for the treatment of wastewater fail in degrading such compounds resulting in their accumulation in the environment, posing a hazard to the plants, animals and humans. Consequently, alternative methods to remove xenobiotic compounds from wastewater are needed. The use of biological degradation is seen as an economic and ecological alternative to remove hazardous compounds from wastewater. Among them, the use of white-rot fungi (WRF) represents a promising approach.

WRF have the unique ability to degrade the bulky, heterogeneous and recalcitrant polymer lignin (Fig. 5.1). This ability is due to the secretion of an extracellular non-specific enzymatic complex during their secondary metabolism (idiophasic), usually under nitrogen depletion. This enzymatic complex is mainly composed of lignin peroxidases (LiPs), manganese-dependent peroxidases (MnPs), versatile peroxidases (VPs) and laccases together with accessory enzymes (mostly H_2O_2-generating oxidases and dehydrogenases) (Mester et al. 2004).

S. Rodríguez-Couto (✉)
Ceit-IK4, Unit of Environmental Engineering, Paseo Manuel de Lardizábal 15, 20018 San Sebastian, Spain
e-mail: srodriguez@ceit.es

S. Rodríguez-Couto
IKERBASQUE, Basque Foundation for Science, Díaz de Haro 3, 48013 Bilbao, Spain

D. Purchase (ed.), *Fungal Applications in Sustainable Environmental Biotechnology*, Fungal Biology, DOI 10.1007/978-3-319-42852-9_5

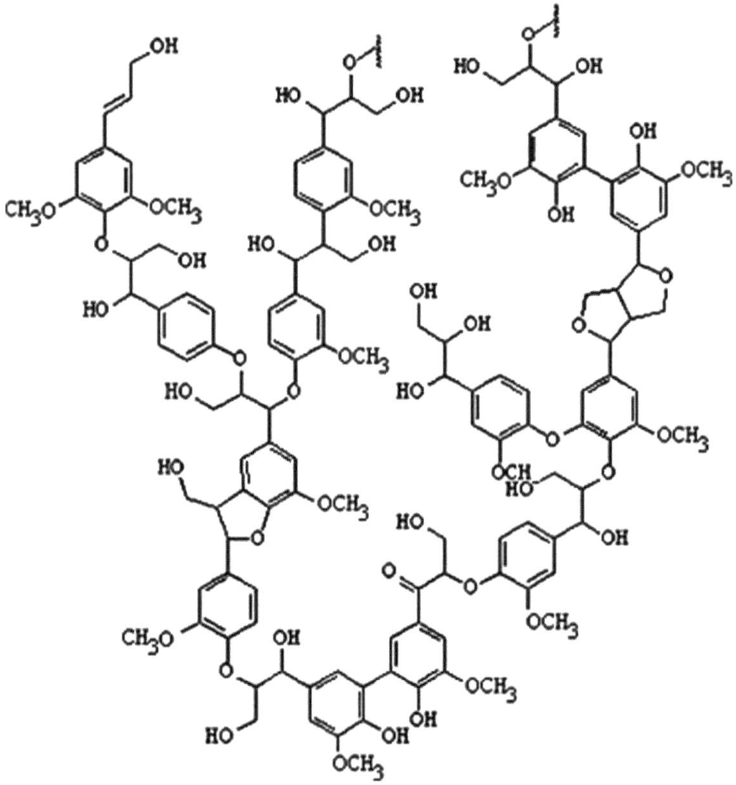

Fig. 5.1 Schematic structure of a lignin molecule. *Source* http://www.research.uky.edu/odyssey/winter07/green_energy.html

The ligninolytic enzymes secreted by the WRF have wide substrate specificity and are able to degrade a wide variety of complex molecules and even a mixture of them. This ability has driven the interest in the development of biotechnology processes based on WRF in the past couple of decades. However, studies dealing with the treatment of real wastewater are scarce. Therefore, in this chapter the latest research on xenobiotic removal from real wastewater by WRF is reviewed.

5.2 White-Rot Fungi

WRF are filamentous wood-degrading fungi, ubiquitous in nature. Most WRF belong to the Basidiomycota *phylum* (Polyporales and Agaricales orders) and together with some related litter-decomposing fungi are the only organisms able to mineralise lignin efficiently (Kirk and Cullen 1998; Hatakka 2001a, b).

Some WRF as grown in nature are shown in Fig. 5.2. The fungus *Phanerochaete chrysosporium* (order Thelephorales) was the first white-rot fungus studied and has become a model fungus for lignin biodegradation studies. The name white-rot derives from the bleached appearance of the wood attacked by these fungi due to the removal of the dark coloured lignin (Fig. 5.3). They grow mostly on hardwoods e.g., birch and aspen, although certain species grow on softwoods such as spruce and pine (Blanchette 1995). Some WRF degrade all wood components (i.e., cellulose, hemicellulose and lignin) simultaneously whereas others degrade lignin selectively. The former are called simultaneous or non-selective white-rot degraders and the latter selective white-rot degraders. The selective white-rot degraders are very interesting from a biotechnological point of view, since they remove lignin leaving the valuable cellulose intact (Dashtban et al. 2010). Simultaneous white-rot occurs mainly on hardwoods, whereas selective white-rot occurs both on hardwood and softwood. The typical characteristics of selective and simultaneous white-rot types are summarised in Table 5.1.

Fig. 5.2 Pictures of the white-rot fungi *Phanerochaete chrysosporium*. *Source* http://botit.botany.wisc.edu/toms_fungi/may97.html, *Trametes* versicolor. Photo Copyright © Michael Wood http://www.mykoweb.com/, *Pleurotus ostreatus*. Photo Copyright © Fred Stevens http://www.mykoweb.com/ and *Bjerkandera adusta*. Photo Copyright © Michael Wood http://www.mykoweb.com/

Fig. 5.3 Photograph of wood attacked by a white-rot fungus. *Source* www.bio.miami.edu/dana/pix/whiterot.jpg

The same mechanism that gives these fungi the potential to degrade lignin also allows them to degrade a wide variety of recalcitrant pollutants. Hence, the WRF are promising and attractive candidates for the bioremediation of xenobiotic compounds.

The mechanism used by the WRF to degrade pollutants gives them several advantages (Christian et al. 2005). For example:

- The WRF are able to mineralise a wide variety of toxic xenobiotics and complex mixtures as their enzymatic system is non-specific, non-stereoselective and based on free radicals.
- The WRF are ubiquitously found in nature.
- The WRF are able to oxidise low soluble compounds at high concentrations due to the extracellular nature of their main enzymatic system.
- The ligninolytic system of the WRF is triggered by nutrient limitation; hence, they do not need any pre-conditioning of the target pollutant.
- The WRF can degrade very low pollutant concentrations to non-detectable levels.
- The WRF can be cultivated on inexpensive substrates like agro and forestry wastes as well as in liquid media and in soil.
- The WRF also produce oxygen radicals (e.g., OH·) which are able to oxidise biomolecules such as proteins and DNA, and help to destroy microbes.
- The WRF are able to adjust the pH of their surrounding environment using the plasma membrane-dependent redox system.

The above-mentioned advantages helped to generate much interest in the development of technologies based on WRF for the biodegradation of hazardous and recalcitrant pollutants.

Table 5.1 Typical characteristics of selective and simultaneous white-rot

	Selective white-rot	Simultaneous white-rot	References
Degraded cell wall components	Initial stages of decay: Hemicellulose and lignin Later stages: Hemicellulose, cellulose and lignin	Cellulose, hemicellulose and lignin	Adasgavek et al. (1995); Fackler et al. (2006)
Anatomical features of decayed wood	Middle lamella dissolved Adjacent wood cells separated	Eroded cell walls, degradation beginning from the secondary wall proceeding to middle lamella	Blanchette (1995)
Lignin loss	Lignin loss diffusive throughout wood cell wall without major degradation of polysaccharides	Lignin loss together with wood cell wall polysaccharides starting progressively from lumen	Blanchette (1995)
Representatives	*Ceriporiopsis subvermispora, Phlebia radiata, Pleurotus* spp., *D. squalens, Ganoderma austral, Phlebia tremellosa, P. cinnabarinus, Phellinus pini*	*Phanerochaete chrysosporium, Fomes fomentarius, Phellinus, robustus, Trametes versicolor, Irpex lacteus, Heterobasidium annosum*	Blanchette (1995); Otjen et al. (1987); Nishida et al. (1988); Martínez et al. (2005)

5.3 Enzymatic System of WRF

In addition to lignin, WRF can oxidise a wide variety of organic compounds with structural similarities to lignin including soil humic substances (Hofrichter et al. 1998), organic pollutants (Tuomela and Hatakka 2011a, b) and synthetic dyes (Glenn and Gold 1983).

WRF usually produce one or more ligninolytic enzymes in different combinations according to which they can be divided into four groups (Hatakka 1994; Tuor et al. 1995; Nerud and Misurcova 1996): (i) laccase, LiP and MnP-producing, (ii) laccase and at least one of the peroxidases, (iii) laccases only and (iv) peroxidases only.

The ligninolytic enzymes most frequently found in the WRF are laccases and MnPs, and the least, LiPs and VPs. The ligninolytic enzymes can act jointly or separately but accessory enzymes (glyoxal oxidase, aryl alcohol oxidase, pyranose 2-oxidase, cellobiose dehydrogenase, etc.) are required to complete the process of lignin or xenobiotic degradation. In addition, intracellular cytochrome P450 monooxygenases as well as low-molecular mass oxidants such as hydroxyl radicals and chelated Mn^{3+} have also shown to be involved in the degradation of lignin and many xenobiotics (ten Have and Teunissen 2001; Hammel et al. 2002; Subramanian and Yadav 2009; Taboada-Puig et al. 2011). Recently dye-decolourising peroxidases (DyPs), involved in the decolouration of high redox potential synthetic dyes and non-phenolic lignin model compounds (Liers et al. 2010), and aromatic peroxygenases (APOs), involved in the catalysis of oxygen transfer reactions resulting in the cleavage of ethers (Hofrichter et al. 2010; Liers et al. 2011), have been found to be part of the ligninolytic system of the WRF. The main ligninolytic enzymes, their substrates and reactions are summarised in Table 5.2.

It is worth pointing out that although a white-rot fungus species can potentially secrete laccase, MnP and LiP, a particular strain may not secrete all of them. Thus, for instance *Trametes versicolor* generally produces all the three enzymes (i.e., laccase, MnP and LiP) but laccase may be predominant in certain strains (Yang

Table 5.2 Ligninolytic enzymes and their main reactions (Hatakka 2001a, b; Harms et al. 2011; Tuomela and Hatakka 2011a, b; Lundell and Mäkelä 2013)

Enzyme and abbreviation	Cofactor	Substrate, mediator	Reaction	Occurrence in fungi
Laccase (EC 1.10.3.2)	O_2	Phenols, mediators e.g., hydroxybenzotriazole or ABTS	Phenols are oxidised to phenoxyl radicals; other reactions in the presence of mediators	Basidiomycota and Ascomycota, in most white-rot fungi and litter-degrading fungi
Lignin peroxidase (EC 1.11.1.4), LiP	H_2O_2	Veratryl alcohol	Aromatic ring oxidised to cation radical	Basidiomycota only in few white-rot fungi
Manganese peroxidase (EC 1.11.1.13), MnP	H_2O_2	Mn, organic acids as chelators, thiols, unsaturated fatty acids	Mn(II) oxidised to Mn(III); chelated Mn(III) oxidises phenolic compounds to phenoxyl radicals; other reactions in the presence of additional compounds	Basidiomycota, common in white-rot fungi and litter-degrading fungi
Versatile peroxidase (EC 1.11.1.16), VP	H_2O_2	Mn, veratryl alcohol, compounds similar to LiP and MnP	Mn(II) oxidised to Mn(III), oxidation of phenolic and non-phenolic compounds, and dyes	Basidiomycota, only in *Pleurotus* sp., *Bjerkandera* sp. and *Trametes versicolor*
Dye-decolourising peroxidase (EC 1.11.1.19), DyP	H_2O_2	Antraquinonic dyes	Oxidation of organic compounds; decolouration of Reactive Blue 5	Basidiomycota and Ascomycota

Table 5.3 Characteristics of the main ligninolytic enzymes (Dashtban et al. 2010; Sigoillot et al. 2012; Liers et al. 2014)

Enzyme	Molecular mass (kDa)	Isolectric point (pI)	Glycosylation	Redox potential (eV)	Localization
Laccase	54–80	3–4	Yes (10–20 %)* N-glycosylated	0.4–0.8	Mostly extracellular
LiP	35–48	3.1–4.7	Yes (up to 20–30 %) N-glycosylated	1.2 (at pH 3.0)	Extracellular
MnP	38–62.5	2.9–7.1	Yes (4–18 %) N-glycosylated	0.8 (at pH 4.5)	Extracellular
VP	40–45	3.4–3.9	Yes	>1	Extracellular
DyP	40–67	3.5–4.3	Yes (9–31 %)	1.1–1.2	Extracellular

*in some cases they can reach up to 49 %

et al. 2013). In addition, the secretion of specific enzymes may also depend on the culture conditions including the composition of the growth medium.

The characteristic of the main ligninolytic enzymes are presented in Table 5.3.

5.3.1 Lignin Peroxidases

Lignin peroxidases (EC 1.11.1.14, 1,2-bis(3,4-dimethoxyphenyl)propane-1,3-diol: hydrogen-peroxide oxidoreductase, family 2 at http://www.cazy.org, LiPs) were first discovered in the white-rot fungus *Phanerochaete chrysosporium* in the mid-1980s. They are considered as true ligninases since they directly catalyse lignin oxidation. LiPs are glycoproteins and contain an iron protoporphyrin IX (heme) as a prosthetic group.

LiPs catalyse the monoelectronic and H_2O_2-dependent oxidation of a wide variety of aromatic compounds through a multistep reaction. These reactions induce the formation of aryl cationic radicals, which further undergo many non-enzymatic reactions generating a number of end products such as glycolate and oxalate. Both the catalytic cycle (Fig. 5.4) and the enzymatic intermediates are similar to those of the other peroxidases. Veratryl alcohol enhances the action of LiP on many substrates, including lignin (Hammel et al. 1993), by acting as a mediator (Harvey et al. 1986) or by protecting the enzyme against inactivation by H_2O_2 (Wariishi and Gold 1989).

5.3.2 Manganese-Dependent Peroxidases

Manganese-dependent peroxidases (EC 1.11.1.13, Mn(II)-hydrogen-peroxide oxidoreductase, family 2 at http://www.cazy.org, MnPs). The first extracellular MnP

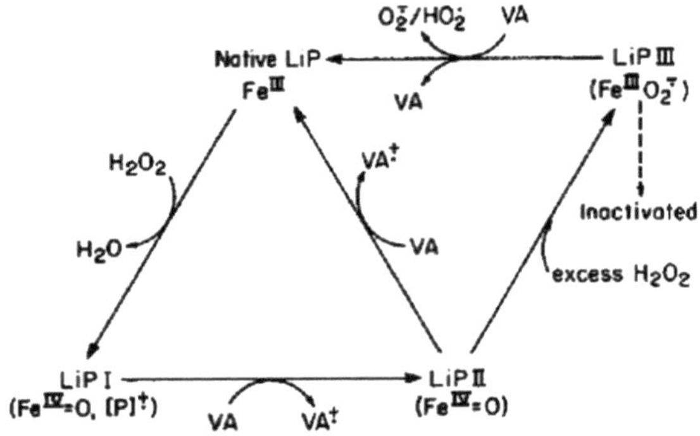

Fig. 5.4 The catalytic cycle of lignin peroxidase (LiP); $VA^{\cdot+}$: veratryl alcohol cation radical. Reprinted from FEBS Letters 243, Copyright Wariishi and Gold (1989), with permission from Elsevier Ltd., UK

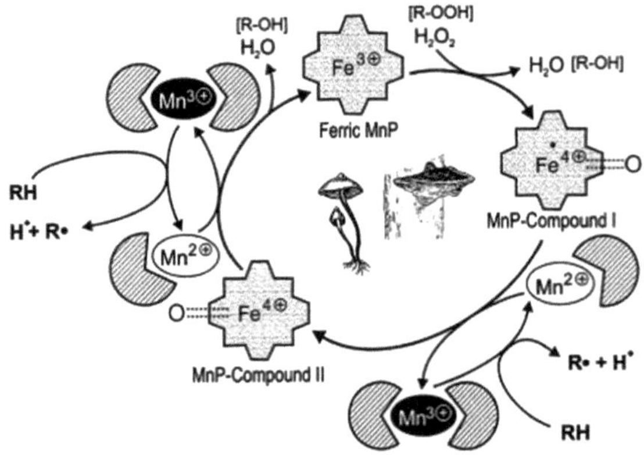

Fig. 5.5 The catalytic cycle of manganese-dependent peroxidase (MnP). Reprinted from Enzyme and Microbial Technology 30, Copyright Hofrichter (2002), *Review* lignin conversion by manganese peroxidase (MnP), 454–466, with permission from Elsevier Ltd., UK

was purified from *P. chrysosporium* and its expression and production showed to be regulated by the presence of Mn(II) in the culture medium (Bonnarme and Jeffries 1990). The catalytic cycle of MnP (Fig. 5.5) is essentially the same as for LiP with the exception that Mn(II) is necessary to complete the cycle.

5.3.3 Laccases

Laccases (EC 1.10.3.2, p-diphenol:oxygen oxidoreductases, lignin oxidases family 1, http://www.cazy.org/Auxiliary-Activities.html) are multi-copper-containing oxidases which catalyse the four-electron reduction of O_2 to water coupled with the oxidation of various organic substrates. They are widely distributed in nature and are found in plants, fungi, bacteria (Dwivedi et al. 2011) and a few insects (Xu 1999).

Laccases cannot directly oxidise all substrates either because of their large size, which hinders their introduction into the enzyme active site, or because of their particular high redox potential. However, it was shown that in the presence of low-molecular weight organic compounds acting as electron transfer mediators,

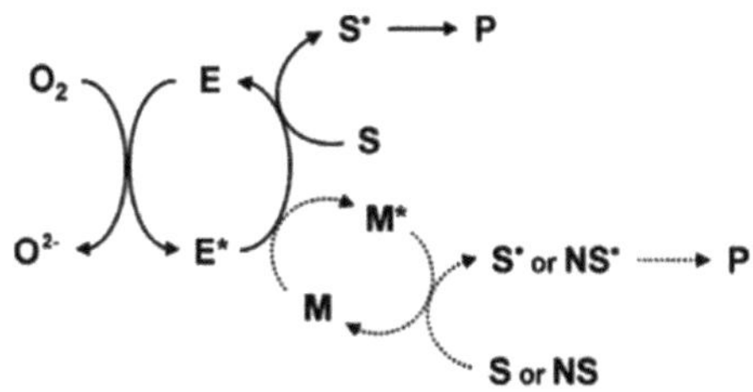

Fig. 5.6 Catalytic cycle of the laccase enzyme; E: native laccase; E^*: oxidised laccase; S: substrate; S·: substrate radical; NS: non-substrate; NS·: non-substrate radical; P: end products; O_2: oxygen; O^{2-}: divalent oxygen; M: mediator; M^*: oxidised mediator. Reprinted from Enzyme and Microbial Technology 41, Copyright Kurniawati and Nicell (2007), with permission from Elsevier Ltd., UK

laccases were also able to oxidise non-phenolic structures (Bourbonnais and Paice 1990; Call and Mücke 1997). The first step of the laccase mediator system (LMS) is the oxidation of the mediator by the laccase enzyme. Then, the oxidised mediator oxidises the bulky or high redox potential substrate. Thus, the mediator acts as an electron shuttle between the substrate and the enzyme (Galli and Gentili 2004; Widsten and Kandelbauer 2008).

Figure 5.6 represents the catalytic cycle of laccase. In typical interactions of laccase with a substrate, the catalytic site of laccase abstracts electrons from the substrate and releases an oxidised product. When a mediator is present, the mediator can be oxidised by laccase and further oxidises another compound that is either a substrate or a non-substrate of laccase resulting in the formation of oxidised product(s) and the mediator regeneration (Banci et al. 1999).

5.3.4 Versatile Peroxidases

Versatile peroxidases (EC 1.11.1.16, hybrid peroxidases, polyvalente peroxidases, family 2 at http://www.cazy.org, VPs) share catalytic properties of both LiP and MnP (Dosoretz and Reddy 2007; Hofrichter et al. 2010). Thus like MnPs, they have high affinity for Mn(II) and catalyse the oxidation of Mn(II) to Mn(III) and oxidise both phenolic and non-phenolic substrates in the absence of Mn(II) like LiPs.

VPs seem to be produced only by fungi from the genera *Pleurotus*, *Bjerkandera* and *Lepista* (Heinfling et al. 1998; Mester and Field 1998; Ruiz-Dueñas et al. 1999; Zorn et al. 2003) and maybe also by *Panus* and *Trametes* species (Martinez 2002; Lisov et al. 2003). In Fig. 5.7 the catalytic cycle of VP is depicted (Pérez-Boada et al. 2005).

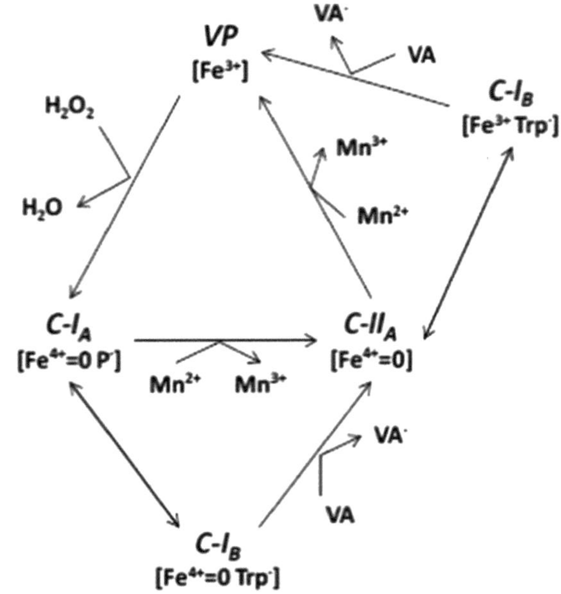

Fig. 5.7 The catalytic cycle of versatile peroxidase. Reprinted from Journal of Molecular Biology 354, Copyright Pérez-Boada et al. (2005), with permission from Elsevier Ltd., UK

5.3.5 *Dye-Decolourising Peroxidases*

Dye-decolourising peroxidases (DyP-type peroxidases; EC 1.11.1.19, DyPs) are glycoproteins having one heme as a cofactor and require H_2O_2 for all enzyme reactions, indicating that they function as peroxidases. They are named after their ability to oxidise a wide range of synthetic dyes, in particular, anthraquinonic dyes, which are poorly oxidised by other peroxidases (Kim and Shoda 1999; Passardi et al. 2005; Sugano 2009). In addition, they function under lower pH conditions than other peroxidases. A very important characteristic of DyPs is that they have a free position for the H_2O_2 binding (Petrides and Nauseef 2000).

Typical peroxidase substrates degraded by DyPs are, for example, 2,2′ azinobis-(3-ethylbenzthiazoline-6-sulphonate and phenolic compounds. DyPs have also been reported to cleave β-carotene and other carotenoids as well as oxidise methoxylated aromatics such as veratryl alcohol and non-phenolic β-O-4 lignin model compounds (van Bloois et al. 2009; Zelena et al. 2009; Liers et al. 2010). However, their physiological function still remains unclear.

5.4 Xenobiotics Degraded by WRF

The ability of the WRF to degrade xenobiotic compounds comes from their ability to degrade lignin, since it resembles the chemical structure of many xenobiotics (Fig. 5.8). Thus, the same mechanisms that give the WRF the ability to degrade

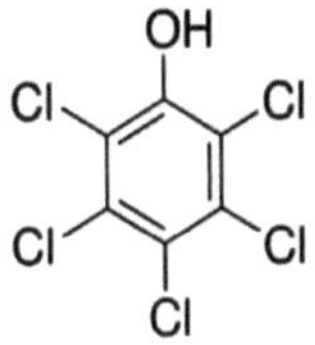

Pentachlorophenol

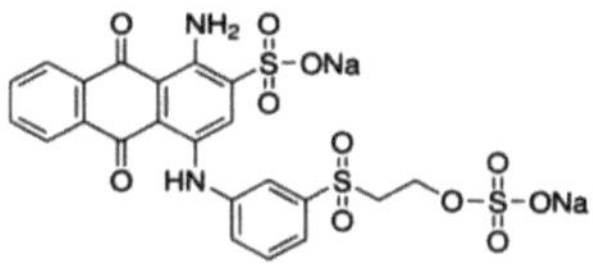

Remazol Brilliant Blue R

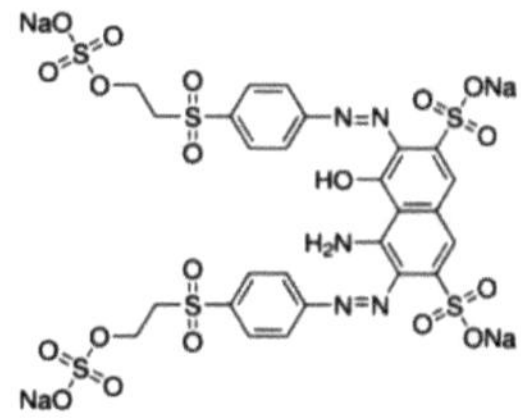

Reactive Black 5

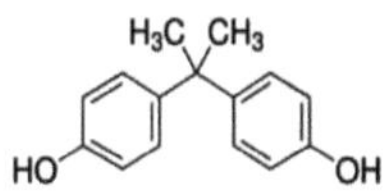

Bisphenol A

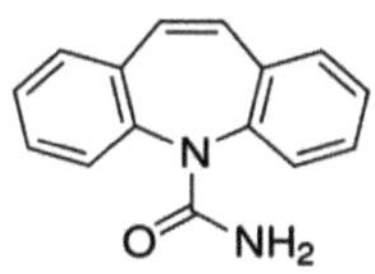

Carbamazepine

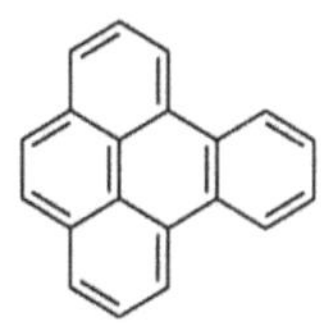

Benzo(a)pyrene

Fig. 5.8 Chemical structures of different xenobiotic compounds

lignin can be used to degrade a wide variety of recalcitrant pollutants. Under ligninolytic conditions, many xenobiotics are oxidised and mineralised to different extents by the WRF (Field et al. 1993).

Several reviews about environmental pollutant degradation by the WRF have already been published (Bumpus et al. 1985; Reddy 1995; Raghukumar et al. 2008; Pointing 2001; Reddy and Mathew 2001; Wesenberg et al. 2003; Chang 2008; Pinedo-Rivlla et al. 2009; Majeau et al. 2010). However, there are few reports focused on the application of WRF in the treatment of real wastewater. In this section, recent reports on xenobiotic removal from real wastewater are reviewed (Table 5.4).

5.4.1 *Pharmaceuticals*[1]

Accinelli et al. (2010) studied the potential of *P. chrysosporium* entrapped in granular bioplastics to remove different pharmaceutical compounds (i.e., the antiviral drug oseltamivir and the antibiotics erythromycin, sulfamethoxazole and ciprofloxacin) from a municipal wastewater treatment plant (WWTP). It was found

[1]For further information on fungal treatment of wastewater containing pharmaceutical products, please refer to Chap. 6—*Fungal bioremediation of emerging micropollutants in municipal wastewaters* and Chap. 8—*Mycoremediation of organic pollutants: principles, opportunities and pitfalls.*

Table 5.4 Degradation of real wastewater containing different xenobiotic compounds by different white-rot fungi in the past years

White-rot fungus	Wastewater source	Xenobiotic	Removal	Reference
Pharmaceuticals				
Phanerochaete chrysosporium	Municipal WWTP (Italy)	Oseltamivir (Tamiflu)	>50 % in 30 days	Accinelli et al. (2010)
		Erythromycin	>80 % in 5 days	
		Sulfamethoxazole	>50 % in 5 days	
		Ciprofloxacin	>70 % in 5 days	
P. chrysosporium	Municipal WWTP (Germany)	Carbamazepine (1 mg/L)	60 % in 100 days	Zhang and Geissen (2012)
T. versicolor	Urban (Spain)	Pharmaceutical compounds	50 %	Cruz-Morató et al. (2013)
T. versicolor	Hospital (Spain)	Pharmaceutical and endocrine disrupting compounds	83.2 % (sterile) in 8 days; 53.3 % (non-sterile) in 8 days	Cruz-Morató et al. (2014)
T. versicolor	Hospital (Spain)	Iopromide	87 % (sterile), 65.4 % (non-sterile) in 8 days	Gros et al. (2014)
		Ofloxacin	98.5 % (sterile), 99 % (non-sterile) in 8 days	
Textile wastewater				
Bjerkandera adusta	Textile (Italy)	Dyes	Up to 84 % during 10 cycles	Anastasi et al. (2010)
B. adusta	Textile (Italy)	Dyes	40 % in 24 h	Anastasi et al. (2011)
P. chrysosporium	Textile (India)	Dyes	84 % in 6 days	Sangeeta et al. (2011)
T. pubescens	Textile (Italy)	Dyes	76 % decolouration in 24 h, COD reduction and toxicity removal (flasks); 30 % decolouration (bioreactor)	Anastasi et al. (2012)

(continued)

Table 5.4 (continued)

White-rot fungus	Wastewater source	Xenobiotic	Removal	Reference
Bjerkandera sp.	Textile (Colombia)	Dyes Everzol Black EDR and Everzol Black EDG	65 % (sterile) and 40 % (non-sterile) in 8 days	Osorio-Echavarría et al. (2012)
P. chrysosporium	Textile (India)	Dyes	80 % (5 g/L glucose); 83 % (10 g/L glucose)	Pakshirajan and Kheria (2012)
Curvularia lunata	Textile (Brazil)	Indigo dye	95 % (non-aerated) and 93 % (aerated) in 10 days	de Miranda et al. (2013)
P. chrysosporium			95 % (non-aerated) and 98 % (aerated) in 10 days	
Pleurotus floridanus	Textile (India)	Dyes	71.2 % colour, 80.5 % COD	Sathian et al. (2013)
B. adusta	Textile effluent from a WWTP (South Korea)	Industrial dyes	71–92 % in 3 weeks	Choi et al. (2014)
Ganoderma sp. En3	Textile (China)	Indigoid and sulphur dyes	85.1 % in 8 days	Ma et al. (2014)
Combination of *P. floridanus, G. lucidum* and *T. pubescens*	Textile (India)	Dyes	71.3 % colour and 79.4 % COD (HRT 5 days)	Sathian et al. (2014)
Olive mills				
Trametes versicolor	Olive mill (Italy)	Phenolics (277 mg/L)	60 % colour, 72 % phenols (shaken flasks) in 216 h; 65 % colour, 89 % phenols (reactor, continuous) in 192 h	Cerrone et al. (2011)

(continued)

Table 5.4 (continued)

White-rot fungus	Wastewater source	Xenobiotic	Removal	Reference
Ganoderma spp.	Olive mill (Greece)	Phenolics (4.9 mg/mL)	40–46 % colour, 64-67 % phenolics in 20 days	Ntougias et al. (2012)
Pleurotus spp.			60–65 % colour, 74–81 % phenolics in 20 days	
Pleurotus ostreatus	Olive mill (Italy)	Polyphenols (5 g/L)	70 % in 4-7 days (batch); 42–68 % for 5 cycles (batch with biomass recycling and nutrient addition)	Olivieri et al. (2012)
Wastewater from other sources				
Trametes pubescens	Distillery (South Africa)	Phenolics (866 mg/L)	86 % in 2 days	Strong (2010)
Ceriporiopsis subvermispora			57 % in 2 days	
Pycnoporus cinnabarinus			69 % in 2 days	
P. chrysosporium			<40 % in 2 days	
P. chrysosporium	Pulp and paper mill (India)		83 % colour in 96 h	Gomathi et al. (2012)
P. ostreatus	Petrochemical (Italy)	Mixture of 2-NSA (2-naphthalene sulfonic acid) polymers	70 % (20–24 % adsorbed by fungal biomass) in 40 days	Palli et al. (2014)

WWTP Wastewater treatment plant
HRT Hydraulic retention time

that the antibiotics were more readily removed by *P. chrysosporium* than the antiviral drug (Table 5.4). DNA analysis showed that fungal growth was mainly confined to the bioplastic carriers making it easy to insert the fungus to the polluted site.

Zhang and Geissen (2012) studied the degradation of carbamazepine in an effluent from a municipal WWTP by *P. chrysosporium* immobilised on polyether foam in a novel plate bioreactor. Carbamazepine (1 mg/L) was removed by 60 % in

100 days of continuous operation provided that additional glucose and nitrogen were supplied.

Cruz-Morató et al. (2013) reported for the first time the degradation of pharmaceutical compounds (PhACs) in urban wastewater by *T. versicolor* pellets in a batch fluidised-bed bioreactor operating under non-sterile conditions where 50 % of the detected PhACs was removed. In addition, a considerable reduction in toxicity was achieved after the fungal treatment. In the following study, Cruz-Morató et al. (2014) reported the removal of PhACs and endocrine disruptor compounds (EDCs) from hospital effluents under sterile and non-sterile conditions using the same approach. They found that the overall load removal was 83.2 % under sterile and 53.3 % under non-sterile conditions after 8 days of treatment. In addition, toxicity tests showed the reduction of wastewater toxicity after the fungal treatment.

Gros et al. (2014) studied the degradation of the X-ray contrast agent iopromide (IOP) and the antibiotic ofloxacin (OFLOX) in hospital wastewater by *T. versicolor* in a 10-L fluidised-bioreactor. They found that within 8 days, IOP and OFLOX were degraded by 87 and 98.5 % respectively, under sterile conditions, and by 65.4 and 99 % respectively, under non-sterile conditions. In addition, toxicity of the treated wastewater was reduced after the fungal treatment.

5.4.2 Textile Wastewater[2]

Anastasi et al. (2010) reported the ability of *Bjerkandera adusta* to treat wastewater from a textile factory in a fixed-bed reactor operated in continuous mode. This fungus was able to decolourise the effluent up to 84 % during 10 cycles under non-sterile conditions. In addition, the chemical oxygen demand (COD) and the toxicity were effectively reduced after the fungal treatment. Subsequently, Anastasi et al. (2011) tested the capacity of the same fungus to degrade wastewater from a textile industry after a secondary treatment and found that the fungal treatment decolourised the effluent by 40 % in 24 h. Further, they (Anastasi et al. 2012) showed that fungal treatment with *Trametes pubescens* followed by activated sludge of wastewater from a cotton dyeing industry led to very good results in terms of decolouration (76 % in 24 h), COD reduction and toxicity removal. However, the scale-up in a 5-L moving-bed bioreactor (working volume 2 L) with *T. pubescens* immobilised on 2 cm^3 cubes of polyurethane foam (PUF) led to lower decolouration (30 %). Therefore, optimisation of the reactor technology is needed before fungal treatment could be successfully applied.

Sangeeta et al. (2011) studied the decolouration of textile wastewater by *P. chrysosporium* in shaken flasks and found that the decolouration of raw

[2]Additional information on treatment of dye using fungi is presented in Chap. 4—*Application of biosorption and biodegradation function of fungi in wastewater and sludge treatment.*

wastewater was negligible. Nevertheless, when wastewater was diluted with medium containing glucose and other nutrients, the decolouration considerably increased (84 % in 6 days).

Osorio-Echavarría et al. (2012) reported the decolouration of textile wastewater by the white-rot fungus anamorph R1 of *Bjerkandera sp.* under sterile and non-sterile conditions. The former led to a decolouration percentage of 65 % in 8 days, whereas the latter led to a decolouration percentage of 40 % for the same period of time. The decolouration under non-sterile conditions was mainly due to dye adsorption onto fungal mycelium since the pH increased affecting both the fungus and the ligninolytic enzymes. They found that the presence of high concentration of salts (i.e. NaCl and Na_2CO_3) in the wastewater favoured the decolouration process. This indicates that the fungus anamorph R1 of *Bjerkandera sp.* is able to grow under hypersaline conditions. This makes this fungus advantageous for the treatment of industrial effluents with high salt concentrations such as those from the textile industries.

Pakshirajan and Kheria (2012) investigated the continuous treatment of textile wastewater by *P. chrysosporium* in a rotating biological contactor reactor operating at an HRT of 48 h. The fungus was able to decolourise the effluent by more than 64 % when diluted with media containing glucose. Maximum decolouration efficiencies of 83 and 80 % were attained with 10 and 5 g/L of glucose respectively.

de Miranda et al. (2013) investigated the decolouration of a textile effluent by the white-rot fungi *Curvularia lunata* and *P. chrysosporium* in static bioreactors under aerated and non-aerated conditions. The effluent was almost totally decolourised within 10 days under both conditions. However, the effluent treated by *P. chrysosporium* contained a mutagenic byproduct from indigo biodegradation that was not found in the effluent treated by *C. lunata.* This indicates that different degradation pathways are used by different ligninolytic fungi and that degradation is not always accompanied by detoxification.

Sathian et al. (2013) studied the decolouration of textile wastewater by *Pleurotus floridanus* in batch culture. After optimisation of different parameters, the fungal treatment achieved 71.2 % decolouration and 80.5 % COD reduction. Furthermore, in studying the ability of the white-rot fungi *Coriolus versicolor*, *P. floridanus*, *Ganoderma lucidum* and *T. pubescens* to decolourise textile wastewater in pure and mixed cultures, Sathian et al. (2014) found that the combination of *P. floridanus*, *G. lucidum* and *T. pubescens* led to the best results (87.2 % decolouration) and this combination was used subsequently in a sequential batch reactor (SBR). When operating at the optimised conditions, a decolouration percentage of 71.3 % and a COD reduction of 79.4 % could be obtained.

Choi et al. (2014) investigated the ability of the white-rot fungi *B. adusta*, *Ceriporia lacerata*, *Phanerochaete calotricha* and *Porostereum spadiceum* to decolourise an untreated textile effluent from a WWTP. They found that only *B. adusta* was able to decolourise the effluent significantly (71–92 % in 3 weeks). In addition, wastewater toxicity decreased after fungal *B. adusta* treatment. These results highlight again the different degrading abilities of different fungal species. Ma et al. (2014) reported that *Ganoderma* sp. En3, a white-rot fungus isolated from a forest in China, was able to decolourise indigo jean dyeing wastewater from a textile factory up to 85.1 % in 8 days.

5.4.3 Olive Mills

Cerrone et al. (2011) evaluated the white-rot fungi *Panus tigrinus*, *Funalia trogii* and *T. versicolor* to treat olive washing wastewater (OWW) and found that *T. versicolor* performed well, reducing colour, COD and phenols by 60, 72 and 87 %, respectively, in 216 h. Also, only this fungus grew well in a bubble-column bioreactor (working volume 1 L) and the treatment of OWW in continuous operation reduced colour, COD and phenols by 65, 73 and 89 %, respectively, after 192 h.

Ntougias et al. (2012) studied the treatment of olive mill wastewater (OMW) by different strains belonging to the *Ganoderma* and *Pleurotus* genera and found that the *Ganoderma* spp removed 40–46 % colour and 64–67 % phenolics and the *Pleurotus* spp removed 60–65 % colour and 74–81 % phenolics within 20 incubation days. This indicates that different fungal species exhibit different degrading abilities.

Olivieri et al. (2012) studied the removal of polyphenols in raw OMW by *P. ostreatus* under controlled non-sterile conditions in flasks and in an internal loop airlift bioreactor (ILAB) operating in batch with biomass recycling and in continuous culture. Biomass recycling with nutrient addition was the most effective configuration, removing 42–68 % of polyphenols for 5 cycles. The continuous treatment in the ILAB was effectively performed provided that OMW was previously aerated to avoid oxygen consumption by endogenous micro-organisms.

5.4.4 Wastewater from Other Sources

Strong (2010) studied the treatment of Amarula distillery wastewater by *T. pubescens*, *Ceriporiopsis subvermispora*, *Pycnoporus cinnabarinus* and *P. chrysosporium*. *T. pubescens* was found to be the most efficient fungus in phenolic removal (86 %) followed by *P. cinnabariunus* (69 %) and *C. subvermispora* (57 %) within 2 cultivation days. However, *P. chrysosporium* removed less than 40 % of the phenolics for the same time period. In addition, *T. pubescens* was also very effective in removing colour and reducing COD. Therefore, this study showed the possibility to treat an effluent containing high COD and high phenolic concentration using the white-rot fungus *T. pubescens*.

Gomathi et al. (2012) reported high decolouration (83 % in 96 h) of a pulp and paper mill effluent by *P. chrysosporium* entrapped in calcium alginate when 1 % sucrose and 1 % ammonium chloride were added to the effluent.

Palli et al. (2014) assessed the ability of *P. ostreatus* to remove 2-naphthalenesulfonic acid polymers (2-NSAP) from petrochemical wastewater. In the presence of an adequate carbon source, the fungus was able to remove about 70 % of the oligomers in 40 days, from which about 20–24 % was adsorbed by the fungal biomass. Furthermore, respirometric tests showed a considerable increase of the BOD/COD ratio (from 9 % up to 57 %) after the fungal treatment which confirmed that the fungus did not mineralise the NSAP but increased their biodegradability.

5.5 Concluding Remarks

WRF hold an enormous potential for the biodegradation of a great variety of xenobiotic compounds due to the secretion of enzymatic complexes with broad substrate specificity. Different WRF show different biodegrading abilities for different xenobiotic compounds mainly due to their different physiology, culture and/or environmental conditions and nature of enzymes secreted. Also, the characteristics of the ligninolytic enzymes from different WRF sources differ considerably.

Despite the promising results reported so far, in order to assess the true technical potential of WRF to biodegrade xenobiotics, more studies under real industrial conditions are needed. However, most studies using real wastewater were performed required some pre-conditioning of wastewater (dilution, pH adjustment, sterilisation, addition of nutrients).

Detailed characterisation of the intermediates and metabolites produced during biodegradation as well as toxicity tests should also be carried out to measure the detoxification of the fungal treated wastewater and prevent accumulation of toxic byproducts. Although some studies regarding the metabolic pathway of xenobiotic degradation by WRF have been performed there is still a gap in the degradation mechanisms of xenobiotics by WRF and their ligninolytic enzymes.

References

Accinelli C, Saccà ML, Batisson I, Fick J, Mencarelli M, Grabic R (2010) Removal of oseltamivir (Tamiflu) and other selected pharmaceuticals from wastewaterusing a granular bioplastic formulation entrapping propagules of *Phanerochaete chrysosporium*. Chemosphere 81:436–443

Adasgavek JE, Gilbertson RL, Dunlap MR (1995) Effects of incubation time and temperature on in vitro selective delignification of Silver leaf oak by *Ganoderma colossum*. Appl Environ Microbiol 61:138–144

Anastasi A, Spina F, Prigione V, Tigini V, Giansanti P, Varese GC (2010) Scaleup of a bioprocess for textile wastewater treatment using Bjerkandera adusta. Bioresour Technol 101:3067–3075

Anastasi A, Parato B, Spina F, Tigini V, Prigione V, Varese GC (2011) Decolourisation and detoxification in the fungal treatment of textile wastewaters from dyeing processes. New Biotechnol 29:38–45

Anastasi A, Spina F, Romagnolo A, Tigini V, Prigione V, Varese GC (2012) Integrated fungal biomass and activated sludge treatment for textile wastewaters bioremediation. Bioresour Technol 123:106–111

Banci L, Ciofi-Baffoni S, Tien M (1999) Lignin and Mn peroxidase-catalyzed oxidation of phenolic lignin oligomers. Biochemistry 38:3205–3210

Blanchette RA (1995) Degradation of the lignocellulose complex in wood. Can J Bot 73:S999–S1010

Bonnarme P, Jeffries TW (1990) Mn (II) regulation of lignin peroxidase and manganese-dependent peroxidase from lignin-degrading white rot fungi. Appl Environ Microb 56:210–217

Bourbonnais R, Paice MG (1990) Oxidation of non-phenolic substrates. An expanded role for laccase in lignin biodegradation. FEBS Lett 267:99–102

Bumpus JA, Tien M, Wright D, Aust SD (1985) Oxidation of persistent environmental pollutants by a white rot fungus. Science 228:1434–1436

Call HP, Mücke I (1997) History, overview and applications of mediated lignolytic systems, especially laccase-mediatorsystems (Lignozym ®-process). J Biotechnol 53:163–202

Cerrone F, Barghini P, Pesciaroli C, Fenice M (2011) Efficient removal of pollutants from olive washing wastewater in bubble-column bioreactor by *Trametes versicolor*. Chemosphere 84:254–259

Chang YS (2008) Recent developments in microbial biotransformation and biodegradation of dioxins. J Mol Microb Biotech 15:152–171

Choi YS, Seo JY, Lee H, Yoo J, Jung J, Kim JJ, Kim GH (2014) Decolorization and detoxification of wastewater containing industrial dyes by *Bjerkandera adusta* KUC9065. Water Air Soil Poll 225:1801–1810

Christian V, Shrivastava R, Shukla D, Modi HA, Vyas BR (2005) Degradation of xenobiotic compounds by lignin-degrading white-rot fungi: Enzymology and mechanisms involved. Indian J Exp Biol 43:301–312

Cruz-Morató C, Ferrando-Climent L, Rodriguez-Mozaz S, Barceló D, Marco-Urrea E, Vicent T, Sarrà M (2013) Degradation of pharmaceuticals in non-sterile urban wastewater by *Trametes versicolor* in a fluidized bed bioreactor. Water Res 47:5200–5210

Cruz-Morató C, Lucas D, Llorca M, Rodriguez-Mozaz S, Gorga M, Petrovic M, Barceló D, Vicent T, Sarrà M, Marco-Urrea E (2014) Hospital wastewater treatment by fungal bioreactor: Removal efficiency for pharmaceuticals and endocrine disruptor compounds. Sci Total Environ 493:365–376

Dashtban M, Schraft H, Syed TA, Qin W (2010) Fungal biodegradation and enzymatic modification of lignin. Int J Biochem Mol Biol 1:36–50

de Miranda RCM, Gomes E, Pereira N, Marin-Morales MA, Machado KMG, de Gusmao NB (2013) Biotreatment of textile effluent in static bioreactor by *Curvularia lunata* URM 6179 and *Phanerochaete chrysosporium* URM 6181. Bioresour Technol 142:361–336

Dosoretz CG, Reddy CA (2007) Lignin and lignin-modifying enzymes. In: Reddy CA, Beveridge TJ, Breznak JA, Marzluf GA, Schmidt TM, Snyder LJ (eds) Methods for general and molecular microbiology. American Society for Microbiology, Washington, pp 611–620

Dwivedi UN, Singh P, Pandey VP, Kumar A (2011) Structure–function relationship among bacterial, fungal and plant laccases. J Mol Catal B-Enzym 68:117–128

Fackler K, Gradingerm C, Hinterstoisser B, Messner K, Schwanninger M (2006) Lignin degradation by white rot fungi on spruce wood shavings during shorttime solid-state fermentations monitored by near infrared spectroscopy. Enzym Microb Tech 39:1476–1483

Field JA, de Jong E, Feijoo Costa G, de Bont JAM (1993) Screening for ligninolytic fungi applicable to the biodegradation of xenobiotics. Trends Biotechnol 11:44–49

Galli C, Gentili P (2004) Chemical messengers: mediated oxidations with the enzyme laccase. J Phys Org Chem 17:973–977

Glenn JK, Gold MH (1983) Decolorization of several polymeric dyes by the lignin degrading basidiomycete *Phanerochaete chrysosporium*. Appl Environ Microb 45:1741–1747

Gomathi Cibichakravarthy V, Ramanjaneya B, Nallapeta AS, Mula R, Jayasimha Rayalu D (2012) Decolourization of paper mill effluent by immobilized cells of *Phanerochaete chrysosporium*. Int J Pl An Environ Sci 2:141–146

Gros M, Cruz-Morato C, Marco-Urrea E, Longrée P, Singer H, Sarrà M, Hollender J, Vicent T, Rodriguez-Mozaz S, Barceló D (2014) Biodegradation of the X-ray contrast agent iopromide and the fluoroquinolone antibiotic ofloxacin by the white rot fungus *Trametes versicolor* in hospital wastewaters and identification of degradation products. Water Res 60:228–241

Hammel KE, Jensen K, Mozuch M, Landucci L, Tien M, Pease E (1993) Ligninolysis by a purified lignin peroxidase. J Biol Chem 268:12274–12281

Hammel KE, Kapich AN, Jensen KA, Ryan ZC (2002) Reactive oxygen species as agents of wood decay by fungi. Enzym Microb Tech 30:445–453

Harms H, Schlosser D, Wick LY (2011) Untapped potential: exploiting fungi in bioremediation of hazardous chemicals. Nat Rev Microbiol 9:177–192

Harvey PJ, Schoemaker HE, Palmer JM (1986) Veratryl alcohol as a mediator and the role of radical cations in lignin biodegradation by*Phanerochaete chrysosporium*. FEBS Lett 195:242–246

Hatakka A (1994) Lignin-modifying enzymes from selected white-rot fungi: production and role in lignin degradation. FEMS Microbiol Rev 13:125–135

Hatakka A (2001a) Biodegradation of lignin. In: Hofrichter M, Steinbüchel A (eds) Lignin, humic substances and coal, vol 1, Wiley-VCH, Weinheim (Germany), pp 129–180

Hatakka A (2001b) Biodegradation of lignin. In: Steinbüchel A, Hofrichter M (eds) Biopolymers, vol 1., Lignin, humic substances and coalWiley-VCH, Weinheim, pp 129–180

Heinfling A, Martinez MJ, Martıinez AT, Bergbauer M, Szewzyk U (1998) Purification and characterization of peroxidases from the dye decolorizing fungus *Bjerkandera adusta*. FEMS Microbiol Lett 165:43–50

Hofrichter M (2002) Review: lignin conversion by manganese peroxidase (MnP). Enzym Microb Tech 30:454–466

Hofrichter M, Scheibner K, Schneegaß I, Fritsche W (1998) Enzymatic combustion of aromatic and aliphatic compounds by manganese peroxidase of *Nematoloma frowardii*. Appl Environ Microb 64:399–404

Hofrichter M, Ulrich R, Pecyna MJ, Liers C, Lundell T (2010) New and classical families of secreted fungal heme peroxidases. Appl Microbiol Biot 87:871–897

Kim SJ, Shoda M (1999) Purification and characterization of a novel peroxidase from *Geotrichum candidum* Dec 1 involved in decolorization of dyes. Appl Environ Microb 65:1029–1035

Kirk TK, Cullen D (1998) Enzymology and molecular genetics of wood degradation by white-rot fungi. In: Young RA, Akhtar M (eds) Environmentally friendly technologies for the pulp and paper industry. Wiley, New York, pp 273–307

Kurniawati S, Nicell JA (2007) Efficacy of mediators for enhancing the laccase-catalyzed oxidation of aqueous phenol. Enzyme Microb. Tech. 41:353–361

Liers C, Bobeth C, Pecyna M, Ullrich R, Hofrichter M (2010) DyP-like peroxidases of the jelly fungus *Auricularia auricula-judae* oxidize nonphenolic lignin model compounds and high-redox potential dyes. Appl Microbiol Biot 85:1869–1879

Liers C, Arnstadt T, Ullrich R, Hofrichter M (2011) Patterns of lignin degradation and oxidative enzyme secretion by different wood- and litter-colonizing basidiomycetes and ascomycetes grown on beech-wood. FEMS Microbiol Ecol 78:91–102

Liers C, Aranda E, Strittmatter E, Piontek K, Plattner DA, Zorn H, Ullrich R, Hofrichter M (2014) Phenol oxidation by DyP-type peroxidases in comparison to fungal and plant peroxidases. J Mol Catal B Enzym 103:41–46

Lisov AV, Leontievsky AA, Golovleva LA (2003) Hybrid Mn-peroxidase from the ligninolytic fungus *Panus tigrinus* 8/18. Isolation, substrate specificity, and catalytic cycle. Biochemistry (Moscow) 68:1027–1035

Lundell T, Mäkelä M (2013) Puunlahottajat (Wood-degrading fungi). In: Timonen S, Valkonen J (eds) Sienten biologia (Biology of fungi). Gaudeamus Oy, Tallinn (Estonian), pp 259–279

Ma L, Zhuo R, Liu H, Yu D, Jiang M, Zhang X, Yang Y (2014) Efficient decolorization and detoxification of the sulfonated azo dye Reactive Orange 16 and simulated textile wastewater containing Reactive Orange 16 by the white-rot fungus *Ganoderma* sp. En3 isolated from the forest of Tzu-chin Mountain in China. Biochem Eng J 82:1–9

Majeau JA, Brar SK, Tyagu RD (2010) Laccases for removal of recalcitrant and emerging pollutants. Bioresource Technol. 101:2331–2350

Martinez AT (2002) Molecular biology and structure–function of lignin-degrading heme peroxidases. Enzyme Microb Tech 30:425–444

Martinez AT, Speranza M, Ruiz-Duenas FJ, Ferreira P, Camarero S, Guillen F, Martinez MJ, Guttirez A, del Rio JC (2005) Biodegradation of lignocellulosics: Microbial, chemical and enzymatic aspects of the fungal attack of lignin. Int Microbiol 8:195–204

Mester T, Field JA (1998) Characterization of a novel manganese peroxidase-lignin peroxidase hybrid isozyme produced by *Bjerkandera* species strain BOS55 in the absence of manganese. J Biol Chem 273:15412–15417

Mester T, Varela E, Tien M (2004) Wood degradation by brown-rot and white-rot fungi. The Mycota II: genetics and biotechnology. Springer-Verlag, Berlin-Heidelberg

Nerud F, Misurcova Z (1996) Distribution of lignilolytic enzymes in selected white-rot-fungi. Folia Microbiol 41:264–266

Nishida T, Kashino Y, Mimura A, Takahara Y (1988) Lignin biodegradation by wood-rotting fungi I. Screening of lignin-degrading fungi. Mokuzai gakkaishi 34:530–536

Ntougias S, Baldrian P, Ehaliotis C, Nerud F, Antoniou T, Merhautová V, Zervakis GI (2012) Biodegradation and detoxification of olive mill wastewater by selected strains of the mushroom genera *Ganoderma* and *Pleurotus*. Chemosphere 88:620–626

Olivieri G, Russo ME, Giardina P, Marzocchella A, Sannia G, Salatino P (2012) Strategies for dephenolization of raw olive mill wastewater by means of *Pleurotus ostreatus*. J Ind Microbiol Biot 39:719–729

Osorio-Echavarría J, Vidal Benavides AI, Quintero Díaz JC (2012) Decolorization of textile wastewater using the white rot fungi anamorph R1 of *Bjerkandera* sp. Revista de la Facultad de Ingeniería de la Universidad de Antioquia 57:85–93

Otjen R, Blanchette RA, Effland M, Leatham G (1987) Assessment of 30 white rot basidiomycetes for selective lignin degradation. Holzforschung 41:343–349

Pakshirajan K, Kheria S (2012) Continuous treatment of coloured industry wastewater using immobilized *Phanerochaete chrysosporium* in a rotating biological contactor reactor. J Environ Manage 101:118–123

Palli L, Gulloto A, Tilli S, Gori R, Lubello C, Scozzafava A (2014) Effect of carbon source on the degradation of 2-naphthalenesulfonic acid polymers mixture by *Pleurotus ostreatus* in petrochemical wastewater. Process Biochem 49:2272–2278

Passardi F, Cosio C, Penel C, Dunand C (2005) Peroxidases have more functions than a Swiss army knife. Plant Cell Rep 24:255–265

Pérez-Boada M, Ruiz-Dueñas J, Pogni R, Basosi R, Choinowski T, Martínez MJ, Piontek K, Martínez AT (2005) Versatile peroxidase oxidation of high redox potential aromatic compounds: site-directed mutagenesis, spectroscopic and crystallographic investigation of three long-range electron transfer pathways. J Mol Biol 354:385–402

Petrides PE, Nauseef WM (2000) The peroxidase multigene family of enzymes, biochemical basis and clinical applications. Springer Verlag, Berlin-Heidelberg

Pinedo-Rivlla C, Aleu J, Collado IG (2009) Pollutants biodegradation by fungi. Curr Org Chem 13:1194–1214

Pointing SB (2001) Feasibility of bioremediation by white-rot fungi. Appl Microbiol Biot 57:20–33

Raghukumar C, D'Souza -Ticlo D, Verma AK (2008) Treatment of colored effluents with lignin-degrading enzymes: an emerging role of marine-derived fungi. Crit Rev Microbiol 34:189–206

Reddy CA (1995) The potential for white-rot fungi in the treatment of pollutants. Curr Opin Biotech 6:320–328

Reddy CA, Mathew Z (2001) Bioremediation potential of white rot fungi. In: Gadd GM (ed) Fungi in bioremediation. Cambridge University Press, London, pp 52–78

Ruiz-Dueñas FJ, Martinez MJ, Martinez AT (1999) Molecular characterization of a novel peroxidase isolated from the ligninolytic fungus *Pleurotus eryngii*. Mol Microbiol 31:223–236

Sangeeta P, Kheria S, Pakshirajan K (2011) Biodecolourization of real textile industry wastewater using white-rot fungus, P*hanerochaete chrysosporium*. J Sci Ind Res India 70:982–986

Sathian S, Radha G, Shanmugapriya V, Rajasimman M, Karthikeyan C (2013) Optimization and kinetic studies on treatment of textile dye wastewater using *Pleurotus floridanus*. Appl Water Sci 3:41–48

Sathian S, Rajasimman M, Radha G, Shanmugapriya V, Karthikeyan C (2014) Performance of SBR for the treatment of textile dye wastewater: Optimization and kinetic studies. Alexandria Eng J 53:417–426

Sigoillot JC, Berrin JG, Bey M, Lesage-Meessen L, Levasseur A, Lomascolo A, Record E, Uzan-Boukhris E (2012) Fungal strategies for lignin degradation. In: Jouanin L, Lapierre C (eds) Lignins: biosynthesis, biodegradation and bioengineering, advances in research. Elsevier, Amsterdam, pp 262–308

Strong PJ (2010) Fungal remediation of Amarula distillery wastewater. World J Microb Biot 26:133–144

Subramanian V, Yadav JS (2009) Role of P450 monooxygenases in the degradation of the endocrine-disrupting chemical nonylphenol by the white rot fungus *Phanerochaete chrysosporium*. Appl Environ Microb 75:5570–5580

Sugano Y (2009) DyP-type peroxidases comprise a novel heme peroxidase family. Cell Mol Life Sci 66:1387–1403

Taboada-Puig R, Lù-Chau T, Eibes G, Moreira MT, Feijoo G, Lema JM (2011) Biocatalytic generation of Mn(III)-chelate as a chemical oxidant of different environmental contaminants. Biotechnol Progr 27:668–676

ten Have R, Teunissen PJM (2001) Oxidative mechanisms involved in lignin degradation by white-rot fungi. Chem Rev 101:3397–3413

Tuomela M, Hatakka A (2011a) Oxidative fungal enzymes for bioremediation. In: Moo-Young M, Agathos S (eds) Comprehensive biotechnology, 2nd edn. Elsevier, Spain, pp 183–196

Tuomela M, Hatakka A (2011b) Oxidative fungal enzymes for bioremediation. In: Moo-Young M, Agathos S (eds) Comprehensive biotechnology. Elsevier, Spain, pp 183–196

Tuor U, Winterhalter K, Fiechter A (1995) Enzymes of white-rot fungi involved in lignin degradation and ecological determinants for wood decay. J Biotechnol 41:1–17

van Bloois E, Torres Pazmino DE, Winter RT, Fraaije MW (2009) A robust and extracellular heme-containing peroxidase from *Thermobifida fusca* as prototype of a bacterial peroxidase superfamily. Appl Microbiol Biot 86:1419–1430

Wariishi H, Gold MH (1989) Lignin peroxidase compound III: formation, inactivation and conversion to the native enzyme. FEBS Lett 243:165–168

Wesenberg D, Kyriakides I, Agathos SN (2003) White-rot fungi and their enzymes for the treatment of industrial dye effluents. Biotechnol Adv 22:161–187

Widsten P, Kandelbauer A (2008) Laccase applications in the forest products industry: a review. Enzym Microb Tech 42:293–307

Xu F (1999) Recent progress in laccase study: properties, enzymology, production, and applications: the encyclopedia of bioprocessing technology: fermentation, biocatalysis, and bioseparation. John Wiley & Sons, New York

Yang S, Hai FI, Nghiem LD, Nguyen LN, Roddick F, Price WE (2013) Removal of bisphenol A and diclofenac by a novel fungal membrane bioreactor operated under non-sterile conditions. Int Biodeter Biodegr 85:483–490

Zelena K, Hardebusch B, H€ulsdau B, Berger RG, Zorn H (2009) Generation of norisoprenoid flavors from carotenoids by fungal peroxidases. J Agr Food Chem 57:9951–9955

Zhang Y, Geissen SU (2012) Elimination of carbamazepine in a non-sterile fungal bioreactor. Bioresour Technol 112:221–227

Zorn H, Langhoff S, Scheibner M, Nimtz M, Berger RG (2003) A peroxidase from Lepista irina cleaves b, b-carotene to flavor compounds. Biol Chem 384:1049–1056

Chapter 6
Fungal Bioremediation of Emerging Micropollutants in Municipal Wastewaters

Federica Spina and Giovanna Cristina Varese

Abbreviations

EPA	Environmental protection agency
EC	European Commission
EDCs	Endocrine disrupting chemicals
BPA	Bisphenol A
PCPs	Personal care products
WTP	Wastewater treatment plant
EEQ	17β-estradiol equivalent concentration
NP	Nonylphenol
TCS	Triclosan
HBT	Hydroxybenzotriazole
ABTS	2,2′-azino-bis(3-etilbenzotiazolin-6-sulfonico)
PEG	Polyethylene glycol
YES	Yeast estrogen screen assay
E-screen test	Human breast cancer cell line (MCF-7 BUS) proliferation test
MELN assay	Luciferase-transfected human breast cancer cell line (MCF-7 BUS) gene reporter assay

F. Spina · G.C. Varese (✉)
Department of Life Sciences and Systems Biology, Mycotheca Universitatis Taurinensis (MUT), University of Turin, Viale Mattioli 25, 10125 Turin, Italy
e-mail: cristina.varese@unito.it

F. Spina
e-mail: federica.spina@unito.it

D. Purchase (ed.), *Fungal Applications in Sustainable Environmental Biotechnology*, Fungal Biology, DOI 10.1007/978-3-319-42852-9_6

6.1 Introduction

Over the past decades, clean and safe waters became a major concern of modern society. International legislations have been developed accordingly, gradually reducing emissions and losses of priority hazardous substances.

Through the years, global industrialization and human activities evolved affecting deeply the condition of surface waters worldwide. A new concept of *priority substances* has to be drawn, including those compounds that have never been considered before; due to advanced techniques that improve the detection limit, many chemicals have been discovered in watercourses. Although no regulatory standard has been set for most of these substances, more and more evidences demonstrated their environmental relevance.

The Environmental Protection Agency (EPA) began to handle this issue in the early-1990s, introducing the term *emerging contaminants*. The European Commission (EC) followed with a few initiatives aimed to monitor the aquatic environment and to assess chemicals risk, leading to the Directive 2013/39/EU. European Commission recently required more detailed information on the effects of up to 30,000 chemicals (Sung et al. 2012). A watch list of compounds including estrogens, herbicides, pesticides, anti-inflammatory drugs, and sunscreen ingredients was outlined. These chemicals are not necessarily new, being part of the industrial and manufacturing chain but their environmental significance emerged only recently.

Why should emerging contaminants be our concern? Public interest has risen due to the possible interference with the physiological functioning of human and wildlife endocrine systems. A primary issue is represented by endocrine disrupting chemicals (EDCs), which are defined as exogenous substances that may provoke changes in the endocrine functions that consequently affect the health of an organism or its progeny. The hazard of EDCs is due to their ability to interact with human estrogenic receptors, altering in this way the levels of hormones that control growth, reproduction, and development. Damages are not restricted to human beings but also to every exposed organism (Snyder et al. 2003; Bicchi et al. 2009). The low-dose to which human beings are exposed should not be underestimated, since it could cause long-term effects as recently demonstrated by Mersha and collaborators (2015). The exposure to low concentration of bisphenol A (BPA) and bisphenol S during embryogenesis altered adulthood behavior causing neural dysfunctioning (Mersha et al. 2015).

Although it is not clear whether the detected concentrations are high enough to compromise human metabolism, trace levels (from μg/L to ng/L) have been recognized to be deleterious to the aquatic ecosystem, e.g., the embryogenesis of fish (Frye et al. 2011; Flint et al. 2012); the presence of BPA, equal to or even lower than 1 pg/L, is responsible of hormonal metabolism alteration at different stages of amphibians and fish development (Flint et al. 2012).

EDCs are mostly man-made and include pesticides, herbicides and fungicides, heat stabilizers, plasticizers, additives in food, pharmaceuticals, and personal care products (PCPs), which are derived from many industrial sources as well as the human daily-life (Wuttke et al. 2010; Frye et al. 2011). Due to their different origins, it is difficult to define a single contamination point, but once the chemicals entered in the water cycle, wastewater treatment plants (WTPs) are considered to be a clear bottleneck. In many cases the 17β-estradiol equivalent concentration (EEQ) was significantly higher in the WTPs influents than in rivers. For instance, rivers closed to Shanghai showed EEQ of 2.48–41.9 ng/L, but this value increased up to 47.7–80.1 ng/L in WTPs influents (Yang et al. 2011). The same was observed in Korean rivers by Ra and collaborators (2011): EEQ contribution of WTPs effluents to river pollution was minimal, mainly due to the dilution effect and self-depuration of the receiving waters. The actual role and effectiveness of WTPs remained uncertain.

Little information is available about the amount of EDCs, their fate and effects in municipal sewage wastewaters (Schilirò et al. 2009; Liu et al. 2015). The detected concentration is strongly affected by a number of seasonal factors that are mainly uncontrollable and unpredictable; for instance rainfalls, water consumption by the area population, drugs uptake for medical uses, industrial procedures, and agriculture practices (Luo et al. 2014). Besides, many compounds are designed to be chemically and biologically stable (Mompelat et al. 2009), limiting the final efficiency of any treatments. Conventional approaches are not specifically designed to act against EDCs (Bicchi et al. 2009; Luo et al. 2014). Many chemical and physical treatments are adopted by WTPs, but none of them can completely solve the issue of micropollutants (Benotti et al. 2009; Luo et al. 2014). Tertiary treatments such as chlorination, ozonation, photocatalytic degradation, and adsorption on activated carbons usually play a more active role in micropollutants removal than primary and secondary treatments (Lee et al. 2008; Oulton et al. 2010; Fatta-Kassinos et al. 2011a; Katsigiannis et al. 2015); particular attention must be given to advanced biological oxidation, though effect in treating micropollutants still has room for significant improvements. For example, along the whole WTPs, the activated sludge system gave the best degradation yields of BPA (Melcer and Klecka 2011). The possibility to apply bacterial-based process is still controversial. Among 14 bacterial strains isolated from activated sludge capable of reducing 17β-estradiol, only one could also degrade estrone (Yu et al. 2007).

Conventional treatments have many drawbacks in terms of (i) energy requirement; (ii) large use of chemicals; and (iii) formation of undesired by-products, which potentially could be even more harmful than the parent compounds (Reungoat et al. 2010; Garcia et al. 2011). A clear example is given by a combined O_3/UV treatment of ketoprofen: the transformation products were much more toxic than ketoprofen itself and this byproduct could be reduced only by increasing the operational time, with consequent repercussion of the economic sustainability of the process (Illés et al. 2014). Due to the limited effectiveness of WTPs toward a complex mixture of EDCs, micropollutants can re-enter the water cycle accumulating even into drinking waters (Benotti et al. 2009; Mompelat et al. 2009; Lundstrom et al. 2010; Touraud et al. 2011; Vulliet et al. 2011). Sun and collaborators (2013) collected data worldwide, associating the major long-term ecological risk of municipal effluents and reclaimed waters to the presence of steroid estrogens and phenolic compounds. Also, agricultures practices are negatively affected by the reuse of still contaminated wastewaters (Fatta-Kassinos et al. 2011b).

Nowadays the removal of emerging contaminants including EDCs, pharmaceutical, PCPs from municipal wastewaters represents a social and technological challenge. WTPs are seeking alternative methodologies, specifically designed to remove micropollutants and to assure the reduction of ecological risk. In addition to concerns arisen within the scientific community, public perception of the risk is also increasing. Future efforts will be aimed to converge the activities of public and private institutions and research centers in order to avoid the uncontrolled propagation of the problem. Any alternative technology will need to focus on breaking the *status quo* of the pollution of water sources.

6.2 Fungal Treatment

Alternative biological treatments are potentially capable of overcoming the disadvantages of conventional approaches and mediating the efficient removal of micropollutants (Cabana et al. 2007b; Domaradzka et al. 2015). Fungal-based technology rose through the years as a perfect candidate to be integrated into WTP process line for the transformation of EDCs and other hazardous emerging pollutants. The coupling of fungi and bacteria for wastewater treatment has already been investigated, reaching the goal to reduce the contaminant content and enhance the water quality for textile wastewaters (Anastasi et al. 2012). So far, this approach has not been investigated for municipal wastewaters.

Fungi have long been recognized for their abilities to transform a broad range of recalcitrant compounds through the use of nonspecific extracellular oxidative enzymes (Kaushik and Malik 2009; Gao et al. 2010). The idea of using fungi for environmental bioremediation sets its basis in the late 1980s (Bumpus and Aust 1987), when a direct connection between the degradation of aromatic compounds and the fungal enzymatic pattern involved in the delignification of wood was hypothesized for the first time.

Having developed a specific mechanism that employs few enzymes and molecules with high oxidizing power, fungi can oxidize aromatic compounds. In particular, oxidoreductases are the most important fungal enzymes involved in the aromatic compounds degradation: laccases (EC 1.10.3.2), peroxidases (EC 1.11.1.x), and peroxygenases (EC 1.11.2.x) are mainly extracellular enzymes. Over the years the actual involvement of these enzymes has often been demonstrated in micropollutants degradation. For instance, effective laccases of *Phlebia tremellosa* were overproduced by genetic transformation in *Irpex lacteus*. The much higher laccase expression led to faster and wider conversion of some EDCs such as BPA, nonylphenol (NP) and two phthalates derivate (Kum et al. 2009).

Fungi, and in particular basidiomycetes, have been efficiently used to treat water samples contaminated with micropollutants. Many reports dealt with the removal of recognized EDCs, PCPs, biological active compounds and pharmaceuticals, but the residual toxicity after the fungal-based treatment has rarely been taken into consideration. However, where the information related to the chemical composition of the treated samples is coupled with the toxicity, a more comprehensive assessment of water quality could be obtained. For example, BPA concentration reduction by *Stereum hirsutum* and *Heterobasidium insulare* was associated with estrogenic activity removal (Lee et al. 2005). *Trametes versicolor* caused the decrease of 99 % of parabens concentration together with a complete detoxification after 2 days (Mizuno et al. 2009). The great potential of *T. versicolor* strains was also confirmed in the effective removal of two sulfonamides, listed among the most widely used antibiotics (Rodriguez-Rodriguez et al. 2012), ciprofloxacin and norfloxacin (Prieto et al. 2011), as well as clofibric acid, ibuprofen and carbamazepine (Marco-Urrea et al. 2009; Jelic et al. 2012). Noteworthy among the four tested strains (*T. versicolor*, *I. lacteus*, *Ganoderma lucidum* and *Phanerochaete chrysosporium*), *T. versicolor* was the only one capable of converting clofibric acid and carbamazepine at high rates (91 and 58 % respectively) (Marco-Urrea et al. 2009).

The transformation rate varies in different fungi; micropollutants degradation is indeed strictly dependent on the fungal species and even on each specific strain. The importance to work with the proper fungal species is underlined by the high variance of degradation percentage observed by different fungi. Eight basidiomycetes have been used to treat several EDCs from aqueous solution. Even though most of them coupled an effective degradation to the estrogenic activity reduction, a great variability in the process yields was observed: the fastest and highest transformation was obtained using *I. lacteus*, *Pycnoporus cinnabarinus*, and *T. versicolor* (Cajthaml et al. 2009). A different scenario as outlined by a rapid screening, focused on 5 fungi (*P. chrysosporium*, *I. lacteus*, *G. lucidum*, *Bjerkandera* sp., and *Penicillium ochrochloron*) and 10 mg/L 17β-estradiol: 77 and 99 % removal was obtained by *P. chrysosporium* and *G. lucidum*, respectively (Blanquez and Guieysse 2008).

Besides, the use of specific strain seemed to be important as well. For example, in two independent studies set up to screen for NP removal using the some common species but different substrate concentration: 3 mg/L (Cajthaml et al. 2009) and 100 mg/L (Soares et al. 2006) showed that *T. versicolor* and *Bjerkandera adusta* have comparable behavior in term of both substrate removal (100 %) and duration required (2 weeks). As regard to the other tested strains, the data did not always correspond, e.g., *P. chrysosporium*: the yields ranged from 100 % after 7 days (Cajthaml et al. 2009) to 50 % after 25 days (Soares et al. 2006).

Up to now most of the researches deal with synthetic and simplified solutions, using single molecules at high concentration and controlling operative conditions, e.g., pH. However, a sterile and controlled environment gives weak information about the capability of the fungus to grow and remain active in very variable and restrictive conditions of municipal wastewater (several micropollutants at low concentration, suspended solids, competition with the autochthonous microflora). Moreover, working at high pollutants concentration (above mg/L), the acquired information is poorly predictive of the fungal performances on environmental water samples (Blanquez and Guieysse 2008; Cajthaml et al. 2009). The applied substrate concentration should be as close to environmentally relevant ones (μg/L–ng/L) as possible. It is noteworthy that *Trametes hirsuta* halved the initial amount of 9 out of 17 pharmaceuticals set at realistic concentrations (20–500 ng/L) (Haroune et al. 2014). The removal of naproxen and diclofenac by *T. versicolor* was almost complete at 10 mg/L and at 45–55 μg/L (Marco-Urrea et al. 2010a, b).

Only few researches have used real wastewaters and operated in harsh working conditions that can be found in WTPs that can ultimately negatively affect the fungal growth and EDCs removal. Hospital and municipal wastewaters have been treated, obtaining positive responses by T. *versicolor* and *P. chrysosporium* (Jelic et al. 2012; Zhang and Geißen 2012; Ferrando-Climent et al. 2015). Interestingly, a waste landfill leachate was successfully treated by *T. versicolor* and *S. hirsutum* that combined adsorption and biodegradation mechanisms. BPA, NP, ethynilestradiol, dimethoate, and linuron were detected in the sample and used as target analytes. With the only exception of dimethoate, *T. versicolor* alone almost completely removed the other contaminants (80–100 %). *S. hirsutum* gave optimal response only for NP (100 %), whereas the other compounds were converted at a lower extent (30–80 %). It should be considered that the addition of synthetic nutrients in the form of potatoes dextrose medium and synthetic adsorbent materials was always necessary to strengthen the fungal activity, reaching better process yields (Castellana and Loffredo 2014).

Regarding the industrial scale feasibility of a process, the set up and the validation of a proper reactor configuration is mandatory. Efforts aimed to develop fungal bioreactors have been made but mostly addressed to textile wastewaters treatment (Anastasi et al. 2010; Baccar et al. 2011; Novotny et al. 2011).

One of the first reports on the scaling up of micropollutants treatment from benchtop assay was carried out by Blanquez and Guieysse (2008) who set up a small (100 mL working volume with substrates concentration varying from 3 to 18.8 mg/L) air-fluidized bioreactor with free biomass of *T. versicolor*. The system continuously operated for 26 days, completely removing 17β-estradiol and 17α-ethinylestradiol by expressing solely laccase activity (156 U/L).

A plate reactor (2 L working volume, recirculating every hour) with *P. chrysosporium* mycelium immobilized on polyurethane foam under non-sterile conditions (Zhang and Geißen 2012) showed that carbamazepine (5 mg/L) was efficiently eliminated by means of the combination of biodegradation and biosorption mechanisms: after the first 3 batches, the process yields lessened at the fourth batch (from 20 to 55 % of residual concentration). Loading the system with municipal wastewaters, carbamazepine elimination was around 60 % and the reactor could run for 100 days, but an adequate nutrients addition was necessary (Zhang and Geißen 2012). A lower concentration of carbamazepine was treated by *T. versicolor* in batch (200 μg/L) and continuous (11.9 μg/g dry weight pellets per day) treatment in an air-pulsed fluidized bed bioreactor (1.5 L working volume, 3 days hydraulic retention time) (Jelic et al. 2012).

Immobilized *P. chrysosporium* was used to set a countercurrent seepage bioreactor (3 L working volume, 2 days hydraulic retention time) for the treatment of a synthetic wastewater spiked with naproxen and carbamazepine: no bacterial contamination was observed in 165 days even though sterility was not controlled. When compared to other configurations (i.e., submerged mode reactor), it could be observed that the 80–100 % degradation was achieved using the specific operational parameters in the countercurrent seepage bioreactor (Li et al. 2015).

It is noteworthy that hospital wastewaters containing 43 pharmaceutical active compounds and 46 EDCs have been treated by a fluidized bed reactor with *T. versicolor* (Cruz-Moratò et al. 2014). The vitality of the fungus even in axenic conditions was indicated by the immediate consumption of the added glucose and the production of 320 U/L laccases. However, without any sterility control, the treatment did not last more than 5 days; a massive bacterial development was observed and coupled with fungal biomass disaggregation.

Although the ability of fungi to degrade emerging contaminants has been confirmed in many studies in the past decades, their involvement along WTPs has not found a clear confirmation so far. One of the main limits of biological treatment based on living organisms is the requirement of specific and controlled operative conditions in order to maintain a durable and efficient system. In addition, fungi are influenced by chemical-physical parameters as well as by the culturing conditions: nutrients presence, pH values, immobilization on different supports, agitation/static growth conditions can strongly affect the fungal oxidative metabolism (Kaushik and Malik 2009).

So far, fungal-based processes have not been transferred from lab scale to industrial level working with municipal wastewaters. Process developments are still necessary to achieve the technical and economic feasibility of fungal treatments. The optimization of chemical-physical parameters (e.g. nutrients addition, pH) and technological features (e.g. carrier selection, design of proper reactor configuration) would allow setting up a whole-cell fungal treatment in WTPs, synergistically working with existing techniques to reduce micropollutants concentration and toxicity.

6.3 Enzymatic Treatment[1]

In this context, a new environmentally friendly technology for industrial biotransformation, with low costs and low consumption of resources, could be very competitive. A promising alternative could be represented by the direct use of fungal crude extract or by purified enzymes capable of maintaining the catalytic yields of the fungal cultures with less technical drawbacks (Torres et al. 2003). The

[1]For further information on enzymatic treatment of wastewater using white-rot fungi, please see Chap. 5—*Potential of white-rot fungi to treat xenobiotic-containing wastewater.*

use of purified enzymes poses some concerns for operational purposes; without other enzymatic isoforms and fungal metabolites, their catalytic efficiency and stability could be compromised. Complex matrixes as municipal wastewaters would benefit from robust, versatile, and effective systems that can be guaranteed by heterogeneous enzymatic fungal extracts.

The potential of enzymes-based methods has been recognized worldwide; to this regard, the Swiss Industrial Biocatalysis Consortium defined oxidative enzymes as the biocatalysts displaying the highest development potential in the next decades (Meyer and Munch 2005). Considerable importance is given to the discovery of novel stable enzymes with wide substrate specificity and industrial potential.

6.3.1 Fungal Peroxidases

Peroxidases include several enzymatic classes found in many fungi: lignin peroxidase (EC 1.11.1.14) and manganese peroxidase (EC 1.11.1.13) were found for the first time in *P. chrysosporium* (Glenn and Gold 1983; Tien and Kirk 1983), versatile peroxidase (EC 1.11.1.16) was described in *Pleurotus* and *Bjerkandera* genera (Martinez 2002), and dye-decolorizing peroxidase (EC 1.11.1.19) was observed in *Auricularia auricula-judae* and *Pleurotus ostreatus* (Faraco et al. 2007; Liers et al. 2010). The expression of these enzymes is not constitutive and depends on the external conditions. Response elements to the presence of xenobiotic molecules have been found on the promoters of the genes coding for lignin peroxidase, while manganese peroxidase and versatile peroxidase are regulated by the presence of metals and thermal stress (Heinfling et al. 1998; Martınez and Martınez 1996).

Besides conventional pollutants, research demonstrated the potential of peroxidase against emerging contaminants. For example, both analytes concentration (octylphenol, BPA, and estrogens in buffer solution) and toxicity were reduced by lignin peroxidases from *Phanerochate sordida* (Wang et al. 2012).

Peroxidases generally show a higher redox potential than laccases, allowing a clear catalytic advantage. Manganese peroxidase from *P. chrysosporium* was able to totally degrade triclosan (TCS), while the best performance of laccases from *T. versicolor* (33.9 % residual TCS) was obtained only in the presence of hydroxybenzotriazole (HBT) as mediator and at double enzymatic concentration (Inoue et al. 2010). Similarly, carbamazepine depletion was close to 14 % with manganese peroxidase but almost null with laccases alone. Laccase's efficiency was significantly enhanced (22 %) only in presence of mediators (Hata et al. 2010).

The main restriction of haem containing enzymes is their poor stability in environmental conditions or due to the presence of intermediate products as well as rapid deactivation by hydrogen peroxide (Buchanan and Nicell 1998; Buchanan and Han 1999; Gasser et al. 2012). Even in model system containing sterile and buffered solutions, 90 % of lignin peroxidase activity was lost after 24 h (Wang et al. 2012). Enzyme inactivation was the major drawback of the treatment of 5

EDCs by an innovative approach based on the high reactive oxidants (i.e., Mn^{3+}-malonate complex) produced by peroxidases of *B. adusta* (Taboada-Puig et al. 2015). In particular, in the versatile peroxidase catalytic cycle, Mn^{3+}-malonate complexes are usually formed to stabilize the oxidized Mn ion and propagate the oxidative reaction. The continuous production of the complex was not possible because of peroxidase inactivation. The authors overcame the problem by designing a two-stage system (complex formation and then EDCs degradation) enhancing the system stability up to 8 h (Taboada-Puig et al. 2015).

At the present state of art, these factors ultimately prevent the use of peroxidase-based technologies at WTP level. In contrast, laccases use oxygen as final electron acceptor, they are not negatively influenced by peroxide deactivation but they need sufficient aeration.

6.3.2 Fungal Laccases

Laccases (EC 1.10.3.2) have been pointed out as *green* catalysts, which work with air, producing water as the only by-product (Riva 2006). Laccases are glycosylated multicopper oxidases whose production is rarely constitutive in fungi, but they are part of secondary metabolisms activated by external stimuli. Gene transcription is regulated by metal ions, various aromatic compounds related to lignin, nitrogen, and carbon: all these factors act synergistically, stimulating laccase production (Piscitelli et al. 2011). Moreover, laccase's stability in a wide range of chemical-physical condition makes them good candidates to be used as biocatalysts for industrial purposes (Riva 2006).

The involvement of laccases in the oxidative cascade triggered by fungi has been confirmed by observations of fungal behavior in the presence of micropollutants: fungi often respond to these chemicals by stimulating the enzymes production and secretion. Agrochemical and industrial pollutants strongly induced the production of laccases in *T. versicolor* (Mougin et al. 2002). A similar induction was noticed after phthalates, BPA and NP addition to a *T. versicolor* culture: the secreted enzyme further degraded the molecules tested (Kim et al. 2008). Cheong and collaborators (2006) monitored the expression levels of laccase genes in *T. versicolor* during the treatment of 2,4,6-trinitrotoluene. A higher gene expression was observed in the cultures containing the substrate in comparison with the control (no substrate), demonstrated the actual involvement of laccases.

Laccase extracts from different fungi (e.g. *T. versicolor*, *Coriolopsis gallica*, *Cerrena unicolor*, *G. lucidum*, *Myceliophthora thermophila*) have been tested on several well-known micropollutants such as BPA, NP, TCS, carbamazepine, anti-inflammatory pharmaceuticals and estrogens (Kim and Nicell 2006a; Auriol et al. 2008; Lloret et al. 2010; Murugesan et al. 2010; Lloret et al. 2012b; Songulashvili et al. 2012; Torres-Duarte et al. 2012; Lloret et al. 2013; Tran et al. 2013). Transformation rates of many compounds were not always consistent among evidences present in literature. This could be explained by looking at the intrinsic biochemical and catalytic properties (e.g., structure of the active site and the region nearby, redox potential) of each laccase, which is strain specific. Any laccase isoform has indeed the potential to differently interact with a substrate. For instance NP was not a substrate for laccases from *Phoma* sp. whereas 90 % of removal was achieved by laccases form *P. ostreatus*. On the contrary, the two laccases crude extracts did not show significant variation treating BPA and 17α-ethinylestradiol (Libardi Junior et al. 2012). These evidences suggest that an enzymatic treatment could be based on a mixture of different enzymes, each with its own substrate specificity, substrate affinity, optimal working conditions (e.g., pH). A multi-enzymatic approach could thus enlarge the range of target compounds and the working conditions compared to single-laccase systems. Amman and collaborators (2014) further investigate this assumption by co-immobilizing laccases of a number of fungal strains: *Thielavia* sp., *Coriolopsis polyzona*, *C. unicolor*, *P. ostreatus,* and *T. versicolor* on silica nanoparticles. Interestingly, the co-immobilization of *Thielavia* sp. and *P. ostreatus* laccases led to a wider operative pH range: the recovery of 70 % of the activity between pH 3 and 7 which was not observed in the single-laccase solutions. As regard to the substrate affinity, *C. polyzona* and *T. versicolor* laccases worked better together than alone: the multi-enzymes system converted the 5 test compounds, including benzophenone-2 and gemfibrozil which were not oxidized by individual laccase produced by *C. polyzona* or *T. versicolor* (Amman et al. 2014). The development of this technology could take into consideration also the co-immobilization of different enzymes typology, e.g., laccases and cytochrome P450 both played an active role in the conversion of diclofenac and naproxen by *T. versicolor* (Marco-Urrea et al. 2010a, b).

The transformation rate of single-compound and mixture solutions can be highly variable. Indeed some molecules and/or the products of their partial degradation can act as radicals and propagate the laccase oxidation toward other micropollutants. The responses derived from single-compound and mixture solutions do not usually correspond as the substrates themselves and their products could influence the oxidation of one another; the effects could not be determined a priori. Diclofenac oxidation (25 %) was enhanced in the presence of BPA (60 %) and mefenamic acid (95 %). On the contrary, the presence of diclofenac decreased BPA and mefenamic acid conversion, probably due to competition phenomena for the active site (Margot et al. 2013).

Estrogens, phthalate, pyrene, and several phenols derivatives (20 mg/L) were treated by the crude laccase extracts from *T. hirsuta*, *Trametes orientalis*, *T. versicolor,* and *P. coccineus*. With the exception of phthalate and pyrene, the micropollutants were concurrently and almost completely (90 %) removed (Sei et al. 2008). Eighteen EDCs, pharmaceuticals and PCPs with known or suspected estrogenic activity were treated by laccases from *Trametes pubescens*: high removal efficiency (>60 %) was achieved for 13 compounds (Spina et al. 2015).

6.3.2.1 *Laccase-Mediated System*

Laccase-catalyzed oxidation includes two distinct paths: (i) the direct interaction with the active site, and/or (ii) the propagation of the oxidation cascade by means of small molecules which diffuse and act towards the target substrates. These low-molecular weight compounds enlarge the spectrum of the potential substrates of laccases, increasing the low redox potential (0.5–0.8 V) up to 1.3 V. The *laccase mediated system* oxidizes non phenolic molecules as well as compounds which, due to their size and steric hindrance, cannot be oxidized by laccases alone (Canas and Camarero 2010; Desai and Nityanand 2011). Many natural mediators produced by the fungal mycelium and synthetic compounds have been evaluated for their capability to propagate the laccases oxidative cascades: at the moment, more than 100 molecules have been analyzed (Desai and Nityanand 2011). The use of 2,2′-azino-bis(3-etilbenzotiazolin-6-sulfonico) (ABTS), HBT, veratryl alcohol, cathecol, syringaldazine, and TEMPO have been often coupled with laccases, greatly enhancing the final oxidation rate (Desai and Nityanand 2011). The strength of the propagated oxidation cascade depends also on the laccase-mediator interaction. The mediators should have a high affinity for the laccase isoform and not interfering with the enzyme active site. Once oxidized, they should be stable and capable of reducing back to their initial form to enter again in the enzymatic catalytic cycle (Morozova et al. 2007).

In many cases, no reactions were observed by laccases alone, making mediators addition mandatory. This phenomenon was true for the degradation of very recalcitrant compounds such as the herbicide dymron (Maruyama et al. 2006) and halogenated pesticides (Torres-Duarte et al. 2009).

ABTS and HBT are among the most studied compounds that are capable of mediating micropollutants conversion (Cabana et al. 2007b). The presence of ABTS-enhanced transformation of dymron (Maruyama et al. 2006), TCS (Kim and Nicell 2006a), sulfonamide antibiotics (Weng et al. 2013), BPA (Kim and Nicell 2006b; Macellaro et al. 2014) and parabens analogues (Macellaro et al. 2014). Whereas HBT was found to be fundamental to laccase treatments of carbamazepine (Hata et al. 2010), iso-butylparaben and *n*-butylparaben (Mizuno et al. 2009), anti-inflammatories and estrogens (Lloret et al. 2010) and tetracycline antibiotics (Suda et al. 2012).

Among the four mediators used to trigger BPA and TCS oxidation by two batches of laccases, ABTS was found to be the best (Arboleda et al. 2013). The role of guaiacol, veratryl alcohol, and HBT depended on the used laccase batch, highlighting the strict connection between laccase isoform, mediator and target substrate (Arboleda et al. 2013). In agreement with this, laccases from another fungus (*G. lucidum*) achieved higher percentage of degradation in presence of HBT and syringaldehyde rather than ABTS (Murugesan et al. 2010).

In particular, dimethoxy-substituted phenols such as syringaldehyde and acetosyringone have a low redox potential favoring their activation as mediators by laccase initial oxidation. They were responsible for strong pesticide transformations, which could be even 100-fold higher than TEMPO-mediated reactions

(Torres-Duarte et al. 2009). Regarding TCS conversion, laccases of *G. lucidum* responded better to syringaldehyde than HBT (Murugesan et al. 2010).

Even considering the great catalytic benefits of a laccase-mediator system, the use of synthetic mediators remains a highly controversial subject. A detailed evaluation of risks and benefits is necessary to carefully consider the following factors: (i) cost of the mediators, (ii) their toxicity, and (ii) the need to be continuously added (Torres et al. 2003; Kunamneni et al. 2008). Moreover, in some cases the oxidized radicals can destabilize the enzymes, leading to an irreversible inactivation (Kunamneni et al. 2008). As an example, 3 mM syringaldehyde inhibited the catalytic activity of *Trametes trogii* laccase, whereas vanillin and coumarate did not (Khlifi-Slama et al. 2012). A significant loss of activity (67 %) was observed in presence of TCS and HBT (Murugesan et al. 2010). The use of crude extracts may be preferred to purified enzymes, since they are cheaper and may be more efficient due to the presence of natural mediators and several soluble molecules that can improve the catalytic efficiency and stability of the enzymes (Morozova et al. 2007; Levin et al. 2010; Majeau et al. 2010; Strong and Claus 2010).

6.3.2.2 Effect of Reaction Conditions on *Micropollutants* Degradation

Technical feasibility of enzymes-based treatment is still a challenge. Enzymes can be destabilized by the chemical-physical conditions of municipal wastewaters, limiting the possibility to hypothesize a new treatment framework for WTPs. Moreover, the operative parameters can affect conversion rates and possibly switch off the system completely. For example, pH, temperature, loaded enzymes concentration, and contact time greatly affected (ranging from 0 to 100 %) the removal of 4 EDCs (Margot et al. 2013).

In detail, pH has a major role in enzymes performances. Laccase-catalyzed system was studied over a pH range 5–9; the authors concluded that pH 6 was optimal for degradation of natural estrogens (estrone, 17β-estradiol and estriol) and synthetic estrogen (17α-ethinylestradiol) (Auriol et al. 2007). The optimal pH range of 4–6 was observed for the transformation of dymron (Maruyama et al. 2006), TCS (Kim and Nicell 2006a; Margot et al. 2013), sulfonamide antibiotics (Weng et al. 2013), anti-inflammatory pharmaceuticals (Margot et al. 2013) and BPA (Kim and Nicell 2006b; Margot et al. 2013).

A factorial design was used to describe the optimal working conditions in terms of pH, temperature and treatment time for the transformation of BPA and TCS by laccases of two forest fungi, *Ganoderma stipitatum* and *Lentinus swartzii* (Arboleda et al. 2013). The reaction was maximal at 40–60 °C and pH 5, in accordance with other data available in literature. Nonylphenol and TCS were degraded by laccases from *C. polyzona* at 50 °C and BPA at 40 °C; acidic environment (pH 5) was optimal for the three micropollutants (Cabana et al. 2007a).

Thermotolerant enzymes should be considered separately. As an example, the best working temperature for laccases of *Trametes* sp. was 70 °C, which sharply decreased of five fold at 30 °C (Maruyama et al. 2006). New industrial scenarios emerge where fungal thermotolerant laccases could be exploited, but it should be carefully taken into consideration the WTP localization and working temperature along the plant.

Operational parameters (i.e., loaded laccase concentration, substrate frequency addition, and aeration) can play a key role. The control of the effective laccase amount is important to constrain the economic feasibility of the process. The use of highly concentrated enzymatic solution does not necessary reflect better conversion yields. For example, the treatment of estrone and estradiol was carried out with laccases from *M. thermophile* using up to 2000 U/L. The minimal effective enzymatic concentration was found to range from 100 to 500 U/L with no mediator need (Lloret et al. 2012b). These values are in agreement with those observed by Spina and collaborators (2015): in the tested range (10–500 U/L) the minimal effective concentration for 18 compounds was usually 100 U/L. Much higher activities (2000–20,000 U/L) were analyzed by Auriol and collaborators (2007): laccases from *T. versicolor* achieved the complete removal of natural estrogens (estrone, 17β-estradiol, estriol) and a synthetic estrogen (17α-ethinylestradiol) but mediators were required. The use of the highest (20,000 U/L) investigated concentration was justified by the fact that 100 % conversion was obtained without mediator use (Auriol et al. 2007).

As oxygen is the cofactor of laccases catalytic cycle, it cannot be missing from the aqueous solution; controlled dosage and air supply technologies have to be carefully taken into consideration to maintain laccases fully operative. Continuous aeration is not always the best methodology to be applied since batch pulses can sustain the biological oxidation, reducing of course cost and energetic requirements (Lloret et al. 2012a). In a continuous enzymatic membrane reactor, oxygenation rate was set as a response parameter as well as enzyme activity and hydraulic residence time. Estrone, estradiol, and ethinylestradiol removal was maximal (71–81 %) at 1000 U/L of laccase, 1 h and 60 mgO_2/h of oxygenation rate (Lloret et al. 2013).

The industrial feasibility of any enzymatic treatment needs to be able to cope with fast processes and integrate into WTPs where hundreds of m^3 of water flux every day. As a consequence, operational time is an important parameter to set. Laccases are optimal biocatalysts and mediate reactions that usually last from minutes to few hours: 4–6 h are needed for the treatment of TCS, BPA, and parabens (Mizuno et al. 2009; Arboleda et al. 2013). A commercial laccase could rapidly convert estrogens (in 15 min), but was much slower (8 h) for more recalcitrant compounds such as naproxen (Lloret et al. 2010). The data acquire great importance if referred to municipal wastewaters. Oxybenzone removal required 2 h (Garcia et al. 2011) in a primary effluent, whereas at 20,000 U/L laccase completely removed estrogens in a wastewater sample within an hour (Auriol et al. 2008). The half-life was lower than 4 h for 10 of the 18 micropollutants, and most of these occurred in the first minutes of the treatment (Spina et al. 2015).

6.3.2.3 Laccases Stability

Although fungal enzymes are known to operate in a wide pH, temperature, and ionic strength range, the harsh conditions of municipal wastewaters may affect the enzymatic catalytic efficiency and stability. The presence of organic compounds did not appear to influence laccase activity (Kim and Nicell 2006b). However, the presence of several salts affected the laccase performance, e.g. cyanide by direct interaction with the active site and further dissociation of the Cu ion. Several chlorine ion forms (NaCl, $CaCl_2$, $MgCl_2$, and NH_4Cl) also resulted in a similar adverse effect on enzymatic activity. The cations Na^+, Mg^{2+}, and NH_4^+ and the anion SO_4^{2-} did not interfere contrarily to Cl^- (Kim and Nicell 2006b). This observation would need deep reflections, mainly in those WTPs that use chlorination as tertiary treatment: laccases could indeed be used in combination with it as advanced oxidation system of the compounds produced or released after the chlorination; but the residual chlorine ions, even in trace quantity, could interfere with the further enzymatic treatment.

The inactivation of the enzymes of interest could be reduced by the presence of additives capable of interacting with the enzyme and/or the reaction products, as demonstrated by Kim and Nicell (2006c). Polyethylene glycol (PEG), polyvinyl alcohol, and the branched hydrophilic polysaccharide Ficoll were applied during BPA transformation by laccase from *T. versicolor* highlighting the great potential of PEG. Unfortunately the toxicity monitored by means of *Vibrio fisheri* increased due probably to the improved solubility of the toxic intermediate products enhanced by the presence of PEG. These results clearly indicated that innovative solutions and process developments could and should be researched, looking at the system efficiency as a whole.

A feasible approach considers the enzymatic immobilization in specific supports, in order to protect the enzymes from denaturing agents: in optimized conditions, the system may remain stable for a long time and the enzymes would not be released in the environment (Desai and Nityanand 2011; Gasser et al. 2014a). Moreover, this technology allows the regeneration and reuse of the biocatalysts avoiding their washing out along a continuously operating WTP. Inert materials are often preferable, avoiding micropollutants adsorption, enzymatic activity alteration and toxic effects.

A number of materials complied these requirements, successfully supported laccases for micropollutants removal: nylon membranes (Diano et al. 2007), silica particles (Songulashvili et al. 2012; Gasser et al. 2014b), Eupergit carriers (Lloret et al. 2012b), TiO_2 nanoparticles (Hou et al. 2014), glass beads (Macellaro et al. 2014), and cross-linked enzyme aggregates (Cabana et al. 2007c) among others.

The selection of carriers manageable for industrial purposes and capable of maintaining enzymes stability and functions as long as possible is a crucial point. Thermostability is generally the first parameter on which immobilization could have positive effects. The half-life at 70 °C of the immobilized laccase on silica beads was threefold higher than free enzyme, and was maintained at least double the activity during the exposure at 80 °C (Songulashvili et al. 2012). Laccase on polyacrylonitrile beads achieved great advantages in terms of pH and thermal stability during bisphenol analogues conversion (Nicolucci et al. 2011). A fluidized bed reactor was set up achieving up to 85 % activity recovery of immobilized laccase after 30 days.

In addition, chemical resilience could be provided by optimized immobilization techniques. Laccase immobilized on isocyanate-terminated prepolymers efficiently transformed few EDCs (Torres-Duarte et al. 2012). Covalent binding of enzymes on diatomaceous earth support was a suitable solution, able to guarantee high laccases recovery in presence of several denaturants (acetone, methanol, EDTA, $CaCl_2$) and the enhancement of the catalytic activity (Cabana et al. 2008). The bio-construct worked continuously over five cycles.

In some cases this technology could even lead to the increase of the initial laccase activity (Hommes et al. 2013; Gasser et al. 2014b). For instance, the immobilization of laccase from *Thielavia* on silica nanoparticles was scaled up to kilogram in weight and the laccase activity recovery significantly increased (132.5 % higher than the loaded activity) (Gasser et al. 2014b). The designed membrane bioreactor was used to treat municipal effluents (BPA as target analyte): 66 % degradation was followed by a significant enzymatic loss (30–40 % recovery). The process effectiveness was evaluated not only in terms of chemical conversion but also in a detailed economic balance; it was found to be both technically and economically competitive with other tertiary wastewater treatments (ozonation and PAC adsorption). In accordance to these findings, membrane reactors have been proven suitable reactor configurations for micropollutants removal (Lloret et al. 2013; Nguyen et al. 2014).

6.3.2.4 *Laccases Versus Municipal Wastewaters*

The inactivation of free laccases may be relevant when challenged with treat real municipal wastewaters. The heterogeneity of their composition includes (i) reducing anions, (ii) organic solvents, (iii) heavy metals, (iv) cyanide, (v) salts, and (vi) suspended particles; when coupled with potentially toxic products of EDCs oxidation pathway they may interfere with the enzymatic catalytic activity and stability. Most of the activity (54 %) was lost after 24 h treatment of municipal wastewater (Spina et al. 2015) and the same effect was observed by Tran and collaborators (2013) who reported that 30 % of laccase activity was lost in only 4 h.

A preliminary indication of the actual laccases potential is given by environmental water matrix spiked with targeted analytes. Groundwater samples were supplemented with BPA, NP, TCS, and 17α-ethinylestradiol. As regards BPA, NP, and 17α-ethinylestradiol the matrix did not affect the laccase catalytic ability: the conversion of the analytes was higher than 89 % both in model and environmental conditions. A significant decrease was instead observed for TCS: the biotransformation yields dropped down from 90 to 55 % using spiked groundwater samples. The authors attributed this phenomenon to the presence of ions that inhibited laccase activity (Garcia-Morales et al. 2015). Estrogens depletion in municipal waters was worse than in spiked solutions using 2000–10,000 U/L of laccases from *T. versicolor* (Auriol et al. 2007). No differences were observed in the presence of higher loaded biocatalysts.

These observations do not necessary imply an ineffective treatment. With the only exception of 2,4-dichlorophenol and BPA, 7 emerging contaminants including biological active compounds as pharmaceutical, pesticides, and herbicides, were converted at high extent (above 72 %) by laccases from *T. pubescens* (Spina et al. 2015) and the toxicological analysis provides the final evidence about the enhanced water quality after the treatment. Commercial fungal laccases worked also on the insect repellent N,N-diethyl-m-toluamide detected in a real wastewater (Tran et al. 2013). It is interesting to notice that degradation rates were higher in real conditions than in theoretical models. One possible explanation could be the presence of phenolic compounds in the environmental samples that act as natural mediators trigging laccase-mediated oxidation to higher extent. Laccase-based system oxidized personal and pharmaceutical care products (oxybenzone as target analyte) from municipal wastewater primary effluent: laccase-mediated system achieved complete removal even in harsh real conditions, but enzymes activity was negatively influenced by the wastewater matrix (Garcia et al. 2011).

Only few studies considered environmentally relevant concentration of micropollutants (few μg/L-ng/L), where indeed laccases may be limited by substrates diffusion and contact time. Those enzymes capable of working in these conditions deserve particular attention. Laccases of *T. versicolor* led to the same results (100 % removal of oxybenzone) at high (1000 μg/L) and low (10 μg/L) concentrations (Garcia et al. 2011). Laccases of *T. pubescens* effectively treated a municipal wastewater containing 9 micropollutants at a concentration of 0.13–67.6 μg/L (Spina et al. 2015).

Advanced in silico technologies open new perspective in this field, clarifying how a certain process could evolve to reach industrial application. To this regard, an intriguing mathematical model was built to describe the behavior of enzymatic membrane reactor in presence of effluents containing tetracycline (Abejon et al. 2015). A large-scale system based on commercial laccases immobilized on ceramic membranes was designed to simulate WTPs in three different case studies (municipal, hospital, and industrial plants). The results for the municipal WTP indicated that 300,000 m^2 of membrane would be needed, meaning 590,000 membrane modules. The technical feasibility was demonstrated but the authors stated that additional analyses are required to enhance the competitiveness of the enzymatic process.

6.4 Toxicity Evaluation

The scenario drawn by the presence of emerging contaminants in water bodies is difficult to define, ultimately limiting the assessment of traditional and innovative treatments effectiveness. Trace levels (from μg/L to ng/L) and the heterogeneity of chemical composition make water risk assessment a true challenge.

First, their detection appears to be strongly influenced by the available analytical techniques, rarely capable of detecting traces of chemicals (Snyder et al. 2003; Mompelat et al. 2009; Lundstrom et al. 2010; Touraud et al. 2011). In 2001, EPA pointed out the lack of sensible and standardized methods for EDCs detection and, at that moment, a dedicated institution, the Endocrine Disruptor Methods Validation Subcommittee, was set up to fill this void (Snyder et al. 2003). The recommended strategy has primarily indicated the test and methodologies able to evaluate the alteration at different stages of the endocrine system (Brown 2010). The identification of EDCs in real water matrixes cannot be taken for granted; routine monitoring programs cannot give a complete chemical characterization of the output waters from a municipal WTP. The occurrence of some target compounds is a first step but it cannot clearly state whether or not water is safe. The Canadian Environmental Protection Act (1999) listed more than 132 molecules, stating that they could alter the biological diversity and affect human health; among them only two (NP and nonylphenol ethoxylates) are recognized EDCs (available from: http://www.ec.gc.ca/lcpe-cepa/). In 2012, EPA launched the Endocrine Disruptor Screening Program, a multiyear program focused to develop computational toxicology methods and high throughput screening; it is aimed to evaluate the potential deleterious effects of chemicals on human health, and more precisely to play an estrogen, androgen, or thyroid-like activity (available from: https://www.epa.gov/endocrine-disruption/endocrine-disruptor-screening-program-edsp-overview).

Further investigation on the biological effects of the released treated effluents is necessary to set water quality protocols and endeavor regulatory limits to follow. A complete understanding of the biological effect of mixed substrates is necessary, considering the synergistic or antagonistic interactions among chemical substances. A mixture of compounds is usually more toxic than expected as a result of the summing effects of each single component (Soupilas et al. 2008). Toxicological analysis by biological assays takes into account the whole effluent; it is a powerful toolbox for the risk evaluation of such complex wastewaters. The long-term ecological hazard associated with the release of municipal wastewaters into the environment found many proofs, allowing the identification of many dangerous chemicals and justifying the request to introduce bioassays as a standard procedure for water quality control (Mendonca et al. 2009; Leusch et al. 2014). Three bioassays with organisms belonging to different trophic levels (algae, zooplanktons, bacteria) highlighted the toxicity of ketoprofen and its derivatives (Illés et al. 2014). The algae *Raphidocelis subcapitata*, formerly known as *Pseudokirchneriella subcapita*, was highly sensitive to the aquatic toxicity of the Sava River, in the Balkans area (Kallqvist et al. 2008). A pan-European screening on 75 WTP effluents has been carried out by the luciferase-transfected human breast cancer cell line (MCF-7 BUS) gene reporter assay (MELN assay). The composition of wastewaters was variable from domestic to many industrial origins; among the 27 samples with significant EEQ (0.53–17.9 ng/L), 16 came from municipal WTPs, confirming the general statement that sees urban areas as hot-spot of pollution (Jarošová et al. 2014a).

Two sides of the same concept need focused analysis. Micropollutants may have deleterious effects at different stages of the life of any microorganisms, being hormone mimicking is only one of the metabolic mechanisms affected by pollutants. High quality waters require evaluation of both ecotoxicity and estrogenic potential. The major issue is still associated with lack of data about the hazard of thousands of chemicals and the uncertainty of the connection exposure-effect (Fuhrman et al. 2015).

Several bioassays have been evaluated for estrogenic activity assessment, using different test organisms or cell lines and end-points: not all of them are based on standard protocols, recognized by water control institutions. Many of them could be hardly defined eligible for routine lab practice: due to high test volume, need of specific equipment, long-term maintenance of living organisms, the economic and technical feasibility is often scars. Besides, their sensitivity to the micropollutants' typology at environmental concentration commonly found in municipal wastewaters needs to be assessed. Though EDCs affinity to human estrogen receptor is lower than the human estrogen 17β-estradiol, some micropollutants such as BPA, NP, and TCS gave positive responses (Torres-Duarte et al. 2012). The affinity decreased after the laccase treatment and the same profile was observed in vivo experiments with zebrafish (Torres-Duarte et al. 2012). The use of animal-based methods has been largely developed through the years, forming the basis of the Endocrine Disruptor Screening Program of EPA. A two-tiered testing program (Tier 1 and 2) has been planned, including in vitro (nonanimal cell lines) and in vivo (animal) approaches (Bishop and Willett 2014). The goal is to assess the toxicity of more than 10,000 chemicals that give negligible effects on human endocrine systems. However, the use of the developed animal-based assays for routine monitoring of environmental samples poses ethical problems that could not underestimated (Purchase 1999).

Easy-to-perform and reliable in vitro tests became attractive solutions. Since their potential to describe estrogenicity is different from in vivo methodologies, their prediction potential and limits should be carefully evaluated (Jarošová et al. 2014b). Among the others, the yeast estrogen screen assay (YES) is a well-known test able to detect the estrogenic activity inherent in the water samples, suitable for monitoring the toxicity profile before and after laccase-based treatments (Mizuno et al. 2009; Arboleda et al. 2013; Chou et al. 2015). The human breast cancer cell line (MCF-7 BUS) proliferation test (*E-screen* test) and the MELN assay have been successfully used to monitor estrogenicity through biological treatments (Lee et al. 2005; Spina et al. 2015).

Many reports included a single test to assess the water quality. Luminescent bacteria *V. fischeri* (UNI EN ISO 11348-3) was used for the conversion of NP by 4 fungi (Soares et al. 2006), BPA and TCS by laccases from *T. versicolor* (Kim and Nicell 2006c), sulfonamide antibiotics by laccases from *Perenniporia* sp. (Weng et al. 2013), naproxen (Marco-Urrea et al. 2010a) and several pharmaceuticals and recognized EDCs by laccases from *T. versicolor* (Cruz-Moratò et al. 2014). Inhibition of *R. subcapitata* growth (UNI EN ISO 8692: 2005 method) was followed in the presence of tetracycline antibiotics (Suda et al. 2012). The YES assay was used to evaluate the estrogenic activity of parabens (Mizuno et al. 2009) and estrogens in municipal wastewaters (Auriol et al. 2008). Methods based of human breast cancer cell line (i.e., MELN assay and *E-screen* test) have been performed to evaluate the actual risk associated to bisphenol treated by two lignin-degrading basidiomycetes (Lee et al. 2005) or a complex mixture of micropollutants (Spina et al. 2015).

However, the sensitivity of each test species depends upon the chemical composition of the sample, and a toxic effect could be detected or not by different model organisms. A series of ecotoxicological tests is recommended, using organisms preferentially belonging to different trophic levels (Soupilas et al. 2008; Tigini et al. 2011). For instance, estrogenic activity of NP and octylphenol treated with laccases was assessed by two bioassays based on the proliferation index of cells (human mesothelioma cell line MPP89) responsive to EDCs and the YES test (Catapane et al. 2013). In this case a common response came out: laccases mediated both chemicals conversion and estrogenic activity was lost.

An extensive toxicological evaluation would ensure good reliability of water quality. The set up of a various toxicological toolbox aimed to describe the ecotoxicity and the estrogenic activity of a certain water sample is a recent but attractive vision (Leusch et al. 2014; Aydin et al. 2015; Wildhaber et al. 2015). Enlarging the groups of tested microorganisms and the targeted end-points would overcome problems associated with the perturbation activity of some micropollutants on certain model organisms that ultimately making some test not applicable. For example, toxicity of TCS and its products could not be determined by the YES assay because the results could be misinterpreted: the substrate interferes with the proper development of the yeast due to its antimicrobial activity, so that it is not possible to detect proper response of the transfected human estrogen receptor (Cabana et al. 2007a); the toxicity profile before and after the enzymatic treatment was instead elucidated by acute toxicity bioassay based on the crustaceous *Daphnia pulex* (Arboleda et al. 2013).

Spina and collaborators (2015) used three ecotoxicological assays (the plant *Lepidium sativum*, the algae *R. subcapitata* and the luminescent bacteria *V. fischeri*) and two tests for estrogenic activity evaluation (*E-screen* test and MELN assay) to evaluate the quality of fungal laccase treated municipal effluents. The out-put data was highly varied: *V. fischeri* was not sensitive, *L. sativum* was not disturbed by the wastewater itself but the laccase treatment provoked a significant biostimulation of the plant, and *R. subcapitata* pointed to the detoxification of the treated effluent. The two estrogenic activity tests showed a common behavior: the moderate estrogenic activity (EEQ 16–18 ng/L) was completely nullified by laccases.

This multifaceted scenario would require clear guidelines that are accepted by the institutions in charge for water control. However, due to the multiple model organisms and end-points, data obtained from bioassays could be difficult to compare. The interpretation of the results could be improved by the introduction of synthetic indices that summarize data coming from a battery of tests. Some of them have shown their potential to describe wastewaters toxicity (Tigini et al. 2011) but municipal effluents have never taken into consideration.

6.5 Conclusion

In conclusion, fungal biocatalysts (fungal biomass and/or enzymes) demonstrated to be powerful tools for wastewaters treatments. However, the actual frontiers of scientific and technological research is to evolve bio-based technology from lab scale to industrial level in WTPs, leading to the set up of competitive and feasible fungal-based treatments.

Considering the actual state-of-art technologies, a number major challenges remain. In order to achieve successful industrial upscaling of a fungal-base wastewater treatment, the following are recommended: (i) use low concentration of micropollutants as close as possible to environmental relevant problematic; (ii) avoid sterility control to assess the fungal capability to compete with autochthonous microflora; (iii) verify the strength of biocatalysts in real variable operative conditions, i.e., municipal wastewaters collected in WTPs; (iv) select the most appropriate carriers able to enhance biocatalysts stability and efficiency at an industrial scale; (v) develop optimal reactor configurations for long operating times; (vi) integrate the optimized technology with those already existing and working in WTPs.

References

Abejon R, De Cazes M, Belleville MP, Sanchez-Marcano J (2015) Large-scale enzymatic membrane reactors for tetracycline degradation in WWTP effluents. Water Res 73:118–131

Amman EM, Gasser CA, Hommes G, Corvini PF-X (2014) Immobilization of defined laccase combinations for enhanced oxidation of phenolic contaminants. Appl Microbiol Biotechnol 98:1397–1406

Anastasi A, Spina F, Prigione V, Tigini V, Giansanti P, Varese GC (2010) Scale-up of a bioprocess for textile wastewater treatment using *Bjerkandera adusta*. Bioresour Technol 101:3067–3075

Anastasi A, Spina F, Romagnolo A, Tigini V, Prigione V, Varese GC (2012) Integrated fungal biomass and activated sludge treatment for textile wastewaters bioremediation. Bioresour Technol 123:106–111

Arboleda C, Cabana H, De Pril E, Peter Jones J, Jiménez GA, Mejía AI et al (2013) Elimination of bisphenol A and triclosan using the enzymatic system of autochthonous colombian forest fungi. ISRN Biotechnol 1–12

Auriol M, Filali-Meknassi Y, Tyagi RD, Adams CD (2007) Laccase-catalyzed conversion of natural and synthetic hormones from a municipal wastewater. Water Res 41:3281–3288

Auriol M, Filali-Meknassi Y, Adams CD, Tyagi RD, Noguerol T-N, Pina B (2008) Removal of estrogenic activity of natural and synthetic hormones from a municipal wastewater: efficiency of horseradish peroxidase and laccase from *Trametes versicolor*. Chemosphere 70:445–452

Aydin ME, Aydin S, Tongur S, Kara G, Kolb M, Bahadir M (2015) Application of simple and low-cost toxicity tests for ecotoxicological assessment of industrial wastewaters. Environ Technol 36(22):2825–2834

Baccar R, Blanquez P, Bouzid J, Feki J, Attiya H, Sarra M (2011) Decolourization of a tannery dye: from fungal screening to bioreactor application. Biochem Eng J 56:184–189

Benotti MJ, Trenholm RA, Vanderford BJ, Holaday JC, Standford BD, Snyder SA (2009) Pharmaceuticals and endocrine disrupting compounds in U.S. drinking water. Environ Sci Technol 43:597–603

Bicchi C, Schilirò T, Pignata C, Fea E, Cordero C, Canale F et al (2009) Analysis of environmental endocrine disrupting chemicals using the *E-screen* method and stir bar sorptive extraction in wastewater treatment plant effluents. Sci Total Environ 407:1842–1851

Bishop PL, Willett CE (2014) The use and acceptance of other scientifically relevant information (OSRI) in the U.S. Environmental Protection Agency (EPA) Endocrine Disruptor Screening Program. Birth Defects Res (Part B) 101:3–22

Blanquez P, Guieysse B (2008) Continuous biodegradation of 17β-estradiol and 17α-ethynylestradiol by *Trametes versicolor*. J Hazard Mater 150:459–462

Brown TR (2010) Androgen receptor binding and transactivation assays identify environmental chemicals as endo disruptors. In: Eldridge JC, Stevens JT (eds) Endocrine toxicology. Informa Healthcare, New York, pp 149–170

Bumpus JA, Aust SD (1987) Biodegradation of environmental pollutants by the white rot fungus *Phanerochaete chrysosporium*: involvement of the lignin degrading system. BioEssays 6:166–170

Buchanan ID, Han YS (1999) Assessment of the potential of arthromyces ramosus peroxidase to remove phenol from industrial wastewaters. Environ Technol 21(5):545–552

Buchanan ID, Nicell JA (1998) Kinetics of peroxidase interactions in the presence of a protective additive. J Chem Technol Biotechnol 72:23–32

Cabana H, Jiwan J-LH, Rozenberg R, Elisashvili V, Penninckx M, Agathos SN, Jones JP (2007a) Elimination of endocrine disrupting chemicals nonylphenol and bisphenol A and personal care product ingredient triclosan using enzyme preparation from the white rot fungus *Coriolopsis polyzona*. Chemosphere 67:770–778

Cabana H, Jones JP, Agathos SN (2007b) Elimination of endocrine disrupting chemicals using white rot fungi and their lignin modifying enzymes: a review. Eng Life Sci 7:429–456

Cabana H, Jones JP, Agathos SN (2007c) Preparation and characterization of cross-linked laccase aggregates and their application to the elimination of endocrine disrupting chemicals. J Biotechnol 132:23–31

Cabana H, Peter Jones J, Agathos SN (2008) Utilization of cross-linked laccase aggregates in a perfusion basket reactor for the continuous elimination of endocrine-disrupting chemicals. Biotechnol Bioeng 102(6):1582–1592

Cajthaml T, Kresinová Z, Svobodová K, Möder M (2009) Biodegradation of endocrine-disrupting compounds and suppression of estrogenic activity by ligninolytic fungi. Chemosphere 75:745–750

Canas AI, Camarero S (2010) Laccases and their natural mediators: biotechnological tools for sustainable eco-friendly processes. Biotechnol Adv 28:694–705

Castellana G, Loffredo E (2014) Simultaneous removal of endocrine disruptors from a wastewater using white rot fungi and various adsorbents. Water Air Soil Pollut 225:1872–1885

Catapane M, Nicolucci C, Menale C, Mita L, Rossi S, Mita DG et al (2013) Enzymatic removal of estrogenic activity of nonylphenol and octylphenol aqueous solutions by immobilized laccase from *Trametes versicolor*. J Hazard Mater 248–249:337–346

Cheong S, Yeo S, Song H-G, Choi HT (2006) Determination of laccase gene expression during degradation of 2,4,6-trinitrotoluene and its catabolic intermediates in *Trametes versicolor*. Microbiol Res 161:316–320

Chou P-H, Lin Y-L, Liu T-C, Chen K-Y (2015) Exploring potential contributors to endocrine disrupting activities in Taiwan's surface waters using yeast assays and chemical analysis. Chemosphere 138:814–820

Cruz-Moratò C, Lucas D, Llorca M, Rodriguez-Mozaz S, Gorga M, Petrovic M et al (2014) Hospital wastewater treatment by fungal bioreactor: removal efficiency for pharmaceuticals and endocrine disruptor compounds. Sci Total Environ 493:365–376

Desai SS, Nityanand C (2011) Microbial laccases and their application: a review. Asian J Biotechnol 98:98–124

Diano N, Grano V, Fraconte L, Caputo P, Ricupito A, Attanasio A et al (2007) Non-isothermal bioreactors in enzymatic remediation of waters polluted by endocrine disruptors: BPA as a model of pollutant. Appl Catal B 69:252–261

Domaradzka D, Guzik U, Wojcieszynska D (2015) Biodegradation and biotransformation of polycyclic non-steroidal anti-inflammatory drugs. Rev Environ Sci Biotechnol 14:229–239

Faraco V, Piscitelli A, Sannia G, Giardina P (2007) Identification of a new member of the dye-decolorizing peroxidase family from *Pleurotus ostreatus*. World J Microbiol Biotechnol 23:889–893

Fatta-Kassinos D, Vasquez MI, Kümmerer K (2011a) Transformation products of pharmaceuticals in surface waters and wastewater formed during photolysis and advanced oxidation processes-Degradation, elucidation of byproducts and assessment of their biological potency. Chemosphere 85:693–709

Fatta-Kassinos D, Kalavrouziotis IK, Koukoulakis PH, Vasquez MI (2011b) The risks associated with wastewater reuse and xenobiotics in the agroecological environment. Sci Total Environ 409:3555–3563

Ferrando-Climent L, Cruz-Morató C, Marco-Urrea E, Vicent T, Sarrà M, Rodriguez-Mozaz S et al (2015) Non conventional biological treatment based on *Trametes versicolor* for the elimination of recalcitrant anticancer drugs in hospital wastewater. Chemosphere 136:9–19

Flint S, Markle T, Thompson S, Wallace E (2012) Bisphenol A exposure, effects, and policy: a wildlife perspective. J Environ Manage 104:19–34

Frye CA, Bo E, Calamandrei G, Calza L, Dessì-Fulgheri F, Fernandez M et al (2011) Endocrine disrupters: a review of some sources, effects, and mechanisms of actions on behaviour and neuroendocrine systems. J Neuroendocrinol 24:144–159

Fuhrman VF, Tal A, Arnon S (2015) Why endocrine disrupting chemicals (EDCs) challenge traditional risk assessment and how to respond. J Hazard Mater 286:589–611

Gao D, Du L, Yang J, Wu W, Liang H (2010) A critical review of the application of white rot fungus to environmental pollution control. Crit Rev Biotechnol 30:70–77

Garcia HA, Hoffman CM, Kinney KA, Lawler DF (2011) Laccase-catalyzed oxidation of oxybenzone in municipal wastewater primary effluent. Water Res 45:1921–1932

Garcia-Morales R, Rodríguez-Delgado M, Gomez-Mariscal K, Orona-Navar C, Hernandez-Luna C, Torres E et al (2015) Biotransformation of endocrine-disrupting compounds in groundwater: bisphenol A, nonylphenol, ethynylestradiol and triclosan by a laccase cocktail from *Pycnoporus sanguineus* CS43. Water Air Soil Pollut 226:251–265

Gasser CA, Hommes G, Schaffer A, Corvini PF-X (2012) Multi-catalysis reactions: new prospects and challenges of biotechnology to valorize lignin. Appl Microbiol Biotechnol 95:1115–1134

Gasser CA, Amman EM, Shahgaldian P, Corvini PF-X (2014a) Laccases to take on the challenge of emerging organic contaminants in wastewater. Appl Microbiol Biotechnol 98:9931–9952

Gasser CA, Yu L, Svojitka J, Wintgens T, Amman EM, Shahgaldian P et al (2014b) Advanced enzymatic elimination of phenolic contaminants in wastewater: a nano approach at field scale. Appl Microbiol Biotechnol 98:3305–3316

Glenn JK, Gold MH (1983) Decolorization of several polymeric dyes by the lignin-degrading basidiomycete *Phanerochaete chrysosporium*. Appl Environ Microbiol 45:1741–1747

Haroune L, Saibi S, Bellenger J-P, Cabana H (2014) Evaluation of the efficiency of *Trametes hirsuta* for the removal of multiple pharmaceutical compounds under low concentrations relevant to the environment. Bioresour Technol 171:199–202

Hata T, Shintate H, Kawai S, Okamura H, Nishida T (2010) Elimination of carbamazepine by repeated treatment with laccase in the presence of 1-hydroxybenzotriazole. J Hazard Mater 181:1175–1178

Heinfling A, Martınez MJ, Martınez AT, Bergbauer M, Szewzyk U (1998) Purification and characterization of peroxidases from the dye-decolorizing fungus *Bjerkandera adusta*. FEMS Microbiol Lett 165:43–50

Hommes G, Gasser CA, Amman EM, Corvini PF-X (2013) Determination of oxidoreductase activity using a high-throughput microplate respiratory measurement. Anal Chem 85:283–291

Hou J, Dong G, Luu B, Sengpiel RG, Ye Y, Wessling M et al (2014) Hybrid membrane with TiO_2 based bio-catalytic nanoparticle suspension system for the degradation of bisphenol-A. Bioresour Technol 169:475–483

Illés E, Szabò E, Takàcs E, Wojnárovits L, Dombi A, Gajda-Schrantz K (2014) Ketoprofen removal by O_3 and O_3/UV processes: kinetics, transformation products and ecotoxicity. Sci Total Environ 472:178–184

Inoue Y, Hata T, Kawai S, Okamura H, Nishida T (2010) Elimination and detoxification of triclosan by manganese peroxidase from white rot fungus. J Hazard Mater 180:764–767

Jarošová B, Erseková A, Hilscherová K, Loos R, Gawlik BM, Giesy JP et al (2014a) Europe-wide survey of estrogenicity in wastewater treatment plant effluents: the need for the effect-based monitoring. Environ Sci Pollut Res 21:10970–10982

Jarošová B, Bláha L, Giesy JP, Hilscherová K (2014b) What level of estrogenic activity determined by in vitro assays in municipal waste waters can be considered as safe? Environ Int 64:98–109

Jelic A, Cruz-Moratò C, Marco-Urrea E, Sarrà M, Perez A, Vicent T et al (2012) Degradation of carbamazepine by *Trametes versicolor* in an air pulsed fluidized bed bioreactor and identification of intermediates. Water Res 46:955–964

Kallqvist T, Milacic R, Smital T, Thomas KV, Vranes S, Tollefsen K-E (2008) Chronic toxicity of the Sava River (SE Europe) sediments and river water to the algae *Pseudokirchneriella subcapitata*. Water Res 42:2146–2156

Kaushik P, Malik A (2009) Fungal dye decolourisation: recent advances and future potential. Environ Int 35:127–141

Katsigiannis A, Noutsopoulos C, Mantziaras J, Gioldasi M (2015) Removal of emerging pollutants through granular activated carbon. Chem Eng J 280:49–57

Khlifi-Slama R, Mechichi T, Sayadi S, Dhouib A (2012) Effect of natural mediators on the stability of *Trametes trogii* laccase during the decolourization of textile wastewaters. J Microbiol 50:226–234

Kim Y, Yeo S, Song H, Choi HT (2008) Enhanced expression of laccase during the degradation of endocrine disrupting chemicals in *Trametes versicolor*. J Microbiol 46:402–407

Kim Y-J, Nicell JA (2006a) Laccase-catalysed oxidation of aqueous triclosan. J Chem Technol Biotechnol 81:1344–1352

Kim Y-J, Nicell JA (2006b) Impact of reaction conditions on the laccase-catalyzed conversion of bisphenol A. Bioresour Technol 97:1431–1442

Kim YJ, Nicell JA (2006c) Laccase-catalyzed oxidation of bisphenol A with the aid of additives. Process Biochem 41:1029–1037

Kum H, Kim MK, Choi HT (2009) Degradation of endocrine disrupting chemicals by genetic transformants in *Irpex lacteus* with an inducible laccase gene of *Phlebia tremellosa*. Biodegradation 20:673–678

Kunamneni A, Camarero S, Garcia-Burgos C, Plou FJ, Ballesteros A, Alcalde M (2008) Engineering and applications of fungal laccases for organic synthesis. Microb Cell Fact 7:1–17

Lee S, Koo B, Choi J, An B, Jeung E, Choi I (2005) Degradation of bisphenol A by white rot fungi, *Stereum hirsutum* and *Heterobasidium insulare*, and reduction of its estrogenic activity. Biol Pharm Bull 28:201–207

Lee Y, Escher BI, Von Gunten U (2008) Efficient removal of estrogenic activity during oxidative treatment of waters containing steroid estrogens. Environ Sci Technol 42(17):6333–6339

Leusch FDL, Khan SJ, Laingam S, Prochazka E, Froscio S, Trinh T et al (2014) Assessment of the application of bioanalytical tools as surrogate measure of chemical contaminants in recycled water. Water Res 49:300–315

Levin L, Meligrani E, Ramos AM (2010) Effect of nitrogen sources and vitamins on ligninolytic enzyme production by some white-rot fungi. Dye decolorization by selected culture filtrates. Bioresour Technol 101:4554–4563

Li Z, Xiang X, Li M, Ma Y, Wang J, Liu X (2015) Occurrence and risk assessment of pharmaceuticals and personal care products and endocrine disrupting chemicals in reclaimed water and receiving groundwater in China. Ecotoxicol Environ Saf 119:74–80

Libardi Junior N, Gern RMM, Furlan SA, Schlosser D (2012) Laccase production by the aquatic ascomycete *Phoma* sp. UHH 5-1-03 and the white rot basidiomycete *Pleurotus ostreatus* DSM 1833 during submerged cultivation on banana peels and enzyme applicability for the removal of endocrine-disrupting chemicals. Appl Biochem Biotechnol 167:1144–1156

Liers C, Bobeth C, Pecyna M, Ullrich R, Hofrichter M (2010) DyP-like peroxidases of the jelly fungus *Auricularia auricula-judae* oxidize nonphenolic lignin model compounds and high-redox potential dyes. App Microbiol Biotechnol 85:1869–1879

Liu Z, Lu G, Yin H, Dang Z (2015) Estimated human excretion rates of natural estrogens calculated from their concentrations in raw municipal wastewater and its application. Environ Sci Pollut Res 22:9554–9562

Lloret L, Eibes G, Lú-Chau TA, Moreira MT, Feijoo G, Lema JM (2010) Laccase-catalyzed degradation of anti-inflammatories and estrogens. Biochem Eng J 51:124–131

Lloret L, Eibes G, Feijoo G, Moreira MT, Lema JM (2012a) Degradation of estrogens by laccase from *Myceliophthora thermophila* in fed-batch and enzymatic membrane reactors. J Hazard Mater 213–214:175–183

Lloret L, Hollmann F, Eibes G, Feijoo G, Moreira MT, Lema JM (2012b) Immobilisation of laccase on Eupergit supports and its application for the removal of endocrine disrupting chemicals in a packed-bed reactor. Biodegradation 23:373–386

Lloret L, Eibes G, Feijoo G, Moreira MT, Lema JM (2013) Application of response surface methodology to study the removal of estrogens in a laccase-mediated continuous membrane reactor. Biocatal Biotransform 31(4):197–207

Lundstrom E, Adolfsson-Erici M, Alsberg T, Bjorlenius B, Eklund B, Laven M et al (2010) Characterization of additional sewage treatment technologies: ecotoxicological effects and levels of selected pharmaceuticals, hormones and endocrine disruptors. Ecotox Environ Safe 73:1612–1619

Luo Y, Guo W, Ngo HH, Nghiem LD, Hai FI, Zahng J et al (2014) A review on the occurrence of micropollutants in the aquatic environment and their fate and removal during wastewater treatment. Sci Total Environ 473–474:619–641

Macellaro G, Pezzella C, Cicatiello P, Sannia G, Piscitelli A (2014) Fungal laccases degradation of endocrine disrupting compounds. Biomed Res Int 2014:1–8

Majeau J, Brar SK, Tyagi RD (2010) Laccases for removal of recalcitrant and emerging pollutants. Bioresour Technol 101:2331–2350

Marco-Urrea E, Perez-Trujillo M, Vicent T, Caminal G (2009) Ability of white-rot fungi to remove selected pharmaceuticals and identification of degradation products of ibuprofen by *Trametes versicolor*. Chemosphere 74:765–772

Marco-Urrea E, Perez-Trujillo M, Blanquez P, Vicent T, Caminal G (2010a) Biodegradation of the analgesic naproxen by *Trametes versicolor* and identification of the intermediates using HPLC-DAD-MS and NMR. Bioresour Technol 101:2159–2166

Marco-Urrea E, Perez-Trujillo M, Cruz-Morato C, Caminal G, Vicent T (2010b) Degradation of the drug sodium diclofenac by *Trametes versicolor* pellets an identification of some intermediates by NMR. J Hazard Mater 176:836–842

Margot J, Maillard J, Rossi L, Barry DA, Holliger C (2013) Influence of treatment conditions on the oxidation of micropollutants by *Trametes versicolor* laccase. New Biotechnol 30(6):803–813

Martınez MJ, Martınez AT (1996) Characterization of MnP isoenzymes of *Pleurotus eryngii* exhibiting Mn-independent activities on 2,6-dimethoxyphenol and veratryl alcohol. In: Messner K, Srebotnik E (eds) Biotechnology in the pulp and paper industry: recent advances in applied and fundamental research. Facultas-Universitatsverlag, Vienna, pp 417–420

Martinez AT (2002) Molecular biology and structure-function of lignin-degrading heme peroxidases. Enzyme Microb Technol 30:425–444

Maruyama T, Komatsu C, Michizoe J, Ichinose H, Goto M (2006) Laccase-mediated oxidative degradation of the herbicide dymron. Biotechnol Prog 22:426–430

Melcer H, Klecka G (2011) Treatment of wastewaters containing bisphenol A: state of science review. Water Environ Res 83(7):650–666

Mendonca E, Picado A, Paixao SM, Silva L, Cunha MA, Leitao S et al (2009) Ecotoxicity tests in the environmental analysis of wastewater treatment plants: case study in Portugal. J Hazard Mater 163:665–670

Mersha MD, Patel BM, Patel D, Richardson BN, Dhillon HS (2015) Effects of BPA and BPS exposure limited to early embryogenesis persist to impair non-associative learning in adults. Behav Brain Funct 11:27

Meyer H, Munch T (2005) Swiss Industrial Biocatalysis Consortium. BioWorld Europe 1:14–16

Mizuno H, Hirai H, Kawai S, Nishida T (2009) Removal of estrogenic activity of iso-butylparaben and n-butylparaben by laccase in the presence of 1-hydroxybenzotriazole. Biodegradation 20:533–539

Mompelat S, Le Bot B, Thomas O (2009) Occurrence and fate of pharmaceutical products and by-products, from resource to drinking water. Environ Int 35:803–814

Morozova OV, Shumakovich GP, Shleev SV, Yaropolov YI (2007) Laccase-mediator systems and their applications: a review. Appl Biochem Microbiol 43:523–535

Mougin C, Kollmann A, Jolivalt C (2002) Enhanced production of laccase in the fungus *Trametes versicolor* by the addition of xenobiotics. Biotechnol Lett 24:139–142

Murugesan K, Chang Y, Kim Y, Jeon J, Kim E, Chang Y (2010) Enhanced transformation of triclosan by laccase in the presence of redox mediators. Water Res 44:298–308

Nguyen LN, Hai FI, Price WE, Leusch FDL, Roddick F, McAdam EJ et al (2014) Continuous biotransformation of bisphenol A and diclofenac by laccase in an enzymatic membrane reactor. Int Biodeterior Biodegradation 95:25–32

Nicolucci C, Rossi S, Menale C, Godjevargova T, Ivanov Y, Bianco M et al (2011) Biodegradation of bisphenols with immobilized laccase or tyrosinase on polyacrylonitrile beads. Biodegradation 22:673–683

Novotny C, Svobodová K, Benada O, Kofronová O, Heissenberger A, Fuchs W (2011) Potential of combined fungal and bacterial treatment for color removal in textile wastewater. Bioresour Technol 102:879–888

Oulton RL, Kohn T, Cwiertny DM (2010) Pharmaceuticals and personal care products in effluent matrices: a survey of transformation and removal during wastewater treatment and implications for wastewater management. J Environ Monit 12:1956–1978

Piscitelli A, Giardina P, Lettera V, Pezzella C, Sannia G, Farmaco V (2011) Induction and transcriptional regulation of laccases in fungi. Curr Genomics 12:104–112

Prieto A, Vallejo A, Zuloaga O, Paschke A, Sellergen B, Schillinger E et al (2011) Selective determination of estrogenic compounds in water by microextraction by packed sorbents and a molecularly imprinted polymer coupled with large volume injection-in-port-derivatization gas chromatography-mass spectrometry. Anal Chim Acta 703:41–51

Purchase IFH (1999) Ethical review of regulatory toxicology guidelines involving experiments of animals: the example of endocrine disrupters. Toxicol Sci 52:141–147

Ra J-S, Lee S-H, Lee J, Kim HY, Lim BJ, Kim SH et al (2011) Occurrence of estrogenic chemicals in South Korean surface waters and municipal wastewaters. J Environ Monit 13:101–109

Reungoat J, Macova M, Escher BI, Carswell S, Mueller JF, Keller J (2010) Removal of micropollutants and reduction of biological activity in a full scale reclamation plant using ozonation and activated carbon filtration. Water Res 44:625–637

Riva S (2006) Laccases: blue enzymes for green chemistry. Trends Microbiol 24:219–226

Rodriguez-Rodriguez CE, Jelic A, Pereira MA, Sousa DZ, Petrovic M, Alves MM et al (2012) Bioaugmentation of sewage sludge with *Trametes versicolor* in solid-phase biopiles produces degradation of pharmaceuticals and affects microbial communities. Environ Sci Technol 46:12012–12020

Schilirò T, Pignata C, Rovere R, Fea E, Gilli G (2009) The endocrine disrupting activity of surface waters and of wastewater treatment plant effluents in relation to chlorination. Chemosphere 75:335–340

Sei K, Takeda T, Soda SO, Fujita M, Ike M (2008) Removal characteristicsn of endocrine-disrupting chemicals by laccase from white-rot fungi. J Environ Sci Health A 43:53–60

Snyder SA, Westerhoff P, Yoon Y, Sedlak DL (2003) Pharmaceuticals, personal care products, and endocrine disruptors in water: implications for the water industry. Environ Eng Sci 20:449–469

Soares A, Jonasson K, Terrazas E, Guieysse B, Mattiasson B (2006) The ability of white-rot fungi to degrade the endocrine-disrupting compound nonylphenol. Appl Microbiol Biotechnol 66:719–725

Songulashvili G, Jimenez-Tobon GA, Jaspers C, Penninckx MJ (2012) Immobilized laccase of *Cerrena unicolor* for elimination of endocrine disruptor micropollutants. Fungal Biol 116:883–889

Soupilas A, Papadimitriou CA, Samaras P, Gudulas K, Petridis D (2008) Monitoring of industrial effluent ecotoxicity in the greater Thessaloniki area. Desalination 224:261–270

Spina F, Cordero C, Schilirò T, Sgorbini B, Pignata C, Gilli G et al (2015) Removal of micropollutants by fungal laccases in model solution and municipal wastewater: evaluation of estrogenic activity and ecotoxicity. J Clean Prod 100:185–194

Strong PJ, Claus H (2010) Laccase: a review of its past and its future in bioremediation. Crit Rev Environ Sci Technol 41:373–434

Suda T, Hata T, Kaway S, Okamura H, Nishida T (2012) Treatment of tetracycline antibiotics by laccase in the presence of 1-hydroxybenzotriazole. Bioresour Technol 103:498–501

Sung E, Turan N, Ho PWL, Ho SL, Jarratt PDB, Waring RH et al (2012) Detection of endocrine disruptors—from simple assays to whole genome scanning. Int J Androl 35:407–414

Sun Y, Huang H, Sun Y, Wang C, Shi X-L, Hu H-Y et al (2013) Ecological risk of estrogenic endocrine disrupting chemicals in sewage plant effluent and reclaimed water. Environ Pollut 180:339–344

Taboada-Puig R, Lu-Chau TA, Eibes G, Feijoo G, Moreira MT, Lema JM (2015) Continuous removal of endocrine disruptors by versatile peroxidase using a two-stage system. Biotechnol Prog 31(4):908–916

Tien M, Kirk TK (1983) Lignin-degrading enzyme from the Hymenomycete *Phanerochaete chrysosporium* Burds. Science 221:661–663

Tigini V, Giansanti P, Mangiavillano A, Pannocchia A, Varese GC (2011) Evaluation of toxicity, genotoxicity and environmental risk of simulated textile and tannery wastewaters with a battery of biotests. Ecotoxicol Environ Saf 74:866–873

Torres E, Bustos-Jaimes I, Le Borgne S (2003) Potential use of oxidative enzymes for the detoxification of organic pollutant. Appl Catal B 46:1–15

Torres-Duarte C, Roman R, Tinoco R, Vazquez-Duhalt R (2009) Halogenated pesticide transformation by a laccase–mediator system. Chemosphere 77:687–692

Torres-Duarte C, Viana MT, Vazquez-Duhalt R (2012) Laccase-mediated transformations of endocrine disrupting chemicals abolish binding affinities to estrogen receptors and their estrogenic activity in zebrafish. Appl Biochem Biotechnol 168(4):864–876

Touraud E, Roig B, Sumpter JB, Coetsier C (2011) Drug residues and endocrine disruptors in drinking water: risk for humans? Int J Hyg Environ Health 214:437–441

Tran NH, Hu J, Urase T (2013) Removal of the insect repellent N, N-diethyl-m-toluamide (DEET) by laccase-mediated systems. Bioresour Technol 147:667–671

Vulliet E, Cren-Olive C, Grenier-Loustalot M (2011) Occurrence of pharmaceuticals and hormones in drinking water treated from surface waters. Environ Chem Lett 9:103–114

Wang J, Majima N, Hirai H, Kawagishi H (2012) Effective Removal of endocrine-disrupting compounds by lignin peroxidase from the white-rot fungus *Phanerochaete sordida* YK-624. Curr Microbiol 64:300–303

Weng S-S, Liu S-M, Lai H-T (2013) Application parameters of laccase–mediator systems for treatmentof sulfonamide antibiotics. Bioresour Technol 141:152–159

Wildhaber YS, Mestankova H, Scharer M, Schirmer K, Salhi E, von Gunten U (2015) Novel test procedure to evaluate the treatability of wastewater with ozone. Water Res 75:324–335

Wuttke W, Jarry H, Seidlovα-Wuttke D (2010) Definition, classification and mechanism of action of endocrine disrupting chemicals. Hormones 9:9–15

Yang M, Wang K, Shen Y, Wu M (2011) Evaluation of estrogenic activity in surface waters and municipal wastewaters in Shanghai, China. Bull Environ Contam Toxicol 87:215–219
Yu C-P, Roh H, Chu K-H (2007) 17β-estradiol-degrading bacteria isolated from activated sludge. Environ Sci Technol 41:486–492
Zhang Y, Geißen S-U (2012) Elimination of carbamazepine in a non-sterile fungal bioreactor. Bioresour Technol 112:221–227

Chapter 7
Application of Microalgae and Fungal-Microalgal Associations for Wastewater Treatment

Ewere Job Egede, Huw Jones, Bryan Cook, Diane Purchase and Aidyn Mouradov

7.1 Introduction

The United Nations Human Development Report identified three rapidly emerging global challenges: pollution (shortages of clean water), energy and food supplies (WWAP 2015). There is an increasing requirement to treat wastewater with greater efficacy using low cost, low energy demanding, near zero maintenance systems while generating a beneficial relationship with the user community. Pollutants in the ecosystem have existed since human life evolved and first learnt how to light fires (Borsos et al. 2003). Although natural events such as geothermal activity have contributed significantly to pollutant problems (e.g. air quality impacts) it has only been since the humanised industrialised era, with the extensive development and use of chemical substances, that the wider impact of human activity has been realised (Khan and Ghouri 2011). Numerous studies have shown that different chemicals including pesticides, organophosphorus compounds, phosphates, hydrocarbons [e.g. aliphatic, aromatic, polycyclic aromatic hydrocarbons, BTEX (benzene, toluene, ethylbenzene, and xylenes)], chlorinated hydrocarbons such as polychlorinated biphenyls, nitrates, ammonia, trichloroethylene-like perchloroethylene, nitroaromatic compounds, solvents and heavy metals have been involved in human-related destructive effects on the soil and aquatic environments (Ritter et al. 1995; Carpenter et al. 1998; Chekroun et al. 2014). Wastewater emanates from industries, agricultural establishments or households as by-products of their manufacturing or via everyday activities. They could occur in the form of treated or raw sewage discharged from

E.J. Egede · H. Jones · D. Purchase (✉)
Department of Natural Sciences, Faculty of Science and Technology,
Middlesex University, The Burroughs, London NW4 4BT, United Kingdom
e-mail: d.purchase@mdx.ac.uk

B. Cook · A. Mouradov
School of Sciences, Royal Melbourne Institute of Technology, University,
Bundoora, VIC 3083, Australia

D. Purchase (ed.), *Fungal Applications in Sustainable Environmental Biotechnology*, Fungal Biology, DOI 10.1007/978-3-319-42852-9_7

municipalities and villages; they might be direct run-offs from agricultural fields, leachates from solid waste disposal sites or industrial and manufacturing plants discharges. Pollutants in wastewaters are naturally occurring substances that are in concentrations or amounts exceeding their normal levels or are xenobiotics (synthetic compounds) released into aquatic systems which can harm biota and affect water quality (Abdel-Raouf et al. 2012). Such contamination or pollution of the environment present health, ecosystem and economic problems globally, and are a real cause for concern to society.

Biological treatment of wastewater has been a topic of interest to many researchers. The role of fungi and their potential to treat wastewater are well established; exemplified by the chapters presented in this book examining the applications of fungi to treat different types of wastewater. Similarly, the use of algae to bioremediate wastewater has long been established. Research has also been carried out to study the potential of bacterial and algal partnership to treat wastewater (e.g., Gutzeit et al. 2005; Medina and Neis 2007; Park and Craggs 2010; Su et al. 2011; Ryu et al. 2014; Unnithan et al. 2014). Much less well known is the role and efficacy of fungal–algal consortia in wastewater treatment. The fungal–algal consortium is a non-mutualistic relationship, different from the association that forms lichens which comprises a phycobiont and a mycobiont in a symbiotic relationship. This chapter aims to examine the fungal–algal consortium and explore its potential as an environmentally friendly wastewater treatment technology. For detailed information on fungal treatment of wastewater, readers are referred to other chapters in *Part II*: *Providing sustainable and environmentally friendly solutions to treat wastewater* of this book; this work will focus on the current knowledge on algal bioremediation and its mechanisms, followed by an overview of fungal–algal treatment and examination of its potential in generating value-added products as well as challenges facing this new technology.

7.2 Microalgal-Assisted Bioremediation of Wastewater

Bioremediation involves the use of microorganisms and plants to break down and remove hazardous chemicals and contamination from the environment. Bioremediation is considered efficient and environmentally safe and is an inexpensive means of decontaminating polluted systems. Aquatic macrophytes and microphytes, which include aquatic plants, microalgae, bacteria and fungi, are often found to be the only life forms able to both survive and grow actively in a polluted environment (Priyadarshani et al. 2011). These microorganisms can degrade many pollutants, using them as an energy source in almost all instances and, as such, the commercial application of this capacity has been the focus of much investigation and development.

Microalgae are unicellular microscopic phytoplanktonic species that include cyanobacteria, green flagellates, diatoms and dinoflagellates of sizes 1–500 μm, based on the International Code of Nomenclature for algae, fungi and plants

(McNeill et al. 2012). DNA sequence data suggest that algal phylogeny includes ten major phyla, the prokaryotic *Cyanophyta* and *Prochlorophyta*, and the eukaryotic *Glaucophyta, Euglenophyta, Cryptophyta, Haptophyta, Dinophyta, Heterocontophyta* (including diatoms, brown algae), *Rhodophyta* (red algae), and *Chlorophyta* (green algae).

The efficiency of microalgae in wastewater treatment and production of biofuel and value-added chemicals has significant advantages when compared to that obtained from plants, yeast and other microorganisms because of: (i) their ability to double biomass every 8–12 h, (ii) ability to produce a large number of value-added chemicals including petrochemicals and (iii) phycoremediation capacity. John (2000) coined the termed phycoremediation to describe remediation by microalgae. From the foregoing, it should be stressed that phycoremediation has several applications: (i) nutrient removal from municipal wastewater and from organics-rich effluent, (ii) removal of xenobiotics and nutrients using algal biosorbents, (iii) transformation, conversion and degradation of xenobiotic compounds such as hydrocarbons, dyes and phenolic compounds, (iv) treatment of acidic wastewaters and metals, (v) CO_2 sequestration achieved by the microalgal biomass simultaneously generated whilst decontaminating effluents/wastewaters, and (vi) detection of xenobiotic/toxic compounds using algae-based biosensors (Souza et al. 2012). Further advantages include their lack of competition with agricultural crops for arable lands, and that their growth is less affected by seasonal changes in climate, making them a feedstock of choice for the production of a number of value-added products (Lam and Lee 2012; Rajkumar et al. 2014; Wu et al. 2014; Simas-Rodrigues et al. 2015).

The use of microalgae in the treatment of municipal and animal wastewaters for pollutant remediation has been extensively investigated on reducing nutrients such as nitrogen (N), phosphorus (P), heavy metals, biochemical and chemical oxygen demand, pathogen reduction and water disinfection (Rawat et al. 2011). Since green microalgae have been shown to thrive in high N- and P-rich conditions, they have been found to be very efficient at removing these nutrients from municipal wastewater (Aslan and Kapdan 2006; Ruiz-Marin et al. 2010). Microalgal species, from the *Chlorophytic, Chlorella* and *Scenedesmus* genera have been well studied in wastewater treatment experiments. They offer both tertiary and quaternary treatment solutions due to their ability to use N and P for their growth (Aslan and Kapdan 2006; Ruiz-Marin et al. 2010; Ahluwalia and Goyal 2007; Ma et al. 2014) and for their capacity to efficiently remove heavy metals as well as some toxic compounds (Deblonde et al. 2011; Fu and Wang 2011; Worku and Sahu 2014). Other microalgae, *Phormidium, Botryococcus, Chlamydomonas* and *Spirulina* have also been shown to be capable of removing/reducing nutrient and heavy metals from wastewater (Christenson and Sims 2011; Li et al. 2011). A study by Chinnasamy et al. (2010) described a mixture of 15 native microalgal isolates all showing >96 % nutrient removal of treated carpet mill effluents. The use of contained and controlled systems such as photobioreactors and algal ability to remove nutrients and metals from secondary wastewater has received much interest where different reactor types have demonstrated their effectiveness in improving wastewater effluent (Muñoz and Guieysse 2006; McGinn et al. 2012; Riaño et al. 2012).

7.2.1 Bioremediation of Heavy Metals

The history of harnessing microalgae for the treatment of wastewaters spans about 60 years. One of the first works was that of Oswald and Gotaas (1957). Several countries including Australia, Thailand, Taiwan and Mexico have intensively tested the technique and cultivated significant interest (Abdel-Raouf et al. 2012). Much of the research has centred on using suspended microalgae cultured in sewage-bearing, shallow, artificial ponds (Hoffman et al. 1998). More recently, however, greater emphasis has been placed on utilising non-suspended microalgae, either as axenic, immobilised cultures on polymeric matrices, or as attached communities growing on surfaces of rotating biological contactors (biodiscs) or in artificial ponds (biofilms).

Microalgae are effective biosorbents due to the advantages they possess over other microorganisms such as fungi and bacteria. Since they are autotrophs they require minimal nutrients to produce large biomass. In addition, they are generally less likely to produce toxins compared to bacteria and fungi. Living and dead microalgae are used as biosorbents for the uptake of metal ions. Non-living microalgal biomass or dead cells have some advantages over living cells. They are unaffected by toxic metal ions, require less space and maintenance and are relatively cheaper. Additionally, dead cells are easily regenerated for reuse. Living cells, on the other hand, are prone to metal toxicity and can thus die. Moreover, living microalgal biomass usually requires nutrients for their wastewater treatment processes, which can increase the Biochemical Oxygen Demand (BOD) and Chemical Oxygen Demand (COD) of the effluent.

According to Das et al. (2008), the process of biosorption involves three components. First, a solid-state biosorbent/sorbent or biological material (microalgae in this case); second, the solvent, usually water (the liquid phase); third, the sorbate or metal ions dissolved in the water. Because the sorbent or microalgae have a higher affinity for the metal ions or sorbate species, the ions are attracted to the sorbent and are removed by various mechanisms including ionic interaction, ion exchange, the formation of complexes, physical interaction, and chemisorption. The greater their affinity for the sorbate (metal) species, the less sorbate will be in solution. Thus, the extent of microalgal affinity for a sorbate determines the concentration and distribution of sorbate between solid and liquid phases. As the biosorption process continues, a point is reached where the microalgae-bound sorbate species are in equilibrium with its remnant dissolved counterpart.

The metal biosorption process is a complex one and is influenced by biosorption process parameters such as pH, the initial concentration of biosorbent and metal ion, equilibrium time, the status of biomass (living or dead) and types of biomaterials (Akhtar et al. 2004a, b; Das et al. 2008). Of these, pH seems to be the most important. It affects the activity of the biomass' functional groups, influences solution chemistry of metal ions and competition of ions. Specific uptake appears to be influenced by biomass concentration via interference with binding sites or concentration of metals through shortage of metals in solution.

In living microalgal cells, metal biosorption is a two-stage process. The first stage (termed passive biosorption) is the adsorption of the metal ions to the microalgal cell's surface when the cell surface's functional groups interact with the metal. These chemical functional groups are metal-binding groups including amines, hydroxyls, carboxylates, imidazoles, phosphates, sulphates and sulfhydryls. They are present in their cell walls and, therefore, provide a range of active sites able to bind metal ions. This first step is rapid, independent of metabolism (energy independent) and occurs via one or combination of coordination, ion exchange, physical adsorption, complexation, or inorganic micro-precipitation mechanisms. Dynamic equilibrium is also observed in passive biosorption with reversible adsorption–desorption.

The second step, active biosorption, sees the metal ions penetrating the cell membrane, gaining entry into the microalgal cell. This stage is metabolism-dependent and occurs via metal transport and deposition. Unlike passive biosorption, active biosorption or active uptake of metals from wastewaters is slower. Also, non-living cells are limited to passive biosorption while living biomass undergo both passive and active uptake.

A number of processes can be used to enhance biosorption including pretreatment by physical methods (such as lyophilisation, drying, heating, freezing/thawing) and chemical methods (such as cross-linking with organic solvents, acid or alkali treatment and washing of biomass with detergents). Both methods have been shown to improve the cell's metal ion-binding ability (Vieira and Volesky 2010; Barange et al. 2014). Table 7.1 presents examples of microalgae in treating wastewater containing heavy metals and the mechanisms involved.

7.2.2 Treatment of Phosphorus (P-) and Nitrogen (N-) Containing Substances

Microalgae have been applied to the treatment of human sewage, livestock wastes, industrial waste, and piggery effluents. Although microalgae have been utilised to remove nutrients from wastewaters for over 60 years, the use of combined and varied microalgal species for tertiary treatment of wastewater commenced only about thirty years ago (de-Bashan et al. 2002). Conventional sewage treatment technologies include: (i) preliminary treatment of sewage to remove large solid materials, (ii) primary treatment of sewage to remove the settleable solids by gravity, (iii) secondary treatment of sewage to reduce the BOD exerted by reducing organic matter, (iv) tertiary treatment of sewage to remove major inorganic ions,

Table 7.1 Summary of selected investigations into metal removal

Organism	Substrate used	Metal (concentration range, in mg L^{-1} where unit not indicated)	Mechanism	Tested Variables	Comments	Reference
Chlorella sorokiniana	Immobilised on loofa sponge/free cells	Lead (II) (10–300)	Biosorption	pH, kinetics, other metal ions	Immobilised cells outperformed free cells in large-scale treatments	Akhtar et al. (2004a)
Chlorella sorokiniana	Loofa sponge/free cells	Nickel (II) (2.5–200)	Biosorption	Biomass concentration, pH, metal concentration	During multiple adsorption/desorption cycles metal uptake was maintained	Akhtar et al. (2004b)
Scenedesmus quadricauda	Immobilised on Ca-alginate	Copper (II), zinc (II), nickel (II) (25 and 600)	Hybrid biosorption	pH, speciation, ionic strength, temperature	Physical/ion exchange	Bayramoglu and Arica (2009)
Chlorella regularis (and *Streptomyces*)	Immobilised on polyacrylamide	Uranium (5, 10, and 20 ppm) in sea and freshwater	Adsorption	Temperature	Recovery study used Na_2CO_3 to desorb uranium	Nakajima et al. (1982)
Chlorella miniata and *Chlorella vulgaris*	Immobilised on alginate beads and free cells	Nickel (30) chromium (20–400) copper (30) zinc (30) in industrial wastewaters	Repeated adsorption–desorption cycles	Live and dead cells	Evaluation of suspended versus immobilised cells	Tam et al. (2009) Lau et al. (1998) Chong et al. (2000) Tam et al. (2001) Han et al. (2006)

(continued)

Table 7.1 (continued)

Organism	Substrate used	Metal (concentration range, in mg L^{-1} where unit not indicated)	Mechanism	Tested Variables	Comments	Reference
Chlamydomonas reinhardtii	Immobilised on alginate beads	Mercury (II) cadmium (II) lead (II) (25–500)	Biosorption/chemisorption	pH, metal concentration, temperature sorption/desorption cycles	Mechanism dominated by second order reaction—95 % recovery of multiple sorption cycles	Bayramoglu et al. (2006)
Spirogyra sp.	*Spirogyra* "biomass"	Chromium (VI) (1–25) in wastewater	Biosorption/physical sorption	pH, cell wall functional groups	*Spirogyra* deemed efficient at removing and recovering Cr (VI) from	Gupta et al. (2001)
Chlorella vulgaris	Free cells in culture medium	Copper (II) nickel (II) (0–2 mM)	Biosorption	Single and combined metal solutions, pH, cell pre-treatment	Competition for sorption of metals was observed. Pre-treatment enhanced sorption	Mehta and Gaur (2005)
Chlorella salina	Immobilised Ca-alginate, free cells	Cobalt, zinc, manganese (0.5–100 μM each)	Biosorption	Temperature, light metabolic inhibitor, pH	Differential metal sorption dependent on light, metal concentration and pH observed	Garnham et al. (1992)
Chlorella miniata	Free cells in liquid (Bristol) culture medium	Chromium (III) chromium (VI) (20, 50, 60, 100 and 200) in contaminated wastewater	Biosorption/bioreduction	pH, biomass, metal concentrations	Functional group present at binding sites, biosorption followed by bioreduction were main processes	Han et al. (2007)

(continued)

Table 7.1 (continued)

Organism	Substrate used	Metal (concentration range, in mg L^{-1} where unit not indicated)	Mechanism	Tested Variables	Comments	Reference
Chlorella pyrenoidosa, Chlamydomonas reinhardtii, Stichococcus bacillaris	Heat killed, dead algal biomass	Multimetal solution (0.1, 0.5, 1, 2, and 4)	Biosorption	pH, speciation, matrix components	Adsorption sites characteristics, pH metal form, concentrations impacted bisorption	Mahan et al. (1989)
Nanochloropsis gaditana	Immobilised on Ca-alginate	Copper (0.64, 1.59 mg L^{-1} or 10, 25 μM respectively) zinc (0.65, 1.63 mg L^{-1} or 10, 25 μM)	Biosorption	Live/dead cells	Higher Zn biosorption explained by metal pumping detoxification	Moreno-Garrido et al. (2002)

mainly ammonium, nitrate and phosphate, and (v) advanced treatment intended for the removal of heavy metals, organic compounds and soluble minerals. The two latter stages, which can include chemical precipitation, ozonation, reverse osmosis or carbon adsorption, are about 4 to 16 times more expensive than that of primary treatment, respectively (de la Noüe et al. 1992; Godos et al. 2009; de-Bashan and Bashan 2010; Hering et al. 2010; Rawat et al. 2011). Apart from high operational cost, the main disadvantages of conventional treatments include variable efficiency of nutrient removal, the possibility of secondary pollution and loss of valuable potential nutrients (N, P) (Godos et al. 2009). Microalgal culture offers a cost-effective approach for tertiary and advanced treatments due to the ability of microalgae to use inorganic nitrogen and phosphorus for their growth (Aslan and Kapdan 2006; Ahluwalia and Goyal 2007; Ruiz-Marin et al. 2010; Ma et al. 2014). Many genera including *Euglena*, *Chlamydomonas*, *Chlorella*, *Dunaliella*, *Scenedesmus* and *Nitzschia* are tolerant to these wastes and can be adapted to contaminated environments to degrade the contaminants (Abdel-Raouf et al. 2012). There are ongoing efforts to use hyperconcentrated microalgal cultures which are very efficient for nutrient removal within brief treatment periods of, for example, up to one hour. This could solve substantial land space requirements of microalgal wastewater treatment systems.

Numerous laboratory and pilot plant studies have been developed, including the use of oxidation ponds which photosynthetically produce oxygen. For example, microalgae in oxidation ponds dramatically reduce the problem of high BOD when they oxidise ammonia (a principal source of nitrogen) to build their cellular materials. Oxidation of ammonium to nitrites and nitrates leads to a reduction of the concentration of oxygen. Microalgae, therefore, provide an effective option for tertiary and quaternary wastewater treatment to solve eutrophication of surface waters.

Biological removal of N and P from wastewaters using microalgae is termed nutrient stripping and is applied to the principal forms of N and P, namely ammonium (NH_4^+), nitrate (NO_3^-), nitrite (NO_2^-) and orthophosphates (PO_4^{3-}). The ions are the main contaminants of wastewaters originating from agro-industries that specialise in dairy production, pig and cattle farming and often occur in large quantities (de-Bashan et al. 2002). The nitrogen in wastewaters originates from interconversion of extra-derived compounds while at least half of their phosphorus content comes from synthetic detergents.

Removal of nitrogen in contaminated wastewaters occurs via several processes. First, the microorganisms (including microalgae) present in the wastewaters take up and assimilate nitrogenous compounds to increase their biomass (de-Bashan and Bashan 2010). Second, ammonium compounds are oxidised to nitric oxide, nitrite and nitrate by nitrifying microorganisms and then, via denitrifying microorganisms, transform to nitrogen which escapes to the atmosphere. Alternatively, ammonium can be oxidised directly into nitrogen via the anaerobic ammonium oxidation (anammox) process (Hu et al. 2013).

For phosphorus removal, microalgae often convert phosphorus ions to a solid fraction. The fraction can consist of a solid-state salt precipitate, a plant biomass in constructed wetlands or microbial mass in activated sludge. The recycling process of removal of phosphorus is often unsustainable since other toxic wastes are removed too. Generally speaking, removal of phosphorus by microalgae is quite slow, less efficient with less than 30 % removed compared to N-removal on a commercial scale (de-Bashan and Bashan 2010). Table 7.2 presents examples of microalgae in nutrient removal and the mechanisms involved.

7.2.3 Treatment of Hydrocarbons

Petroleum-related activities include oil spillage, processing and storage activities of oil, and the clean-outs and wash-outs from such activities (Chavan and Mukherji 2008). In the environment, they can exert toxic, mutagenic and carcinogenic effects on aquatic organisms and man. Exposure may be direct or via bioaccumulation in living organisms and via biomagnification along food chains. Aromatics, particularly the polycyclic aromatic hydrocarbons (PAHs), are recalcitrant and often persist in sediments of aquatic systems that receive hydrocarbon-bearing wastewaters. They are therefore considered as priority pollutants. Treatment of these wastewaters is a necessity to mitigate the environmental and health implications they cause.

It has been claimed that the use of algae for removing hydrocarbons from wastewaters is cost-effective, does not require oxygenation, are easy to culture, and have the capacity to transform PAHs and other hydrocarbons via various metabolic pathways to non-toxic metabolites (Semple et al. 1999). The removal of toxicants including hydrocarbons is usually via their autotrophic growth and has been applied in several wastewater treatment processes (Tam et al. 2009). Microalgae utilise the carbon in hydrocarbons and other toxicants, and nitrate and phosphates in nutrients to build up their cells and accumulate biomass (autotrophic growth). Microalgae with greater biomass are therefore better able to absorb, degrade or accumulate the xenobiotics from wastewaters.

Microalgae employ varied mechanisms to remove hydrocarbons from wastewater. Algae can directly biotransform hydrocarbons, or could enhance hydrocarbon-degrading capacities of coexisting microorganisms. Microalgae can also use some dioxygenase enzyme systems to form dihydrodiol intermediates which are subsequently converted to glucoside conjugates. Upon biotransforming hydrocarbons to form intermediate products or metabolites, the microalgae could subsequently accumulate the intermediate metabolic products in their biomass (Semple et al. 1999). In some instances, they do not participate in the metabolism of the hydrocarbon but rather accumulate it in their cell wall (Lei et al. 2002). Others can biodegrade and mineralise hydrocarbons with the help of photooxidation processes. However, complete PAH mineralisation is normally rare (Takáčová et al. 2014). Table 7.3 presents examples of microalgae in hydrocarbon remediation using microalgae and the mechanisms involved.

Table 7.2 Summary of selected investigations into nutrient removal

Organism	Substrate	Treated matrix/nutrients (Concentration range, in mg L^{-1} where unit not indicated)	Variables	Comments	Reference
Scenedesmus sp.	Gel-immobilised cells	Ammonium–nitrogen (22), orthophosphate (1.8)	Sequential starvation treatment cycles, real secondary wastewaters and artificial effluent	Immobilization, successive starvation cycles, use of real wastewaters increased removal	Zhang et al. (2008)
Chlorella vulgaris, Chlorella kessleri, Scenedesmus quadricauda	Polyurethane foam	Ammonium–nitrogen and orthophosphate in raw sewage (RS) and pretreated cattle manure (PCM) (i) RS: ammonium nitrogen (31 ± 5), orthophosphate (6 ± 2) (ii) PCM: ammonium nitrogen (237 ± 43), orthophosphate (34 ± 3)	Flow pattern, natural light, substrate	Differential nutrient removal was observed for various species and strains	Travieso et al. (1996)
Chlorella vulgaris	Immobilised in beads or spherical gels	Ammonium (30) and phosphate (6)	pH, aeration, Ca^{2+} ions	Optimisation of bead concentration	Tam and Wong (2000)
Dunaliella salina	Ca-alginate beads	Phosphate (50-142 μg L^{-1}, ammonium (260-450 μg L^{-1}), nitrate (180 to 210 μg L^{-1})	pH, alginate biomass	Uptake of PO_4^{3-}, NH_4^+ and NO_3^{2-} was higher in the immobilised cell than free-living cells; pH significantly affects the nutrient uptake	Thakur and Kumar (1999)

(continued)

Table 7.2 (continued)

Organism	Substrate	Treated matrix/nutrients (Concentration range, in mg L^{-1} where unit not indicated)	Variables	Comments	Reference
Chlorella vulgaris	Immobilised	Primary treated effluents containing ammonium nitrogen (13–36.2) and phosphates (3.6–9.8)	Variable cell densities	Adsorption/utilisation by cells deemed important. Nitrification and ammonia volatilisation non-significant	Tam et al. (1994)
Chlorella vulgaris	Immobilised carrageenan and alginate matrices	Primary settled domestic wastewater—ammonium nitrogen (37.5–47.5) and phosphate (3.1–4.3)	Cell metabolism, Ca^{2+} ions	Cell metabolism, interaction with polysaccharide matrices enhanced nutrient removal	Lau et al. (1997)
Scenedesmus spp.	Chitosan–immobilised	Nitrate (44) and phosphate (6)	pH, carbon availability	Carbon limitation impacted nitrate removal, pH and immobilization matrix impacted phosphate removal	Fierro et al. (2008)
Chlorella vulgaris *Scenedesmusrubescens*	Self-adhesion to ultrathin, wet microporous layer	In real municipal wastewaters: nitrate (3.7–6.2), ammonium (0.8), phosphate (0.2–0.7) For synthetic wastewater, 3, 20 and 3 respectively	Algal culture conditions, substrate type, wastewater type	The substrate had greater phosphate removal than carrageenan or alginate. Wastewater type, algal species, culture conditions differentiated removal	Shi et al. (2007)
Chlorella sorokiniana *Chlorella vulgaris* *Rhodoturula spharoides*, *Spirulina platenis*	None	"High strength" Synthetic wastewater of ammonia (50-400), nitrate (700), phosphate (100)	Light/dark, aerobic/anaerobic, mono/mixed cultures	Mixed cultures were shown to be much more effective than monocultures in nutrient removal	Ogbonna et al. (2000)
Scenedesmus bicellularis	Immobilised algal system	Municipal wastewater: ammonium (57.4), orthophosphate (5.5)	Intermittent CO_2 enrichment	Nutrient uptake accelerated in response to CO_2 inputs	Kaya et al. (1996)

Table 7.3 Summary of selected investigations into hydrocarbon/organic compound removal

Organism	Substrate	Treated matrix/nutrients (Concentration range, in mg L^{-1} where unit not indicated)	Mechanisms/Variables	Comments	Reference
Scenedesmus abundans, Spirulina sp.	Mixed batch systems	Lubricating and diesel oil (0.5–1.0 % v/v)	Biosorption	Similar results to oil spill clean-up sorbents	Mishra and Mukherji (2012)
Selenastrum capricornutum	Alginate-immobilised and free cells	PAHs (phenanthrene 1, fluorathene 0.25 and pyrene 0.1) from agro-industrial wastewaters	Biosorption/biodegradation	Immobilised algae performed better than free cells	Tam et al. (2010)
Seven microbial species	Live and dead cells	Pyrene (0.1)	Physico-chemical binding and bioaccumulation	Passive physico-chemical biosorption was chief initial mechanism followed by bioaccumulation. *Selenastrum capricornutum* was most efficient remover, *C. vulgaris* the least	Lei et al. (2002)
Chlorella vulgaris, Ankistrodesmus sp., *Scenedesmus* sp.	Axenic cultures	Naphthalene (0.1 mM) in aqueous media and wastewaters	Biotransformation	Biotransformation to 1-naphthol and dihydroxylated derivatives observed—full mechanism not elucidated	Todd et al. (2002)

(continued)

Table 7.3 (continued)

Organism	Substrate	Treated matrix/nutrients (Concentration range, in mg L^{-1} where unit not indicated)	Mechanisms/Variables	Comments	Reference
Prototheca zopfii	Foam cube-immobilised algae in batch cultures and free cells	n-alkanes (3 % v/v)	Biodegradation	Synergy observed between cells and immobiliser leading to more effective biodegradation than free cells	Yamaguchi et al. (1999)
Pseudokirchneriella subcapitata,	Live/dead cells—varying cell densities	Phenanthrene (1), fluoranthene (0.25), pyrene (0.1)	Physical adsorption/biodegradation	Passive adsorption (dead and live cells) followed by accumulation/degradation (live cells only)	Chan et al. (2006)
Prototheca zopfii	Immobilised thermotolerant strain and non-thermotolerant strains	Hydrocarbon mixture (1 % v/v) of PAHs (6 mg/0.2 mL) and n-alkanes (150 mg/0.2 mL)	Biodegradation under varying temperatures	Thermotolerant strains performed better	Ueno et al. (2006)
Prototheca zopfii	Batch experiments	Mixed hydrocarbon substrate comprised of n-alkanes (250 ml) and PAHs (750 mg) in 1 % v/v	Biodegradation/bioaccumulation	Preferential biodegradation of n-alkanes followed by selective retrieval of PAHs	Ueno et al. (2008)

7.2.4 *Treatment of Dye Compounds*

Dyes—water-soluble and insoluble types—are released in wastewaters due to their improper uptake and the extent of fixation on the substrate in textile manufacture processes (Singh and Arora 2011). A large volume of water and chemicals is consumed by the textile industry during the wet processing stages. The dye-containing wastewater from textile industries is highly variable in constituents and may have high colour intensity. It is, therefore, difficult to treat satisfactorily. It has been claimed that textile and dye manufacturing industries generate more toxic wastewaters than any other industry (Singh and Arora 2011). The primary reason is that they have high BOD and COD demands; their characteristic colour is a great problem as they reduce light penetration and photosynthetic processes in the aquatic environment, and dye components pose direct toxic effects on biota (Aksu and Tezer 2005).

A major class of colourant is azo dye. Azo dye compounds are the largest and most diverse classes of dyes. Due to their aromatic and heterocyclic moieties and their synthetic origin, they are environmental hazards. They are toxic to aquatic biota and mutagenic to man and are persistent in the ecosystem. Their disintegration poses greater health hazards (Pinheiro et al. 2004).

Microorganisms including algae often decolourise dyes via biosorption/adsorption mechanisms. Microalgae also utilise reductive and bioconversion mechanisms. They degrade dyes to obtain nitrogen; at the same time, they sequester carbon dioxide (Elkassas and Mohamed 2014). The pH of the medium, initial dye concentration, dosage of adsorbent and temperature have been found to be the key factors that affect the adsorption capacity of microalgal adsorbents (Singh and Arora 2011).

Azo reductases of algae have been implicated in degradation of azo dyes to form aromatic amines usually involving breakage of the azo linkage (Forgacs et al. 2004). Since most dyes are stable and resistant to microalgal attack, the isolation of new algal strains or adaptation of existing strains to decompose dyes could increase their degradation capacity.

Table 7.4 presents examples of microalgae in dye removal in wastewater and the mechanisms involved.

7.3 Challenges of Using Microalgae for Treatment of Wastewaters

Having considered the viability, potential and varied application of microalgae to sustainably remediate wastewaters, our discussion would be incomplete without delineating the drawbacks of utilising them for environmental clean-up purposes. Microalgae alone usually cannot completely mineralise or remove certain xenobiotics even under optimal conditions. A number of factors need to be considered to

Table 7.4 Summary of selected investigation into the removal of selected dye compounds

Organism	Substrate	Treated matrix/nutrients (Concentration range in mg L^{-1})	Mechanisms/Variables	Comments	Reference
Chlamydomonas reinhardtii	Immobilised by polysulfone nanofibrous web	Remazol black 5 (RB5) and Reactive Blue 221 (RB221) in effluents (10, 25 and 50)	Decolourisation, adsorption, degradation	Free cells and repeat cycles	Keskin et al. (2015)
Chlorella vulgaris, Chlamydomonas sajao, Nitzschia hantzschiana, Anabaena cylindrica	Live/dead cells	Aniline (0.02, 0.04, 0.05, 0.08 and 0.10 mmol L^{-1})	Bioaccumulation/biodegradation	Dead cells found to have greater photodegradative capacity than live cells	Wang et al. (2007)
Chlorella vulgaris	Live cells and COD	Tectilon yellow 2G (TY2G) (50, 200 and 400)	Bioconversion to aniline	High COD removal was recorded alongside initial low dye removal	Acuner and Dilek (2004)
Chlorella vulgaris	Dried microalgal biomass	Remazol Black (RB) Remazol Red (RR) Remazol Golden Yellow (RGY) (10–800)	Biosorption	pH and temperature were observed to substantially affect biosorption	Aksu and Tezer (2005)
Gonium sp.	Batch cultures	Reactive blue (RB220) in wastewater (26.2, 49.7, 74.3 and 109)	Biodegradation	Bioaccumulation, passive uptake, biosorption deemed not to be involved in removal	Boduroglu et al. (2014)
Cosmarium sp.	Batch cultures	Malachite green (MG; 0-15 ppm)—a triphenylmethane dye—from wastewaters	Biodegradation, with some bioconversion	Initial dye concentration, pH, temperature and initial algal concentrations all influenced the decolourisation process	Daneshvar et al. (2007)

(continued)

Table 7.4 (continued)

Organism	Substrate	Treated matrix/nutrients (Concentration range in mg L^{-1})	Mechanisms/Variables	Comments	Reference
Chlorella, Chlorococcum, and *Chlamydomonas* sp.	Mixed cultures	Pulping mill effluent, by removing its color (1245, 770, 420 and 172 PtCo) and adsorbable organic halides (AOX) (6, 3, 2.5 and 1.5 g m^{-3})	Adsorption and metabolism	Color removal was metabolism rather than adsorption; mixed cultures were effective	Dilek et al. (1999)
Chlorella vulgaris	Axenic batch culture	Textile waste effluent containing the azo dyes Metanil yellow, Fast Orange, Fast Red, Direct Blue, Acid Fast Red, Direct Fast Scarlet, Congo Red and Acid Fast N Blue (5, 8.5, 17.5, 26.5 and 30 % of real concentrations) from dyeing industry to treat the effluents	Biological assimilation and chemical precipitation	Efficient decolourisation, detoxified and degradation observed together with phosphate removal	Elkassas and Mohammed (2014)
Chlorella vulgaris, Volvox aureus and *Lyngbya lagerlerimi*	Mixed algal cultures	Methyl red, orange II, G-Red (FN-3G), basic fuchsin, basic cationic dyes (5, 10 and 20 ppm or mg L^{-1})	A partial mechanism involved in the process was degradation	Degradation/decolourisation of high ratios of azo dyes by the algae via varied mechanisms and partially by certain enzymes(azo reductase) induced and effective in the degradation process	El-Sheekh et al. (2009)

(continued)

Table 7.4 (continued)

Organism	Substrate	Treated matrix/nutrients (Concentration range in mg L^{-1})	Mechanisms/Variables	Comments	Reference
Scenedesmus quadricauda	Axenic cultures	Remazol Brilliant Blue (CI 612200, Reactive Blue 19, RBBR; 25–200)	Physical adsorption/biosorption reactions	Initial dye concentration and pH significantly influenced the biosorption process	Ergene et al. (2009)
Spirogyra sp.	Batch cultures	Simulated azo dyes direct and reactive blue (25–100)	Biosorption, biodegradation and biocoagulation	Biosorption and further removal via release of exopolymers during algal stress	Mohan et al. (2004)
Spirogyra sp.	Dried, crushed, acid-treated, particle-sized (2 mm) algal biomass	Direct Brown dye (15) in aqueous media	Biosorption/particle diffusion	Low pH and increased temperature significantly increased adsorption rate	Sivarajasekar et al. (2009)
Pithophora sp.	Batch experiments	Malachite green (MG; 20–100)	Multilayer sorption	Processes were pH dependent; biosorption process was controlled by surface diffusion	Kumar et al. (2006), Kumar et al. (2005)
Chlorella vulgaris Chlorella pyrenoidosa	Axenic cultures of microalgae	Azo dyes (10–20 ppm)	Degradation	Azoreductase enzyme of algae was responsible for the degradation	Jinqi and Houtian (1992)
Chlorella pyrenoidosa	Immobilised algae	Direct brown NM (20, 30 and 50)	Degradation	Introduction of excess CO_2 enhanced dye degradation	Guolan et al. (2000)
Scenedesmus bijugatus, Scenedesmus obliquus and diatom, *Nitzschia perminuta*	Axenic cultures of algae	Two dyes—Tartrazine and Ponceau (5, 10 and 20 ppm)	Degradation	Azoreductase enzyme identified as involved in biodegradation/reduction	Omar (2008)

overcome the challenges to improve the accessibility and acceptability of this technology.

First, the compatibility of microalgal biosorbent for the treatment of real industrial wastewaters should be considered. Currently, biosorption by microalgae cannot compete with conventional processes for large-scale environmental applications (Volesky and Naja 2005; Wang and Chen 2009). Many industrial effluents are comprised of a multicomponent mixture of heavy metals, hydrocarbons, dye compounds, phenolic compounds and nutrients, but there is a paucity of knowledge on multi-xenobiotic remediation.

Second, although there is information on technicalities of microalgal treatment options which compare favorably to conventional treatments as has been previously shown, such information/research is relatively scarce. This might hinder the scaling up of microalgal treatment for treating large volumes of wastewater. The scaling up of microalgal-based wastewater treatment methods have not been successful, unlike conventional treatment methods. For example, although microalgae require and use a high amount of nitrogen and phosphorus to build their morphology and physiology, and as a result are efficient nutrient strippers in wastewaters, they have not been demonstrated to be suitable for large-scale nutrient stripping in wastewater treatment (Rawat et al. 2011).

Third, microalgal treatment of wastewater is still a niche technology that requires specialised expertise. The technology itself is imprecise and dependent on environmental factors, quantities of the pollutant and retention times in treatment systems. The combination of these factors makes engineers increasingly reluctant to develop microalgae as a potentially sustainable wastewater treatment technology (de-Bashan and Bashan 2010).

Fourth, the presence of microalgal contaminants in effluents of treated wastewaters has a strong negative impact on animal and human health. Clearly, contaminated waters cannot be used for irrigation and for drinking. On the other hand, collected microalgal biomass represents a potentially valuable feedstock for production of a number of value-added chemicals. One of the important challenges that microalgal-based industries are facing is associated with the high costs of harvesting and dewatering of the microalgal cells.

7.3.1 Microalgae Harvesting Strategies

One of the pressing problems when using microalgae for wastewater treatment, production of biofuel and value-added chemicals is that current harvesting/dewatering methods are not economically feasible, either requiring the addition of chemicals or excessive energy demand. The harvesting of microalgae can contribute to 20–50 % of the total cost (Leite et al. 2013; Pragya et al. 2013). This cost is acceptable for high-value products, but for the mass production of low-value products, such as biofuels, the current methods for microalgae are not cost effective or energy efficient (Leite et al. 2013). There are a number of different techniques used to harvest microalgae, all of which have their flaws and advantages. The methods most widely used are: centrifugation, flocculation, filtration, flotation and gravity sedimentation (Pragya et al. 2013). However, some of these techniques cannot be scaled up economically (Christenson and Sims 2011; Pragya et al. 2013; Vandamme et al. 2013).

Filtration is a process in which microalgae are captured by a filter or membrane as the medium passes through, forming a thick mass of microalgae (Pragya et al. 2013). There are numerous different filtration methods, including different sized filters, the exertion of pressure on the solution, and use of filter aids, to allow for the easier harvest of smaller particles. Larger microalgae such as *Coelastrum proboscideum* and *Spirulina platensis*, are simpler to filter, but the filters fail to recover microalgae when they approach the smaller bacterial dimensions (Molina Grima et al. 2003). Filtration methods can give a good harvest recovery, but the filter needs to be replaced frequently due to fouling. The cost of filter replacement and operating the filtration assembly and pressure pumps all adds to the cost of production, and on a larger scale becomes uneconomical (Molina Grima et al. 2003; Pragya et al. 2013). Centrifugation can harvest about 90 % of the microalgae, but this is associated with high energy input costs, especially with a low-value product such as biofuel (Uduman et al. 2010). Floatation technology is based on attachment of algal cells to air bubbles or flocculants which carry them to the surface (Chinnasamy et al. 2010; de-Bashan and Bashan 2010; Li et al. 2011).

7.3.1.1 Flocculation

Flocculation is the process when microalgae form clumps, pellets or pellet-like compounds called flocs, sometimes referred to as aggregation, coagulation or bio-flocculation. Microalgae have a negative surface charge which is why they do not self-flocculate (Pragya et al. 2013). A number of methods can achieve microalgal flocculation: auto-flocculation, biological flocculation, combined flocculation, chemical flocculation, inorganic flocculation, organic flocculation, or by inducing an electrical impulse to alter the negative charge (Chen et al. 2011). Auto-flocculation describes the process of microalgal self-flocculating and can be achieved by changing the microalgae growth conditions. Changes in pH levels arising from movement in carbonate/bicarbonate balance caused by photosynthesis can sometimes bring about auto-flocculation (Chen et al. 2011). After flocculation, the algae are either removed by gravity sedimentation or filtration.

7.3.1.2 Chemical Flocculation

The chemical flocculation process involves adding a chemical to the culture causing the microalgae to form flocs. There are a number of different chemicals and compounds which can be used. The problems associated with chemical flocculation are that the use of chemicals can be costly, and these chemicals can also make the resultant biomass unusable for certain value-added products. Alum and ferric chloride, including other metal salts, are often used as flocculating agents in some industries, such as mining waste and wastewater. Unfortunately, as Vandamme et al. (2013) demonstrated, the metals remaining in the biomass interfere with the use of algae as an animal feed, making chemical flocculation economically unfeasible. Also, the amount of metal salts needed is considerable which results in significant amounts of sludge formation and is unsuitable for some strains of microalgae (Chen et al. 2011).

A study by Uduman et al. (2010) looked at the use of organic compounds such as polymers to flocculate microalgae. They investigated the effects of different polymers such as anionic, cationic and non-ionic polyelectrolytes and demonstrated that all the polyelectrolytes were effective, with cationic forms having the greatest flocculation ability. The use of other organic flocculants including chitosan can be effective as well but are species-dependent, and the results also depend on temperature and pH of the growth culture (Uduman et al. 2010). Biomolecules such as chitosan, starch and polyglutamic acid, although effective for freshwater microalgae, undergo coiling at high ionic strengths and consequently are less effective for marine microalgae (Bilanovic et al. 1988). Chemical flocculation using alum has been reported as negatively affecting the fermentative production of acetone, butanol and ethanol (Anthony et al. 2013).

Another form of flocculation is electro-flocculation. This process involves the use of electricity to dissolve sacrificial metals to supply the ions required for flocculation (Lee et al. 2013; Vandamme et al. 2013). A study by Lee et al. (2013) investigated the economics behind electro-flocculation where they found that previous studies only looked at the cost of replacing the electrode and not the total cost, including energy. They found that using electro-flocculation resulted in a cost of \$0.19 kg^{-1} of dry microalgae biomass. While these results look promising, it is not feasible for biofuel production since the costs of extraction and refinement were not described.

7.3.1.3 Bio-flocculation

Bio-flocculation involves the use of biological compounds or organisms for microalgal harvesting. In submerged cultures, filamentous microorganisms, including bacteria and fungi, tend to aggregate and grow as granules or pellets. Bio-flocculation shows potential to resolve the major challenges facing algal biotechnology. Apart from efficient harvesting of algae, co-flocculants can demonstrate their synergistic activity on total biomass, total oil production and treatment of wastewaters by recovering their primary nutrients, N and P, and microelements.

For some microalgae species, self-flocculation can occur under certain conditions. Some investigations have demonstrated that it is possible to mix such naturally bio-flocculating microalgae with other species to elicit flocculation (Leite et al. 2013; Vandamme et al. 2013). Other microorganisms that have been shown to instigate bio-flocculation in microalgae are bacteria and fungi (Salim et al. 2011; Leite et al. 2013). Some microalgae grown in wastewater were able to form large colonies (50–200 μm) (Park et al. 2011). An efficient bio-flocculant has been isolated from the auto-flocculating *Chlorella vulgaris* and *Scenedesmus* microalgae when they were grown in wastewater (Reed 1981; Manheim and Nelson 2013). The marine microalga *Nannochloropsis oceanica* sp. demonstrated a flocculation efficiency of up to 90 % with the bacteria *Solibacillus silvestris* and *Bacillus* sp. (Powell and Hill 2013; Ratha et al. 2013). Bacterial bio-flocculants have been shown to be effective in the flocculation of the microalgae *Chlorella* and *Pleurochrysis carterae* (Molina Grima et al. 2003). It was found that organic carbon content in the growth medium is required to enable bacterial growth to flocculate the microalgae (Lee et al. 2009). During bio-flocculation, bacteria have been found to make up to 30 % of the biomass (Sournia 1978) and to have a high proportion of the mixed microalgal biomass grown in wastewater (Park et al. 2011). Sedimentation has been found to be an effective means to separate algae when incorporated into biomass flocs and has been demonstrated in symbiotic algal-bacterial wastewater treatment (Medina and Neis 2007). The use of flue gas from a coal power plant has been shown to settle microalgal bacterial floc from the secondary treatment of sewage, removing 97.5 % of the biomass from the sewage effluent within 30 min and producing sediment of 2 % bacterial–microalgal dry biomass (Van den Hende et al. 2011).

7.3.2 *Fungal-Assisted Algal Bio-flocculation*

7.3.2.1 Fungal Pelletization

Fungal cell self-pelletization has been extensively studied for applications in wastewater treatment, bio-flocculation and the production of a wide spectrum of pharmaceuticals, enzymes and chemical molecules which can be converted to a number of petrochemicals (Hiruta et al. 1996; Liao et al. 2007a; Liu et al. 2008; Gultom and Hu 2013). Self-pelletized fungal cells are commonly used for wastewater treatment to remove pollutants (Papagianni 2004; Papanikolaou et al. 2004; Liao et al. 2007b). Self-pelletization is more widely observed for filamentous fungal strains, which are widely used to produce many bio-products and biofuels (Liao et al. 2007b; Xia et al. 2011). In submerged cultures for industrial applications, filamentous fungi such as *Mortierella isabellina, Mucor circinelloides* and *Cunninghamella echinulata* can grow into two different morphologies: filamentous and pelletized. The most commonly observed pellet morphologies include loose and dense spherical aggregates which originate from a single cell (Cox et al. 1998; Gibbs et al. 2000). The structural properties and the pelletization process are very dependent on cultivation conditions (Krull et al. 2010, 2013; Gultom and Hu 2013). In general, filamentous fungi can form pellets through coagulative and non-coagulative processes. Coagulating pellet formation through aggregation of fungal conidia occurring during the early stage of cultivation has been described for *Aspergillus* representatives as well as *Phanerochaete chrysosporium* (Metz and Kossen 1977; Znidarsic and Pavko 2001; Grimm et al. 2004; Krull et al. 2010). In the non-coagulating process, pellets are formed out of a single spore and have been reported for fungi belonging to *Rhizopus* sp. and *Mucor* sp. (Metz and Kossen 1977) and actinomycetes from the genus *Streptomyces* (Vechtlifshitz et al. 1990; Znidarsic and Pavko 2001).

A number of factors have been reported to affect pellet formation including pH, culture medium, agitation, inoculum concentration, and the addition of nuclei and polymer (Liu et al. 2008). These authors used pH adjustment to induce the formation of fungal cell pelletization, providing a simplified method by which to facilitate the cell harvest of oleaginous cells. Fungal cell pelletization is strain-specific, and not all filamentous fungal strains can form pellets during growth. A correlation was found between fungal cell capacity for self-pelletization and the accumulation of a family of low molecular weight amphipathic, hydrophobic proteins (hydrophobins) accumulated on the hyphal surface (Linder 2009). These hydrophobic proteins are potentially involved in hyphae adherence to solid substrates (Feofilova 2010).

7.3.2.2 Filamentous Fungal-Based Bio-flocculation

The natural symbiosis between filamentous fungi and algae in the form of lichens (Fig. 7.1) has existed since plants evolved from green algae more than 400 million years ago and currently lichens cover 6 % of the Earth's land surface (Taylor et al. 1995) . In this symbiotic relationship fungi consume the sugars and nutrients produced through the photosynthetic activity of algae; in return, the fungi offer protection to the algae by retaining water, offering a larger capture area for mineral nutrients and can also provide minerals obtained from the substrate (Zoller and Lutzoni 2003). Fungal-assisted bio-flocculation of microalgae has shown high efficiency when tested against a large number of microalgal and cyanobacterial strains representing photoautotrophic and heterotrophic, freshwater and marine, unicellular and multicellular, small (3 μm) and giant (300 μm), motile and non-motile strains (Wrede et al. 2014; Miranda et al. 2015; Muradov et al. 2015). Flocculation of some microalgal strains with *Aspergillus fumigatus* is shown in Fig. 7.2. In some cases, flocculation rate is very high. For example, the mycelium of *Aspergillus nomius* added to cultured freshwater *C. vulgaris* and marine *Nannochloropsis* sp was precipitated within 60 min after mixing, resulting in up to 97 % flocculation efficiency of these strains.

The detailed mechanisms of fungal–algal interactions remain unclear, but it is broadly accepted that the interaction between oppositely charged surfaces may enable microalgal attachment to the fungal cell wall (Gultom and Hu 2013; Wrede et al. 2014; Muradov et al. 2015). Microscopic analysis of algal–fungal pellets showed that algal cells are not only trapped within the scaffolds of fungal filaments but are clearly attached to them (Wrede et al. 2014; Muradov et al. 2015; Zhou

Fig. 7.1 Examples of lichen form and structure. **a** Lichens growing on rocks and tree barks. **b** Schematic representation of lichens illustrating fungal hyphae and microalgal components

et al., 2015). Microalgae possess a negative surface charge due to the presence of proton-active carboxylic, phosphoric, phosphodiester, hydroxyl and amine functional groups. The zeta potential of microalgae has been found to be within the range −10 to −35 mV (Henderson et al. 2008). Using coagulating agents, it was shown that reduction of the magnitude of the zeta potential to approximately ten mV and below is required for removal of four algae species *Asterionella formosa*, *Melosira* sp., *Melosira aeruginosa, and C. vulgaris* (Henderson et al. 2008). The electrostatic charge distribution across the surface of the cells of fungal cells depends on the age of the conidia/spores and pH. For *Beuaveria bassiana* zeta, potential for aerial conidia varied from +22 to −30 mV for pH 3 to 9 (Holder et al. 2007). For submerged conidia the net surface charge ranged from +10 to −13 mV and much less variation (+4 to −4 mV) was observed for spores. The charge difference between *Aspergillus flavus* (+46.1 mV) and microalgae cells (−23.7 mV) was suggested to be essential for their flocculating interaction (Rajab 2007). The stability of these microalgal suspensions depends on the forces that interact between the cells themselves and between the cells and water (Uduman et al. 2010).

It is unclear whether the charge difference alone is enough to keep motile microalgal strains such as *Chlamydomonas reinhardtii, Dunaliella tertiolecta* and *Tetraselmis chuii* attached to the fungal filaments (Wrede et al. 2014; Muradov et al. 2015). Moreover, observations that some microalgal cells not only attach to fungal filaments but also to each other suggest that they have lost their negative charges. This may be explained by the fact that secretion of extracellular enzymes

(a)
(b)
1
2
3
4
5
6
7
8
9
10
(c)

◀ **Fig. 7.2** Bio-flocculation of microalgal strains by *Aspergillus fumigatus*. **a** Bio-flocculation of *Raphidocelis subcapitata* (formerly known as *Pseudokirchneriella subcapitata*) by *A. fumigatus*. *A. fumigatus* pellets (*left*); *A. fumigatus* and *R. subcapitata* co-culture, t = 0 (*middle*) and t = 24 h (*right*). **b** Bio-flocculation of different microalgal representatives by *A. fumigatus*. Left: microalgal monocultures; Right: bio-flocculation by *A. fumigatus*. Microalgal cultures: 1, 6. *C. vulgaris*; 2, 7. *Isochrysis sp;* 3, 8. *N. oculata*; 4, 9. *Pavlova* sp; 5, 10. *Haematococcus* sp. **c** *A. fumigatus/C. vulgaris* pellets (*left*); *A. fumigatus* pellets (*right*)

digests microalgal cell walls enabling protoplasts to stick to each other (Wrede et al. 2014; Muradov et al. 2015). Another explanation for algal attachment to fungal cells comes from the ability of fungal cells to secrete a cocktail of concentrated exopolysaccharide molecules during interaction with other microorganisms (Seviour et al. 1992; Selbmann et al. 2003; Wrede et al. 2014). This suggests that as an alternative/additional scenario, algal cells can be entrapped by cocktails of 'sticky' exopolysaccharides secreted by *A. fumigatus*. Metabolomic analysis of media after co-cultivation of *A. fumigatus* with *Synechocystis* with microalgal strains could reveal more information about the biochemistry of fungal-assisted flocculation.

7.4 Associations of Microalgae with Other Microorganisms for Wastewater Treatment

Microalgae can be associated and immobilised with other microorganisms for wastewater treatment. Advantages of this association in wastewater treatment may be summarised as follows: (i) improved physical separation from wastewaters, (ii) consortia can be easily reutilised and manipulated, (iii) improved features of microalgae such as thermal stability, productivity, inhibition and kinetic constants, and (iv) valuable biomass can be preserved for extended usage.

7.4.1 Microalgal–Bacterial Associations

Positive advancements have been made to cope with the challenges associated with microalgal-mediated wastewater treatment described in Sect. 7.3, even though more developments are still needed with a view to large-scale applications (de-Bashan et al. 2004). Co-culturing microalgae with hardier, pollutant-devouring microorganisms such as bacteria have yielded moderate success. Table 7.5 presents some examples of successful co-cultivation of algal-bacterial microcosms in treating some pollutants. There are obvious benefits of such co-culture. Toxicity problems encountered by the algae can be alleviated by the co-immobilising bacterial detoxification mechanisms. Attached bacteria also increase cellular uptake rates for sugars and amino acids (Schweitzer et al. 2001), and they can release amino acids to be utilised by the planktonic community. The CO_2–O_2 cycle generates further enhancement of the mechanism of detoxifying toxic recalcitrant organics and xenobiotics from wastewaters (Borde et al. 2003). Microalgal growth could also increase by microalgae growth-promoting bacteria (MGPB) (de-Bashan et al. 2002). Significant improvement in the removal of xenobiotics from wastewaters has been observed during algae-bacteria co-culture/co-immobilization. For instance, increased and improved cellular metabolism, cell density and cell size of microalgae induced by MGPB evidenced in chlorophyll pigment, cell size, cell cytology, lipid content, microalgal population size and variety of fatty acid has been reported, which resulted in enhanced removal of ammonium and phosphate from wastewaters (de-Bashan et al. 2002, 2004; Table 7.5).

It has been claimed that unlike biological treatment processes based on bacteria only, the algal–bacterial partnership does not produce NH_4^+ because the photoautotrophic microalgae efficiently assimilate NH_4^+ as their most preferred nitrogen source (Perez-Garcia et al. 2011). In addition, as some microalgae can accumulate significantly higher lipids than typical oleaginous crops, the biomass resulting from wastewater treatment can also serve as a biodiesel feedstock (Chisti 2007). Thus, algal–bacterial-based technologies are attracting increasing attention for the treatment of various types of wastewater, including municipal sewage, sludge concentrates, leachates, agricultural anaerobic digested effluent, and industrial effluent (Unnithan et al. 2014).

7.4.2 Photosynthetic Biofilms

Microalgal biofilms, intact or attached to solid surfaces, are gaining increasing attention as an alternative strategy for the concentration of microalgal biomass. In general, these biofilms represent microecosystems with representatives of unicellular and multicellular photosynthetic organisms: green algae, diatoms, cyanobacteria along with some non-photosynthetic microorganisms. Most of these components have the ability to secrete a sticky extracellular polymeric substance (EPS) adhering to each other and to a surface. These three-dimensional structures can be used as a matrix for attachment of commercially valuable algal and diatom cells which can be used for the production of value-added products including biofuels. The ability of biofilms to grow in freshwaters and saline environments as well as efficiently up-taking main nutrients (N and P), trace- and heavy metals provides a low energy alternative to conventional methods for purification of municipal, animal and mining wastewaters. Multicellular biofilms can be specifically designed from photosynthetic components (microalgae, cyanobacteria, diatoms) isolated from different biofilms for production of specific value-added products, such as biofuels, health-promoting oils (Omega 3), carotenoids and/or for efficient treatment of wastewaters displaying synergistic or additive effects of their multiple components (Boelee et al. 2011; Sheng et al. 2012; Berner et al. 2015).

Table 7.5 List of contaminants effectively removed via co-immobilization and co-culturing compared to use of microalgae alone

Contaminants treated	Algal/bacterial strain used	Effectiveness (percentage of xenobiotics removed)	References
Phenol and nonylphenol	*Scenedesmus obliquus, Chlorella vulgaris/Raoultella terrigena, Pantoea Agglomerans*	>99 %	Maza-Márquez et al. (2014)
Hydrocarbons and total petroleum hydrocarbon (TPH)	*Phormidium, Oscillatoria, Chroococcus/Burkholderia cepacia*	98.6–99.4 %	Chavan and Mukherji (2008)
Ammonium and phosphorus ions	*Chlorella vulgaris/Azospirillum brasilense*	100 % for NH_4^+ versus 35, 83 % for PO_4^{3-} versus < 20 %	de-Bashan et al. (2002)
Ammonium, nitrate and phosphorus	*Chlorella vulgaris, C. sorokiniana/Azospirillum brasilense*	100 % for NH_4^+ versus 75, 36 % for PO_4^{3-} versus 19, 15 % nitrate versus 6 %	de-Bashan and Bashan (2004)
Phenol, oil and hydrocarbon, metals	*Chlorella* sp. ES-13, *Chlorella* sp. ES-30, *Scenedesmus obliquus* ES-55, *Stichococcus/Rhodococcus* sp. Ac-1267, *Kibdelosporangium aridum* 754	62–85 %	Safonova et al. (2004)
Phenol, phenanthrene and salicylate	For salicylate: *Chlorella vulgaris/Ralstonia Basilensis*; for phenol: *Chlorella vulgaris/Acinetobacter haemolyticus*; for phenanthrene: *Chlorella vulgaris/Pseudomonas migulae* and *Chlorella vulgaris/Sphingomonas yanoikuyae*	>85 %	Borde et al. (2003)
Thiocyanate (SCN^-)	*Chlorella protocoides or Ettlia* sp./activated sludge (mixed bacteria)	100 % after 100 h	Ryu et al. (2014)

7.4.3 *Fungal–Microalgal Associations*

Both fungi and microalgae have been extensively used for biological treatment of different types of wastewaters using their main nutrients, nitrogen and phosphorus for growth (e.g. Swami and Buddhi 2006; Wang et al. 2010; Li et al. 2011; de Boer et al. 2012). Efficient wastewater treatment with fungal–algal pellets has been described by a number of research groups (Zhou et al. 2011; Wrede et al. 2014; Miranda et al., 2015; Muradov et al. 2015). *Aspergillus* sp./*C. vulgaris* pellets harvested by filtration were used for the treatment of two types of wastewater resources namely 'Centrate', representing a liquid stream generated through centrifugation of activated sludge and diluted swine wastewater (Zhou et al. 2011). The former contained around 50 mg L^{-1} of both ammonia and phosphates while the latter was high in ammonia (89 mg L^{-1}) but low in phosphates (1.8 mg L^{-1}). Most of the ammonia and phosphates were removed after the first 24 h. The brown color of wastewater represented by phenolic substances was also removed after 24 h producing a practically colorless solution.

Co-pellets produced by *A. fumigatus* in association with heterotrophic algae, *Chlorella protothecoides* and sea water algae *Tetraselmis suecica* were used for the treatment of anaerobically digested swine wastewater using its ammonia and phosphate as an alternative source of nutrients (Muradov et al. 2015). For these experiments, the swine wastewater was diluted with tap water for *A. fumigatus*/*C. protothecoides* and with seawater for *A. fumigatus*/*T. suecica* to a final concentration of 10 % and 25 % of the original wastewater. Both types of pellets showed up to 73.9 % uptake of ammonia and up to 55.6 % uptake of phosphates. Growing in different dilutions of swine wastewater in seawater, *A. fumigatus* showed no obvious differences in nutritional uptake compared to swine wastewater diluted with tap water. Interestingly absorption of key wastewater nutrients, ammonia and phosphates led to 1.7- and 1.6-fold increase in biomass production after 48 h of treatment for *A. fumigatus*/*C. protothecoides* and *A. fumigatus*/*T. suecica* pellets, respectively (Muradov et al. 2015).

Similar results were obtained after assessment of the ability of *A. fumigatus*/*Thraustochytrid* sp. and *A. fumigatus*/*T. chuii* pellets to uptake ammonia and phosphate from swine wastewater using diluted swine wastewater prepared from swine lagoon wastewaters. Up to 86 % uptake of ammonia and 69 % of phosphate were reported after 48 h of wastewater treatment. This was also correlated with an increase in biomass of the pellets (Wrede et al. 2014).

Growth rates of cyanobacterial representatives were also increased in the presence of wastewater nutrients (N and P). Genetically modified *Synechocystis* sp. PCC 6803 designed for expression of lactate dehydrogenase gene involved in d-lactate biosynthesis as a feedstock for food, pharmaceutical and plastic industries was successfully grown on BG11 supplemented with alternative sources of N and P from wastewater from municipal sludge subjected to anaerobic digestion (Brightman and Seaward 1977). Co-cultivation of *A. fumigatus*- *Synechocystis*-SD100 pellets produced a synergistic effect on absorption of ammonium and phosphates from diluted swine wastewater (Miranda et al. 2015). After 48 h of growth of *A. fumigatus*-SD100 pellets in 25 % wastewater the concentration of NH_4^+-N reduced from 164.3 to 18.2 mg L^{-1} (89 %) and the concentration of PO_4^{3-}-P reduced from 38.7 to 9.8 mg L^{-1} (75 %). This removal efficiency was higher than achieved separately by *Synechocystis*-SD100 (30 % for NH_4^+-N and 26 % for PO_4^{3-}-P) and by *A.*

fumigatus (52 and 45 %, for NH_4^+-N and PO_4^{3-}-P, respectively). Nutrient uptake by *A. fumigatus*-SD100 pellets led to a 2.3-fold increase in their biomass production after 48 h of treatment and this correlated with a 1.5-fold increase in lipid yield.

Application of lignocellulosic waste for large-scale fungal and algal biomass production has been explored extensively. Fungal representatives are well known for their ability to digest lignocellulosic biomass, secreting a cocktail of very active enzymes including cellulases, hemicellulases, pectinases, laccase manganese peroxidase and lignin peroxidase (Kumar et al. 2008; Zhou et al. 2012; Xie et al. 2013). Representatives of *A. fumigatus* sp. growing in the presence of rice straw showed induced endoglucanase, exoglucanase, beta-glucosidase, laminarinase, lichenase, xylanase and pectin lyase activities (Liu et al. 2013). To improve the economics of fungal-assisted algal bio-flocculation *A. fumigatus* was grown on alternative carbon sources, such as wheat straw. Growing in liquid media containing wheat straw as a sole source of carbon fungal pellets showed similar growth rates and sizes as if grown on glucose (Wrede et al. 2014; Miranda et al. 2015; Muradov et al. 2015). However, the efficiency of algal flocculation was lower when *A. fumigatus* strains were grown on wheat straw than when glucose was used as a source of carbon. Inhibition of algal growth by products of enzymatic digestion of lignocellulosic wheat straw's biomass can be one of the reasons explaining this effect. The application of wheat straw biomass for fungal-assisted flocculation needs to be optimised to reduce the production of antialgal chemicals and to increase the lipid content in fungal cells grown on a straw biomass (Wrede et al. 2014).

7.5 Production of Renewable Bioenergy from Fungal–Algal Pellets[1]

Because of high biomass, growth rates, lipid yield and the composition of essential fatty acids, components of fungal–algal systems, as well as their co-pellets, can be used as feedstock for production of sustainable and renewable petrochemicals. The biomass of *C. protothecoides, A. fumigatus* and *A. fumigatus/C. protothecoides* pellets collected after bioremediation of wastewater were subjected to pyrolysis (Muradov et al. 2015). This led to the production of three main pyrolysis products, biogas, bio-oil, bio-char in ratios 6–9 %, 30–38 %, and 50–55 %, respectively. Production of similar pyrolytic products was observed for *C. protothecoides* grown in monoculture (Miao and Wu 2004; Aguirre et al. 2013). Analysis of liquid fractions showed that phytol (3,7,11,15-tetramethyl-2-hexadecen-1-ol), is the most abundant pyrolysis product in *C. protothecoides* and *A. fumigatus/C. protothecoides*. Because

[1]Additional information on the role of fungi and their enzymes on biofuel production can be found in Part IV—*Biotechnology to reduce reliance on fossil fuels: from biomass to biofuel.*

of the absence of chlorophyll molecules, phytol is not present in the bio-oil produced by *A. fumigatus*.

Phytol is being explored by some bioenergy companies, such as Argonne National Laboratory (USA) for the production of drop-in biofuels (Argonne National Laboratory 2013). The physical and chemical properties of phytol such as density, cetane number and heat of combustion are close to those of diesel fuel.

Lipid level and composition of microalgae and fungi have been extensively investigated for their potential to be used as feedstocks for biodiesel production. Fatty acids in *A. fumigatus* are represented mainly by palmitate, C16:0, oleate, C18:1 and linoleate, C18:2 (Wrede et al. 2014; Muradov et al. 2015). Most of the microalgal strains showed a high proportion of palmitate, C16:0, and palmitoleate, C16:1, stearate, oleate and linoleate (C18:0, C18:1 and C18:2, respectively). Lipid production in the fungal–algal pellets showed complex profiles reflecting at least three main factors: (i) total biomass production, (ii) lipid concentrations in fungal and algal cells before and during co-cultivation and (iii) the harvesting efficiencies of algal cells by fungal cells (Wrede et al. 2014; Miranda et al. 2015; Muradov et al. 2015). In most of the analyzed fungal–algal pellets, total lipid level and their composition showed additive values in comparison to those observed when both components were grown separately in monocultures. Analysis of the fungal–algal pellet's fatty acids composition suggested that it can be tailored and optimised through co-cultivating different algae and fungi without the need for genetic modification.

7.6 Conclusion

Microalgae have been successfully used for wastewater treatment for decades. Generated biomass has been used as a sustainable and renewable feedstock for production of a number of value-added products. Harvesting costs, however, represent a fundamental barrier to the industrial production of these products from microalgae. Fungal-assisted bio-flocculation as a biological way of harvesting has been shown to be highly efficient, not requiring added chemicals and having a low energy input requirement. Associations of fungal–algal representatives are attracting increasing attention because of a number of important features including (i) additive effects both of components on wastewater treatment and generation of biomass, which can be used as feedstock for production of biofuels, (ii) possibility for targeted tailoring of fatty acids and other value-added chemicals through co-cultivating different algae and fungi, (iii) improvement in the economics of value-added chemicals production by providing alternative sources of carbon from lignocellulosic wastes and nitrogen/phosphorous from wasterwater to enhance fungal and algal growth, and (iv) the ability of fungi to secrete a family of hydrolytic enzymes that can convert microalgal cells into cell wall-free protoplasts which can reduce the requirements for organic solvents for lipid extraction.

References

Abdel-Raouf N, Al-Homaidan AA, Ibraheem IBM (2012) Microalgae and wastewater treatment. Saudi J Biol Sci 19(3):257–275

Acuner E, Dilek FB (2004) Treatment of tectilon yellow 2G by *Chlorella vulgaris*. Process Biochem 39(5):623–631

Aguirre AM, Bassi A, Saxena P (2013) Engineering challenges in biodiesel production from microalgae. Crit Rev Biotechnol 33:293–308

Ahluwalia SS, Goyal D (2007) Microbial and plant derived biomass for removal of heavy metals from wastewater. Bioresour Technol 98:2243–2257

Akhtar N, Iqbal J, Iqbal M (2004a) Enhancement of lead (II) biosorption by microalgal biomass immobilized onto loofa (*Luffa cylindrica*) sponge. Eng Life Sci 4(2):171–178

Akhtar N, Iqbal J, Iqbal M (2004b) Removal and recovery of nickel (II) from aqueous solution by loofa sponge-immobilized biomass of *Chlorella sorokiniana*: characterization studies. J Hazard Mater 108(1):85–94

Aksu Z, Tezer S (2005) Biosorption of reactive dyes on the green alga *Chlorella vulgaris*. Process Biochem 40(3):1347–1361

Anthony RJ, Ellis JT, Sathish A, Rahman A, Miller CD, Sims RC (2013) Effect of coagulant/flocculants on bioproducts from microalgae. Bioresour Technol 149:65–70

Argonne National Laboratory (2013). http://www.transportation.anl.gov/engines/multi_dim_model_biofuels.html. Accessed 12 April 2016

Aslan S, Kapdan IK (2006) Batch kinetics of nitrogen and phosphorus removal from synthetic wastewater by algae. Ecol Eng 28:64–70

Barange M, Srivstava A, Srivastava JK, Palsania J (2014) Biosorption of heavy metals from wastewater by using microalgae. Int J Chem Phys Sci 3(6):67–81

Bayramoğlu G, Arıca MY (2009) Construction a hybrid biosorbent using *Scenedesmus quadricauda* and Ca-alginate for biosorption of Cu (II), Zn (II) and Ni (II): kinetics and equilibrium studies. Bioresour Technol 100(1):186–193

Bayramoğlu G, Tuzun I, Celik G, Yilmaz M, Arica MY (2006) Biosorption of mercury (II), cadmium (II) and lead (II) ions from aqueous system by microalgae *Chlamydomonas reinhardtii* immobilized in alginate beads. Int J Miner Process 81:35–43

Berner F, Heimann K, Sheehan M (2015) Microalgal biofilms for biomass production. J Appl Phycol 27:1793–1804

Bilanovic D, Shelef G, Sukenik A (1988) Flocculation of microalgae with cationic polymers—effects of medium salinity. Biomass 17:65–76

Boduroğlu G, Kılıç NK, Dönmez G (2014) Bioremoval of Reactive Blue 220 by *Gonium* sp. biomass. Environ Technol 35(19):2410–2415

Boelee NC, Temmink H, Janssen M, Buisman CJN, Wijffels RH (2011) Nitrogen and phosphorus removal from municipal wastewater effluent using microalgal biofilms. Water Res 45: 5925–5933

Borde X, Guieysse B, Delgado O, Muñoz R, Hatti-Kaul R, Nugier-Chauvin C et al (2003) Synergistic relationships in algal–bacterial microcosms for the treatment of aromatic pollutants. Bioresour Technol 86(3):293–300

Borsos E, Makra L, Béczi B, Vitányi B, Szentpéteri V (2003) Anthropogenic air pollution in the ancient times. Acta Climatol Chrolog 36–37:5–15

Brightman FH, Seaward MRD (1977) Lichens of man-made substrates. In: Seaward MRD (ed) Lichen Ecology. Academic Press, London, pp 253–293

Carpenter SR, Caraco NF, Correll DL, Howarth RW, Sharpley AN, Smith VH (1998) Nonpoint pollution of surface waters with phosphorus and nitrogen. Ecol Appl 8:559–568

Chan SMN, Luan T, Wong MH, Tam NFY (2006) Removal and biodegradation of polycyclic aromatic hydrocarbons by *Selenastrum capricornutum*. Environ Toxicol Chem 25(7): 1772–1779

Chavan A, Mukherji S (2008) Treatment of hydrocarbon-rich wastewater using oil degrading bacteria and phototrophic microorganisms in rotating biological contactor: effect of N: P ratio. J Hazard Mater 154(1):63–72
Chekroun KB, Sánchez E, Baghour M (2014) The role of algae in bioremediation of organic pollutants. Int Res J Public Environ Health 1:19–32
Chen C-Y, Yeh K-L, Aisyah R, Lee D-J, Chang J-S (2011) Cultivation, photobioreactor design and harvesting of microalgae for biodiesel production: a critical review. Bioresour Technol 102:71–81
Chinnasamy S, Bhatnagar A, Hunt RW, Das KC (2010) Microalgae cultivation in a wastewater dominated by carpet mill effluents for biofuel applications. Bioresour Technol 101:3097–3105
Chisti Y (2007) Biodiesel from microalgae. Biotechnol Adv 25(3):294–306
Chong AMY, Wong YS, Tam NFY (2000) Performance of different microalgal species in removing nickel and zinc from industrial wastewater. Chemosphere 41:251–257
Christenson L, Sims R (2011) Production and harvesting of microalgae for wastewater treatment, biofuels, and bioproducts. Biotechnol Adv 29:686–702
Cox PW, Paul GC, Thomas CR (1998) Image analysis of the morphology of filamentous micro-organisms. Microbiology 144:817–827
Daneshvar N, Ayazloo M, Khataee AR, Pourhassan M (2007) Biological decolorization of dye solution containing Malachite Green by microalgae *Cosmarium* sp. Bioresour Technol 98 (6):1176–1182
Das N, Vimala R, Karthika P (2008) Biosorption of heavy metals—an overview. Ind J Biotechnol 7(2):159–169
de-Bashan LE, Bashan Y (2004) Recent advances in removing phosphorus from wastewater and its future use as fertilizer (1997–2003). Water Res 38(19):4222–4246
de-Bashan LE, Bashan Y (2010) Immobilized microalgae for removing pollutants: review of practical aspects. Bioresour Technol 101:1611–1627
de-Bashan LE, Moreno M, Hernandez J-P, Bashan Y (2002) Removal of ammonium and phosphorus ions from synthetic wastewater by the microalgae *Chlorella vulgaris* coimmobilized in alginate beads with the microalgae growth-promoting bacterium *Azospirillum brasilense*. Water Res 36:2941–2948
de-Bashan LE, Hernandez JP, Morey T, Bashan Y (2004) Microalgae growth-promoting bacteria as "helpers" for microalgae: a novel approach for removing ammonium and phosphorus from municipal wastewater. Water Res 38:4222–4246
de Boer K, Moheimani NR, Borowitzka MA, Bahri PA (2012) Extraction and conversion pathways for microalgae to biodiesel: a review focused on energy consumption. J Appl Phycol 24:1681–1698
de la Noüe J, Laliberté G, Proulx D (1992) Algae and waste water. J Appl Phycol 4:247–254
Deblonde T, Cossu-Leguille C, Hartemann P (2011) Emerging pollutants in wastewater: a review of the literature. Int J Hygiene Environ Health 214:442–448
Dilek FB, Taplamacioglu HM, Tarlan E (1999) Colour and AOX removal from pulping effluents by algae. Appl Microbiol Biotechnol 52(4):585–591
Elkassas HY, Mohamed LA (2014) Bioremediation of the textile waste effluent by *Chlorella vulgaris*. Egyp J Aqua Res 40:301–308
El-Sheekh MM, Gharieb MM, Abou-El-Souod GW (2009) Biodegradation of dyes by some green algae and cyanobacteria. Int Biodeter Biodegrad 63(6):699–704
Ergene A, Ada K, Tan S, Katırcıoğlu H (2009) Removal of Remazol Brilliant Blue R dye from aqueous solutions by adsorption onto immobilized *Scenedesmus quadricauda*: equilibrium and kinetic modeling studies. Desalination 249(3):1308–1314
Feofilova EP (2010) The fungal cell wall: modern concepts of its composition and biological function. Microbiology 79:711–720
Fierro S, del Pilar Sanchez-Saavedra M, Copalcua C (2008) Nitrate and phosphate removal by chitosan immobilized *Scenedesmus*. Bioresour Technol 99(5):1274–1279
Forgacs E, Cserhati T, Oros G (2004) Removal of synthetic dyes from wastewaters: a review. Environ Int 30(7):953–971

Fu F, Wang Q (2011) Removal of heavy metal ions from wastewaters: a review. J Environ Manage 92:407–441

Garnham GW, Codd GA, Gadd GM (1992) Accumulation of cobalt, zinc and manganese by the estuarine green microalga *Chlorella salina* immobilized in alginate microbeads. Env Sci Tech 26(9):1764–1770

Gibbs PA, Seviour RJ, Schmid F (2000) Growth of filamentous fungi in submerged culture: problems and possible solutions. Crit Rev Biotechnol 20:17–48

Godos Id, Blanco S, García-Encina PA, Becares E, Muñoz R (2009) Long-term operation of high rate algal ponds for the bioremediation of piggery wastewaters at high loading rates. Bioresour Technol 100:4332–4349

Grimm LH, Kelly S, Hengstler J, Göbel A, Krull R, Hempel DC (2004) Kinetic studies on the aggregation of *Aspergillus niger* conidia. Biotechnol Bioeng 87:213–218

Gultom SO, Hu B (2013) Review of microalgae harvesting via co-pelletization with filamentous fungus. Energies 6:5921–5939

Guolan H, Hongwen S, Li CL (2000) Study on the physiology and degradation of dye with immobilized algae. Art Cells Blood Substit Immobil Biotechnol 28(4):347–363

Gupta VK, Shrivastava AK, Jain N (2001) Biosorption of chromium (VI) from aqueous solutions by green algae *Spirogyra* species. Water Res 35(17):4079–4085

Gutzeit G, Lorch D, Weber A, Engels M, Neis U (2005) Biofocculent algal-bacterial biomass improves low-cost waterwater treatment. Water Sci Technol 52(12):9–18

Han X, Wong YS, Tam NFY (2006) Surface complexation mechanism and modelling in Cr (III) biosorption by a microalgal isolate, *Chlorella miniata*. J Colloid Interf Sci 303:365–371

Han X, Wong YS, Wong MH, Tam NFY (2007) Biosorption and bioreduction of Cr (VI) by a microalgal isolate *Chlorella miniata*. J Hazard Mater 146(1):65–72

Henderson RK, Parsons SA, Jefferson B (2008) Successful removal of algae through the control of zeta potential. Sep Sci Technol 43:1653–1666

Hering D, Borja A, Carstensen J, Carvalho L, Elliott M, Feld CK et al (2010) The European water framework directive at the age of 10: a critical review of the achievements with recommendations for the future. Sci Total Environ 408:4007–4019

Hiruta O, Futamura T, Takebe H, Satoh A, Kamisaka Y, Yokochi T et al (1996) Optimization and scale-up of gamma-linolenic acid production by *Mortierella ramanniana* MM 15-1, a high gamma-linolenic acid producing mutant. J Ferment Bioeng 82:366–370

Hoffmann JP (1998) Wastewater treatment with suspended and nonsuspended algae. J Phycol 34 (5):757–763

Holder DJ, Kirkland BH, Lewis MW, Keyhani NO (2007) Surface characteristics of the entomopathogenic fungus *Beauveria* (Cordyceps) *bassiana*. Microbiology 153:3448–3457

Hu Z, Tommaso L, van Loosdrecht M, Kartal B (2013) Nitrogen removal with the anaerobic ammomium oxidation process. Biotechonol Lett 35(8):1145–1154

Jinqi L, Houtian L (1992) Degradation of azo dyes by algae. Environ Pollut 75:273–278

John J (2000) A self-sustainable remediation system for acidic mine voids. In: 4th International conference of diffuse pollution. International Association of Water Quality, Bangkok, pp 506–511

Kaya V, Goule, de la Noüe J, Picard G (1996) Effect of intermittent CO_2 enrichment during nutrient starvation on tertiary treatment of wastewater by alginate-immobilized *Scenedesmus bicellularis*. Enzyme Microb Technol 18(8):550–554

Keskin NOS, Celebioglu A, Uyar T, Tekinay T (2015) Microalgae immobilised by nanofibrous web for removal of reactive dyes from wastewater. Ind Eng Chem Res 54:5802–5809

Khan MA, Ghouri AM (2011) Environmental pollution: its effects on life and its remedies. Res World: J Arts Sci Commer 2:276–285

Krull R, Cordes C, Horn H, Kampen I, Kwade A, Neu TR et al (2010) Morphology of Filamentous fungi: linking cellular biology to process engineering using *Aspergillus niger*. Adv Biochem Eng Biotechnol 121:1–21

Krull R, Wucherpfennig T, Esfandabadi ME, Walisko R, Melzer G, Hempel DC et al (2013) Characterization and control of fungal morphology for improved production performance in biotechnology. J Biotechnol 163:112–123

Kumar KV, Ramamurthi V, Sivanesan S (2006) Biosorption of malachite green, a cationic dye onto *Pithophora* sp., a fresh water algae. Dyes Pigments 69:102–107

Kumar KV, Sivanesan S, Ramamurthi V (2005) Adsorption of malachite green onto *Pithophora* sp., a fresh water algae: Equilibrium and kinetic modelling. Process Biochem 40:2865–2872

Kumar R, Singh S, Singh OV (2008) Bioconversion of lignocellulosic biomass: biochemical and molecular perspectives. J Ind Microbiol Biotechnol 35:377–391

Lam MK, Lee KT (2012) Microalgae biofuels: a critical review of issues, problems and the way forward. Biotechnol Adv 30:673–690

Lau A, Wong YS, Tong Z (1998) Metal removal studied by a laboratory scale immobilized microalgal reactor. J Environ Sci. 10(3):474–478

Lau PS, Tam NFY, Wong YS (1997) Wastewater nutrients (N and P) removal by carrageenan and alginate immobilized *Chlorella vulgaris*. Environ Technol 8(9):945–951

Lee A, Lewis D, Ashman P (2009) Microbial flocculation, a potentially low-cost harvesting technique for marine microalgae for the production of biodiesel. J Appl Phycol 21:559–567

Lee AK, Lewis DM, Ashman PJ (2013) Harvesting of marine microalgae by electroflocculation: the energetics, plant design, and economics. Appl Energ 108:45–53

Lei AP, Wong YS, Tam NF (2002) Removal of pyrene by different microalgal species. Water Sci Technol 46(11–12):195–201

Leite GB, Abdelaziz AEM, Hallenbeck PC (2013) Algal biofuels. Challenges and opportunities. Bioresour Technol 145:134–141

Li Y, Zhou W, Hu B, Min M, Chen P, Ruan RR (2011) Integration of algae cultivation as biodiesel production feedstock with municipal wastewater treatment: Strains screening and significance evaluation of environmental factors. Bioresour Technol 102:10861–10867

Liao W, Liu Y, Chen SL (2007a) Studying pellet formation of a filamentous fungus *Rhizopus oryzae* to enhance organic acid production. Appl Biochem Biotechnol 137:689–701

Liao W, Liu Y, Frear C, Chen SL (2007b) A new approach of pellet formation of a filamentous fungus—*Rhizopus oryzae*. Bioresour Technol 98:3415–3423

Linder MB (2009) Hydrophobins: proteins that self-assemble at interfaces. Curr Opin Colloid In. 14:356–363

Liu D, Li J, Zhao S, Zhang R, Wang M, Miao Y et al (2013) Secretome diversity and quantitative analysis of cellulolytic *Aspergillus fumigatus* Z5 in the presence of different carbon sources. Biotechnol Biofuels 6(1):149. doi:10.1186/1754-6834-6-149

Liu Y, Liao W, Chen S (2008) Co-production of lactic acid and chitin using a pelletized filamentous fungus Rhizopus oryzae cultured on cull potatoes and glucose. J Appl Microbiol 105:1521–1528

Ma X, Zhou W, Fu Z, Cheng Y, Min M, Liu Y et al (2014) Effect of wastewater-borne bacteria on algal growth and nutrients removal in wastewater-based algae cultivation system. Bioresour Technol 67:8–13

Mahan CA, Majidi V, Holcombe JA (1989) Evaluation of the metal uptake of several algae strains in a multicomponent matrix utilizing inductively coupled plasma emission spectrometry. Anal Chem 6(1):624–627

Manheim D, Nelson Y (2013) Settling and bioflocculation of two species of algae used in wastewater treatment and algae biomass production. Environ Prog Sustain Energy 32:946–954

Maza-Márquez P, Martinez-Toledo MV, Fenice M, Andrade L, Lasserrot A, Gonzalez-Lopez J (2014) Biotreatment of olive washing wastewater by a selected microalgal-bacterial consortium. Int Biodeter Biodegrad 88:69–76

McGinn PJ, Dickinson KE, Park KC, Whitney CG, MacQuarrie SP, Black FJ et al (2012) Assessment of the bioenergy and bioremediation potentials of the microalga *Scenedesmus* sp. AMDD cultivated in municipal wastewater effluent in batch and continuous mode. Algal Res 1:155–165

McNeill J, Barrie FR, Buck WR, Demoulin V, Greuter W, Hawksworth DL et al (2012) International nomenclature code for Algae, Fungi and Plants, (Melbourne code). In: Adopted by the 18th International Botanical Congress. Koeltz Scientific Books, Melbourne, Australia

Medina M, Neis U (2007) Symbiotic algal bacterial wastewater treatment: effect of food to microorganism ratio and hydraulic retention time on the process performance. Water Sci Technol 55(11):165–171

Mehta SK, Gaur JP (2005) Use of algae for removing heavy metal ions from wastewater: progress and prospects. Cri Rev Biotechnol 25(3):113–152

Metz B, Kossen NWF (1977) The growth of molds in the form of pellets–a literature review. Biotechnol Bioeng 19:781–799

Miao XL, Wu QY (2004) High yield bio-oil production from fast pyrolysis by metabolic controlling of *Chlorella protothecoides*. J Biotechnol 110:85–93

Miranda AF, Taha M, Wrede D, Morrison P, Ball AS, Stevenson T et al (2015) Lipid production in association of filamentous fungi with genetically modified cyanobacterial cells. Biotechnol Biofuels 8:179–197

Mishra PK, Mukherji S (2012) Biosorption of diesel and lubricating oil on algal biomass. 3 Biotechnol 2(4):301–310

Mohan SV, Bhaskar YV, Karthikeyan J (2004) Biological decolourisation of simulated azo dye in aqueous phase by algae *Spirogyra* species. Int J Environ Pollut 21(3):211–222

Molina Grima E, Belarbi EH, Acién Fernández FG, Robles Medina A, Chisti Y (2003) Recovery of microalgal biomass and metabolites: process options and economics. Biotechnol Adv 20:491–515

Moreno-Garrido I, Codd GA, GaddGM, Lubián LM (2002) Cu and Zn accumulation by calcium alginate immobilized marine microalgal cells of *Nannochloropsis gaditana* (Eustigmatophyceae). Cienc Mar 28(1):107–119

Muñoz R, Guieysse B (2006) Algal–bacterial processes for the treatment of hazardous contaminants: a review. Water Res 40:2799–2815

Muradov N, Taha M, Miranda AF, Wrede D, Kadali K, Gujar A et al (2015) Fungal-assisted algal flocculation: application in wastewater treatment and biofuel production. Biotechnol Biofuels 8:24. doi:10.1186/s13068-015-0210-6

Nakajima A, Horikoshi T, Sakaguchi T (1982) Recovery of uranium by immobilized microorganisms. Eur J Appl Microbiol Biotechnol 16(2–3):88–91

Ogbonna JC, Yoshizawa H, Tanaka H (2000) Treatment of high strength organic wastewater by a mixed culture of photosynthetic microorganisms. J Appl Phycol 12:277–284

Omar HH (2008) Algal decolorization and degradation of monoazo and diazo dyes. Pak J Biol Sci 1(10):1310–1316

Oswald WJ, Gotaas HB (1957) Photosynthesis in sewage treatment. Trans Am Soc Civil Eng 122:73–105

Papagianni M (2004) Fungal morphology and metabolite production in submerged mycelial processes. Biotechnol Adv 22:189–259

Papanikolaou S, Komaitis M, Aggelis G (2004) Single cell oil (SCO) production by *Mortierella isabellina* grown on high-sugar content media. Bioresour Technol 95:287–291

Park JB, Craggs RJ (2010) Wastewater treatment and algal production in high rate algal ponds with carbon dioxide addition. Water Sci Technol 61(3):633–639. doi:10.2166/wst.2010.951

Park JBK, Craggs RJ, Shilton AN (2011) Wastewater treatment high rate algal ponds for biofuel production. Bioresour Technol 102:35–42

Perez-Garcia O, Escalante FME, de-Bashan LE, Bashan Y (2011) Heterotrophic cultures of microalgae: metabolism and potential products. Water Res 45(1):11–36

Pinheiro HM, Touraud E, Thomas O (2004) Aromatic amines from azo dye reduction: status review with emphasis on direct UV spectrophotometric detection in textile industry wastewater. Dyes Pigments 61:121–139. doi:10.1016/j.dyepig.2003.10.009

Pinto G, Pollio A, Previtera L, Temussi F (2002) Biodegradation of phenols by microalgae. Biotechnol Lett 24(24):2047–2051

Powell RJ, Hill RT (2013) Rapid aggregation of biofuel-producing algae by the bacterium *Bacillus* sp strain RP1137. Appl Environ Microbiol 79:6093–6101

Pragya N, Pandey KK, Sahoo PK (2013) A review on harvesting, oil extraction and biofuels production technologies from microalgae. Renew Sustain Energy Rev 24:159–171

Priyadarshani I, Sahu D, Rath B (2011) Microalgal bioremediation: current practices and perspectives. J Biochem Technol 3:299–304

Rajab AH (2007) Micro-algae removal in domestic wastewater using *Aspergillus flavus* soft pellets as a bio-coagulant. Universiti Putra Malaysia, Faculty of Engineering

Rajkumar R, Yaakob Z, Takriff MS (2014) Potential of the micro and macro algae for biofuel production: a brief review. Bioresources 9:1606–1633

Ratha SK, Babu S, Renuka N, Prasanna R, Prasad RBN, Saxena AK (2013) Exploring nutritional modes of cultivation for enhancing lipid accumulation in microalgae. J Basic Microbiol 53:440–450

Rawat I, Kumar RR, Mutanda T, Bux F (2011) Dual role of microalgae: phycoremediation of domestic wastewater and biomass production for sustainable biofuels production. Appl Energy 88(10):3411–3424

Reed TB (1981) Biomass gasification: principles and technology. NJ Noyes Data Corporation, Park Ridge

Riaño B, Hernández D, García-González MC (2012) Microalgal-based systems for wastewater treatment: Effect of applied organic and nutrient loading rate on biomass composition. Ecol Eng 49:112–117

Ritter L, Solomon KR, Forget J, Stemeroff M, O'Leary C (1995) Persistent organic pollutants. An assessment report on DDT, aldrin, dieldrin, endrin, chlordane, heptachlor, hexachlorobenzene, mirex, toxaphene, polychlorinated biphenyls, dioxins and furans. International Programme on Chemical Safety (IPCS). In: Persistent organic pollutants. An assessment report on DDT, aldrin, dieldrin, endrin, chlordane, heptachlor, hexachlorobenzene, mirex, toxaphene, polychlorinated biphenyls, dioxins and furans. International Programme on Chemical Safety (IPCS). PCS 95, City, vol 38, p 43

Ruiz-Marin A, Mendoza-Espinosa LG, Stephenson T (2010) Growth and nutrient removal in free and immobilized green algae in batch and semi-continuous cultures treating real wastewater. Bioresour Technol 101:58–64

Ryu BG, Kim J, Faroog W, Han JI, Yang JW, Kim W (2014) Algal-bacterial process for the simultaneous detoxification of thiocyanate-containing wastewater and maximized lipid production under photoautotrophic/photoheterotrophic conditions. Bioresour Technol 162:70–79. doi:10.1016/j.biortech.2014.03.084

Safonova E, Kvitko KV, Iankevitch MI, Surgko LF, Afti IA, Reisser W (2004) Biotreatment of industrial wastewater by selected algal-bacterial consortia. Eng Life Sci 4(4):347–353

Salim S, Bosma R, Vermuë MH, Wijffels RH (2011) Harvesting of microalgae by bio-flocculation. J Appl Phycol 23:849–855

Schweitzer B, Huber I, Amann R, Ludwig W, Simon M (2001) Alpha- and betaproteobacteria control the consumption and release of amino acids on lake snow aggregates. Appl Environ Microbiol 67:632–645

Selbmann L, Stingele F, Petruccioli M (2003) Exopolysaccharide production by filamentous fungi: the example of *Botryosphaeria rhodina*. Anton Van Leeuwenhoek 84:135–145

Semple KT, Cain R, Schmidt S (1999) Biodegradation of aromatic compounds by microalgae. FEMS Microbiol Lett 170(2):291–300. doi:10.1111/j.1574-6968.1999.tb13386.x

Seviour RJ, Stasinopoulos SJ, Auer DPF, Gibbs PA (1992) Production of pullulan and other exopolysaccharides by filamentous fungi. Crit Rev Biotechnol 12:279–298

Sheng Y, Chen F, Sheng G, Fu J (2012) Water quality remediation in a heavily polluted tidal river in Guangzhou, South China. Aquat Ecosyst Health 15:219–226

Shi J, Podola B, Melkonian M (2007) Removal of nitrogen and phosphorus from wastewater using microalgae immobilized on twin layers: an experimental study. J Appl Phycol 19(5):417–423

Simas-Rodrigues C, Villela HDM, Martins AP, Marques LG, Colepicolo P, Tonon AP (2015) Microalgae for economic applications: advantages and perspectives for bioethanol. J Exp Bot 66(14):4097–4108. doi:10.1093/jxb/erv130

Singh K, Arora S (2011) Removal of synthetic textile dyes from wastewaters: a critical review on present treatment technologies. Crit Rev Environ Sci Technol 41(9):807–878

Sivarajasekar N, Baskar R, Balakrishnan V (2009) Biosorption of an azo dye from aqueous solutions onto *Spirogyra*. J Uni Chem Technol Metal 44(2):157–164

Sournia A (1978) Phytoplankton manual. UNESCO, Paris

Souza PO, Ferreira LR, Pires NR, Duarte FA, Pereira CM, Mesko MF (2012) Algae of economic importance that accumulate cadmium and lead: a review. Revista Brasil Farmacog 22(4): 825–837

Su Y, Mennerich A, Urban B (2011) Municipal wastewater treatment and biomass accumulation with a water-born and settleable algal-bacterial culture. Water Res 45(11):3351–3358. doi:10.1016/j.watres.2011.03.046

Swami D, Buddhi D (2006) Removal of contaminants from industrial wastewater through various non-conventional technologies: a review. Int J Environ Pollut 27:324–346

Takáčová A, Smolinská M, Ryba J, Mackuľak T, Jokrllová Hronec P et al (2014) Biodegradation of benzo[a]pyrene through the use of algae. Cent Eur J Chem 12(11):1133–1143

Tam NFY, Wong YS (2000) Effect of immobilized microalgal bead concentrations on wastewater nutrient removal. Environ Pollut 107:145–151

Tam NFY, Lau PS, Wong YS (1994) Wastewater inorganic N and P removal by immobilized *Chlorella vulgaris*. Water Sci Technol 30:369–374

Tam NFY, Wong JPK, Wong YS (2001) Repeated use of two *Chlorella* species, *C. vulgaris* and WW1 for cyclic nickel biosorption. Environ Pollut 114:85–92

Tam NFY, Wong YS, Wong MH (2009) Novel technology in pollutant removal at source and bioremediation. Ocean Coast Manage 52:368–373

Tam NFY, Chan MN, Wong YS, Popov V, Itoh H, Mander U et al (2010) Removal and biodegradation of polycyclic aromatic hydrocarbons by immobilized microalgal beads. WIT Trans Ecol Environ 140:391–402. doi:10.2495/WM100351

Taylor TN, Hass H, Remy W, Kerp H (1995) The oldest fossil lichen. Nature 378:244

Thakur A, Kumar HD (1999) Nitrate, ammonium, and phosphate uptake by the immobilized cells of *Dunaliella salina*. B Environ Contam Tox 62(1):70–78

Todd SJ, Cain RB, Schmidt S (2002) Biotransformation of naphthalene and diaryl ethers by green microalgae. Biodegradation 13(4):229–238

Travieso L, Benitez F, Weiland P, Sanchez E, Dupeyron R, Dominguez AR (1996) Experiments on immobilization of microalgae for nutrient removal in wastewater treatments. Bioresour Technol 55(3):181–186

Uduman N, Qi Y, Danquah MK, Hoadley AFA (2010) Marine microalgae flocculation and focused beam reflectance measurement. Chem Eng J 162:935–940

Ueno R, Wada S, Urano N (2006) Synergetic effects of cell immobilization in polyurethane foam and use of thermotolerant strain on degradation of mixed hydrocarbon substrate by *Prototheca zopfii*. Fish Sci 72:1027–1033

Ueno R, Wada S, Urano N (2008) Repeated batch cultivation of the hydrocarbon-degrading, micro-algal strain *Prototheca zopfii* RND16 immobilized in polyurethane foam. Can J Microbiol 54(1):66–70

Unnithan VV, Unc A, Smith GB (2014) Mini-review: a priori considerations for bacteria-algae interactions in algal biofuel sustems receiving municipal wastewaters. Algal Res 4:35–40. doi:10.1016/j.algal.2013.11.009

Van den Hende S, Vervaeren H, Desmet S, Boon N (2011) Bioflocculation of microalgae and bacteria combined with flue gas to improve sewage treatment. New Biotechnol 29:23–31

Vandamme D, Foubert I, Muylaert K (2013) Flocculation as a low-cost method for harvesting microalgae for bulk biomass production. Trends Biotechnol 31:233–239

Vechtlifshitz SE, Magdassi S, Braun S (1990) Pellet formation and cellular aggregation in *Streptomyces tendae*. Biotechnol Bioeng 35:890–896

Vieira RH, Volesky B (2010) Biosorption: a solution to pollution? Int Microbiol 3(1):17–24
Volesky B, Naja G (2005) Biosorption: application strategies. In: 16th International Biotechnology Symposium. Compress Co., Cape Town, South Africa
Wang J, Chen C (2009) Biosorbents for heavy metals removal and their future. Biotechnol Adv 27 (2):195–226
Wang L, Zhang C, Wu F, Deng N (2007) Photodegradation of aniline in aqueous suspensions of microalgae. J Photochem Photobio B: Bio 87:49–57
Wang LA, Min M, Li YC, Chen P, Chen YF, Liu YH et al (2010) Cultivation of green algae *Chlorella* sp in different wastewaters from municipal wastewater treatment plant. Appl Biochem Biotechnol 162:1174–1186
Worku A, Sahu O (2014) Reduction of heavy metal and hardness from ground water by algae. J Appl Environ Microbiol 2:86–89
WWAP (United Nations World Water Assessment Programme) (2015) World water development report 2015: water for a sustainable world. UNESCO, Paris. http://unesdoc.unesco.org/images/0023/002318/231823E.pdf. Accessed 14 Apr 2016
Wrede D, Taha M, Miranda AF, Kadali K, Stevenson T, Ball AS et al (2014) Co-cultivation of fungal and microalgal cells as an efficient system for harvesting microalgal cells, lipid production and wastewater treatment. PLoS ONE 9:e113497. doi:10.1371/journal.pone.0113497
Wu YH, Hu HY, Yu Y, Zhang TY, Zhu SF, Zhuang LL et al (2014) Microalgal species for sustainable biomass/lipid production using wastewater as resource: a review. Renew Sustain Energ Rev 33:675–688
Xia C, Zhang J, Zhang W, Hu B (2011) A new cultivation method for microbial oil production: cell pelletization and lipid accumulation by *Mucor circinelloides*. Biotechnol Biofuels 4:15. doi:10.1186/1754-6834-4-15
Xie SX, Sun S, Dai SY, Yuan JS (2013) Efficient coagulation of microalgae in cultures with filamentous fungi. Algal Res 2:28–33
Yamaguchi T, Ishida M, Suzuki T (1999) An immobilized cell system in polyurethane foam for the lipophilic micro-alga *Prototheca zopfii*. Process Biochem 34(2):167–171
Zhang E, Wang B, Wang Q, Zhang S, Zhao B (2008) Ammonia–nitrogen and orthophosphate removal by immobilized *Scenedesmus* sp. isolated from municipal wastewater for potential use in tertiary treatment. Bioresour Technol 99(9):3787–3793
Zhou W, Cheng YL, Li Y, Wan YQ, Liu YH, Lin XY et al (2012) Novel fungal pelletization-assisted technology for algae harvesting and wastewater treatment. Appl Biochem Biotechnol 167:214–228
Zhou W, Li YC, Min M, Hu B, Chen P, Ruan R (2011) Local bioprospecting for high-lipid producing microalgal strains to be grown on concentrated municipal wastewater for biofuel production. Bioresour Technol 102:6909–6919
Zhou W, Ruan R, Wang J (2015) Bio-flocculation of microalgae: status and prospects. Curr Biotechnol 4(4):448–456
Znidarsic P, Pavko A (2001) The morphology of filamentous fungi in submerged cultivations as a bioprocess parameter. Food Technol Biotechnol 39:237–252
Zoller S, Lutzoni F (2003) Slow algae, fast fungi: exceptionally high nucleotide substitution rate differences between lichenized fungi *Omphalina* and their symbiotic green algae *Coccomyxa*. Mol Phylogenet Evol 29:629–640

Part III
Mycoremediation: Fungi as Prolific Degraders to Control Waste and Reduce Pollution

Chapter 8
Mycoremediation of Organic Pollutants: Principles, Opportunities, and Pitfalls

Stephano Covino, Tatiana Stella and Tomáš Cajthaml

Abbreviations

LEs	Ligninolytic enzymes
LiP	Lignin peroxidase
MnP	Manganese-dependent peroxidase
VP	Versatile peroxidases
DyP	Dye-degrading peroxidases
Lac	Laccase
Tyr	Tyrosinase
UPO	Unspecific peroxigenases
CYP450	Cytochrome P450
CDH	Cellobiose dehydrogenase
SRFs	Soft rot fungi
BRFs	Brown rot fungi
WRFs	White rot fungi
LDFs	Litter-decomposing fungi
PCBs	Polychlorobiphenyls
PCDDs	Polychlorinated dioxins
DDT	1,1-bis(4-chlorophenyl)-2,2,2-trichloroethane
BaP	Benzo[*a*]pyrene
POPs	Persistent organopollutants
PAHs	Polycyclic aromatic hydrocarbons
LFs	Ligninolytic fungi
NLFs	Non-ligninolytic fungi

S. Covino · T. Stella · T. Cajthaml (✉)
Institute for Environmental Studies, Faculty of Science, Charles University in Prague, Benátská 2, CZ-128 01 Prague 2, Czech Republic
e-mail: cajthaml@biomed.cas.cz

S. Covino · T. Stella · T. Cajthaml
Institute of Microbiology, Academy of Sciences of the Czech Republic V.V.I., Vídeňská 1083, CZ-142 20 Prague 4, Czech Republic

D. Purchase (ed.), *Fungal Applications in Sustainable Environmental Biotechnology*, Fungal Biology, DOI 10.1007/978-3-319-42852-9_8

ROS	Reactive oxygen species
CBAs	Chlorobenzoic acids
EDs	Endocrine disruptors
ECMs	Ectomycorrhizal fungal species
NADPH	Reduced nicotinamide adenine dinucleotide phosphate
PCDFs	Polychlorinated dibenzofurans
diCDD	Dichlorodibenzo-dioxin
OCDD	Octachlorodibenzo-*p*-dioxin
MPs	Micropollutants
PPCP	Pharmaceuticals and personal care products
NPs	Nonylphenols
BPA	Bisphenol A
DEHP	Diethylhexyl phthalate
EE2	17α-ethinylestradiol
WWTP	Wastewater treatment plant
HBT	Hydroxybenzotriazole
NSAID	Non-steroidal anti-inflammatory drugs
IBU	Ibuprofen
CA	Clofibric acid
CBZ	Carbamazepine
DBQ	Dimethoxy-1,4-benzoquinone
CLEAs	Cross-linked enzyme aggregates
TNT	Trinitrotoluene
TPH	Total petroleum hydrocarbons
qPCR	Quantitative PCR
PCP	Pentachlorophenol
SMC	Spent Mushroom Compost

8.1 Introduction

Fungi are a diverse group of microorganisms ubiquitously present in natural ecosystems and they are regarded as the major decomposers of complex biomaterials. Fungi can thrive in a huge variety of habitats, ranging from surface waters to terrestrial environments—including extreme zones of the biosphere like deserts or polar regions—and they use air to disperse their propagules. Regardless of their taxonomic classification, all fungi are essentially heterotrophic microorganisms, i.e., they assimilate nutrients by absorption from the extracellular environment. In general terms, they can be subdivided into biotrophic and saprotrophic organisms based on their ecophysiology. The first group encompasses symbionts and pathogens of other species, while the second is the group of "decomposers" of biomass.

Although several studies have highlighted the ecological importance of some biotrophic fungi (for example mycorrhizas associated with plant roots in polluted soils), for the rationale of this chapter, which deals with the importance of fungi in the decontamination of polluted environments, saprotrophic fungi are more significant.

Saprophytic fungi are able to decompose and transform a wide array of organic (and inorganic) substrates. In contrast to bacteria, which need to transport nutrients into their intracellular compartment prior to their utilization, fungal mycelia can be seen as "externalized stomachs" which secrete their hydrolytic enzymes and organic acids into the extracellular environment and transport digested substances and chelates inside their cell wall. Although bacteria are, to some extent, faster and more efficient than fungi in utilizing readily assimilable substrates (proteins, lipids, starch and free sugars), fungi have evolved extremely efficient enzymatic machineries for the degradation of complex nonprotein polymers, such as ligno-cellulose and chitin, which are inaccessible and/or recalcitrant to the majority of prokaryotes. Due to these unique capabilities, fungi are considered to play key roles in a number of biogeochemical processes and, more generally, in the cycling of carbon and other elements.

8.1.1 Lignocellulolytic Fungi and Their Enzymes[1]

Lignocellulose is the most abundant renewable biomass on earth. All woody materials are composed mainly of complex carbohydrates (cellulose and hemicellulose) and lignin, the most complex and heterogeneous biopolymer known till date. Thanks to the production of lignocellulolytic enzymes, different saprotrophic fungi can degrade and utilize plant polymers. In general, extracellular lignocellulolytic enzymes from fungi can be subdivided into two major groups. The first one comprises a series of hydrolytic enzymes capable of breaking down polysaccharide constituents of the plant cell wall, namely cellulases (e.g., glucanases, glucosidases, cellobiohydrolases) and hemicellulases (xylanases, mannanases, xylosidase). The other group, referred to as ligninolytic system, promotes the radical-mediated non-specific decomposition of lignin besides cleaving the aromatic moieties thereof. It includes heme peroxidases (lignin peroxidase, LiP, E.C. 1.11.1.14; manganese-dependent peroxidase, MnP, E.C.1.11.1.13; versatile peroxidases, VP, E.C. 1.11.1.16; dye-degrading peroxidases, DyP, E.C. 1.11.1.19), phenol oxidases (laccase, Lac, E.C. 1.10.3.2; tyrosinase, Tyr, E.C. 1.14. 18.1) and a series of accessory

[1]Additional information on decomposition of lignocellulase redsidues and waste by white and brown rot fungi is presented in Chap. 9—*White and brown rot fungi as decomposers of lignocellulosic materials and their role in waste and pollution control.*

enzymes for the production of H_2O_2 (e.g., glyoxal oxidase, glucose oxidase and aryl alcohol oxidase). During the last decade, a new type of H_2O_2-requiring enzyme has been described for the first time in the Basidiomycete *Agrocybe aegerita* (Ullrich et al. 2004) and, hence, in other fungi (Hofrichter and Ullrich 2014). These proteins are heme-thiolate haloperoxidases, nowadays classified as fungal unspecific peroxigenases (UPO, E.C. 1.11.2.1). Their catalytic cycle combines the typical pathways of extracellular heme peroxidases and intracellular monooxygenases (unspecific cytochrome P450 system, CYP450, E.C. 1.14.14.1), thus enabling them to catalyze an outstanding array of reactions (Hofrichter and Ullrich 2014).

By reason of the different strategy to decompose wood polymers—and owing to the appearance of the wood following the decay process—it is possible to distinguish soft rot (SRFs), brown rot (BRFs), and white rot fungi (WRFs). An additional group ecophysiologically close to the white rot is formed by litter-decomposing fungi (LDFs), which inhabit the organic horizon of forests and grassland soil. These basidiomycetes produce ligninolytic oxidases and peroxidases as WRFs, but their typical niche is soil, where they cause the decay of the leaf-litter and other plant-derived debris.

Ligninolysis per se does not support fungal growth, but fungi can cause a certain deterioration of the amorphous structure of lignin to access the carbohydrates intertwined in the lignin structure of the plant cell walls. To this end, SRFs almost exclusively produce cellulases that diffuse in the lignocellulose structure providing the fungus with nourishment, without significantly altering the structure of the cell wall. Ubiquitous SRFs include species of *Chaetomium*, *Fusarium*, *Paecilomyces* and other common soil genera—not typically wood-inhabiting fungi. Any timber which is in contact with the soil (for example, fence posts or telegraph poles) is ultimately attacked by these fungi, especially under condition of high moisture. SRFs such as *Trichoderma* spp. and *Chaetomium* spp. preferably attack plant biomass characterized by low lignin content. BRFs, instead, use a Fenton-type catalytic system to disrupt the lignin and holocellulose structure in wood (Rabinovich et al. 2004; Baldrian and Valášková 2008), thus facilitating the penetration of hydrolases in the plant cell wall. Strong and diffusible $OH^{\cdot}$ radicals are produced starting from reduced iron (Fe^{2+}) and H_2O_2, which are, in turn, produced either endogenously or *via* cellobiose dehydrogenase (CDH) catalyzed reactions, quinone-redox cycling or low molecular mass glycopeptides (Rabinovich et al. 2004; Baldrian and Valášková 2008). WRFs are also known to use Fenton-type reactions (Baldrian and Valášková 2008; Gómez-Toribio et al. 2009; Dashtban et al. 2010), although the distinctive feature of this fungal group is the production of extracellular high redox potential ligninolytic enzymes (LEs), which are responsible for the so called "enzymatic combustion" of lignin (Kirk and Farrell 1987). LDFs also produce ligninolytic enzymes and attack lignocellulose in a fashion very similar to that of wood-inhabiting white rots, although their niche is the superficial organic layer of soil.

8.1.2 Mycoremediation: Origin and Principles

The link between bioremediation potential of fungi and the decay of wood relies on the similarity between the components of the lignin macromolecules and the majority of aromatic pollutants. Indeed, the catabolic enzymes that fungi have evolved to access the cellulose and hemicellulose fibrils embodied in the wood structure were also found to be active towards a vast range of aromatic pollutants. Concerning this, an overview of the mechanism that ligninolytic and non-ligninolytic fungi use to attack both lignocellulose and organic contaminants is provided in Fig. 8.1.

The first evidence about fungal degradation of aromatic pollutants dates back more than 50 years (Anastasi et al. 2013), when Lyr (1963) reported the chlorophenol-removing potential of a wood-inhabiting fungus and awareness of fungal degradation of wood preservative agents had developed (Duncan and Deverall 1964). In the same period, the "kerosene fungus" *Cladosporium resinae* was also described for the first time upon its isolation from fuel tanks (Hendey 1964). However, the actual driving force which led to the birth of a new branch of bioremediation was the study of the model white rot fungus *Phanerochaete chrysosporium*. In fact, the ligninolytic machinery of *P. chrysosporium* became the focus of several researchers' attention in the early 1980s and a ligninase from the same fungus was characterized for the first time in 1984 (Tien and Kirk 1984). Driven by the capability of *P. chrysosporium* to degrade chlorinated lignin-derived by-products of the kraft pulping process, Bumpus and colleagues demonstrated one year later, in 1985, the ability of the same fungus to degrade (and partially mineralize) a set of persistent organopollutants, namely polychlorobiphenyls (PCBs), dioxins (PCDDs), lindane, 1,1-bis(4-chlorophenyl)-2,2,2-trichloroethane (DDT)

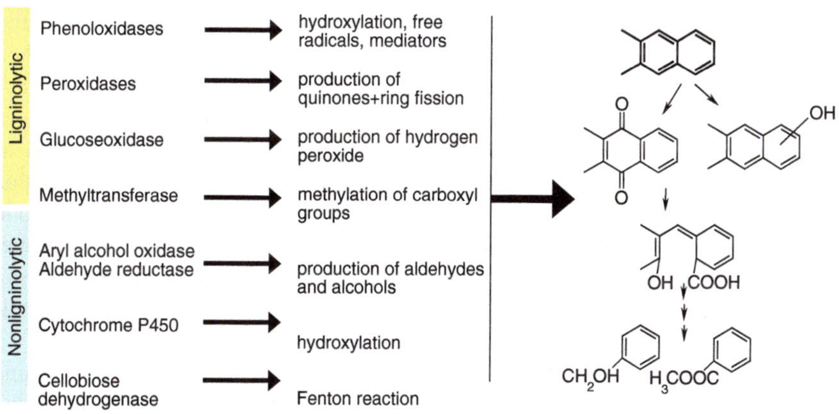

Fig. 8.1 Outline of the mechanisms used by ligninolytic and non-ligninolytic fungi to decompose lignocellulose and organic pollutants

and benzo[*a*]pyrene (BaP) (Bumpus et al. 1985). Since then, an intense period of research on the bioremediation potential of *P. chrysosporium* and other wood-degrading fungi has started (Hammel 1995; Rabinovich et al. 2004; Harms et al. 2011). Concerning this issue, the term "Mycoremediation" has been coined by the eminent mycologist Paul Stamets to specifically address the use of fungal mycelia and their enzymes in bioremediation. The majority of studies concerning fungal remediation have mainly considered the use of lignin degrading fungi (WRFs and LDFs) with typical target matrices being soils and sediments contaminated by persistent aromatic pollutants, such as wood preservatives, oil-derived products, explosives, dielectric fluids, pesticides and other man-made chemical products (Mougin 2002; Baldrian 2008; Cerniglia and Sutherland 2006; Cajthaml and Svobodová 2012).

Several physiological and biochemical features of ligninolytic fungi make them potential candidates for soil bioremediation strategies. As already mentioned, they produce a large number of oxidative enzymes, mainly Lac, LiP and MnP, which exhibit very low substrate specificity and, being active in the extracellular environment, are able to reach and attack scarcely bioavailable contaminants by non-specific radical-based reactions. In addition to the extracellular one, WRFs possess an intracellular enzymatic system involving CYP450 monooxygenases (Črešnar and Petrič 2011). This intracellular pathway occurs in all eukaryotic organisms and it mainly regulates the bioconversion of hormones and the detoxification of drugs and xenobiotics (Bernhardt 2006). In wood rotting fungi, cytochrome P450 is supposed to cooperate with the ligninolytic system in the general mechanism of xenobiotic degradation (van den Brink et al. 1998). Moreover, white rots develop a spatially extensive hyphal growth, enabling them to penetrate across air-filled soil pores, air–water interfaces and even rock matrices (Bornyasz et al. 2005) and to act as dispersion vectors for bacteria (Kohlmeier et al. 2005; Bonfante and Anca 2009). Fungi can also tolerate high concentrations of organic contaminants and heavy metals with limited or no deleterious effects on their own enzymatic activities (Baldrian et al. 2000; Baldrian 2003; Tuomela et al. 2005). In spite of their extraordinary degrading capabilities, the vast majority of fungi cannot assimilate contaminants as a source of carbon and energy like bacteria, except in some cases of non-ligninolytic fungi. For this reason, lignocellulosic residues are used as amendants to support the fungal growth and, thus to improve the mycoremediation performances (Singh 2006). Furthermore, it is well known that fungi are involved in soil humification process: in this respect, the use of these organisms in soil remediation could lead, not only to the decontamination, but also to the reuse of the soil for agricultural purposes once the remediation goals are met (Bollag 1992; Michels 1998).

In the following sections, we present the principles and mechanisms behind fungal degradation of persistent organic pollutants (POPs) and other emerging organic micropollutants (MPs).

8.2 Polycyclic Aromatic Hydrocarbons

Polycyclic aromatic hydrocarbons (PAHs) are contaminants of great concern due to their worldwide distribution. Although they are natural compounds, PAHs also originate from incomplete combustion of fossil fuels and biomasses. For this reason, human activities are considered as the major source of PAH contamination of the environment. Chemically, PAHs consist of two or more fused benzene rings arranged either in a linear or a cluster mode and several members of this class—e.g., benzo[*a*]anthracene, chrysene and benzo[*a*]pyrene—have been included among priority pollutants owing to their toxic, mutagenic and carcinogenic properties (Haritash and Kaushik 2009). The persistence of PAHs in the environment is mainly due to their hydrophobicity, toxicity and recalcitrance to microbial degradation, which largely increase as the molecular weight of PAHs increases.

PAHs are, by far, the most extensively studied contaminants with respect to their susceptibility to fungal biodegradation. Owing to the chemical stability of PAHs, which is favored by the dense cloud of delocalized π electrons on the coplanar structure, the critical step for aerobic organisms is the initial incorporation of oxygen on the aromatic ring. In general, the way fungi use to attack PAHs is the same as that of mammalian and other eukaryotes, i.e., oxidation of PAHs to PAH *trans*-dihydrodiols by Phase I intracellular enzymes (CYP450 monooxygenases and epoxide hydrolase) followed by Phase II conjugation reactions mediated by transferases. This detoxification reaction was first described in *Cunninghamella elegans* by Cerniglia and Gibson (1977). Cerniglia (1997) also used this member of Mucoraceae as a model organism for subsequent PAH degradation studies, which prompted him to identify high molecular weight PAH-dihydrodiols, quinones and phenols metabolites. However, the above-cited study of Bumpus et al. (1985) about the xenobiotic degradation capabilities of *P. chrysosporium*, shifted the attention of many researchers towards ligninolytic fungi (LFs) and their powerful extracellular machineries. Since then, several other genera of LFs (mainly WRFs) such as *Trametes*, *Pleurotus*, *Bjerkandera*, *Irpex*, *Phlebia*, *Nematoloma*, and *Lentinus* have been demonstrated to metabolize PAHs efficiently under model liquid culture conditions or in the soil (Sack et al. 1997; Pointing 2001; Giubilei et al. 2009; Novotný et al. 2009; Baldrian 2008; Covino et al. 2010a, b, c). Additionally, it was proved that PAHs can be oxidized under in vitro conditions using Lac and peroxidases from LFs (Hammel et al. 1986; Majcherczyk et al. 1998; Eibes et al. 2006; Baborová et al. 2006; Covino et al. 2010c) and that the use of redox mediators can improve the rate of PAH degradation as well as expand the spectrum of substrates oxidized by such catalysts (Sack et al. 1997; Johannes and Majcherczyk 2000; Camarero et al. 2008; Cañas and Camarero 2010; Covino et al. 2010c). Besides, the involvement of CYP450 monooxygenase-epoxide hydrolase system was shown to be active in the initial degradation of PAHs also in LFs (Bezalel et al. 1997). Using ^{14}C-labeled compounds, it was proved that LFs are able to completely mineralize PAHs to carbon dioxide (Bumpus et al. 1985; Bezalel et al. 1996a; Wolter et al. 1997; Sack et al. 1997). However, evidence for PAH ring-cleavage by fungi are poorly described. Hammel et al. (1991) showed that *P. chrysosporium* was able to decompose anthracene to phthalic acid while Bezalel et al. (1996b) presented the mechanism of 2,2′-diphenic acid production from phenanthrene. These authors suggested that CYP450 of *Pleurotus ostreatus* was responsible for the attack to phenanthrene, enabling further ring opening reactions. Moen and Hammel (1994) reported the formation of 2,2′-diphenic acid from phenanthrene as a result of lipid peroxidation by MnP, while other authors found several ring-cleavage products of acenaphthylene and acenaphthene after incubation with *Trametes versicolor* Lac as well as a laccase-mediator system (Johannes et al. 1998; Majcherczyk et al. 1998). A few years later, Cajthaml et al. (2002) identified several PAH metabolites in *Irpex lacteus* axenic cultures spiked with phenanthrene, anthracene, fluoranthene and pyrene: the presence of ring-cleavage products and PAH-dihydrodiols reinforced the hypothesis that ligninolytic fungi use both the CYP450 system and extracellular ligninases. Baborová et al. (2006) not only confirmed that purified MnP from *I. lacteus* can attack recalcitrant representative of PAHs, but also that it can cleave the benzene ring to form 2-(2′-hydroxybenzoyl)-benzoic acid. A pathway similar to that proposed for anthracene has been also hypothesized by Cajthaml and colleagues (2006) for benzo(*a*)anthracene, with emphasis on intermediates possessing two aromatic rings, e.g., 1,4-naphthalenedione, 1,4-naphthalenediol, and 1,2,3,4-tetrahydro-1-hydroxynaphthalene. The breakdown products in the metabolic pathway of benzo[*a*]anthracene by *I. lacteus* are shown in Fig. 8.2.

The WRF *P. ostreatus*, well known as oyster mushroom, is one of the most efficient degrader of PAHs (Bezalel et al. 1996b; Wolter et al. 1997). When it was grown in the presence of several PAHs (benzo[*a*]pyrene, pyrene, fluorene, phenanthrene, anthracene) metabolization and mineralization was shown to occur. The main PAH degradation products identified were: phenanthrene trans-9,10-dihydrodiol and 2,2′-diphenic acid, pyrene trans-4,5-dihydrodiol, anthracene trans-1,2-dihydrodiol, and 9,10-anthraquinone. For instance, the fungus was able to decompose phenanthrene, anthracene and pyrene by 50, 92 and 35 % in 5 days, respectively, in bran flakes media (Pickard et al. 1999). Schützendübel et al. (1999) found that *Bjerkandera adusta* removed 56 and 38 % of fluorene and anthracene 3 days after spiking these PAH in the fungal cultures, while *P. ostreatus* degraded 43 and 60 % of these compounds; other PAHs were degraded to a lower extent. All PAHs were removed uniformly during the period of incubation by *P. ostreatus* cultures, but fluorene and anthracene were degraded faster than other compounds in basidiomycete’s rich media. The detected intermediates were mostly keto compounds (Schützendübel et al. 1999).

Another member of the Polyporales with remarkable PAH-degrading capabilities is the white rot fungus *Lentinus tigrinus* (Valentin et al. 2006; Covino et al. 2010a, c). Regardless of the N-content in the culture media, this fungus was able to remove up to 97 % of a mixture of 7 PAHs from 3 to 5 fused benzene rings. In addition, the major ligninolytic isoenzymes (Lac and MnP) from *L. tigrinus* were shown to efficiently degrade individual PAH congeners under in vitro conditions, either in the presence or in the absence of low molecular weight redox mediators, although MnP showed a wider PAH substrate range and faster oxidation rates than

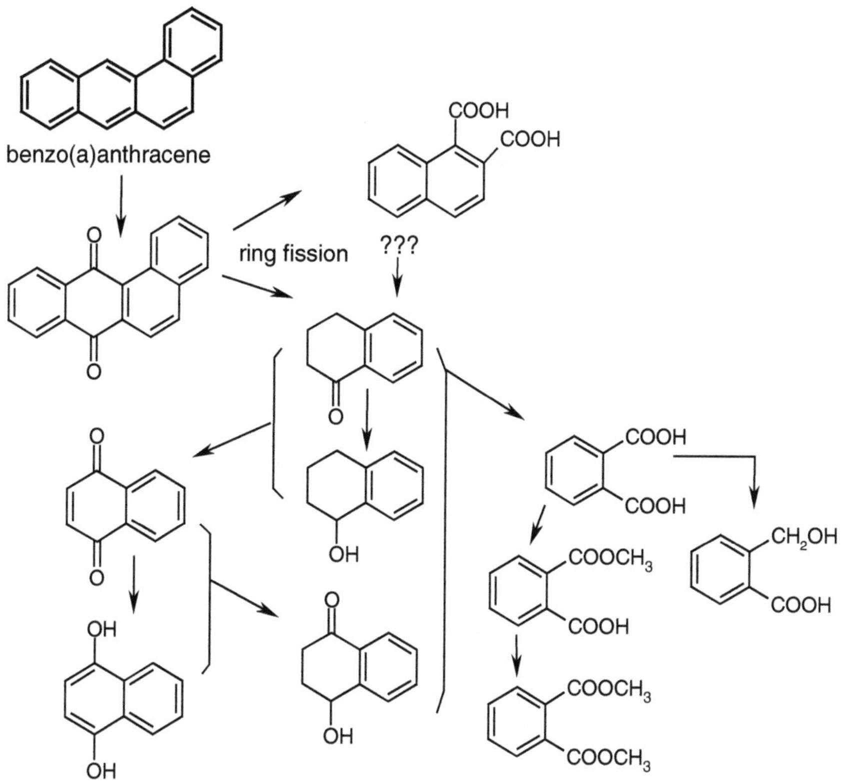

Fig. 8.2 Benzo[*a*]anthracene degradation by *I. lacteus*—adapted from Cajthaml et al. (2006)

the Lac produced by the same strain. When *L. tigrinus* was bioaugmented in a PAH-contaminated soil and in creosote wood, it removed PAHs to a higher extent than *I. lacteus* (Covino et al. 2010a). Moreover, Valentin et al. (2006) reported that *L. tigrinus*, *I. lacteus* and *B. adusta* were capable of growing and degrading PAHs under the typical halophilic conditions of a marsh soil.

Litter-decomposing fungi (LDFs) were claimed to have a greater potential than white rots in the mycoremediation of contaminated soil, mainly due to their innate capability to compete with natural soil microflora (Šašek 2003; Steffen et al. 2007). In addition to an early report on the BaP-degradation by *Marasmiellus troyanus* (Wunch et al. 1999), the first extensive screening of the PAH degrading potential of LDFs was performed by Gramss et al. (1999). Outstanding results in term of removal of a mixture of 5 PAHs were reported for two fungi, namely *Gymnophilus sapineus* and *Agrocybe praecox* (92 and 90 %, respectively, after 14 d). Steffen et al. (2002) reported the bioconversion of PAHs by nine litter decomposers in model liquid systems and concomitantly assessed the extent of ^{14}C-benzo[*a*]pyrene (BaP) mineralization. The results showed that all fungi were capable of oxidizing PAHs, with two strains belonging to the genus *Stropharia* (*S. coronilla* and

S. rugosoannulata) that were more efficient than others. The authors also highlighted the positive correlation between Mn^{2+} supplementation, expression of MnP activity and overall degradation/mineralization of PAHs (Steffen et al. 2002). As a consequence, *S. coronilla* MnP was further purified and incubated under in vitro condition with ^{14}C-BaP. The results confirmed the involvement of this enzyme in the mineralization of BaP and a 12-fold higher increase in $^{14}C\text{-}CO_2$ production upon Mn^{2+} supplementation. In addition, the two *Stropharia* spp. mentioned above were the most efficient PAHs degraders among other litter decomposers in an artificially contaminated soil (Steffen et al. 2007). Similarly, Lac from the oak-litter decomposer *Marasmius quercophilus* was shown to degrade PAHs with ionization potential lower than 7.55 eV, i.e., anthracene and BaP (Farnet et al. 2009). As for the PAH degradation in real contaminated matrices, the shaggy mane mushroom *Coprinus comatus* was used in a laboratory-scale mycoremediation test targeting historically creosote-contaminated soil and wood. Regardless of lignocellulose support selected as growth substrate, the PAH degradation performances of this fungus were similar to that of representative white rots (*Dichomitus squalens* and *P. ostreatus*), and some high molecular weight PAHs were degraded by *C. comatus* beyond their threshold of bioavailability (Covino et al. 2010b).

Comparatively, BRFs were less investigated than ligninolytic fungi with respect to PAH degradation. However, in spite of early studies claiming the inadequacy of brown rots for mycoremediation applications (Martens and Zadrazil 1998), recent reports have highlighted the PAH biotransformation potential of the button mushroom *Agaricus bisporus* and the possible use of *A. bisporus* spent compost in soil decontamination (Mayolo-Deloisa et al. 2011; García-Delgado et al. 2015).

Soft-rotting fungi include common soil saprotrophs that usually grow on substrates other than wood, but they can occasionally cause its decay. Indeed, Ascomycetes such as *Trichoderma*, *Chaetomium*, *Fusarium*, *Aspergillus*, *Penicillium* and *Paecilomyces* are still regarded by many authors as non-ligninolytic fungi (Rabinovich et al. 2004; Marco-Urrea et al. 2015) although several SRFs genera are known to produce Lac (Verdin et al. 2004; Baldrian 2006; Wu et al. 2010; Cázares-García et al. 2013; Viswanath et al. 2014) and ligninolytic peroxidases (Yang et al. 2003, 2013; Silva et al. 2009). The redox potential of these enzymes, however, is lower than those typically produced by wood-rotting fungi and their role in lignin and PAH biotransformation has not yet been fully understood (Baldrian 2006). In general, WRFs and LDFs have attracted more attention than soft-rotters for their possible applicability in bioremediation. In spite of this, soil saprotrophs are frequently isolated from crude oil- and PAH-contaminated soils, where they constitute a large proportion of the fungal community (Rafin et al. 2000, 2012; Covino et al. 2015; Bourceret et al. 2015: Cébron et al. 2015). Very recently, an interesting review on NLFs with potential application in bioremediation of organopollutants has been published (Marco-Urrea et al. 2015) and an exhaustive list of species with reported PAH degrading capability can be found in the literature (Gadd 2001; Cerniglia and Sutherland 2006; Fernández-Luqueño et al. 2011; Harms et al. 2011; Marco-Urrea et al. 2015). The mechanism by which these fungi attack PAHs is the same as that described for *C. elegans*, i.e., *via* CYP450 monooxygenase-epoxide hydrolase system, which yields hydroxylated, (*trans-*) di-hydroxylated and transient dihydro-epoxide derivatives. These activated structures, that are capable of binding to nucleophilic groups of DNA, can be also transformed into sulfate-, methyl-, glucose-, xylose-, or glucuronic acid conjugates by Phase II enzymes (e.g., aryl sulfotransferase, glutathione S-transferase, UDP-glucuronosyltransferase and UDP-glucosyltransferase, methyl transferases). This step, followed by the excretion of these compounds into the extracellular environment, results in a significant detoxification with respect to both parent molecules and mutagenic dihydroxy-epoxides, as it is known to occur in higher eukaryotes (Cerniglia and Sutherland 2006; Marco-Urrea et al. 2015). Interestingly, some members of this group, e.g., the yeasts *Pichia anomala* and *Rhodotorula glutinis* or the hyphomycetes *Fusarium solani*, *Trichoderma* and *Penicillium*, possess the capability to use PAHs with up to 5 fused benzene rings as sole carbon source (Romero et al. 1998, 2002; Rafin et al. 2000; Pan et al. 2004; Verdin et al. 2004; Cerniglia and Sutherland 2006), and some of them can do so even in the presence of microaerobic ($5 < O_2 < 15$ %) or very-low oxygen conditions ($1\ \% < O_2$) (Silva et al. 2009). Detailed studies on BaP-degradation by *F. solani* demonstrated that this fungus (i) produces a surfactant-like hydrogel (polysaccharides and glycoproteins) to increase the BaP solubility and uptake; (ii) transports BaP in the intracellular environment and accumulate it in lipid vesicles; (iii) degrades the substrates by means of either CYP450 or reactive oxygen species (ROS) and (iv) mineralizes BaP to an extent similar to that reported for the white rot fungus *P. ostreatus* (Verdin et al. 2004, 2005).

8.3 Chlorinated Aromatic Pollutants

8.3.1 Polychlorinated Biphenyls

Polychlorinated biphenyls (PCBs) are synthetic compounds the structure of which consists of a biphenyl molecule (two aromatic rings linked by a C–C bond) carrying from one to ten chlorine atoms. Theoretically, the entire set of PCB homologs comprises 209 congeners. In the past, PCB mixtures were widely used for a number of industrial applications due to their thermal and chemical stability, flame resistance and dielectric properties. As a consequence, their use led to extensive environmental contamination either through accidental releases and/or inappropriate methods of disposal. Owing to their aforementioned inertness, PCBs are therefore still present in a number of areas where their production had been carried out for decades, although their application was drastically restricted long time ago. Nowadays, PCBs are considered among the most hazardous contaminants in the world, placing them at the forefront of public health concern. Indeed, the teratogenic, carcinogenic and endocrine disrupting effects of these xenobiotics have been widely reported in the literature (Crinnion 2011; Helmfrid et al. 2012; Kramer et al.

2012; El Majidi et al. 2013) especially for coplanar PCBs (not substituted at the ring positions *ortho-*) which exhibit "dioxin-like" toxicity features (US Environmental Protection Agency 2000; Van den Berg et al. 2006).

In the light of all these considerations, the cleaning-up of PCB-contaminated sites has become a global priority. Among diverse remediation approaches, the use of biological systems represents an efficient, cost-effective, and environmentally friendly alternative to the more commonly used thermal and physico-chemical technologies (Passatore et al. 2014).

Bacteria play a key role in PCB biodegradation processes: anaerobes can use highly chlorinated biphenyls as electron acceptors and convert them into less chlorinated congeners, while aerobic bacteria can co-metabolize lower chlorinated biphenyls (Pieper 2005; Field and Sierra-Alvarez 2008). However, the co-metabolism of PCBs *via* the biphenyl pathway leads to the formation of chlorobenzoic acids (CBAs), which tend to be accumulated as dead-end products. The build-up of these metabolites exerts an inhibitory feedback effect on the upper biphenyl degradation pathway resulting in the impediment or slowing down of the PCB biotransformation process as a whole (Adebusoye et al. 2008; Furukawa and Fujihara 2008). Therefore, it is worthwhile to search for other microorganisms that are able to mineralize PCBs completely.

Due to their wide metabolic capabilities, a large number of ligninolytic fungi belonging to the *Basidiomycota* phylum were tested in laboratory-scale model liquid systems for their ability to degrade technical PCB mixtures or single PCB congeners, notably: *P. chrysosporium* (Eaton 1985; Thomas et al. 1992; Vyas et al. 1994; Yadav et al. 1995; Kamei et al. 2006a) *T. versicolor* (Zeddel et al. 1993; Vyas et al. 1994; Cloete and Celliers 1999), *Lentinus edodes* (Ruiz-Aguilar et al. 2002), *P. ostreatus* (Kubátová et al. 2001; Čvančarová et al. 2012), *Grifola frondosa* (Seto et al. 1999), *Coriolopsis polyzona* (Vyas et al. 1994; Novotný et al. 1997), *Phlebia brevispora* (Kamei et al. 2006b), *I. lacteus* (Čvančarová et al. 2012); *B. adusta* (Beaudette et al. 1998; Čvančarová et al. 2012), *Pycnoporus cinnabarinus* (Čvančarová et al. 2012), *Phanerochaete magnoliae* (Čvančarová et al. 2012).

Generally, all these studies confirmed that the extent of degradation significantly decreases with increasing chlorine content. For instance, a negligible mineralization of individual tetra- and hexa-chlorobiphenyls was reported for *P. chrysosporium* (Dietrich et al. 1995), whereas the level of degradation remarkably increased for three- and di-chlorinated congeners. Similar trends were also observed with technical PCB mixtures. For example, Yadav et al. (1995) showed that *P. chrysosporium* was able to remove 60.9, 30.5, and 17.6 % of Aroclor 1242, 1254, and 1260 PCB mixtures (42, 54 and 60 % chlorinated), respectively, in 30 days.

Moreover, some fungal strains (e.g., *P. ostreatus*) have shown to selectively remove PCB congeners with a preference for compounds with chlorine atom in *ortho* > *meta* > *para* position (Kubátová et al. 2001), while others (e.g., *P. chrysosporium*) did not exhibit any noticeable specificity for the position of chlorine substitutions (Yadav et al. 1995).

A number of studies have dealt with the interpretation of PCB degradation mechanisms in fungi. A few works indicated that purified fungal extracellular phenoloxidases and peroxidases were unable to oxidize PCB congeners (Beaudette et al. 1998; Krčmar et al. 1999; Takagi et al. 2007). For instance, Krčmar et al. (1999) performed a biodegradation experiment with technical PCB mixtures containing low and high chlorinated congeners (Delor 103 and Delor 106) using *P. chrysosporium* mycelium, crude extracellular liquid and crude extract enriched in MnP and LiP activities. A decrease in the PCB concentration after 44 h of treatment with mycelium (74 %) or crude extracellular liquid (60 %) was observed whereas MnP and LiP isolated from the extracellular liquid did not catalyze any degradation.

Afterwards, based on the identification of PCB transformation products in liquid fungal cultures, the involvement of both extracellular ligninolytic system and intracellular CYP450 monooxygenases system in the degradation process of PCBs was hypothesized (Kamei et al. 2006a, b; Čvančarová et al. 2012). In this respect, other studies proved that the ligninolytic enzymes were able to breakdown some of the PCB degradation intermediates such as their hydroxylated derivatives, which are produced by CYP450 system (Keum and Li 2004; Takagi et al. 2007; Kordon et al. 2010).

Kamei et al. (2006a) investigated the transformation products of 4,4′-dichlorobiphenyl in *P. chrysosporium* in the attempt to determine its degradation pathway. Methoxylated- and hydroxylated-PCB derivatives were detected in *P. chrysosporium* liquid cultures and, thus the involvement of cytochrome CYP450 in the degradation process was supposed. The addition of piperonyl butoxide, a well-known CYP450 inhibitor, to fungal cultures prevented the formation of hydroxylated metabolites, further supporting the initial hypothesis. Chlorobenzoic acids, chlorobenzaldehydes and chlorobenzylalcohols were also identified as PCB degradation products. In particular, Cvancarová et al. (2012) reported the formation of chlorobenzoates from hydroxylated PCBs and hypothesized the further transformation *via* reductive pathways. Once CYP450 oxidizes the aromatic structure of PCBs, ring fission reaction can be mediated by other enzymatic systems (i.e., LEs) (Cajthaml et al. 2006; Čvančarová et al. 2012); subsequently, a reductive mechanism can operate on the carboxyl group of CBAs leading to the formation of chlorinated aldehydes and alcohols (Muzikář et al. 2011; Stella et al. 2013). An overview of the PCB degradation pathway in ligninolytic fungi, including the formation and further biotransformation of CBAs, is provided in Fig. 8.3.

Despite the promising outcomes achieved in liquid cultures, few studies investigated the ability of white rot fungi to degrade chlorinated biphenyls either in artificially contaminated (Zeddel et al. 1993; Kubátová et al. 2001) or in actual PCB-contaminated soils (Borazjani et al. 2005; Federici et al. 2012). With regard to historically PCB-polluted matrices, a recent study demonstrated that a bioaugmentation approach based on the use of maize stalk-immobilized fungus *L. tigrinus* brought about 33.6 % of Aroclor 1260 depletion and 23.2 % of dechlorination in 60 days (Federici et al. 2012).

Compared to PCB biotransformation processes mediated by basidiomycetes, few studies investigated the PCB biodegradation potential of NLFs. One of the first

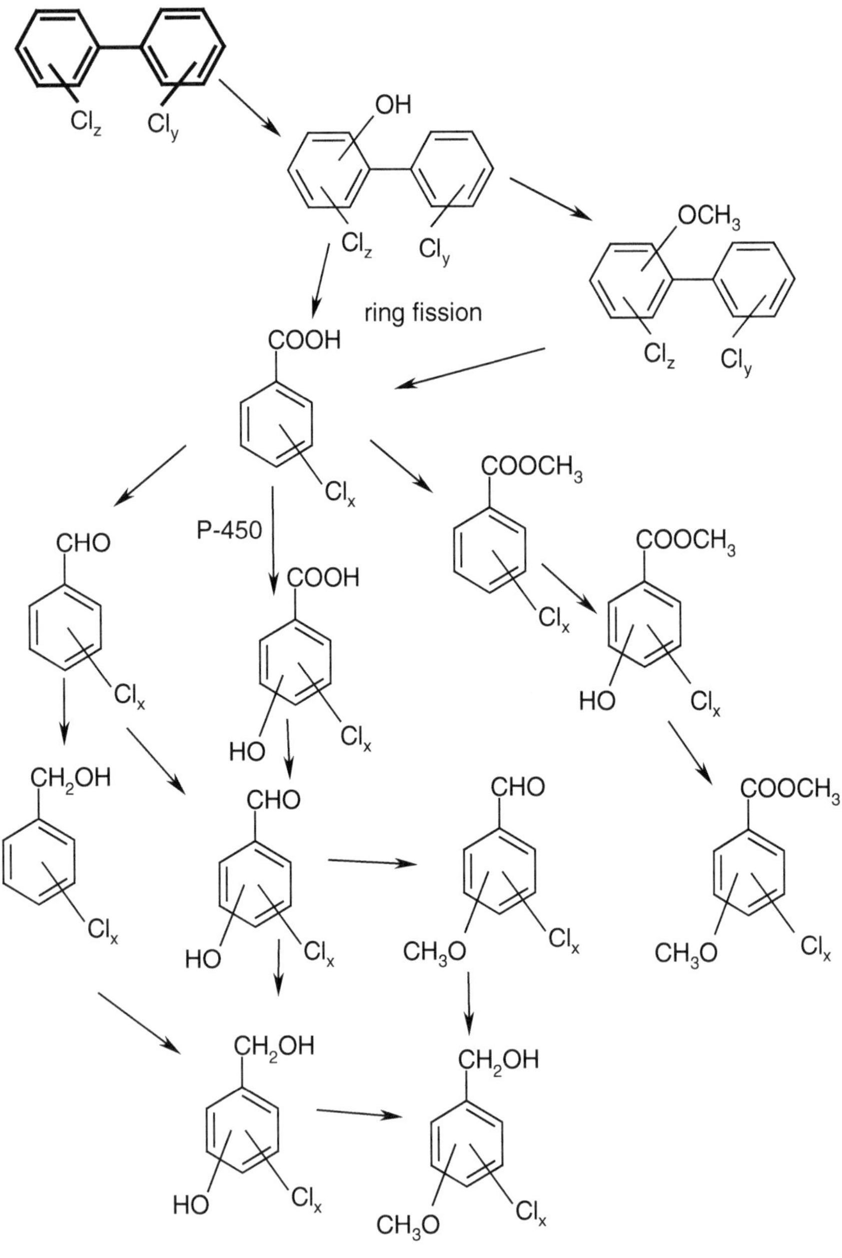

Fig. 8.3 Degradation pathway of PCBs and CBAs by *P. ostreatus*—adapted from Čvančarová et al. (2012) and Muzikář et al. (2011)

report (Dmochewitz and Ballschmiter 1988) revealed that the ascomycete *Aspergillus niger* was able to degrade a technical mixture of PCBs (Chlophen A) with a low chlorine content (42 %), while PCB mixtures with higher chlorination levels (54 and 60 %) remained untransformed. Recently, six fungal strains belonging to *Aspergillus*, *Penicillium*, *Fusarium* and *Scedosporium* genera were isolated from an aged contaminated soil (0.9 g kg^{-1} PCB mixture Aroclor 1260) and tested for their ability to degrade three target congeners (2-monoCB, 4,4′-diCB and 2,2′,5,5′-tetraCB) in liquid mineral medium in the presence of glucose (Tigini et al. 2009). The degradation experiments demonstrated that PCBs were cometabolically transformed by all the isolates at a similar rate, regardless of the number of halogenated substituents. A similar experimental approach was adopted by Mouhamadou et al. (2013). The PCB degradation ability of autochthonous fungal strains isolated from former industrial sites highly contaminated with PCBs was tested in liquid cultures. Among all the isolates, the ascomycetes *Doratomyces nanus*, *Doratomyces purpureofuscus*, *Doratomyces verrucisporus*, *Myceliophthora thermophila*, *Phoma eupyrena,* and *Thermoascus crustaceus* showed remarkable degradation ability (>70 %). As for the previous study, the number of chlorine substituents on the biphenyl nucleus did not affect the degradation rate. Thereafter, a consortium of the aforementioned fungal strains was used for the remediation of both PCB-contaminated soil and sediment (Sage et al. 2014). After 6 months of treatment, 18.7 and 33.3 % of 15 target PCBs were degraded in soil and sediment mesocosms, respectively. The fungal strains were re-inoculated and incubated for additional 2 months to further promote the biodegradation of PCBs. However, no additional PCB depletion was observed. This result suggested that the bioavailable fraction of PCB was completely removed after 6 months and the remaining fraction was inaccessible to fungi.

Research regarding the PCB degradation pathways in other non-white rot fungi, revealed that no dechlorination occurred during the transformation of 4-chlorobiphenyl by *Paecilomyces lilacinus*, but five chlorinated metabolites, including ring fission products, were identified (Sietmann et al. 2006). Another study by Tigini et al. 2009, where mitosporic fungi were incubated in the presence of PCBs, showed that no chloride ions were released. Moreover, peroxidase activities were not detected in the liquid cultures, either in the absence or in the presence of PCBs, providing further evidence that these extracellular enzymes were not involved in the biotransformation process. Only Lac activity was detected at significant levels, irrespective of the presence of PCBs, but the role played by this enzyme in the PCBs degradation process was not clarified.

In conclusion, despite the large number of studies regarding the mycoremediation of PCBs, further research is needed to explore the physiological and biochemical characteristics of fungi prior to any practical application. Moreover, as suggested in a recent study (Stella et al. 2015), a comprehensive evaluation of both physico-chemical factors (i.e., interaction of PCBs with soil organic matter) and biological factors (i.e., interaction with native microorganisms) is strongly required to develop an effective fungal-based remediation approach for the treatment of PCB contaminated matrices.

8.3.2 Chlorobenzoic Acids

Chlorobenzoic acids (CBAs) are a class of relevant environmental pollutants consisting of benzoic acid with different chlorination degrees of the aromatic ring. CBAs primarily result from bacterial biodegradation of polychlorinated biphenyls (PCBs) under aerobic conditions as reviewed by Field and Sierra-Alvarez (2008). During this process, CBAs tend to accumulate as dead-end products, acting as inhibitors for the upper biphenyl pathway and therefore restricts further PCB transformation (Pieper 2005; Adebusoye et al. 2008). Due to their relatively high solubility in water, CBAs are characterized by a mobility that is several orders of magnitude higher than that of PCBs. In addition, some of CBA isomers are highly toxic for aquatic organisms (Lee and Chen 2009), exhibit genotoxicity toward higher plants (Gichner et al. 2008) and act as endocrine disruptors (EDs) (Svobodová et al. 2009).

The number of publications dealing with the fungal degradation of CBAs is very limited. Four white rot fungi (*P. chrysosporium*, *P. ostreatus*, *T. versicolor*, *Heterobasidion annosum*) and two ectomycorrhizal (ECMs) fungal species (*Paxillus involutus*, *Suillus bovinus*) were tested for their capability to grow on different concentrations of 3-chlorobenzoic acid (0.1, 1, and 3 mM) and to degrade it in 4 weeks (Dittmann et al. 2002). Even low concentrations of 3-CBA led to a remarkable reduction in the growth of both ECMs and *P. chrysosporium*, whereas that of *P. ostreatus*, *T. versicolor* and *H. annosum* was not affected. This study showed, in particular, that the great ability of the latter group of white rot fungi to grow in the presence of 3-CBA correlated to the highest extent of degradation achieved at the end of incubation period.

Muzikár et al. (2011) performed a more comprehensive study dealing with fungal degradation of CBA. Eight ligninolytic fungal strains (*I. lacteus*, *B. adusta*, *P. chrysosporium*, *P. magnoliae*, *P. ostreatus*, *T. versicolor*, *P. cinnabarinus* and *D. squalens*) were tested for their ability to degrade a mixture of 12 CBAs (mono-, di- and tri-chlorobenzoates) in both model liquid systems and contaminated soils. In liquid media, *I. lacteus*, *P. cinnabarinus* and *D. squalens* were the most effective fungi in terms of CBA removal and toxicity reduction, whereas *I. lacteus* and *P. ostreatus* were the most efficient in an artificially ("ad hoc") contaminated soil. Analysis of the degradation products revealed that methoxy and hydroxy derivatives were produced along with reduced forms of the original acids (benzaldehydes and benzyl alcohols) and a general degradation pathway was proposed. However, the mechanism by which individual enzymes can degrade CBAs was not fully clarified. Subsequently, the white rot fungus *L. tigrinus* was selected for the evaluation of its efficiency in degrading a mixture of chlorinated benzoates (Stella et al. 2013). The in vivo degradation experiment showed that this organism was able to deplete most of the target CBA isomers within 20 days of incubation, with the exception of 2,6-diCBA, 2,3,6-triCBA and 2,4,6-triCBA. The authors suggested that the recalcitrance of these isomers might be due to the steric hindrance of the two chlorine substituents adjacent to the carboxyl moiety and to the general electron-withdrawing effect of halides. Furthermore, to gain insights into the CBA degradation pathway of *L. tigrinus*, in vitro experiments were performed with both extracellular LEs (Lac and MnP) and intracellular enzymatic systems (microsomal fraction containing CYP450 monooxygenases) extracted from *L. tigrinus* cultures. Lac and MnP were shown not to be involved in the initial transformation steps of CBAs, even in the presence of redox mediators, whereas CYP450 were able to oxidize chlorobenzoates. The key role played by CYP450 in the degradation of CBAs under in vitro condition was inferred by the NADPH-dependency of the reaction and by its susceptibility to piperonyl butoxide and carbon monoxide, two well-known CYP450 inhibitors (Stella et al. 2013). Additionally, the detection of a hydroxylated chlorobenzoic acid further supported the pivotal role of CYP450 system in the initial CBA bioconversion steps by *L. tigrinus*.

8.3.3 Chlorinated Dioxins and Furans

Polychlorinated dibenzodioxins (PCDDs) are a group of polyhalogenated organic compounds which consist of two benzene rings joined by two oxygen bridges (dibenzo-1,4-dioxin) with one or several of the hydrogens replaced by chlorines and polychlorinated dibenzofurans (PCDFs) are a group of organopollutants with properties and chemical structures similar to polychlorinated dibenzodioxins. PCDDs and PCDFs are released in the environment as by-products of several industrial processes, including incineration of waste, chemical and manufacturing of pesticides. Because of their lipophilic properties, they get accumulated in humans and wildlife causing developmental disorders and cancer as already reported several decades ago (Huff et al. 1980). For this reason, they were listed among the most persistent hazardous organic pollutants in the Stockholm convention. A large number of white rot fungi, namely *P. chysosporium*, *Panellus stypticus*, *Phlebia* spp., and *Bjerkandera* spp., were assessed as potential dioxin degraders. The capability of *P. chysosporium* to mineralize PCDDs was demonstrated in 1985 (Bumpus et al. 1985) without the elucidation of the metabolic pathway. Later, Valli et al. (1992) demonstrated that *P. chrysosporium* degraded 50 % of 2,7-dichlorodibenzo-*p*-dioxin (2,7-diCDD) under ligninolytic conditions while only 10 % was degraded under non-ligninolytic conditions, implying the involvement of LiP and MnP. The proposed pathway for 2,7-diCDD transformation involved oxidative cleavage of 2,7-diCDD by LiP resulting in the production of 4-chloro-1,2-benzoquinone and 2-hydroxy-1,4-benzoquinone. Afterwards, hydroquinones or catechols were produced via LiP and/or MnP mediated oxidation followed by methylation leading to the formation of methoxybenzenes. Interestingly, though the white rot fungus *Panellus stypticus* does not produce either LiP or MnP, it could transform 2,7-diCDD. The formation of 4-chlorocatechol as an intermediate and the inhibition of the degradation process upon addition of piperonyl butoxide, a well-known CYP450 inhibitor, suggested a degradation pathway involving CYP450 different from that described for *P. chrysosporium* (Sato et al. 2002). The same degradation mechanism was proposed for several *Phlebia* species that mineralized 6.5 % of [^{14}C]-2,7-diCDD producing hydroxylated and methoxylated intermediates (Mori and Kondo 2002a, b; Kamei and Kondo 2005). Moreover, Kamei et al. (2005) demonstrated that different *Phlebia* species were also able to degrade higher chlorinated dioxins, such as 2,3,7-triCDD, 1,2,8,9-tetraCDD and 1,2,6,7-tetraCDD. Several strains of the genus *Bjerkandera* were also screened for their ability to transform PCDDs and the production of key enzymes that catalyze the conversion of these pollutants was evaluated in either nitrogen-sufficient or -limited conditions (Manji and Ishihara 2004). Among the tested strains, *Bjerkandera* MS325, that could produce high levels of both MnP and LiP even under non-ligninolytic conditions, was proved to be the most efficient degrader of 1,3,6,8-tetraCDD (21 % of the original content removed in 10 days). Other WRFs were also studied for their ability to degrade dibenzo-*p*-dioxins as reviewed by Pinedo-Rivilla et al. (2009). In this respect, *Trametes* sp. CH2, *Irpex* sp. W3 and *Pleurotus pulmonarius* were reported for their capability to hydroxylate and methoxylate PCDDs (Yamaguchi et al. 2007; Nam et al. 2008).

In contrast to the above, evidence of PCDDs degradation by NLFs is limited to a few ascomycetes, but the enzymatic systems involved in such biotransformations remain unknown. A dioxin-degrading organism was isolated from the activated sludge of a landfill treatment plant and identified as *Acremonium* sp., (Nakamiya et al. 2002). This fungus (later identified as *Pseudallescheria boydii*) was able to remove 82 % of a mixture containing tetrachlorinated up to octachlorinated dioxins (2 ng/mL) under denitrifying conditions in 1 day. The incubation with octachlorodibenzo-*p*-dioxin (OCDD, 100 ng) under aerobic conditions led to the formation of 1- and 2-hydroxydibenzo-*p*-dioxin indicating that OCDD was transformed to non-chlorinated dibenzo-*p*-dioxins through a dechlorination process. However, the degradation mechanism of OCDD and other highly chlorinated dioxins was not described in detail. In another study, *P. boydii* was tested in a slurry bioreactor system for the treatment of dioxin-contaminated soils (Ishii and Furuichi 2007). Surprisingly, in spite of its great capability to remove dioxins, *P. boydii* failed to lower the dioxins level to the environmental standard in this particular experimental setup and, thus the biological treatment was coupled with a physico-chemical process (ethanol extraction) to reach the required quality standard. Two years after, the same authors clarified that *P. boydii* use glucose as carbon source to grow and degrade PCDDs during its logarithmic growth phase regardless of the concentration of dioxins (Ishii et al. 2009). Another study showed that the cyclic ether degrading fungus, *Cordyceps sinensis*, was able to degrade dibenzo-*p*-dioxin (DD), 2,3,7-trichlorodibenzo-*p*-dioxin (2,3,7-triCDD) and octachlorodibenzo-*p*-dioxin (OCDD) using either glucose or 1,4-dioxane as a carbon source (Nakamiya et al. 2005). Less chlorinated catechols and catechols themselves along with *cis, cis*-muconates were identified as by-products, this leading to the postulation of a new dioxin degradation pathway.

Considering the fungal degradation studies on PCDFs, the number of publications dealing with furans is very limited. A mixture of 10 of tetra- to octachlorodibenzofurans with chlorine substituents on the aromatic rings at 2-, 3-, 7-, and 8- positions was partially degraded by *Phanerochaete sordida* YK-624 stationary cultures in low-nitrogen medium and 4,5-dichlorocatechol and tetrachlorocatechol were formed from the degradation of 2,3,7,8-tetra- and octaCDD, respectively (Takada et al. 1996). The white rot fungus *Phlebia lindtneri* was able to degrade 2,8-diCDF producing hydroxyl-diCDF as an intermediate differently from *P. sordida* (Mori and Kondo 2002b). A yeast-like fungal strain was isolated from a dioxin-contaminated soil under selective conditions: the soil sample was suspended in a mineral salt medium supplemented with 1 g L^{-1} of dibenzofuran, a model compound for highly chlorinated dibenzofurans (Hammer et al. 1998). The yeast strain, identified as *Trichosporon mucoides*, metabolized dibenzofuran at fast rate (more than 50 % of the compound added was degraded within 8 h) and water-soluble intermediates were detected during the incubation period. In detail, different isomers of mono- and di-hydroxylated dibenzofurans were observed, along with the ring-cleavage product (2-(1-carboxy methylidene)-2,3-dihydrobenzo [*b*]furanylidene glycolic acid). The ability of the imperfect soil fungus *Paecilomyces lilacinus* to transform dibenzofurans was also investigated and 14 degradation intermediates were identified by UV spectroscopy, mass spectrometry, and proton nuclear magnetic resonance spectroscopy (Gesell et al. 2004). According to this study, the biotransformation of dibenzofuran was initiated by two separate hydroxylation steps leading to the accumulation of monohydroxylated- and dihydroxylated-dibenzofurans. Further oxidation yielded to the ring-cleavage of dibenzofuran structure, which was not previously described for filamentous fungi. The ring fission products were identified as benzo[*b*]furo[3,2-d]-2-pyrone-6-carboxylic acid and [2-(1-carboxy-methylidene)-benzofuran-3-ylidene]-hydroxyacetic acid and its derivatives hydroxylated at 7 and 8- positions on the non-cleaved ring. Riboside conjugates of 2-hydroxydibenzofuran and 3-hydroxydibenzofuran were also detected. These results suggested the involvement of CYP450 in the initial steps of the furans transformation process as already reported for other chlorinated environmental pollutants; the formation of the riboside conjugates that increased with the dibenzofuran concentration supported the common opinion that different types of conjugation (glycosylation, sulfate, or glutathion conjugation) are used by eukaryotic cells to encounter toxic hydroxylated compounds.

8.4 Emerging Organic Micropollutants: Endocrine Disruptors, Pharmaceuticals and Personal-Care Products[2]

In recent years, the presence of organic micropollutants (MPs) in the aquatic environment (i.e., surface waters, groundwaters and drinking water reservoirs) has become an issue of worldwide concern (Luo et al. 2014). The development of powerful and sensitive analytical techniques contributed to the detection and quantification of a large number of organic chemicals in the environment, thus compelling the scientific community to consider this type of contamination as a subject that merits attention (Santos et al. 2010). Such MPs, also termed as "emerging" organic contaminants (Gasser et al. 2014), encompass a vast and ever-expanding array of substances, mainly of anthropogenic origin, that are detected in surface waters at trace concentrations (ng to μg per liter). Among these substances, pharmaceuticals and personal care products (PPCPs) such as antibiotics, anti-inflammatories, cytostatic drugs, disinfectants, β-blockers, UV-filters, fragrances, and other compounds are common water MPs (Verlicchi et al. 2010; Luo et al. 2014). The other group is represented by endocrine disrupting compounds, i.e., chemical compounds that interfere with the endocrine system and therefore could be detrimental to human and wildlife health. The most studied and detected EDs are the natural hormones estrone and 17β-estradiol, the synthetic estrogen 17α-ethinylestradiol, industrial compounds and additives, nonylphenols (NPs), bisphenol A (BPA) and also the personal care product triclosan (Cajthaml et al. 2009a). Besides, endocrine disrupting effects have been reported for PAHs (Arcaro et al. 1999), PCBs, CBAs, PCDDs and other POPs (Cajthaml 2014).

The occurrence of emerging organic MPs in the environment has been associated with a number of negative effects, mainly regarding the aquatic ecosystems (see below). Till date, however, discharge guidelines and standards for most of the MPs do not exist. For example, environmental quality standards for a minority of MPs (e.g., NPs, BPA, diethylhexyl phthalate-DEHP and diuron) were stipulated in Directive 2008/105/EC (European Parliament and The Council 2008). Nonylphenol and nonylphenol ethoxylates were also recognized as toxic substances by the Canadian government (Canadian Environmental Protection Act 1999); but other potent MPs, such as PPCP and EDs have not been included yet in the list of regulated substances (Luo et al. 2014).

[2]Additional information on fungal treatment of emerging pollutant in municipal wastewater is presented in Chap. 5—*Fungal bioremediation of emerging micropollutants in municipal wastewater.*

8.4.1 Toxicity of Emerging Micropollutants

The adverse effects associated with the occurrence of MPs in the aquatic environment include short-term and long-term toxicity, endocrine disrupting effects and induced antibiotic resistance of microorganisms—although the latter phenomenon is still largely disputed (Luo et al. 2014). Some damaging health effects that occur in human and wildlife are due to the exposure to particular organic pollutants, many of which could be detected in all human beings, which cause damage or impairment of neural, immune and reproductive systems (Luo et al. 2014). Although the majority of EDs and PPCP are not considered as recalcitrant as other POPs (e.g., PAHs, PCBs, PCDDs, etc.), the widespread usage and continuous delivery of low concentration of contaminants to the receiving ecosystems might give rise to toxicity even without high rates of persistence (Santos et al. 2010). It has been demonstrated that environmental concentrations of EDs induce the "feminization" of male tissues and/or the presence of female reproductive organs in males, along with other dysfunctions of the reproductive systems in many aquatic animals (Liu et al. 2009; Santos et al. 2010; Eggen et al. 2014). Since the 1940s, decline in semen quality and increasing incidence of testicular cancer in humans has been observed, although a direct link between exposure to EDs and manifestation of negative effects could be proved only in model organisms (UNEP 2012), which is also applicable to a number of scientific studies that reported the thyroid-disrupting effects of EDs and other environmental MPs. (e.g., BPA, phthalates, and perfluorinated chemicals) (Boas et al. 2012).

The challenges facing environmental scientists are significant. However, it is essential to conduct research that increases our scientific knowledge on the wide range of water MPs, including a detailed understanding of their environmental cycling and toxicological impact. To understand the dimension and the importance of such problem, it is interesting to look at the strategies adopted by the leading EU countries. As an example, the Swiss Federal Office for the Environment (FOEN) conducted the project "Strategy Micropoll" in the period 2006–2010 in order to develop a strategy to reduce the input of organic MPs from domestic wastewater into Swiss surface waters. Results from this project led to the development of the VSA-Platform "Verfahrenstechnik Mikroverunreinigungen," for which the Swiss Federation is going to grant ca. 1 billion euro funds in the following years.

8.4.2 Fungal Biodegradation of EDs

The number of scientific contributions dealing with fungal degradation of EDs and other water MPs has been growing steadily through the last decades. The majority of such studies, although not all, focused on the groups of wood-rotting fungi, mainly because of their capability to produce unspecific, high redox potential LEs (Ashger et al. 2008; Cajthaml et al. 2009a). These potent biocatalysts can, bring about the attack on the phenolic moieties present in the chemical structure of many of the MPs, resulting in the formation of phenoxyl radicals, which subsequently undergo oxidative coupling. Once dimers, tetramers or oligomers are formed, the overall biological and endocrine disrupting activity is significantly reduced or completely removed (Cabana et al. 2007b; Cajthaml 2014). Intracellular enzymatic complexes like CYP450 were also shown to be involved in the degradation of several compounds by the intact fungal cells, as well as Fenton-like reactions (Marco-Urrea et al. 2010d; Cajthaml 2014). Due to the number and diversity of compounds in the MPs family, we will only select some of the compounds which were extensively studied and provide a brief overview of the most significant results reported in the scientific literature.

Typical man-made EDs with estrogen-like action include NPs, BPA and 17α-ethinylestradiol (EE2). NPs isomers mainly occur in the environment as degradation products of nonylphenol-polyethoxylates which are widely used as nonionic surfactants in many industrial processes. BPA is a key building block for polycarbonate plastics and epoxy resins, but it is also used in a number of other materials and applications. Synthetic estrogens, such as EE2, are used worldwide as oral contraceptives. Non-metabolized EE2 and its conjugates are first excreted into the urban sewage systems and, after incomplete degradation in the wastewater treatment processes, they reach the surface water ecosystems. The capability of the above-cited compounds to bind human receptors and cause endocrine disruptions has been documented (Cabana et al. 2007b).

NPs: NPs were found to be very efficiently degraded by LF cultures and purified LEs (see Corvini et al. 2006; Cabana et al. 2007b; Cajthaml et al. 2009a; Cajthaml 2014). The intermediate steps of NPs biotransformation by LFs, including (C–O) coupling reactions and oxidation at the terminal carbon of the alkyl chain, are summarized in Fig. 8.4. Several studies reported the formation of dimeric and holigomeric NP transformation products using Lac of *Coriolopsis gallica* and *C. polyzona* (Cabana et al. 2007a; Torres-Duarte et al. 2012), a mechanism that is known to suppress the estrogenic activity of the parent compounds. Tsutsumi et al. (2001) suggested similar transformation products following oxidation with MnP of *P. chrysosporium*. This coupling metabolite is probably produced as a result of the action of LEs on phenolic hydroxyl groups that results in the formation of phenoxyl radicals which subsequently undergo different spontaneous reactions, including oxidative coupling (Hofrichter 2002; Dec et al. 2003). Dubroca et al. (2005) proposed the formation of a direct C–C bond between the phenolic rings and the analogous structure was finally identified by Wang et al. (2012) with respect to *p*-t-octylphenol. Production of NP oligomers was similarly observed by other authors who tested, e.g., the degradation abilities of Lac from the aquatic fungus *Clavariopsis aquatica* (Junghanns et al. 2005). Syed et al. (2011, 2013) investigated the involvement of CYP450 (CYP5136A3 and CYP63A2) from *P. chrysosporium* in the NP transformation mechanisms and discovered a very versatile function of these intracellular monooxygenases. They documented the oxidation of NP and other alkylphenols at the terminal alkyl chain carbon (omega oxidation) *via* alcohols resulting in aldehydes. Based on the literature and known pathway in

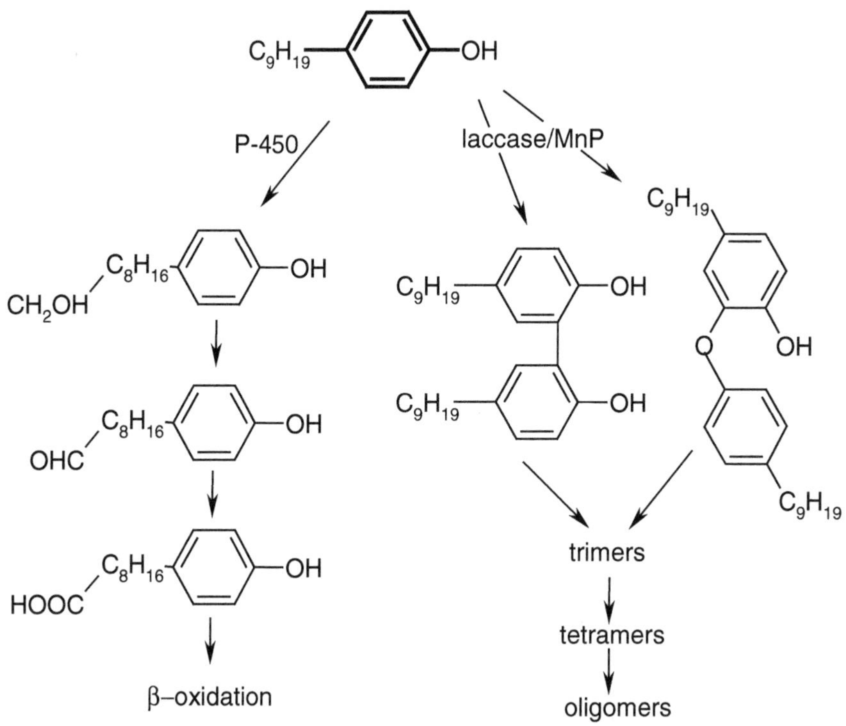

Fig. 8.4 Nonylphenols biotransformation by LFs—adapted from Cajthaml (2014)

yeasts/fungi for the degradation of an alkylphenol moiety, the authors concluded that the terminal oxidation of the alkyl side-chain is followed by the removal of the terminal carbons via the β-oxidation of NPs *via* carbon–carbon (C–C) and carbon–oxygen pathway as noted in other fungi, e.g., the yeast *Candida aquatextoris* (Vallini et al. 2001; Corvini et al. 2006; Rozalska et al. 2010). The possible entry into the basal metabolism is also supported by a study of Dubroca et al. (2005), who observed mineralization of NPs by a culture of *T. versicolor* using ^{14}C-labeled NP. Surprisingly, the potentially anticipated metabolites following direct hydroxylation of NPs by CYP450 of LFs have not been detected yet (Corvini et al. 2006). It was shown that the production of the coupled NPs leads to a reduction of estrogenic activity (Tsutsumi et al. 2001), an aspect which has not been investigated with respect to alkylphenol carboxylic acids so far.

BPA: BPA has also been shown to be susceptible to biodegradation by several fungi and fungal enzymes. Fukuda et al. (2001) demonstrated that purified Lac from *Trametes villosa* efficiently transformed BPA under in vitro conditions without any requirement for electron-shuttling mediator. Structural analysis of the BPA degradation products using nuclear magnetic resonance spectroscopy indicated that the oligomers of BPA were formed as a result of successive oxidation–condensation entailing C–C coupling (Fukuda et al. 2004). The presence of oligomeric fragments, each including a phenol moiety, suggested the cleavage of the oligomeric products to release 4-isopropenylphenol or simultaneous fission of the original BPA molecule. Neither the soluble nor the insoluble fractions of the BPA reaction products had estrogenic activity, even at high concentrations (Fukuda et al. 2004). Similar metabolites were also detected by Michizoe et al. (2005) who identified 4-isopropylphenol following the transformation of BPA by Lac (*Trametes* sp.) in a reverse micelle system in organic media, and by Cabana et al. (2007a), who detected BPA polymeric structures. Huang et al. (2005) proposed that the formation of C–O coupled polymer occurred *via* radical-mediated mechanisms triggered by horseradish peroxidase and reported the formation of 4-isopropenylphenol. Wang et al. (2012, 2013a) tested LiP from *P. sordida* and identified dimerized BPA with C–C coupling and also with a C–O bridge. The same authors also detected hydroxylated BPA and suggested the involvement of CYP450 in the transformation (Wang et al. 2013b). Hirano et al. (2000) used a MnP isoenzyme from *P.ostreatus* and detected mostly breakdown products such as phenol, *p*-cresol, 4-isopropenylphenol and its dimer. Structures akin to these were reported by Torres-Duarte et al. (2012) following the treatment of BPA with Lac of *C. gallica*. It is noteworthy that in all the studies where the authors monitored the estrogenic potency of samples after treatment, the removal of BPA by either purified LEs or fungal cultures was always accompanied by a simultaneous depletion of the estrogenic activity (Tsutsumi et al. 2001; Cabana et al. 2007b; Cajthaml et al. 2009a; Torres-Duarte et al. 2012).

EE2: Synthetic and natural estrogens are considered to be major contributors to the estrogenic activity associated with wastewater treatment plant (WWTP) effluents due to their incomplete removal in conventional wastewater treatment process (Clouzot et al. 2008). EE2 was generally reported to be more resistant to biodegradation than other natural estrogens (Cajthaml et al. 2009b; Tran et al. 2013). Numerous authors reported the high efficiency of LFs and LEs in biodegradation or transformation of EE2 (see reviews Cabana et al. 2007a; Cajthaml et al. 2009b; Demarche et al. 2012b) and effective EE2 degradation by whole cultures of several LFs has been described (Blánquez and Guieysse 2008; Cajthaml et al. 2009a). Estrogenic activity generally decreased with the progression of degradation. However, during EE2 biodegradation tests with *I. lacteus*, *P. ostreatus*, *P. cinnabarinus* and *P. magnolia*, residual or increased estrogenic activity was observed, suggesting the production of estrogen-like degradation products (Cajthaml et al. 2009a). Suzuki et al. (2003) investigated the removal of the steroidal hormones E2 and EE2 in organic solvent under biphasic conditions using MnP and the Lac-mediator (hydroxybenzotriazole—HBT) systems from *P. chrysosporium* ME-446 and *T. versicolor* IFO-6482. They also monitored the evolution of estrogenic activity and documented the high potential of LEs for the biodegradation of EE2. Similar results were obtained by other authors (Tanaka et al. 2001) who worked with Lac from *Trametes* sp. and *Pycnoporus coccineus* in a rotating reactor. Auriol et al. (2007) studied the *T. versicolor* Lac-catalyzed conversion of natural oestrogens and EE2 and showed that the Lac-catalyzed system was not significantly affected by the presence of a real municipal wastewater matrix.

Moreover, when HBT was used as a mediator, the Lac-catalyzed system showed an improved efficiency. However, limited information is available about the fungal biotransformation products of EE2. Kresinová et al. (2012) published a study including a broad set of in vivo and in vitro experiments using sub-cellular and cellular fractions as well as purified LEs from *P. ostreatus* and suggested that the EE2 degradation mechanisms could involve intracellular microsomal enzymes, mycelium-associated laccase-like isoenzymes and extracellular ligninases (MnP and Lac). The synergic contributions of the individual transformation steps were documented and each phase of transformation was reported to individually reduce estrogenic activity. The authors also detected a set of intermediates clearly indicating that in spite of its recalcitrance, EE2 is efficiently transformed by the ligninolytic system and also by the CYP450 monooxygenase system. The intermediates included methoxylated estron (E1) and dioxo 17β-estradiol (E2), which were probably formed by the action of Lac, and dehydrogenated and hydroxylated EE2. Metabolites similar to the latter ones were also described by Choudhary et al. (2004) and Della et al. (2008) following the incubation of EE2 with the mucoromycete *Cunninghamella elegans* and microalgae, respectively. In a previous study, Tanaka et al. (2001) detected dimeric products of EE2 after treatment with Lac from *Trametes* sp.and observed a significant decrease in estrogenic activity but they did not perform any further analysis of the intermediates in detail. Other authors studied EE2 and natural estrogen biotransformation by Lac from the ascomycete *Myceliphthora thermophila* in a membrane bioreactor (Lloret et al. 2013) and detected two dimers and trimers of EE2 along with several transformation products of 17β-estradiol (E2). They identified E1 following Lac oxidation of E2 and characterized two C–C and two C–O dimers as well as four trimers with various coupling combinations and, concluding that the Lac transformation of E2 is analogous to that of EE2. The authors generally observed a decrease in the estrogenic activity during the course of the transformation. Similar results were achieved using Lac from *Myceliophthora* and *Trametes pubescens*, both adsorbed on glass beads. The enzymes were used for the transformation of natural estrogen E2 (Nicotra et al. 2004) and the reaction products were also tentatively identified as C–C dimeric compounds with coupling between either 4′-4 or 2′-2 and C–O (phenolic ring) at the 4′ or 2′ positions.

8.4.3 Fungal Degradation of Pharmaceuticals and Personal Care Products (PPCP)

The ability of fungi, especially belonging to the group of wood-inhabiting Basidiomycetes, to degrade PPCP detected in surface and process waters and sludge biosolids, has been demonstrated in recent years. In relation to this issue, non-steroidal anti-inflammatory drugs (NSAID) (e.g., diclofenac, naproxen, ibuprofen, ketoprofen), blood lipid regulators (clofibric acid), β-blockers (e.g., propanolol, atenolol), anti-epileptic drugs, (carbamazepine) and antibiotics (e.g., sulfamethoxazole, ciprofloxacin, erythromycin) are the most intensively studied pharmaceutical compounds (Marco-Urrea et al. 2009, 2010a, b, c, d; Accinelli et al. 2010; Rodarte-Morales et al. 2011, 2012; Domaradzka et al. 2015).

Marco-Urrea et al. (2009) screened four white rot fungi (*T. versicolor*, *I. lacteus*, *G. lucidum* and *P. chrysosporium*) for their ability to co-metabolize ibuprofen (IBU), clofibric acid (CA) and carbamazepine (CBZ) in model (glucose-supplemented) liquid systems. All strains efficiently removed IBU while the turkey-tail fungus *T. versicolor* was the most efficient in the biotransformation of the recalcitrant CA and CBZ. Failure of PPCP degradation by Lac and Lac-mediator systems under in vitro condition, reduced biotransformation rates in the presence of CYP450 inhibitors and the detection of hydroxylated and di-hydroxylated derivatives of IBU prompted the authors to conclude that CYP450 monooxygenase system is responsible for the initial oxidative attack to this NSAID. Further experiments in which *T. versicolor* was incubated in the presence of other NSAIDs (naproxen, ketoprofen and diclofenac) showed the involvement of intracellular (CYP450) and to a lesser extent, extracellular (Lac) enzymatic systems in pharmaceuticals degradation by this fungus (Marco-Urrea et al. 2010a, b, c). In all cases, the time course of the degradation of NSAIDs was accompanied by a significant drop in toxicity as assessed by the test with the luminescent bacterium *Vibrio fisheri*, and also by the presence of transient biotransformation products, the chemical structure of which had been elucidated by nuclear magnetic resonance techniques. The same authors took care to demonstrate whether *T. versicolor* could use mechanisms other than those reported above (CYP450-Lac) to oxidize representative PPCP (Marco-Urrea et al. 2010d). To rule this out, the fungus was incubated under conditions triggering the production of ROS (e.g., $OH^{\cdot}$) *via* quinone-redox cycling (see also Gómez-Toribio et al. 2009). The degradation of CBZ, CA, and the β-blockers atenolol and propranolol occured (80 % of 10 mg L^{-1} compounds initially added was removed after 6 h) only under 2,6-dimethoxy-1, 4-benzoquinone (DBQ) redox cycling conditions (i.e., fungus mediated Fenton-like mechanism) although the NSAID-bioconversion process was short-lived due to the depletion of DBQ by the same $OH^{\cdot}$. Several hydroxylated metabolites where reported for the first time, although the authors noted the complete disappearance of degradation products at the end of the incubation period (Marco-Urrea et al. 2010d).

Scale-up processes of *T. versicolor*-mediated degradation of PPCP from flasks to bubble column reactors were successful under both batch and continuous mode of operation (Blánquez and Guieysse 2008; Jelić et al. 2012), even in the presence of non-sterile real wastewaters from WWTP (Cruz-Morató et al. 2013), although sources of C and N (glucose and ammonium tartrate) were necessary for the growth of the fungus. In some cases, however, the production of hydroxylated metabolites with increased toxicity with respect to parent molecules has also been reported (Cruz-Morató et al. 2013).

8.4.3.1 White Rot Inocula for the Treatment of Sludge Biosolids[3]

Depending on their hydrophobicity, water MPs tend to partition in wastewater processes and adsorb onto sludges and therefore, considerable concentrations of PPCP and EDs are often detected in sludge biosolids (Cabana et al. 2007b; Clarke and Smith 2011). In consideration of this, Rodriguez-Rodriguez and colleagues (2012) tested a sludge-bioaugmentation approach using *T. versicolor* mycelia to remove the load of organic MPs therein. The fungus, pre-grown on lignocellulose-based pellets, was mixed (38 % w/w dry weight) with sewage sludge and control microcosms were set up with non-inoculated, sterile lignocellulose. The fungus was deemed responsible of a faster disappearance of some MPs (ranitidine and fenofibrate) and enhanced the removal of atorvastatin, diclofenac and hydrochlorothiazide in the initial phase of incubation. However, *T. versicolor* was rapidly overgrown by members of the sludge microflora and the overall degradation of several compounds at the end of the incubation period remained similar to that observed in the non-inoculated control. The authors therefore concluded that the bioaugmentation with white rot inoculant is feasible and advantageous with respect to the rates of contaminants removal but the biostimulation of the sludge microbiota, as assessed *via* denaturing gradient gel electrophoresis (DGGE), merits equal attention in the light of future applications (Rodríguez-Rodríguez et al. 2012).

8.4.3.2 Enzyme Aggregates and Immobilized Forms for the Treatment of Process Waters

The pollutant degradation potential of LEs was widely evidenced in the previous sections of this chapter. However, the application of fungal catalysts in free solution is unfeasible due to the denaturing and/or inactivating conditions (temperature, pH, metal ions etc.) often encountered in sewage or waste waters (Cabana et al. 2007b; Demarche et al. 2012a, b; Kües 2015). Therefore, for the optimal retention of catalytic properties and to promote an enduring reuse of such biocatalysts in continuous-flow processes, immobilization, encapsulation, and insolubilization are techniques of great interest nowadays.

Immobilization of Lac for its potential use in a variety of applications, including bioremediation, has been the subject of intensive research in the past decades (see the review Durán et al. 2002; D'Annibale et al. 1999, 2000). Among the carriers for the immobilization/encapsulation pre-activated for ionic or covalent binding with the enzyme, glass and mesoporous silica beads, ceramics and clay minerals, natural (chitin, chitosan, agarose, cellulose nanofibers) and synthetic (polystyrene, nylon 6, poly acrylamide) polymers were successfully used in a number of studies and

[3]Further information on the use of white rot and brown rot fungi in wastewater and sludge treatment can be found in Chap. 4—*Application of biosorption and biodegradation function of fungi in wastewater and sludge treatment.*

applications (Durán et al. 2002; Ba et al. 2013; Kües 2015). The only technical disadvantage of such immobilized or encapsulated forms is the low ratio of enzyme to support in the overall volume of biocatalyst (Cabana et al. 2007b; Ba et al. 2014). To overcome this, carrier-free insolubilization of laccase and other enzymes can be achieved by producing cross-linked enzyme aggregates (CLEAs). The technique relies on the precipitation of enzymes (e.g., Lac) and the subsequent addition of cross-linking agents, which bind together amino-acid residues of proteins to form stable and insoluble aggregates. The most widely used cross-linking agents are homobifunctional aldehydes (dialdehydes) such as glutaraldehyde, glyoxal and dextran polyaldehydes, although these chemical agents can have adverse effect on aquatic organisms (and humans) besides causing a certain loss of enzymatic activity during biocatalyst preparation (Ba et al. 2013). Chitosan represents a valuable alternative to dialdehydes as cross-linkers, provided that the carboxylic groups of the protein of interest are pre-activated by carbodiimides such as 1-ethyl-3-(3-dimethylaminoisopropyl) carbodiimide hydrochloride (EDC) (Zhang et al. 2009; Arsenault et al. 2011; Cabana et al. 2011; Ba et al. 2013).

Several studies reported that regardless of the cross-linker used, Lac- or peroxidases-based CLEAs are biocatalysts with improved properties (i.e., stability, catalytic efficiency and reiterate use) as compared to enzymes in their free-form and offer greater potential for applicability in the treatment of wastewater MPs (Cabana et al. 2007b; Taboada-Puig et al. 2011; Ba et al. 2013; Kües 2015). In an attempt to broaden the range of substrates oxidized by these biocatalysts, some authors have recently tested the co-aggregation of different type of enzymes, i.e., combi-CLEAs. For instance, Ba et al. (2014) characterized combi-CLEAs of *T. versicolor* Lac and mushroom-derived Tyr and then tested its efficiency for the removal of acetaminophen. Taboada-Puig et al. (2011) instead optimized the co-insolubilization of *B. adusta* VP and *A. niger* glucose oxidase (GOD), the latter serving as ancillary H_2O_2-producing enzyme. A good example of combi-CLEAs is represented by the concomitant aggregation of Lac, VP, and GOD from *T. versicolor, B. adusta* and *A. niger*, respectively (Touahar et al. 2014). This versatile combi-CLEA was able to oxidize a wider spectrum of MPs (pharmaceuticals) than Lac-based one, although the requirement of co-substrates (e.g., GOD substrates and Mn^{2+}) for its optimal activity might generate additional costs in case of practical applications.

Different types of reactors bearing immobilized or insolubilized enzymes have been tested for the removal of MPs from artificial or real wastewaters, although most of the research work has been carried out at bench-scale. Packed bed, fluidized bed, perfusion basket, and membrane reactors (Cabana et al. 2007a, 2009a, b; Taboada-Puig et al. 2011; Lloret et al. 2012a, b; Nair et al. 2013) were successfully used in batch or continuous mode of operation. Among these, membrane reactors appear to be the most promising for the treatment of micropollutant-containing waters under continuous-flow conditions since the porosity of the membrane allows the retention of biocatalyst while letting the permeate to flow through (Ba et al. 2013; Kües 2015).

The pilot-scale experiment of Gasser et al. (2014) demonstrated that a fluidized bed reactor (460 L total working volume) equipped with *Thielavia* (*Myceliophthora thermophila*) laccase immobilized on fumed silica nanoparticles (fsNP, 0.5 kg) and an ultra-filtration membrane unit (UF porosity 0.04 μm) could effectively and continuously remove BPA from wastewater treatment plant effluents. The experiment was conducted under field conditions, at a Swiss wastewater treatment plant, over a period of 45 days and the reactor was fed with WWTP effluents and operated continuously in a cyclic filtration mode which included membrane backwash in order to prevent clogging (net permeate production78 L h^{-1}). At the end of the test, the Lac-fsNP biocatalyst retained around 30–40 % of its initial activity and approximately 66 % of BPA in the reactor's inlet was removed (Gasser et al. 2014). The authors also estimated the overall economic feasibility of this novel enzymatic treatment at WWTP and concluded that the costs for the Lac-fsNP bioreactor process (0.130 € m^{-3}) are just slightly higher than other suitable treatments, namely ozonation (0.078 € m^{-3}) and powdered activated carbon adsorption (0.114 € m^{-3}). Further improvement of the treatment efficiency and additional reduction of costs might be achieved by optimizing some of the running parameters of the process (e.g., mixing conditions, nanobiocatalyst load, and hydraulic retention time). Moreover, as already stated above for combi-CLEAs, the co-immobilization of enzyme mixtures on the same nano-carrier could widen the range of wastewater MPs removed by such bioreactor treatment, thus enhancing the overall applicability potential of enzymatic treatment technologies at WWTP.

8.5 Pilot- and Field-Scale Mycoremediation

Pilot- and full-scale applications are essential to evaluate the effectiveness and economic viability of fungal technologies. However, till date, there are few reports concerning the use of fungi in the treatment of contaminated matrices at field-scale and thus many technical and engineering challenges remained in the area of its application. For instance, the model WRF *P. chrysosporium* was very efficiently degrading organic pollutants in laboratory-scale experiments but never showed similar effectiveness in any field-scale tests.

In the early 1990s, *P. chrysosporium* was used in a pilot-scale treatability study at a trinitrotoluene (TNT) contaminated-site—a former USA Naval submarine base (Bangor, Washington). The initial TNT concentration of 1844 ppm was reduced to 1087 ppm in 120 days (degradation of 41 % of the original TNT content). However, the final concentration was still significantly above the target level of 30 ppm and thus the test was considered a failure (US EPA Handbook 1993).

A decade later, a field-scale trial for the remediation of crude oil-contaminated soil was performed using composting biopiles technology where the efficiency of the traditional process (soil aerated and amended with bulking agents, such as straw, sawdust, wood chips, etc.) was compared with that of a bioaugmentation approach (Li et al. 2002). In particular, three fungal strains characterized by high lipase activities (*Mucor* sp., *Cunninghamella* sp. and *Fusarium* sp.) were isolated and reintroduced into the contaminated matrices either with or without *P. chrysosporium.* The results showed that the inoculum composed of only indigenous fungi remarkably increased the degradation rate of total petroleum hydrocarbons (TPH) with respect to non-inoculated soil, while the introduction of the allochthonous fungus did not improve the efficiency of TPH removal process.

A similar outcome was also achieved a few years later in a pilot-scale experiment for the treatment of soil contaminated by TNT, PAHs, PCDDs and PCDFs (Tuomela et al. 2012). Pine bark was selected as a substrate to support the growth of the litter-decomposing fungus *Phanerochaete velutina.* Once the inoculum was prepared, the allochthonous fungus was introduced into soil and the extent of degradation of the target pollutants was evaluated. *P. velutina* growth was completely inhibited when introduced in soil with high concentration of PAHs (5000 mg kg^{-1}) and TNT (>10 g kg^{-1}) and no degradation occurred, moreover, the presence of molds such as *Trichoderma* clearly prevented its growth. On the other hand, dilution of PAH and TNT-contaminated soil with garden compost significantly enhanced the degradation process. However, since the growth of both *P. velutina* and indigenous microorganisms was clearly stimulated under such conditions, the role of the litter-decomposing soil fungus in the removal of PAHs and TNT could not be clarified. These results were also confirmed two years later when a field-scale experiment (2 t) for the treatment of a PAH-contaminated soil was carried out (Winquist et al. 2014). At first, the degradation of PAHs was assessed in a laboratory-scale experiment where soil was mixed with composted green waste (1:1) and incubated with or without *P. velutina*. The higher extent of degradation (96 % for 4-ring PAHs and 39 % for 5- and 6-ring PAHs) observed in microcosms inoculated with the fungus after three months of incubation as compared to the biostimulated ones prompted the researchers to scale-up the process. Unfortunately, even if the percentage of degraded PAHs in soil under field condition was extremely high, the results obtained in the case of *P. velutina*-inoculated treatment (bioaugmentation) or non-inoculated treatment (biostimulation) were comparable. This again suggested that the degradation of PAHs could not be ascribed univocally to the exogenous fungus.

P. velutina inoculated on pine bark was also used to remediate a TNT-contaminated soil diluted with composted green waste (1:20) at pilot-scale level (0.3 t) (Anasonye et al. 2015). TNT was efficiently degraded up to 80 % in 107 days. The quantification of the fungal ITS region DNA copy number by quantitative PCR (qPCR) using specific primers for *P. velutina* proved that the fungus grew in such contaminated soil withstanding high concentration of TNT and autochthonous microorganisms. However, due to economical and practical issues, no control with non-inoculated soil was prepared and thus the real efficiency of the fungus in the removal of TNT could not be compared with the degradation ability of indigenous microbiota present in the green waste.

The capability of the white rot fungus *T. versicolor* to remediate a pentachlorophenol (PCP)-contaminated soil at field-scale level was also evaluated (Walter et al. 2005). The contaminated soil was collected at a timber treatment plant where PCP was widely used as a wood preservative. In a previous study performed at laboratory scale, PCP was removed from both biostimulated (soil mixed only with a lignocellulosic substrate) and bioaugmentated microcosms (*T. versicolor*-colonized lignocellulosic substrate) when the contaminated soil was diluted with uncontaminated soil. The role of the fungal inoculum versus indigenous microorganisms in the PCP degradation was not completely clarified, but complete mineralization of PCP was observed only in the presence of the allochthonous fungus. In view of this last result, soil "cells" (final volume approx. 500 L) were designed to carry out exclusively a *T. versicolor*-based treatment. The contaminated soil (1000 mg kg^{-1}) was diluted with non-contaminated soil and mixed with wood chips obtained from cuttings and shreds. Thereafter, such blended pile of material was mixed with a specific formulation composed of sawdust, cornmeal and starch previously inoculated with *T. versicolor* to obtain a final ratio of soil: woodchips: fungal inoculum of 40: 20: 40 or 60: 20: 20, respectively. After a rapid decline in PCP concentration within the first 25 weeks of incubation, the degradation rate remarkably decreased. Nonetheless, after 2.5 years, more than 90 % of the original PCP content was removed in both cells (20 or 40 % of fungal inoculum used) indicating that PCP levels at the end of treatment were not statistically different with respect to the differences in the amount of fungal inoculum.

P. ostreatus is one of the best candidates for potential applicability in mycoremediation of soils, therefore several field-scale tests based on the use of such were designed. This WRF was used for the filed-scale remediation of explosive-contaminated soil from the Yorktown Naval Weapons Station (Virginia, USA) where munitions manufacturing process was carried out for decades (Axtell et al. 2000). Two plots (6 cubic yards) of contaminated soil were blended with 3 cubic yards of a substrate mixture previously colonized by *P. ostreatus*. Concentrations of the target explosives, namely TNT, HMX (octogen) and RDX (cyclonite), were noticeably reduced during a 62-day incubation period. Nonetheless, in soil amended only with the substrate mixture, the concentrations of TNT, HMX, and RDX were also reduced substantially during the same period suggesting that the addition of amendments enhanced the growth and activity of indigenous microorganisms and thus sufficient to promote the degradation of these compounds in soil. However, the role of the *P.ostreatus* in this field-scale test was not elucidated.

Further application of fungal bioaugmentation was evaluated for the remediation of creosote-contaminated soil at a wood-preserving site in West Virginia (Lamar et al. 2002). After a preliminary bench-scale evaluation of the degradation capabilities of two WRFs, namely *P. ostreatus* and *I. lacteus*, *P. ostreatus* from a commercial spawn provider was selected to scale up the process. Two pilot-scale biocells comprising the contaminated soil, a sterile substrate (wood splinters or wood splinters with sawdust) and *P. ostreatus* spawn were monitored for 276 days. Concentration of all the 16 PAHs was reduced in both biocells below the industrial risk-based level with the exception of benzo[*a*]pyrene (Lamar et al. 2002). However, the involvement of indigenous microorganisms in the degradation process was not assessed.

The influence of *P. ostreatus* mycelia on the degradation of selected PAHs in soil was investigated under field conditions also in the Northern temperate zone (Hestbjerg et al. 2003). Homogenized *P. ostreatus* refuse from commercial mushroom production was added to PAH-contaminated soils collected from two sources: a former shipyard and coal tar storage from an asphalt factory. Treatments (soil control, soil mixed with autoclaved sawdust substrate and soil mixed with *P. ostreatus* refuse) were set up in concrete cylinders for both soils. In the case of the soil from the coal tar storage, the addition of the oyster mushroom significantly enhanced the degradation rate of 4-ring PAHs compared to the rate in either soil control or biostimulated cylinder, while 5- and 6-ring PAHs were only slightly degraded and no significant difference was observed between the two treatments. On the contrary, the growth and the activity of *P. ostreatus* were completely prevented when inoculated in the other soil, although the factors that resulted in its inefficacy were not clarified.

8.6 Mycoremediation of Soils: Opportunities and Pitfalls[4]

As it was evidenced in the previous sections, there is no doubt that both LFs and NLFs are potent biological tools for the degradation of persistent soil organopollutants and other emerging water MPs. Considering the soil compartment, fungal inoculants were successfully used since the 1980s for remediation studies at laboratory scale but only a limited number of field-scale trials, which were attempted in the past two decades, showed remarkable limitations with regard to the practical application of technologies based on the use of LFs. As a consequence, all the attempts to promote soil mycoremediation on the market and propose it as a valuable on-site bioremediation option have failed (Šašek 2003; Baldrian 2008).

The reasons that lie behind this failure are diverse, as it was already observed by several authors. The first and most important is, without any doubt, the competition between introduced fungi and the resident soil microflora. The idea that a single species, in most cases allochthonous to the soil, should bring about the complete restoration of a contaminated matrix is absolutely unrealistic and unsustainable. Fungi, like bacteria and plant roots introduced into the soil, are influenced by the type and activity of other resident soil organisms and this has a great influence on their colonization capacity. Moreover, the inoculation support onto which the fungi are immobilized prior to their introduction in non-sterile soil (e.g., straw, hay, corn cobs, seeds and seeds husks, wood chips or shavings, bark, etc.) represents a source of nutrients for the autochthonous soil microbiota, this generating divergent effects: an ensuing positive impact on bioremediation efficiency as has been reported in some cases (Lang et al. 2000; Steffen et al. 2007; Federici et al. 2012) and

[4]Additional information on using fungi and mycoaugmented compost to treat polyaromatic hydrocarbon can be found in Chap. 1—*Fungi in composting*.

growth-inhibition and mycophagy in other circumstances (de Boer and van der Wal 2008; Baldrian 2008).

The strict growth conditions required by most white rot fungi represent another significant constraint for the application of fungal technologies at field-scale. For example, *P. chrysosporium* requires high temperature to grow (37 °C) and to produce LEs (30 °C) making its application inadequate under actual environmental conditions (i.e., due to daily and seasonal fluctuations of temperatures) (Hestbjerg et al. 2003). On the other hand, several studies demonstrated that *P. ostreatus* was less affected by temperature and remained active in the degradation of organic pollutants even at very low temperatures (up to 8 °C) (Eggen and Sveum 1999; Lang et al. 2000; Hestbjerg et al. 2001). Moreover, in contrast to *P. chrysosporium* and other white rot fungi such as *D. squalens*, *Pleurotus* sp.was less affected by the presence of soil organisms and being highly competitive with soil microbiota, was more successful at colonizing the contaminated matrix (in der Wiesche et al. 1996; Lang et al. 1997, 1998).

One of the parameters which deeply influence the success of bioremediation application is the bioavailability of contaminants. In turn, this parameter depends on soil organic matter content, texture (e.g., percentage of clay minerals), structure (e.g., peds size, microporosity, etc.) and hydrophobicity of the contaminants. It has been claimed several times that biodegradation hardly ever reaches the complete removal of POPs, for example PAHs and PCBs, this being particularly valid in the case of historically contaminated matrices with high proportion of non-bioavailable compounds. It should be borne in mind that thorough extractions of lipophilic contaminants under harsh conditions that allow the quantification of total contaminant load in a defined matrix (soil) would also include the portion of "aged" pollutants tightly bound to soil components which are inaccessible to microbes in reality. Therefore, the concept of contaminant bioavailability is more readily accepted by decision makers and regulatory organizations (Naidu et al. 2013). Several authors in fact assert that bioavailability should be the underlying basis for risk assessment and in setting the remediation goals at contaminated sites (Latawiec et al. 2011; Naidu et al. 2013). In view of this, establishing contaminant bioavailability and careful testing of the treatability of contaminated matrices prior to field-scale application might help to clarify the feasibility of certain bioremediation interventions. Moreover, accurate compound-specific and ecotoxicological analyses (during and after mycoremediation trials) can help to assess whether the oxidized and more water-soluble derivatives are likely to exert a higher toxicity than parent molecules themselves, or if the residual portion of contaminants that has been immobilized in the soil humus, is inaccessible. In case polar metabolites are formed, as in the case of oxidized high molecular weight (HMW)-PAHs, it is of utmost importance to identify the members of the bacterial community which are able to cooperate with fungi and to consider their contribution.

Another critical factor that requires consideration while dealing with mycoremediation approaches is the formulation and the amount of inoculum which can favor an efficient and long-lasting colonization of the soil matrix. Ligninolytic fungi, both white rots and litter decomposers, are not endowed with intrinsic capabilities to grow in soil unless an external substrate is also provided. Therefore, lignocellulose pre-colonized by the fungus constitutes the inoculum/amendment for soil. However, using large amount of lignocellulose-immobilized fungi is not economically feasible and would significantly increase the volume of the soil after treatment. A valid alternative to bypass the high costs associated with the production of fungal inoculum is the reuse of Spent Mushroom Compost (SMC) which is a cheap and readily available source of both fungal mycelia and enzymes (Phan and Sabaratnam 2012). In fact, in recent years, the worldwide market of commercially grown edible fungi (i.e., *P. ostreatus*) has been growing exponentially and spent residues from mushroom industries are still considered as a waste to be disposed. In this respect, *P. ostreatus* refuse from commercial mushroom production was exploited as a source of viable mycelia to remediate contaminated soils under field conditions as reported in the previous section.

8.7 Conclusions

The potential of fungi in the remediation of contaminated matrices (soil, sediments, surface water) were highlighted and the shortcomings that make this technology not suitable yet for large-scale application were also analysed in this chapter.

A deeper understanding of the ecology of fungi in their natural environment and a detailed knowledge of the physiology of mycoremediation process, with specific regard to the condition when fungi come in contact with the recalcitrant organic pollutants, could enable us to take a step forward in the application of fungal-based technology for the treatment of polluted environmental matrices.

In this respect, advances in genomics have triggered a revolution in understanding even the most complex biological systems. Indeed, the composition of microbial community and its dynamics during the fungal colonization of contaminated solid matrices as well as the interactions between exogenously introduced mycelia and resident soil microorganisms can be described in depth by means of innovative molecular biology techniques.

Transcriptomics and proteomics are steadily progressing field of science and their application to study mycoremediation processes could clarify how the fungal gene expression and the production of enzymes involved in the degradation process are regulated in response to the variations of environmental condition (i.e., chemicals/pollutants concentration, physiological and nutritional stress condition, etc.). However, due to some methodological biases and the high costs associated to their applications, the analysis of metatranscriptome and/or metaproteome in contaminated/treated matrices is still limited to a few studies at bench-scale level. In the near future, as soon as these limitations could be surmounted, these approaches might be used as a tool for understanding the real potential of fungal inocula in the clean up of polluted environmental compartments.

Concerning this, mycoremediation cannot be considered as a method of choice nowadays, but it still deserves attention because the combination of fungal treatment with bacteria-based approaches or with phytoremediation could overcome the limitations associated with their stand-alone application. Moreover, despite the failure in the bioaugmentation of allochthonous species in soil, the importance of fungi in some widely accepted bioremediation approaches (e.g., co-composting of polluted matrices, biopiling, etc.) should not be overlooked. In some specific cases, such as low permeability clay soils where conventional (e.g., in situ) technologies are not applicable or ineffective, the mere addition of lignocellulose might represent a valuable alternative to stimulate the growth of resident mycoflora, and, in turn, bacterial populations that could attack the target pollutants.

On the other hand, the application of high redox potential fungal enzymes (laccases, peroxidases and peroxigenases) for the treatment of micropollutant-contaminated wastewaters appears more realistic and feasible than other approaches. In this respect, the development of bioreactors with immobilized/insolubilized biocatalysts appears to be at a technologically mature stage. In the authors' opinion, this is the most promising and likely fungal-derived technology that have a chance to be practically applied in the near future. The work of Gasser et al. (2014), along with a plethora of studies targeting the implementation of tertiary treatments at wastewater treatment plants, reinforces this assertion. Nonetheless, such biocatalyst-based bioreactors should not be viewed exclusively as an end-of-pipe (tertiary) treatment unit at urban WWTP because similar technologies might also be applied to reduce the chemical burden arising from critical point source of MPs contamination, such as chemical and pharmaceutical production plants effluents.

Acknowledgments This work was supported by the Grant No. 15-02328S of the Czech Science Foundation.

References

Accinelli C, Saccà ML, Batisson I, Fick J, Mencarelli M, Grabic R (2010) Removal of oseltamivir (Tamiflu) and other selected pharmaceuticals from wastewater using a granular bioplastic formulation entrapping propagules of *Phanerochaete chrysosporium*. Chemosphere 81: 436–443

Adebusoye SA, Picardal FW, Ilori MO, Amund OO (2008) Influence of chlorobenzoic acids on the growth and degradation potentials of PCB-degrading microorganisms. World J Microb Biot 24:1203–1208

Anasonye F, Winquist E, Räsänen M, Kontro J, Björklöf K, Vasilyeva G, Jørgensen K, Steffen K, Tuomela M (2015) Bioremediation of TNT contaminated soil with fungi under laboratory and pilot-scale conditions. Int Biodeter Biodegr 105:7–12

Anastasi A, Tigini V, Varese GC (2013) The bioremediation potential of different ecophysiological groups of fungi. In: Goltapeh EM et al (eds) Fungi as Bioremediators, Soil Biology 32, Springer, Berlin Heidelberg. pp 29–49

Arcaro KF, O'Keefe PW, Yang Y, Clayton W, Gierthy JF (1999) Antiestrogenicity of environmental polycyclic aromatic hydrocarbons in human breast cancer cells. Toxicology 133:115–127

Arsenault A, Cabana H, Jones JP (2011) Laccase-Based CLEAs: Chitosan as a Novel Cross-Linking Agent. Enzyme Res Artcle ID: 376015

Ashger M, Bhatti HN, Ashraf M, Legge RL (2008) Recent developments in bio-degradation of industrial pollutants by white rot fungi and their enzyme system. Biodegradation 19:771–783
Auriol M, Filali-Meknassia Y, Tyagi RD, Adams CD (2007) Laccase-catalyzed conversion of natural and synthetic hormones from a municipal wastewater. Water Res 41:3281–3288
Axtell C, Johnston CG, Bumpus JA (2000) Bioremediation of soil contaminated with explosives at the naval weapons station yorktown. J Soil Contam 9:537–548
Ba S, Arsenault A, Hassani T, Jones JP, Cabana H (2013) Laccase immobilization and insolubilization: from fundamentals to applications for the elimination of emerging contaminants in wastewater treatment. Crit Rev Biotechnol 33:404–418
Ba S, Haroune L, Cruz-Morato C, Jacquet C, Touahar IE, Bellenger J-P, Legault CY, Jones JP, Cabana H (2014) Synthesis and characterization of combined cross-linked laccase and tyrosinase aggregates transforming acetaminophen as a model phenolic compound in wastewaters. Sci Total Environ 487:748–755
Baborová P, Moder M, Baldrian P, Cajthamlová K, Cajthaml T (2006) Purification of new manganese peroxidase of the white-rot fungus *Irpex lacteus*, and degradation of polycyclic aromatic hydrocarbons by the enzyme. Res Microbiol 157:248–253
Baldrian P (2003) Interactions of heavy metals with white-rot fungi. Enzyme Microb Technol 32:78–91
Baldrian P (2006) Fungal laccases- occurrence and properties. FEMS Microbiol Rev 30:215–242
Baldrian P (2008) Wood-inhabiting ligninolytic basidiomycetes in soils: ecology and constraints for applicability in bioremediation. Fungal Ecol 1:4–12
Baldrian P, in der Wiesche C, Gabriel J, Nerud F, Zadrazil F (2000) Influence of cadmium and mercury on activities of ligninolytic enzymes and degradation of polycyclic aromatic hydrocarbons by *Pleurotus ostreatus* in soil. Appl Environ Microbiol 66:2471–2478
Baldrian P, Valášková V (2008) Degradation of cellulose by basidiomycetous fungi. FEMS Microbiol Rev 32:501–521
Beaudette LA, Davies S, Fedorak PM, Ward OP, Pickard MA (1998) Comparison of gas chromatography and mineralization experiment for measuring loss of selected polychlorinated biphenyl congeners in cultures of white rot fungi. Appl Environ Microbiol 64:2020–2025
Bernhardt R (2006) Cytochromes as versatile biocatalists. J Biotechnol 124:128–145
Bezalel L, Hadar Y, Cerniglia CE (1996a) Mineralization of polycyclic aromatic hydrocarbons by the white rot fungus *Pleurotus ostreatus*. Appl Environ Microbiol 62:292–295
Bezalel L, Hadar Y, Fu PP, Freeman JP, Cerniglia CE (1996b) Metabolism of phenanthrene by the white rot fungus *Pleurotus ostreatus*. Appl Environ Microbiol 62:2547–2553
Bezalel L, Hadar Y, Cerniglia CE (1997) Enzymatic mechanisms involved in phenanthrene degradation by the white rot fungus *Pleurotus ostreatus*. Appl Environ Microbiol 63: 2495–2501
Blánquez P, Guieysse B (2008) Continuous biodegradation of 17β-estradiol and 17α-ethynylestradiol by Trametes versicolor. J Hazard Mater 50:459–462
Boas M, Feldt-Rasmussen U, Main KM (2012) Thyroid effects of endocrine disrupting chemicals. Mol Cell Endocrinol 355:240–248
Bollag JM (1992) Decontaminating soil with enzymes. Environ Sci Technol 26:1876–1881
Bonfante P, Anca IA (2009) Plants, mycorrhizal fungi, and bacteria: a network of interactions. Annual Rev Microbiol 63:363–383
Borazjani H, Wiltcher D, Diehl S (2005) Bioremediation of polychlorinated biphenyl and petroleum contaminated soil. In: Lyon WG, Hong JJ, Reddy RK (eds) Proceedings of the international conference on environmental science and technology. American Science Press, New Orleans, USA, pp 502–507
Bornyasz MA, Graham RC, Allen MF (2005) Ectomycorrhizae in a soil-weathered granitic bedrock regolith: linking matrix resources to plants. Geoderma 126:141–160
Bourceret A, Cébron A, Tisserant E, Poupin P, Bauda P, Bequiristain T, Leyval C (2015) The bacterial and fungal diversity of an aged PAH- and heavy metal-contaminated soil is affected by plant cover and edaphic parameters. Soil Microbiol Microb Ecol Springer, US, pp 1–14

Bumpus JA, Tien M, Wright D, Aust SD (1985) Oxidation of persistent environmental pollutants by a white rot fungus. Science 228:1434–1436

Cabana H, Jiwan JLH, Rozenberg R, Elisashvili V, Penninckx M, Agathos SN, Jones JP (2007a) Elimination of endocrine disrupting chemicals nonylphenol and bisphenol A and personal care product ingredient triclosan using enzyme preparation from the white rot fungus *Coriolopsis polyzona*. Chemosphere 67:770–778

Cabana H, Jones JP, Agathos SN (2007b) Elimination of endocrine disrupting chemicals using white rot fungi and their lignin modifying enzymes: a review. Eng Life Sci 7:429–456

Cabana H, Alexandre C, Agathos SN, Jones JP (2009a) Immobilization of laccase from the white rot fungus *Coriolopsis polyzona* and use of the immobilized biocatalyst for the continuous elimination of endocrine disrupting chemicals. Bioresour Technol 100:3447–3458

Cabana H, Jones JP, Agathos SN (2009b) Utilization of cross-linked laccase aggregates in a perfusion basket reactor for the continuous elimination of endocrine-disrupting chemicals. Biotechnol Bioeng 102:1582–1592

Cabana H, Ahamed A, Leduc R (2011) Conjugation of laccase from the white rot fungus *Trametes versicolor* to chitosan and its utilization for the elimination of triclosan. Bioresour Technol 102:1656–1662

Cajthaml T (2014) Biodegradation of endocrine-disrupting compounds by ligninolytic fungi: mechanisms involved in the degradation. Environ Microbiol. doi:10.1111/1462-2920.12460

Cajthaml T, Möder M, Kačer P, Šašek V, Popp P (2002) Study of fungal degradation products of polycyclic aromatic hydrocarbons using gas chromatography with ion trap mass spectrometry detection. J Chromatogr A 974:213–222

Cajthaml T, Erbanová P, Šašek V, Moeder M (2006) Breakdown products on metabolic pathway of degradation of benzo(*a*)anthracene by a ligninolytic fungus. Chemosphere 64:560–564

Cajthaml T, Křesinová Z, Svobodová K, Möder M (2009a) Biodegradation of endocrine disrupting compounds and suppression of estrogenic activity by ligninolytic fungi. Chemosphere 6: 745–750

Cajthaml T, Křesinová Z, Svobodová K, Sigler K, Řezanka T (2009b) Microbial transformation of synthetic estrogen 17α-ethinylestradiol. Environ Pollut 157:3325–3335

Cajthaml T, Svobodová K (2012) Biodegradation of aromatic pollutants by ligninolytic fungal strains. In: Singh SN (ed) Microbial degradation of xenobiotics. Springer-Verlag, pp 291–316

Camarero S, Cañas AI, Nousiainen P, Record E, Lomascolo A, Martínez MJ, Martínez ÁT (2008) *p*-hydroxycinnamic acids as natural mediators for laccase oxidation of recalcitrant compounds. Environ Sci Technol 42:6703–6709

Canadian Environmental Protection Act (1999). Priority substances list assessment report, "nonylphenol and its ethoxylates"

Cañas AI, Camarero S (2010) Laccases and their natural mediators: biotechnological tools for sustainable eco-friendly processes. Biotechnol Adv 28:694–705

Cázares-García SV, Vázquez-Garcidueñas MS, Vázquez-Marrufo G (2013) Structural and phylogenetic analysis of laccases from *Trichoderma*: a bioinformatic approach. PLoS ONE 8:e55295

Cébron A, Beguiristain T, Bongoua-Devisme J, Denonfoux J, Faure P, Lorgeoux C, Ouvrard S, Parisot N, Peyret P, Leyval C (2015) Impact of clay mineral, wood sawdust or root organic matter on the bacterial and fungal community structures in two aged PAH-contaminated soils. Environ Sci Pollut Res 22:13724–13738

Cerniglia CE (1997) Fungal metabolism of polycyclic aromatic hydrocarbons: past, present and future applications in bioremediation. J Ind Microbiol Biotechnol 19:324–333

Cerniglia CE, Gibson DT (1977) Metabolism of naphthalene by *Cunninghamella elegans*. Appl Environ Microbiol 34:363–370

Cerniglia CE, Sutherland JB (2006) Relative roles of bacteria and fungi in polycyclic aromatic hydrocarbon biodegradation and bioremediation of contaminated soils. In: Geoffrey Michael Gadd (ed) Fungi in biogeochemical cycles. Cambridge University Press, pp 182–211

Choudhary MI, Musharraf SG, Ali RA, Atif M, Atta-ur-Rahman (2004) Microbial transformation of antifertility agents, norethisterone and 17α-ethinylestradiol. Z. Naturforschung 59b:319–323

Clarke BO, Smith SR (2011) Review of 'emerging' organic contaminants in biosolids and assessment of international research priorities for the agricultural use of biosolids. Environ Int 37:226–247

Cloete TE, Celliers L (1999) Removal of Aroclor 1254 by the white rot fungus *Coriolus versicolor* in the presence of different concentration of Mn (IV) oxide. Int Biodeterior Biodegrad 44: 243–253

Clouzot L, Marrot B, Doumenq P, Roche N (2008) 17α-Ethinylestradiol: an endocrine disrupter of great concern. Analytical methods and removal processes applied to water purification. A review Environ Prog 27:383–396

Corvini PFX, Schaffer A, Schlosser D (2006) Microbial degradation of nonylphenol and other alkylphenols—our evolving view. Appl Microbiol Biotechnol 72:223–243

Covino S, Čvančarová M, Muzikar M, Svobodová K, D'Annibale A, Petruccioli M, Federici F, Křesinová Z, Cajthaml T (2010a) An efficient PAH-degrading *Lentinus* (*Panus*) *tirinus* strain: effect of inoculum formulation and pollutant bioavailability in solid matrices. J Hazard Mater 183:669–676

Covino S, Svobodová K, Čvančarová M, D'Annibale A, Petruccioli M, Federici F, Křesinová Z, Galli E, Cajthaml T (2010b) Inoculum carrier and contaminant bioavailability affect fungal degradation performances of PAH-contaminated solid matrices from a wood preservation plant. Chemosphere 79:855–864

Covino S, Svobodová K, Křesinová Z, Petruccioli M, Federici F, D'Annibale A, Čvančarová M, Cajthaml T (2010c) In vivo and in vitro polycyclic aromatic hydrocarbon degradation by *Lentinus (Panus) tigrinus* CBS 577.79. Bioresour Technol 101:3004–3012

Covino S, D'Annibale A, Stazi SR, Cajthaml T, Čvančarová M, Stella T, Petruccioli M (2015) Assessment of degradation potential of aliphatic hydrocarbons by autochthonous filamentous fungi from a historically polluted clay soil. Sci Total Environ 505:545–554

Črešnar B, Petrič Š (2011) Cytochrome P450 enzymes in the fungal kingdom. Review Biochim Biophys Acta 1814:29–35

Crinnion WJ (2011) Polychlorinated biphenyls: persistent pollutants with immunological, neurological and endocrinological consequences. Altern Med Rev 16(1)

Cruz-Morató C, Ferrando-Climent L, Rodriguez-Mozaz S, Barceló D, Marco-Urrea E, Vicent T, Sarrà M (2013) Duran Degradation of pharmaceuticals in non-sterile urban wastewater by *Trametes versicolor* in a fluidized bed bioreactor. Water Res 47:5200–5210

Čvančarová M, Křesinová Z, Filipová A, Covino S, Cajthaml T (2012) Biodegradation of PCBs by ligninolytic fungi and characterization of the degradation products. Chemosphere 88: 1317–1323

D'Annibale A, Stazi SR, Vinciguerra V, Di Mattia E, Sermanni GG (1999) Characterization of immobilized laccase from *Lentinula edodes* and its use in olive mill wastewater treatment. Process Biochem 34:697–706

D'Annibale A, Stazi SR, Vinciguerra V, Di Mattia E, Sermanni GG (2000) Oxirane-immobilized *Lentinula edodes* laccase: stability and phenolics removal efficiency in olive mill wastewater. J Biotechnol 77:265–273

Dashtban M, Schraft H, Syed TA, Qin W (2010) Fungal biodegradation and enzymatic modification of lignin. Int J Biochem Mol Biol 1:36–50

de Boer W, van der Wal A (2008) Chapter 8 interactions between saprotrophic basidiomycetes and bacteria. British Mycological Society Symposia Series. Ecol Saprotrophic Basidiomycetes 28:143–153

Dec J, Haider K, Bollag J (2003) Release of substituents from phenolic compounds during oxidative coupling reactions. Chemosphere 52:549–556

Della Greca M, Pinto G, Pistillo P, Pollio A, Previtera L, Temussi F (2008) Biotransformation of ethinylestradiol by microalgae. Chemosphere 70:2047–2053

Demarche P, Junghanns C, Mazy N, Agathos SN (2012a) Design-of-experiment strategy for the formulation of laccase biocatalysts and their application to degrade bisphenol A. New Biotechnol 30:96–103

Demarche P, Junghanns CH, Nair RR, Agathos SN (2012b) Harnessing the power of enzymes for environmental stewardship. Biotechnol Adv 30:933–953

Dietrich D, Hickey WJ, Lamar R (1995) Degradation of 4,4′-dichlorobiphenyl, 3,3′,4,4′-tetrachlorobiphenyl, and 2,2′,4,4′,5,5′-hexachlorobiphenyl by the white rot fungus *Phanerochaete chrysosporium*. Appl Environ Microbiol 61:3904–3909

Dittmann J, Heyser W, Bücking H (2002) Biodegradation of aromatic compounds by white rot and ectomycorrhizal fungal species and the accumulation of chlorinated benzoic acid in ectomycorrhizal pine seedlings. Chemosphere 49:297–306

Dmochewitz S, Ballschmiter K (1988) Microbial transformation of technical mixtures of polychlorinated biphenyls (PCB) by the fungus *Aspergillus niger*. Chemosphere 17:111–121

Domaradzka D, Guzik U, Wojcieszyńska D (2015) Biodegradation and biotransformation of polycyclic non-steroidal anti-inflammatory drugs. Rev Environ Sci Biotechnol 14:229–239

Dubroca J, Brault A, Kollmann A, Touton I, Jolivalt G, Kerhoas L, Mougin C (2005) Biotransformation of nonylphenol surfactants in soils amended with contaminated sewage sludges. In: Lichtfouse E, Dudd S, Robert D (eds) Environmental chemistry: green chemistry and pollutants in ecosystems. Springer, Berlin, Germany, pp 305–315

Duncan CG, Deverall FJ (1964) Degradation of wood preservatives by fungi. Appl Microbiol 12:57–62

Durán N, Rosa MA, D'Annibale A, Gianfreda L (2002) Applications of laccases and tyrosinases (phenoloxidases) immobilized on different supports: a review. Enzyme Microb Tech 31: 907–931

Eaton DC (1985) Mineralization of polychlorinated biphenyls by *Phanerochaete chrysosporium*: a ligninolytic fungus. Enzyme Microb Tech 7:194–196

Eggen RIL, Hollender J, Joss A, Schärer M, Stamm C (2014) Reducing the discharge of micropollutants in the aquatic environment: the benefits of upgrading wastewater treatment plants. Environ Sci Technol 8:7683–7689

Eggen T, Sveum P (1999) Decontamination of aged creosote polluted soil: The influence of temperature, white rot fungus *Pleurotus ostreatus*, and pretreatment. Int Biodeterior Biodegrad 43:125–133

Eibes G, Cajthaml T, Moreira MT, Feijoo G, Lema JM (2006) Enzymatic degradation of anthracene, dibenzothiophene and pyrene by manganese peroxidase in media containing acetone. Chemosphere 64:408–414

El Majidi N, Bouchard M, Carrier G (2013) Systematic analysis of the relationship between standardized prenatal exposure to polychlorinated biphenyls and mental and motor development during follow-up of nine children cohorts. Regul Toxicol Pharmacol 66:130–146

European Parliament and The Council (2008) Directive 2008/105/EC of the European Parliament and of the Council of 16 December 2008 on environmental quality standards in the field of water policy, amending and subsequently repealing Council Directives 82/176/EEC, 83/513/EC, 84/156/EEC, 84/491/EEC, 86/280/EEC and amending Directive 2000/60/EC of the European Parliament and of the Council. Off J Eur Union L348:84–97

Farnet AM, Gil G, Ruaudel F, Chevremont AC, Ferre E (2009) Polycyclic aromatic hydrocarbon transformation with laccases of a white-rot fungus isolated from a mediterranean schlerophyllous litter. Geoderma 149:267–271

Federici E, Giubilei M, Santi G, Zanaroli G, Negroni A, Fava F, Petruccioli M, D'Annibale A (2012) Bioaugmentation of a historically contaminated soil by polychlorinated biphenyls with *Lentinus tigrinus*. Microb Cell Fact 11:35

Fernández-Luqueño F, Valenzuela-Encinas C, Marsch R, Martínez-Suárez C, Vázquez-Núñez E, Dendooven L (2011) Microbial communities to mitigate contamination of PAHs in soil-possibilities and challenges: a review. Environ Sci Pollut Res 18:12–30

Field JA, Sierra-Alvarez Reyes (2008) Microbial transformation and degradation of polychlorinated biphenyls. Rev Environ Pollut 155:1–12

Fukuda T, Uchida H, Suzuki M, Miyamoto H, Morinaga H, Nawata H, Uwajima T (2004) Transformation products of bisphenol A by a recombinant *Trametes villosa* laccase and their estrogenic activity. J Chem Technol Biotechnol 79:1212–1218

Fukuda T, Uchida H, Takashima Y, Uwajima T, Kawabata T, Suzuki M (2001) Degradation of bisphenol A by purified laccase from *Trametes villosa*. Biochem Biophys Res Commun 284:704–706
Furukawa K, Fujihara H (2008) Microbial degradation of polychlorinated biphenyls: biochemical and molecular features. J Biosc Bioeng 105:433–449
Gadd GM (2001) Fungi in bioremediation (No 23). Published for the British Mycological Society. Cambridge University Press. ISBN 0521781191
García-Delgado C., D'Annibale A., Pesciaroli L., Yunta F., Crognale S., Petruccioli M., Eymar E. (2015). Implications of polluted soil biostimulation and bioaugmentation with spent mushroom substrate (*Agaricus bisporus*) on the microbial community and polycyclic aromatic hydrocarbons biodegradation. *Sci. Total Environ.* 508: 20–28
Gasser CA, Yu L, Svojitka J, Wintgens T, Ammann EM, Shahgaldian P, Corvini PF-X, Hommes G (2014) Advanced enzymatic elimination of phenolic contaminants in wastewater: a nano approach at field scale
Gesell M, Hammer E, Mikolasch A, Schauer F (2004) Oxidation and ring cleavage of dibenzofuran by the filamentous fungus *Paecilomyces lilacinus*. Arch Microbiol 182:51–9
Gichner T, Lovecká P, Vrchotová B (2008) Genomic damage induced in tobacco plants by chlorobenzoic acids metabolic products of polychlorinated biphenyls. Mutat Res 657:140–145
Giubilei MA, Leonardi V, Federici E, Covino S, Šašek V, Novotný C, Federici F, D'Annibale A, Petruccioli M (2009) Effect of mobilizing agents on mycoremediation and impact on the indigenous microbiota. J Chem Technol Biotechnol 84:836–844
Gómez-Toribio V, García-Martín AB, Martínez MJ, Martínez AT, Guillén F (2009) Induction of extracellular hydroxyl radical production by white-rot fungi through quinone redox cycling. Appl Environ Microbiol 75:3944–3953
Gramss G, Kirsche B, Voigt KD, Gunther T, Fritsche W (1999) Conversion rates of five polycyclic aromatic hydrocarbons in liquid cultures of fifty-eight fungi and the concomitant production of oxidative enzymes. Mycol Res 103:1009–1018
Hammel KE (1995) Mechanisms for polycyclic aromatic hydrocarbon degradation by ligninolytic fungi. Environ Health Perspect 103:41–43
Hammel KE, Kalyanaraman B, Kirk TK (1986) Oxidation of polycyclic aromatic hydrocarbons and dibenzo[*p*]-dioxins by *Phanerochaete chrysosporium* ligninase. J Biol Chem 261: 16948–16952
Hammel KE, Green B, Gai WZ (1991) Ring fission of anthracene by a eukaryote. Proc Natl Acad Sci USA 88:10605–10608
Hammer E, Krowas D, Schafer A, Specht M, Francke W, Schauer F (1998) Isolation and characterization of a dibenzofuran-degrading yeast: identification of oxidation and ring cleavage products. Appl Environ Microbiol 64:2215–2219
Haritash AK, Kaushik CP (2009) Biodegradation aspects of polycyclic aromatic hydrocarbons (PAHs): a review. J Hazard Mater 169:1–15
Harms H, Schlosser D, Wick LY (2011) Untapped potential: exploiting fungi in bioremediation of hazardous chemicals. Nat Rev Microbiol 9:177–192
Helmfrid I, Berglund M, Löfman O, Wingren G (2012) Health effects and exposure to polychlorinated biphenyls (PCBs) and metals in a contaminated community. Environ Intern 44:53–58
Hendey NI (1964) Some observations on *Cladosporium resinae* as a fuel contaminant and its possible role in the corrosion of aluminium alloy fuel tanks. T Brit Mycol Soc 47:467–475
Hestbjerg H, Brinch UC, Nielsen T (2001) Potential of *Pleurotus ostreatus* for bioremediation: Temperature requirements and competitiveness. Abstracts, SETAC Organic Soil Contaminants, Copenhagen, Denmark, September 2–5, p 24
Hestbjerg H, Willumsen PA, Christensen M, Andersen O, Jacobsen CS (2003) Bioaugmentation of tar-contaminated soils under field conditions using *Pleurotus ostreatus* refuse from commercial mushroom production. Environ Toxicol Chem 22:692–698
Hirano T, Honda Y, Watanabe T, Kuwahara M (2000) Degradation of bisphenol A by the lignin-degrading enzyme, manganese peroxidase, produced by the white rot basidiomycete, *Pleurotus ostreatus*. Biosci Biotechnol Biochem 64:1958–1962

Hofrichter M (2002) Review: lignin conversion by manganese peroxidase (MnP). Enzyme Microb Technol 30:454–466

Hofrichter M, Ullrich R (2014) Oxidations catalyzed by fungal peroxygenases. Curr Opin Chem Biol 19:116–125

Huang Q, Walter J, Weber JR (2005) Transformation and removal of bisphenol A from aqueous phase via peroxidase-mediated oxidative coupling reactions: efficacy, products, and pathways. Environ Sci Technol 39:6029–6036

Huff JE, Moore JA, Saracci R, Tomatis L (1980) Long-term hazards of polychlorinated dibenzodioxins and polychlorinated dibenzofurans. Environ Health Persp 36:221–240

in der Wiesche C, Martens R, Zadrazil F (1996) Two-step degradation of pyrene by white-rot fungi and soil microorganisms. Appl Microbiol Biotechnol 46:653–659

Ishii K, Furuichi T (2007) Development of bioreactor system for treatment of dioxin-contaminated soil using *Pseudallescheria boydii*. J Hazard Mater 148:693–700

Ishii K, Furuichi T, Tanikawa N, Kuboshima M (2009) Estimation of the biodegradation rate of 2,3,7,8-tetrachlorodibenzo-*p*-dioxin by using dioxin-degrading fungus, *Pseudallescheria boydii*. J Hazard Mater 162:328–332

Jelić A, Gros M, Petrović M, Ginebreda A, Barceló D (2012) Occurrence and elimination of pharmaceuticals during conventional wastewater treatment. Emerging and Priority Pollutants in Rivers. The Handbook of Environmental Chemistry, Springer US, vol 19, pp 1–23

Johannes C, Majcherczyk A (2000) Natural mediators in the oxidation of polycyclic aromatic hydrocarbons by laccase mediator systems. Appl Environ Microbiol 66:524–528

Johannes C, Majcherczyk A, Hüttermann A (1998) Oxidation of acenaphthene and acenaphthylene by laccase of *Trametes versicolor* in a laccase-mediator system. J Biotechnol 61:151–156

Junghanns C, Moeder M, Krauss G, Martin C, Schlosser D (2005) Degradation of the xenoestrogen nonylphenol by aquatic fungi and their laccases. Microbiology 151:45–57

Kamei I, Kondo R (2005) Biotransformation of dichloro-, trichloro-, and tetrachlorodibenzo-*p*-dioxin by the white rot fungus *Phlebia lindtneri*. Appl Microbiol Biotechnol 68:560–566

Kamei I, Suhara H, Kondo R (2005) Phylogenetical approach to isolation of white rot fungi capable of degrading polychlorinated dibenzo-*p*-dioxin. Appl Microbiol Biotechnol 69: 358–366

Kamei I, Kogura R, Kondo R (2006a) Metabolism of 4,4′-dichlorobiphenyl by white-rot fungi *Phanerochaete chrysosporium* and *Phanerochaete* sp. MZ142 *Appl*. Microbiol Biotechnol 72:566–575

Kamei I, Sonoki S, Haraguchi K, Kondo R (2006b) Fungal bioconversion of toxic polychlorinated biphenyls by white-rot fungus, *Phlebia brevispora*. Appl Microbiol Biotechnol 73:932–940

Keum YS, Li QX (2004) Fungal laccase-catalyzed degradation of hydroxy polychlorinated biphenyls. Chemosphere 56:23–30

Kirk TK, Farrell RL (1987) Enzymatic "combustion": the microbial degradation of lignin. Annu Rev Microbiol 41:465–505

Kohlmeier S, Smits TMH, Ford RM, Keel C, Harms H, Lukas YW (2005) Taking the fungal highway: mobilization of pollutant-degrading bacteria by fungi. Environ Sci Technol 39: 4640–4646

Kordon K, Mikolasch A, Schauer F (2010) Oxidative dehalogenation of chlorinated hydroxy-biphenyls by laccases of white-rot fungi. Intern Biodet Biodegrad 64:203–209

Kramer S, Moller Hikel S, Adams K, Hinds D, Moon K (2012) Current status of the epidemiologic evidence linking polychlorinated biphenyls and non-Hodgkin lymphoma and the role of immune dysregulation. Environ Health Perspect 120(8)

Krčmar P, Kubátová A, Votruba J, Erbanová P, Novotný C, Sašek V (1999) Degradation of polychlorinated biphenyls by extracellular enzymes of *Phanerochaete chrysosporium* produced in a perforated plate bioreactor. World J Microbiol Biotechnol 15:269–276

Křesinová Z, Moeder M, Ezechiáš M, Svobodová K, Cajthaml T (2012) Mechanistic study of 17α-ethinylestradiol biodegradation by *Pleurotus ostreatus*: tracking of extracelullar and intracelullar degradation mechanisms. Environ Sci Technol 46:13377–13385

Kubátová A, Erbanová P, Eichlerová I, Homolka L, Nerud F, Šašek V (2001) PCB congener selective biodegradation by the white-rot fungus *Pleurotus ostreatus* in contaminated soil. Chemosphere 43:207–215

Kües U (2015) Fungal enzymes for environmental management. Curr Opin Biotechnol 33: 268–278

Lamar RT, White RB, Ashley KC (2002) Evaluation of white-rot fungi for the remediation of creosote-contaminated soil. Remediation J 12:97–106

Lang E, Eller G, Zadrail F (1997) Lignocellulose decomposition and production of ligninolytic enzymes during interaction of white rot fungi with soil microorganisms. Microb Ecol 34:1–10

Lang E, Nerud F, Zadrazil F (1998) Production of ligninolytic enzymes by *Pleurotus* sp. and *Dichomitus squalens* in soil and lignocellulose substrate as influenced by soil microorganisms. FEMS Microb Lett 167:239–244

Lang E, Gonser A, Zadrazil F (2000) Influence of incubation temperature on activity of ligninolytic enzymes in sterile soil by *Pleurotus* sp. and *Dichomitus squalens*. Basic Microbiol 40:33–39

Latawiec AE, Swindell AL, Simmons P, Reid BJ (2011) Bringing bioavailability into contaminated land decision making: the way forward? Crit Rev Env Sci Tec 41:52–77

Lee PY, Chen CY (2009) Toxicity and quantitative structure–activity relationships of benzoic acids *to Pseudokirchneriella subcapitata*. J Hazard Mater 165:156–161

Li P, Sun T, Stagnitti F, Zhang C, Zhang H, Xiong X, Allinson G, Ma X, Allinson M (2002) Field-scale bioremediation of soil contaminated with crude oil. Environ Eng Sci 19:277–289

Liu ZH, Kanjo Y, Mizutani S (2009) Removal mechanisms for endocrine disrupting compounds (EDCs) in wastewater treatment—physical means, biodegradation, and chemical advanced oxidation: a review. Sci Total Environ 407:731–748

Lloret L, Eibes G, Feijo G, Moreira MT, Lema JM (2012a) Continuous operation of a fluidized bed reactor for the removal of estrogens by immobilized laccase on Eupergit supports. J Biotechnol 182:404–406

Lloret L, Hollmann F, Eibes G, Feijoo G, Moreira MT, Lema JM (2012b) Immobilisation of laccase on Eupergit supports and itsapplication for the removal of endocrine disrupting chemicals in a packed-bed reactor. Biodegradation 23:373–386

Lloret L, Eibes G, Moreira MT, Feijoo G, Lema JM (2013) Removal of estrogenic compounds from filtered secondary wastewater effluent in a continuous enzymatic membrane reactor. Identification of biotransformation products. Environ Sci Technol 47:4536–4543

Luo Y, Guo W, Ngo HH, Nghiem LD, Hai FI, Zhang J, Liang S (2014) A review on the occurrence of micropollutants in the aquatic environment and their fate and removal during wastewater treatment. Sci Total Environ 473–474:619–641

Lyr H (1963) Enzymatische detoxification chlorieter phenole. Phytopathol Z 47:73–83

Majcherczyk A, Johannes C, Huttermann A (1998) Oxidation of polycyclic aromatic hydrocarbons (PAH) by laccase of *Trametes versicolor*. Enzyme Microb Technol 22:335–341

Manji S, Ishihara A (2004) Screening of tetrachlorodibenzo-*p*-dioxin-degrading fungi capable of producing extracellular peroxidases under various conditions. Appl Microbiol Biotechnol 63:438–444

Marco-Urrea E, Pérez-Trujillo M, Vicent T, Caminal G (2009) Ability of white-rot fungi to remove selected pharmaceuticals and identification of degradation products of ibuprofen by *Trametes versicolor*. Chemosphere 74:765–772

Marco-Urrea E, Pérez-Trujillo M, Blánquez P, Vicent T, Caminal G (2010a) Biodegradation of the analgesic naproxen by *Trametes versicolor* and identification of intermediates using HPLC-DAD-MS and NMR. Bioresour Technol 101:2159–2166

Marco-Urrea E, Pérez-Trujillo M, Cruz-Morató C, Caminal G, Vicent T (2010b) Degradation of the drug sodium diclofenac by Trametes versicolor pellets and identification of some intermediates by NMR. J Hazard Mater 176:836–842

Marco-Urrea E, Pérez-Trujillo M, Cruz-Morató C, Caminal G, Vicent T (2010c) White-rot fungus-mediated degradation of the analgesic ketoprofen and identification of intermediates by HPLC-DAD-MS and NMR. Chemosphere 78:474–481

Marco-Urrea E, Radjenović J, Caminal G, Petrović M, Vicent T, Barceló D (2010d) Oxidation of atenolol, propranolol, carbamazepine and clofibric acid by a biological Fenton-like system mediated by the white-rot fungus *Trametes versicolor*. Water Res 44:521–532

Marco-Urrea E, García-Romera I, Aranda E (2015) Potential of non-ligninolytic fungi in bioremediation of chlorinated and polycyclic aromatic hydrocarbons. New Biotechnol 32: 620–628

Martens R, Zadrazil F (1998) Screening of white-rot fungi for their ability to mineralize polycyclic aromatic hydrocarbons in soil. Folia Microbiol 43:97–103

Mayolo-Deloisa K, Machín-Ramírez C, Rito-Palomares M, Trejo-Hernández MR (2011) Oxidation of polycyclic aromatic hydrocarbons using partially purified laccase from residual compost of *Agaricus bisporus*. Chem Eng Technol 34:1368–1372

Michels J (1998) Processes for the bioremediation of soil. Editor: Federal Environmental Agency BMBF Project Management Agency for Waste Management and Remediation of Hazardous Abandoned Sites (PT AWAS). Section III 3.6

Michizoe J, Ichinose H, Kamila N, Maruyama T, Goto M (2005) Biodegradation of phenolic environmental pollutants by a surfactant-laccase complex in organicmedia. J Biosci Bioeng 99:642–647

Moen MA, Hammel KE (1994) Lipid peroxidation by the manganese peroxidase of *Phanerochaete chrysosporium* is the basis for phenanthrene oxidation by the intact fungus. Appl Environ Microbiol 60:1956–1961

Mori T, Kondo R (2002a) Degradation of 2, 7-dichlorodibenzo-*p*-dioxin by wood-rotting fungi, screened by dioxin degrading ability. FEMS Microbiol Lett 213:127–131

Mori T, Kondo R (2002b) Oxidation of dibenzo-*p*-dioxin, dibenzofuran, biphenyl, and diphenyl ether by the white rot fungus *Phlebia lindtneri*. Appl Microbiol Biotechnol 60:200–205

Mougin C (2002) Bioremediation and phytoremediation of industrial PAH-polluted soils. Polycyclic Aromat Compd 22:1–33

Mouhamadou B, Faure M, Sage L, Marçais J, Souard F, Geremia RA (2013) Potential of autochthonous fungal strains isolated from contaminated soils for degradation of polychlorinated biphenyls. Fungal Biol 117:268–274

Muzikář M, Křesinová Z, Svobodová K, Filipová A, Čvančarová M, Cajthamlová K, Cajthaml T (2011) Biodegradation of chlorobenzoic acids by ligninolytic fungi. J Hazard Mater 196: 386–394

Naidu R, Channey R, McConnell S, Johnston N, Semple KT, McGrath S, Dries V, Nathanail P, Harmsen J, Pruszinski A, MacMillan J, Palanisami T (2013) Towards bioavailability-based soil criteria: past, present and future perspectives. Environ Sci Pollut Res Int 22:8779–8785

Nair RR, Demarche P, Agathos SN (2013) Formulation and characterization of an immobilized laccase biocatalyst and its application to eliminate organic micropollutants in wastewater. New Biotech 30:814–823

Nakamiya K, Furuichi T, Ishii K (2002) Isolation of a fungus from denitrifying activated sludge that degrades highly chlorinated dioxins. J Mater Cycles Waste Manage 4:127–134

Nakamiya K, Hashimoto S, Ito H, Edmonds JS, Yasuhara A, Morita M (2005) Degradation of dioxins by cyclic ether degrading fungus. Cordyceps sinensis FEMS Microbiol Lett 248:17–22

Nam I, Kim Y, Murugesan K, Jeon J, Chang Y (2008) Bioremediation of PCDD/Fs-contaminated municipal solid waste incinerator fly ash by a potent microbial biocatalyst. J Hazard

Nicotra S, Intra A, Ottolina G, Riva S, Danieli B (2004) Laccase-mediated oxidation of the steroid hormone 17b-estradiol in organic solvents. Tetrahedron Asymmetry 15:2927–2931

Novotný C, Cajthaml T, Svobodová K, Susla M, Šašek V (2009) *Irpex lacteus*, a white rot fungus with biotechnological potential—review. Folia Microbiol 54:375–390

Novotný C, Vyas BRM, Erbanová P, Kubátová A, Šašek V (1997) Removal of PCBs by various white-rot fungi in liquid cultures. Folia Microbiol 42:136–140

Pan F, Yang Q, Zhang Y, Zhang S, Yang M (2004) Biodegradation of polycyclic aromatic hydrocarbons by *Pichia anomala*. Biotechnol Lett 26:803–806

Passatore L, Rossetti S, Juwarkar AA, Massacci A (2014) Phytoremediation and bioremediation of polychlorinated biphenyls (PCBs): state of knowledge and research perspectives. J Hazard Mater 278:189–202
Phan CW, Sabaratnam V (2012) Potential uses of spent mushroom substrate and its associated lignocellulosic enzymes. Appl Environ Microbiol 96:863–873
Pickard MA, Roman R, Tinoco R, Vazquez-Duhalt R (1999) Polycyclic aromatic hydrocarbon metabolism by white rot fungi and oxidation by *Coriolopsis gallica* UAMH 8260 laccase. Appl Environ Microbiol 65:3805–3809
Pieper DH (2005) Aerobic degradation of polychlorinated biphenyls. Appl Microbiol Biotechnol 67:170–191
Pinedo-Rivilla C, Aleu J, Collado IG (2009) Pollutants biodegradation by fungi. Curr Org Chem 13:1194–1214
Pointing S (2001) Faesibility of bioremediation by white-rot fungi. Appl Microbiol Biotechnol 57:20–33
Rabinovich ML, Bolobova AV, Vasilchenko LG (2004) Decomposition of natural aromatic structures and xenobiotics by fungi. Appl Biochem Microbiol 40:5–23
Rafin C, Potin O, Veignie E (2000) Degradation of benzo[a]pyrene as sole carbon source by a non white-rot fungus, *Fusarium solani*. Polycycl Aromat Compd 21:311–329
Rafin C, de Foucault B, Veignie E (2012) Exploring micromycetes biodiversity for screening benzo[a]pyrene degrading potential. Environ Sci Pollut Res Int 20:3280–3289
Rodarte-Morales AI, Feijoo G, Moreira MT, Lema MJ (2011) Degradation of selected pharmaceutical and personal care products (PPCPs) by white-rot fungi. World J Microbiol Biotechnol 27:1839–1846
Rodarte-Morales AI, Feijoo G, Moreira MT, Lema MJ (2012) Biotransformation of three pharmaceutical active compounds by the fungus *Phanerochaete chrysosporium* in a fed batch stirred reactor under air and oxygen supply. Biodegradation 23:145–156
Rodríguez-Rodríguez CE, Barón E, Gago-Ferrero P, Jelić A, Llorca M, Farré M, Díaz-Cruz MS, Eljarrat E, Petrović M, Caminal G, Barceló D, Vicent T (2012) Removal of pharmaceuticals, polybrominated flame retardants and UV-filters from sludge by the fungus *Trametes versicolor* in bioslurry reactor. J Hazard Mater 233–234:235–243
Romero MC, Cazau MC, Giorgieri S, Arambarri AM (1998) Phenanthrene degradation by microorganisms isolated from a contaminated stream. Environ Pollut 101:355–359
Romero MC, Salvioli ML, Cazau MC, Arambarri AM (2002) Pyrene degradation by yeasts and filamentous fungi. Environ Pollut 117:159–163
Rozalska S, Szewczyk R, Dlugonski J (2010) Biodegradation of 4-n-nonylphenol by the non-ligninolytic filamentous fungus *Gliocephalotrichum* simplex: a proposal of a metabolic pathway. J Hazard Mater 180:323–331
Ruiz-Aguilar GML, Fernandez-Sanchez JM, Rodnguez-Vazquez R, Poggi-Varaldo H (2002) Degradation by white-rot fungi of high concentrations of PCB extracted from a contaminated soil. Adv Environ Res 6:559–568
Sack U, Heinze TM, Deck J, Cerniglia CE, Martens R, Zadrazil F, Fritche W (1997) Comparison of phenanthrene and pyrene degradation by different wood-decaying fungi. Appl Environ Microbiol 63:3919–3925
Sage L, Périgon S, Faure M, Gaignaire C, Abdelghafour M, Mehu J, Geremia RA, Mouhamadou B (2014) Autochthonous ascomycetes in depollution of polychlorinated biphenyls contaminated soil and sediment. Chemosphere 110:62–69
Santos LH, Araújo AN, Fachini A, Pena A, Delerue-Matos C, Montenegro MC (2010) Ecotoxicological aspects related to the presence of pharmaceuticals in the aquatic environment. J Hazard Mater 175:45–95
Šašek V (2003) Why mycoremediations have not yet come into practice. The utilization of bioremediation to reduce soil contamination: problems and solutions. NATO Science Series. 19:247–266
Sato A, Watanabe T, Watanabe Y, Harazono K, Fukatsu T (2002) Screening for basidiomycetous fungi capable of degrading 2, 7-dichlorodibenzo-*p*-dioxin. FEMS Microbiol Lett 213:213–217

Schützendübel A, Majcherczyk A, Johannes C, Hüttermann A (1999) Degradation of fluorene, anthracene, phenanthrene, fluoranthene, and pyrene lacks connection to the production of extracellular enzymes by *Pleurotus ostreatus* and *Bjerkandera adusta*. Int Biodeterior Biodegrad 43:93–100

Seto M, Nishibori K, Masai E, Fukuda M, Ohdaira Y (1999) Degradation of polychlorinated biphenyls by a 'Maitake' mushroom, *Grifola frondosa*. Biotechnol Lett 21:27–31

Sietmann R, Gesell M, Hammer E, Schauer F (2006) Oxidative ring cleavage of low chlorinated biphenyl derivatives by fungi leads to the formation of chlorinated lactone derivatives. Chemosphere 64:672–685

Silva IS, Grossman M, Durrant LR (2009) Degradation of polycyclic aromatic hydrocarbons (2-7 rings) under microaerobic and very-low-oxygen conditions by soil fungi. Int. Biodeterior. Biodegrad. 63:224–229

Singh H (2006) Mycoremediation: fungal bioremediation. Wiley-Interscience

Steffen KT, Hatakka A, Hofrichter M (2002) Removal and mineralization of polycyclic aromatic hydrocarbons by litter-decomposing basidiomycetous fungi. Appl Microbiol Biotechnol 60:212–217

Steffen KT, Schubert S, Tuomela M, Hatakka A, Hofrichter M (2007) Enhancement of bioconversion of high-molecular mass polycyclic aromatic hydrocarbons in contaminated non-sterile soil by litter-decomposing fungi. Biodegradation 18:359–369

Stella T, Covino S, Křesinová Z, D'Annibale A, Petruccioli M, Cvančarová M, Cajthaml T (2013) Chlorobenzoic acid degradation by *Lentinus* (*Panus*) *tigrinus*: In vivo and in vitro mechanistic study-evidence for P-450 involvement in the transformation. J Hazard Mater 260:975–983

Stella T, Covino S, Burianová E, Filipová A, Křesinová Z, Voříšková J, Větrovský T, Baldrian P, Cajthaml T (2015) Chemical and microbiological characterization of an aged PCB-contaminated soil. Sci Total Environ 533:177–186

Suzuki K, Hirai H, Murata H, Nishida T (2003) Removal of estrogenic activities of 17α-estradiol and ethinylestradiol by ligninolytic enzymes from white rot fungi. Water Res 37:19721975

Svobodová K, Placková M, Novotná V, Cajthaml T (2009) Estrogenic, androgenic activity of PCBs, their chlorinated metabolites and other endocrine disruptors estimated with two in vitro yeast assays. Sci Total Environ 407:5921–5925

Syed K, Porollo A, Ying WL, Yadav JS (2011) A fungal P450 (CYP5136A3) capable of oxidizing polycyclic aromatic hydrocarbons and endocrine disrupting alkylphenols: role of Trp129 and Leu324. PLoS ONE 6:1–14

Syed K, Porollo A, Lam YW, Grimmett PE, Yadav JS (2013) CYP63A2, a catalytically versatile fungal P450 monooxygenase capable of oxidizing higher-molecular-weight polycyclic aromatic hydrocarbons, alkylphenols, and alkanes. Appl Environ Microbiol 79:2692–702

Taboada-Puig R, Junghanns C, Demarche P, Moreira MT, Feijoo G, Lema JM, Agathos SN (2011) Combined cross-linked enzyme aggregates from versatile peroxidase and glucose oxidase: production, partial characterization and application for the elimination of endocrine disruptors. Bioresour Technol 102:6593–6599

Takada S, Nakamura M, Matsueda T, Kondo R, Sakai K (1996) Degradation of polychlorinated dibenzo-*p*-dioxins and polychlorinated dibenzofurans by the white rot fungus *Phanerochaete sordida* YK-624. Appl Environ Microbiol 62:4323–4328

Takagi S, Shirota C, Sakaguchi K, Suzuki J, Sue T, Nagasaka H, Hisamatsu S, Sonoki S (2007) Exoenzymes of *Trametes versicolor* can metabolize coplanar PCB congeners and hydroxy PCB. Chemosphere 67:54–57

Tanaka T, Tonosaki T, Nose M, Tomidokoro N, Kadomura N, Fujii T, Taniguchi M (2001) Treatment of model soils contaminated with phenolic endocrine disrupting chemicals with laccase from *Trametes* sp. in a rotating reactor. J Biosci Bioeng 92:312–316

Thomas DR, Carswell KS, Georgiou G (1992) Mineralization of biphenyl and PCBs by the white rot fungus *Phanerochaete chrysosporium*. Biotechnol Bioeng 40:1395–1402

Tien M, Kirk TK (1984) Lignin-degrading enzyme from *Phanerochaete chrysosporium*: purification, characterization, and catalytic properties of a unique H_20_2-requiring oxygenase. Proc Natl Acad Sci 81:2280–2284

Tigini V, Prigione V, Di Toro S, Fava F, Varese GC (2009) Isolation and characterisation of polychlorinated biphenyl (PCB) degrading fungi from a historically contaminated soil. Microb Cell Fact 8:5

Torres-Duarte C, Teresa VM, Vazquez-Duhalt R (2012) Laccase-mediated transformations of endocrine disrupting chemicals abolish binding affinities to estrogen receptors and their estrogenic activity in zebrafish. Appl Biochem Biotechnol 168:864–876

Touahar IE, Haroune L, Ba S, Bellenger JP, Cabana H (2014) Characterization of combined cross-linked enzyme aggregates from laccase, versatile peroxidase and glucose oxidase, and their utilization for the elimination of pharmaceuticals. Sci Total Environ 481:90–99

Tran NH, Urase T, Ngo HH, Hu JY, Ong SL (2013) Insight into metabolic and cometabolic activities of autotrophic and heterotrophic microorganisms in the biodegradation of emerging trace organic contaminants. Bioresour Technol 146:721–731

Tsutsumi Y, Haneda T, Nishida T (2001) Removal of estrogenic activities of bisphenol A and nonylphenol by oxidative enzymes from lignin-degrading Basidiomycetes. Chemosphere 42:271–276

Tuomela M, Steffen KT, Kerko E, Hartikainen H, Hofrichter M, Hatakka A (2005) Influence of Pb contamination in boreal forest soil on the growth and ligninolytic activity of litter-decomposing fungi. FEMS Microbiol Ecol 53:179–186

Tuomela M, Jørgensen K, Winquist E, Björklöf K, Schultz E, Anasonye F, Häkkinen L, Räsänen M, Sorvari J, Hartikainen ES, Steffen K, Valentin L (2012) Mycoremediation of contaminated soil in field scale. Conference abstract. Environ Eng Manag J 11:S38

Ullrich R, Nüske J, Scheibner K, Spantzel J, Hofrichter M (2004) Novel haloperoxidase from the agaric basidiomycete *Agrocybe aegerita* oxidizes aryl alcohols and aldehydes. Appl Environ Microbiol 70:4575–4581

UNEP, United Nations Environment Programme (2012) Annual report

US Environmental Protection Agency (2000) Exposure and human health.reassessment of 2,3,7,8-tetrachlorodibenzo-p-dioxin (TCDD) and related compounds. Part I: Estimating exposure to dioxin-like compounds. Vol. 3EPA/600/P-00/001

US EPA Handbook (1993) Approaches for the remediation of federal facility sites contaminated with explosive and radioactive wastes. EPA/625/R-93/013

Valentin L, Feijoo G, Moreira MT, Lema JM (2006) Biodegradation of polycyclic aromatic hydrocarbons in forest and salt marsh soils by white-rot fungi. Int Biodeterior Biodegrad 58:15–21

Valli K, Wariishi H, Gold MH (1992) Degradation of 2,7-dichlorodibenzo-*p*-dioxin by the lignin degrading basidiomycete *Phanerochaete chrysosporium*. J Bacteriol 174:2131–2137

Vallini G, Frassinetti S, D'Andrea F, Catelani G, Agnolucci M (2001) Biodegradation of 4-(1-nonyl)phenol by axenic cultures of the yeast *Candida aquaetextoris*: identification of microbial breakdown products and proposal of a possible metabolic pathway. Int Biodeterior Biodegrad 47:133–140

Van den Berg M, Birnbaum LS, Denison M, De Vito M, Farland W, Feeley M (2006) The 2005 World Health Organization reevaluation of human and mammalian toxic equivalency factors for dioxins and dioxin-like compounds. Toxicol Sci 93:223–241

van den Brink HM, van Gorcom RF, van den Hondel CA, Punt PJ (1998) Cytochrome P450 enzyme systems in fungi. Fungal Genet Biol 23:1–17

Verdin A, Sahraoui ALH DR (2004) Degradation of benzo[a]pyrene by mitosporic fungi and extracellular oxidative enzymes. Int Biodeterior Biodegrad 53:65–70

Verdin A, Sahraoui ALH, Newsam R, Robinson G, Durand R (2005) Polycyclic aromatic hydrocarbons storage by *Fusarium solani* in intracellular lipid vesicles. Environ Pollut 133:283–291

Verlicchi P, Galletti A, Petrovic M, Barceló D (2010) Hospital effluents as a source of emerging pollutants: an overview of micropollutants and sustainable treatment options. J Hydrol 389:416–428

Viswanath B, Rajesh B, Janardhan A, Kumar AP, Narasimha G (2014) Fungal laccases and their applications in bioremediation. Enzyme Res. doi:10.1155/2014/163242

Vyas BRM, Šašek V, Matucha M, Bubner M (1994) Degradation of 3,3',4,4'- tetrachlorobiphenyl by selected white rot fungi. Chemosphere 28:1127–1134

Walter M, Boyd-Wilson K, Boul L, Ford C, McFadden D, Chong B, Pinfold J (2005) Field-scale bioremediation of pentachlorophenol by *Trametes versicolor Int.* Biodeter Biodegr 56:51–57

Wang J, Majima N, Hirai H, Kawagishi H (2012) Effective removal of endocrine-disrupting compounds by lignin peroxidase from the white-rot fungus *Phanerochaete sordi*da YK-624. Curr Microbiol 64:300–303

Wang J, Yamamoto Y, Hirai H, Kawagishi H (2013a) Dimerization of bisphenol by hyper lignin-degrading fungus *Phanerochaete sordida* YK-624 under ligninolytic condition. Curr Microbiol 66:544–547

Wang J, Yamamoto R, Yamamoto Y, Tokumoto T, Dong J, Thomas P, Hirai H, Kawagishi H (2013b) Hydroxylation of bisphenol A by hyper lignin-degrading fungus *Phanerochaete sordida* YK-624 under non-ligninolytic condition. Chemosphere 93:1419–1423

Winquist E, Björklöf K, Schultz E, Räsänen M, Salonen K, Anasonye F, Cajthaml T, Steffen K, Jørgensen K, Tuomela M (2014) Bioremediation of PAH-contaminated soil with fungi- From laboratory to field scale. Int Biodeter Biodegr 86:238–247

Wolter M, Zadrazil F, Martens R, Bahadir M (1997) Degradation of eight highly condensed polycyclic aromatic hydrocarbons by *Pleurotus* sp. *florida* in solid wheat straw substrate. Appl Microbiol Biotechnol 48:398–404

Wu YR, Luo ZH, Chow KKR, Vrijmoed LLP (2010) Purification and characterization of an extracellular laccase from the anthracene-degrading fungus *Fusarium solani* MAS2. Bioresource Technol 101:9772–9777

Wunch KG, Alworth WL, Bennett JW (1999) Mineralization of benzo[a]pyrene by *Marasmiellus troyanus*, a mushroom isolated from a toxic waste site. Microbiol Res 154:75–79

Yadav JS, Quensen JF III, Tiedje JM, Reddy CA (1995) Degradation of polychlorinated biphenyl mixtures (aroclors 1242, 1254, and 1260) by the white rot fungus *Phanerochaete chrysosporium* as evidenced by congener-specific analysis. Appl Environ Microbiol 61:2560–2565

Yamaguchi M, Kamei I, Nakamura M, Takano M, Sekiya A (2007) Selection of *Pleurotus pulmonarius* from domestic basidiomycetous fungi for biodegradation of chlorinated dioxins as environmentally persistent organopollutants. Shinrin Sogo Kenkyusho Kenkyu Hokoku 6: 231–237

Yang Q, Yang M, Pritsch K, Yediler A, Hagn A, Schloter M, Kettrup A (2003) Decolorization of synthetic dyes and production of manganese-dependent peroxidase by new fungal isolates. Biotechnol Lett 25:709–713

Yang Q, Zhang H, Li X, Wang Z, Xu Y, Ren S, Chen X, Xu Y, Hao H, Wang H (2013) Extracellular enzyme production and phylogenetic distribution of yeasts in wastewater treatment systems. Bioresour Technol 129:264–273

Zeddel A, Majcherczyk A, Hüttermann A (1993) Degradation of polychlorinated biphenyls by white-rot fungi *Pleurotus ostreatus* and *Trametes versicolor* in a solid state systém. Toxicol Environ Chem 40:255–266

Zhang J, Xu Z, Chen H, Zong Y (2009) Removal of 2,4-dichlorophenol by chitosan-immobilized laccase from *Coriolus versicolor*. Biochem Eng J 45:54–59

Chapter 9
White and Brown Rot Fungi as Decomposers of Lignocellulosic Materials and Their Role in Waste and Pollution Control

Tripti Singh and Adya P. Singh

9.1 Introduction

Lignocellulosic materials are derived from the cell walls of secondary xylem tissues (wood) of forest trees and other woody plants. The primary xylem of some grass species, such as palm and bamboo, are also a rich source of lignocellulosic biomass. Certain fungi and bacteria can degrade lignocellulosic cell walls (reviewed in Blanchette et al. 1990; Kim and Singh 2000; Hatakka and Hammel 2010; Singh 2012; Arantes and Goodell 2014; Daniel 2014, 2015; Singh et al. 2015). Among these, basidiomycete white and brown rot fungi are the most efficient degraders of lignocellulosic materials, and play an important role in the recycling of carbon in nature, adding value to our environment. The ability of white and brown rot fungi to produce a variety of enzymes and extensively degrade wood cell wall polymers has been exploited in a wide range of biotechnological applications, from biopulping for pulp and paper production, pretreatment of lignocellulosic biomass to improve the yield of sugars for biofuels, bioremediation via detoxification of harmful environmental pollutants to targeted processing for improvements in food and health sectors. Therefore, this review focuses on white and brown rot basidiomycete fungi, outlining the processes associated with the decomposition of lignocellulosic materials by these fungi and then presenting an overview of biotechnological applications of these fungi and their enzymes relevant to waste and pollution control.

T. Singh (✉) · A.P. Singh
Scion (Formerly, Forest Research Institute), Te Papa Tipu Innovation Park, 49 Sala Street, Rotorua 3046, New Zealand
e-mail: Tripti.singh@scionresearch.com

D. Purchase (ed.), *Fungal Applications in Sustainable Environmental Biotechnology*, Fungal Biology, DOI 10.1007/978-3-319-42852-9_9

9.2 Lignocellulosic Materials

Lignocellulosic materials include secondary xylem tissues of tree species and other woody plants, such as shrubs and horticultural plants. Lignified fibres present in the secondary phloem of some plants, such as kenaf, and vascular bundles of palm and bamboo stems, also contribute significantly to lignocellulosic biomass on earth. Of these, the secondary xylem (wood) from forest trees is the most abundant lignocellulosic material and therefore the focus of this review. The term wood and lignocellulosic material will be used interchangeably.

The cell walls of wood, an abundant renewable plant biomass on our planet, consist of three main polymers: cellulose, hemicelluloses and lignin. Versatility in the application and high performance of wood-based products derives from the unique structure of wood cell walls, in which the three polymers form a complex association and are organised in specific patterns across the cell wall. Nature cycles its lignocellulosic biomass with the help of microorganisms, such as fungi and bacteria, which can independently and cooperatively degrade lignocellulosic cell walls (breakdown bonds within and between polymers), releasing carbon and other nutrients. Fungi can degrade cell walls more rapidly than bacteria and thus are considered to play a greater role in the recycling of carbon; despite their slower rate of degradation bacteria also make significant contributions, particularly under conditions unfavourable to the growth and activity of white and brown rot fungi, such as depleted oxygen environments (e.g. waterlogged terrestrial environments) and timbers containing high amounts of toxic preservatives and extractives (Kim and Singh 2000). White and brown rot fungi are the most dominant species among wood degrading microorganisms, as they can depolymerise all cell wall components, including the highly recalcitrant lignin (see Sect. 4).

9.3 Cell Wall Polymers of Lignocellulosic Materials

The major constituents of lignocellulosic cell walls are cellulose, hemicelluloses and lignin. Cellulose among these is the most abundant natural polymer on earth, constituting about 40 % of the dry weight of woody biomass (Sjöström 1993). Structurally, wood cell wall is a layered composite, consisting of compound middle lamella (middle lamella + primary cell wall) and the three layered secondary cell wall. Cellulose is predominantly located in the secondary cell wall. Cellulose is a homopolymer, consisting of β-D-glucopyranoside units. Cellulose chains are formed from interlinking of these units. Many such chains form a structural entity called microfibrils, which can be visualised by high resolution microscopes, such as transmission electron microscope (TEM), field-emission scanning electron microscope (FE-SEM) and atomic force microscope (AFM). Cellulose is largely crystalline, with some regions displaying semi-crystalline or non-crystalline characteristics, proportions of which vary among wood types and plant species.

Short chain heteropolymer hemicellulose consists of glucose, xylose, mannose, galactose and arabinose. Mannans (galactoglucomannans) are the main hemicelluloses of softwoods, which may also contain some xylans (arabinoglucuronoxylans). The major hemicelluloses present in hardwood cell walls are xylans. Hemicelluloses account for 15–30 % of the dry weight of wood, depending upon wood type and plant species. Hemicelluloses are linked to cellulose via hydrogen bonding, to lignin via ester linkages and to each other via covalent and hydrogen bonds. Cellulose chains are held tightly by numerous hydrogen bonds.

Lignin is a complex amorphous polymer, which forms a three dimensional network and is the main recalcitrant factor in lignocellulosic cell walls. Lignin is associated with hemicelluloses but is not considered to form a direct link with cellulose. Lignin consists of three different phenylpropane monomers. Three major lignin types have been recognised: syringyl (S), guaiacyl (G) and para-hydroxyphenyl (H) (Vanholme et al. 2010). Hardwood cell walls predominantly consist of S lignin, although certain cell types, e.g. vessels, are rich in G lignin. Softwood cell walls largely contain G lignin. Generally, G lignin exhibits greater resistance to wood degrading microorganisms than S lignin. Because of its complexity isolation of lignin in its native form is difficult and therefore it has not been possible to fully characterise lignins of lignocellulosic cell walls.

9.4 Degradation of Lignocellulosic Materials by White and Brown Rot Fungi

9.4.1 Natural Habitat

Of all wood rotting basidiomycete fungi more than 90 % are white rot fungi (Gilbertson 1980). Their abundance in nature coupled with the ability of some white rot fungi to completely mineralise all lignocellulosic cell wall polymers and convert into carbon dioxide and water makes them important players in the recycling of carbon in nature. Although brown rot fungi constitute only about 7 % of wood rotting basidiomycete fungi, they are also valuable recyclers of lignocellulosic biomass, particularly in softwood forests and man-made wooden structures exposed to external environments (Eriksson et al. 1990).

White rot fungi are widely present in terrestrial situations, with some species also present in aquatic environments. Because of their requirement for oxygen for growth and lignocellulosic cell wall degradation, they are largely excluded from oxygen deficient habitats, such as waterlogged land masses. Hardwood forest floors are particularly rich in white rot fungi, as they are more effective in degrading hardwoods than softwoods. The presence of whitish fibrous stumps in harvested hardwood forests is a common sight. In comparison, brown rot fungi are more common in softwood forests. Like the white rotters, brown rot fungi have a preference for terrestrial habitats and are largely absent from water saturated environments.

9.4.2 Colonisation of Lignocellulosic Materials

White and brown rot fungi deploy similar strategies for colonising and penetrating lignocellulosic materials, such as wood. Initially, these fungi gain entry into wood tissues via rays, which contain nutrient rich parenchyma cells in both hard and softwoods. The storage materials in parenchyma cells are readily metabolisable. Also, parenchyma cell walls in many wood species do not contain the more recalcitrant lignin. Pits, which occur in the common walls between neighbouring cells, serve as an important pathway for fungal movement within wood tissues. Although pits vary in the structure and composition of their partitioning membrane with wood and wood tissue types, pit membranes in the sapwood consist of pectins, cellulose and hemicelluloses and are not lignified. Thus, pit membranes are readily degraded by penetrating hyphae. White and brown rot fungi have also evolved a mechanism to directly penetrate through the double wall between wood elements for their movement within wood tissues. Fungi accomplish this by producing slender hyphae which can bore through the cell wall by localised cell wall degradation. Whilst rays are the main pathways for fungal entry into wood tissues in both soft and hardwoods, hardwood vessels are more readily and selectively colonised by fungal hyphae compared to other cell types (Blanchette et al. 1990). Their larger diameter can accommodate greater hyphal mass in the lumen, accelerating longitudinal fungal penetration of wood tissues and hyphal movement within the lignocellulosic material.

9.4.3 Degradation of Lignocellulosic Materials

White and brown rot fungi differ in their strategies for the degradation of lignocellulosic cell walls. White rotters can degrade both polysaccharides and lignin, but brown rotters while capable of rapidly and extensively depolymerising cellulose, can only partially modify lignin. Possessing greater capacity to degrade syringyl lignin than guaiacyl lignin, white rot fungi are more common to hardwood forests than softwood forests. Based on (i) their ability to degrade wood cell wall polymers and (ii) the micromorphological patterns produced during cell wall degradation, white rot fungi are categorised as: simultaneous (non-selective) and preferential degraders. Simultaneous degraders colonise cell lumen and attack exposed faces of the cell wall, resulting in cell wall erosion from simultaneous removal of polysaccharides and lignin (Eriksson et al. 1990). With progressive degradation, cell wall erosion deepens and expands from coalescence of eroded regions. All cell wall regions, including the more highly lignified cell corner middle lamella (CML), are eventually lost, with CML being the last structure to disappear. TEM studies have demonstrated presence of extracellular slime within cell walls undergoing degradation (Daniel 2014), which suggests that the extracellularly produced slime has a role in the degradation process, perhaps by channelling cell wall loosening radicals and other small molecular weight substances from the hyphae present in the cell lumen into the cell wall. This may be a mechanism to minimise loss of these potent molecules through uncontrolled diffusion. Preferential (selective) white rot degraders remove lignin from lignocellulosic cell walls to a greater extent than cellulose. The stumps in hardwood forests degraded by preferential white rotters typically have a stringy or whitish fibrous appearance, resulting mainly from the selective loss of lignin from larger diameter vessel elements. This type of fungal attack has also been described as white-pocket or white-mottled rot for some white rot fungi (Blanchette 1995).

In comparison to the white rotters, brown rot fungi mainly degrade polysaccharide components of lignocellulosic cell walls, with only slight modification of lignin. Brown rot fungi cause rapid depolymerisation of cellulose, because of which degraded wood tissues loose much of their strength (Cowling 1961), and thus can readily crumble under pressure, particularly in a dry state. Brown rotted wood can be readily distinguished from white rotted wood both macroscopically and microscopically. Softwood tree stumps degraded by brown rot fungi have a brownish appearance, and can disintegrate if touched or forced with a hard object. Superficially, microscopic views of brown rotted wood tissues are deceptive, as degraded wood tissues have a normal appearance when sections are viewed by light microscopy without staining for lignin or at low magnification under SEM. However, when viewed using polarisation mode of light microscope, dark appearing degraded regions can be readily distinguished from bright appearing (birefringent) sound tissue regions. Also, cell walls of degraded tissues stain more intensely with lignin contrasting stains than uncompromised tissues. TEM, in conjunction with staining of ultrathin sections with potassium permanganate to contrast lignin, has also been employed to confirm the presence largely of lignin in degraded cell walls as well as to characterise increases in cell wall porosity due to removal of polysaccharides (Singh et al. 2002). Production of abundant low molecular weight agents, such as hydroxyl radicals, is a key mechanism by which brown rot fungi loosen the microstructure of lignocellulosic cell walls. The greater porosity of the cell walls developed facilitates enzymatic attack on cellulose. The mechanism was proposed several decades ago (Cowling 1961; Koenigs 1974) and later expanded by many investigators (reviewed in Arantes et al. 2012).

9.5 Application of Fungi and Their Enzymes in Waste and Pollution Minimisation

Basidiomycetes and their enzymes have been widely used in waste minimisation and pollution control. Bioremediation and industrial applications of enzymes from the two important groups of basidiomycete fungi are reviewed.

White and brown rot fungi produce extracellular enzymes or exoenzymes, which are synthesised inside the cell and then secreted outside the cell. These enzymes are employed in a number of industries for pollution control. For example, laccases which can be produced by white rot fungi are capable of oxidising several environmental pollutants (reviewed by Viswanath et al. 2014). Cellulase, the other most commonly produced enzyme by basidiomycestes fungi, catalyses the conversion of carbohydrates to starch and is particularly used in pharmaceutical, food and beverage industries. The global market for enzymes was approximately US $4500 million in 2013 and it is expected to increase to over US $7000 million in the next decade (BBC Research Report 2011). Following topics will provide examples of industrial and other uses of key enzymes from white and brown rot fungi.

9.5.1 Pulp and Paper

Fungal pretreatment is an effective way of energy saving during mechanical pulping; fungal species and treatment duration directly influence the extent of energy saving. It has been shown that significant improvement in pulp strength and also substantial savings on energy use can be achieved from pretreatment of wood chips with lignin degrading fungi, their enzymes and enzyme-mediated systems (Bar-Lev et al. 1982; Abuhasan et al. 1988; Akhtar et al. 1998; Mansfield 2002; Chandra and Ragauskas 2005; Liu et al. 2012). In particular, white rot fungi have shown great promise. For example, Mardones et al. (2006) reported that the pretreatment of *Eucalyptus nitens* with the white rot fungus *Ceriporiopsis subvermispora* resulted in a loss of more than 13 % in lignin in 15 days; and a lower amount of alkali was needed to reach the target Kappa number (a measure of residual lignin). In addition, the pulp yield and strength properties of the pulp also markedly increased. Using white rot fungus *Trametes hirsute* to modify eucalyptus chemi-thermomecahnical pulp (CTMP) fibres, Yang et al. (2008) showed that fibre internal bond strength increased by 32 %, and that handsheet-tensile and tear properties were significantly improved.

In a study where energy saving was compared after treatment with different white rot fungi and a brown rot fungus for four weeks, Leatham et al. (1990) showed that greatest saving (>35 % to produce 100 ml Canadian Standard Freeeness (CSF) pulp) was achieved from treatment with *Phlebia* species. Treatment with the brown rot fungus, *Wolfiporia cocos* resulted in 10 % energy saving; however, there was no improvement in the strength properties of the pulp.

Extractives (pitch) in wood can cause serious technical and environmental problems in pulp and paper manufacture, particularly when the pitch is rich in lipophylic contents, such as resin acids, fatty acids and trigycerides (Hillis and Sumimoto 1989). The pitch can coat and clogg machines, reducing the paper quality (Beatson et al. 1999; Del Rio et al. 1999). Additionally, the presence of certain pitch components, such as resin acids in pulp mill effluents causes aquatic toxicity (Munkittrick and Sandstrom 2003), and the removal of pitch can enhance the effluent quality.

The use of selected fungi and enzymes has shown promise among the environmentally compatible methods attempted for pitch removal (reviewed in Gutiérrez et al. 2009). For their nutrition, sapstaining fungi mainly depend on the contents of rays and resin canals, where pitch is most abundant, and earlier studies employed these fungi for pitch control (Farrell et al. 1993). Enzymes, such as lipases, have been effective, particularly for pulping softwoods with high resin acid content. Use of wood degrading basidiomycetes has been explored in more recent developments. Because brown rot fungi extensively degrade cellulose, their use in pitch control is limited. However, white rot fungi have been successfully used, as they degrade lignin (particularly the selective degraders) and can also remove lipophylic substances. There is more efficient removal of certain pitch components, such as sterols and resin acids, by white rot fungi than sapstaining fungi (Martinez-Inigo et al. 1999). However, in selecting fungi for pitch control, toxicity of certain extractives is also an important factor to consider (Eberhardt et al. 1994). The use of white rot fungi is gaining favour because pitch control can be combined with the removal of residual lignin from the pulp (bio-bleaching), which can lead to considerable energy saving during pulping (Akhtar et al. 2000; Bajpai et al. 2001; Gutiérrez et al. 2006) while minimising the risk of chemical pollution of the environment. It is important, however, that these processes are fine-tuned to avoid degradation of cellulose when using white rot fungi for pitch control so that fibre strength is not compromised.

Effectiveness of a range of enzymes, such as lipase, resinase, lipoxigenases, laccases, has also been tested for pitch control (reviewed in Gutiérrez et al. 2009), laccases being the most effective and the preferred choice for the application. While laccases are generally most effective on phenolic compounds, laccase from a *Trametes* species has some activity on polyunsaturated fatty acids and conjugated resin acids (Karlsson et al. 2001). Laccase appears to be effective also against a range of lipophilic extractives (Buchert et al. 2002; Dubé et al. 2008; Zhang et al. 2005). Laccase mediator system, which can efficiently remove lipophylic extractives from a wide range of wood substrates, has been employed. For example, a study by Gutiérrez et al. (2006) showed that in the presence of mediator HBT (1-hydroxybenzotriazole) laccase from the basidiomycete *Pycnoporus cinnabarinus* removed free and conjugated sterols, triglycerides, resin acids and sterols by up to 100 %, and fatty alcohols, alkanes and sterols by 40–100 % from various pulp systems.

9.5.2 Timber Treatment (CCA Remediation)

Metal-based compounds have been employed for a long time to preserve wood. These compounds are toxic and can adversely affect human health and the environment (Decker et al. 2002). Among different chemicals and their combinations employed, copper-chromium-arsenate (CCA) is favoured and the most commonly used preservative. CCA preservatives have been in use world-wide for some

70 years. This preservative is highly effective against fungi and termites, and is being used in a wide range of outdoor applications, including: (i) housing, (ii) vineyard and other utility poles, (iii) decks, (iv) playground structures and (v) landscaping. However, disposal or recycling of CCA-treated wood after its service life poses a tough technological challenge, and effective processes have to be developed for the detoxification of CCA. While much of CCA is fixed in the wood, some amount of arsenic, chromium and copper is often released due to the exposure of treated wood to rain and irrigation water (Jamberck et al. 2006) and can contaminate soils and water. Also of increasing concern is the disposal of treated wood after service life, particularly via placing underground (land filling), because toxic components can persist in the ground for a long time when they are not readily degraded by microorganisms. It is therefore important that after service preservative components are extracted from the treated wood and detoxified prior to disposal or recycling.

Public and scientific awareness on managing wastes from treated wood at the end of service life has stimulated attempts at remediation by extraction of preservative components using organic acid and microorganisms. Bacteria have been the main force for the majority of commercial scale bioremediation operations, based on degradation/detoxification of toxic chemicals, as certain types of bacteria are highly tolerant to toxic chemicals, and may even rely on them for growth (Wolfe-Simon et al. 2011). Use of wood rotting fungi and their enzymes for this purpose has attracted attention in recent years.

Metal leaching from CCA-treated wood is promoted by a wide range of microorganisms (Clausen and Smith 1998; Kartal et al. 2004). High levels of oxalic acid and other polycarboxylic acids are produced by brown rot fungi. Being chelating and reducing agents, these acids can be readily oxidised as a fungal product during fermentation; they can be used for the removal of toxic metal elements via biochelation or bioleaching (Kartal and Kose 2003). Evaluation of bioremediation of CCA-treated wood by three different brown rot fungi indicated although the extent of CCA leaching varied (Kartal et al. 2004), the three fungi shared a common feature: the process was mediated by the production of oxalic acid, which increased the acidity of the substrate and solubility of metals. Solid-state fermentation with Cu-tolerant brown rot fungi has proven to be a promising approach for the remediation of decommissioned wood treated with Cu, Cr and/or As (Sierra-Alvarez 2009). Thus, preservative tolerant microorganisms are of great interest and can be used for bioremediation and bioconversion of CCA (Schubert et al. 2011).

Owing to their ability to oxidise diverse range of compounds, white rot fungi may offer advantages over other microorganisms. The ability of white rot fungi to produce lignin degrading/modifying enzymes and low molecular mass-mediators may also enhance bioavailability of pollutants. Thus, the use of white rot fungi offers the possibility to expand the substrate range via biodegradation of pollutants that cannot be readily removed by bacteria- or chemical-based methods.

9.5.3 Coal Industry

Low-grade coal such as peat and lignite has low fuel value due to its relatively low heat content. However, their degradation by microorganisms, can lead to better utilisation of lignite resources for both fuel and nonfuel applications. Lignin degrading enzymes, such as laccase, liginin peroxidases and manganese peroxidase (MnP), can potentially solubilise coal. Coal solubilisation for the production of liquid lignite could become an alternative to thermo-chemical operations that require high temperature and pressure, and result in the reduction of environmental pollutants such as sulphur, nitrogen oxide and others (Faison 1991).

The process of solubilising low-grade coal with fungi has been termed 'bio-solubilisation'. Various studies have shown that in bio-solubilisation procedure, dissolution of colloidal coal occurs through the action of enzymes produced by a number of basidiomycetes, such as *Phanerochaete chrysosporium* and *Nematoloma frowardii* (Catcheside and Mallet 1991; Hofrichter and Fritsche 1997; Gokcay et al. 2001). A recent study by Sekhola and Igbinigie (2013) also looked at the biological properties of bio-solubilisation products and found the presence of oxidative and reducing capabilities and other biological activities. The precise role of these enzymes in the process of solubilisation is not clear. However, specific enzyme and mutant studies by Ralph and Catcheside (1997) suggested that coal can be both polymerised and depolymerised by MnP, without any change in its absorbance at 400 nm. Lignin peroxidase decolourises coal without apparently altering its molecular mass. Coal-derived monomers were not recovered from cultures of *P. chrysosporium*, suggesting that they were taken-up by fungal cells. This indicates that cellular transformations may permit conversion of diverse catabolic products derived from low rank coal to specific low molecular mass compounds.

Substrates obtained from fungal degradation of low-grade coals have been investigated for their potential as soil conditioning agents and as substrates, which could be subjected to subsequent processing for the generation of alternative fuel, like methane (Haider et al. 2015).

9.5.4 Food Industry

Food industry has utilised fungi and their metabolites since ancient times in a variety of ways. Selected enzymes from basidiomycetes fungi are used for processing of food for desirable attributes, such as nutrimental elements, colour or flavour (Minussi et al. 2002). For example, laccases has been used to modify and enhance the colour of food or beverage by eliminating the undesirable phenolics, responsible for haze formation and browning during alcohol or certain food production (Rodriguez and Toca 2006; Selinheimo et al. 2006). Enzymes lignin peroxidase and MnP produced by a white rot fungus *P. chrysosporium* are shown to be effective in enhancing flavour through vanillin production (Barbosa et al. 2006).

Residues from agricultural and horticultural industries constitute a large quantity of waste. Much of this waste is lignocellulosic biomass, which presents a great potential as a feedstock for value-added products, such as chemicals, enzymes, ruminant feed and various applications in food industry. For example, Zhi and Wang (2014) showed degradation of up to 25 % lignin from wheat straw using a white rot fungus *P. chrysosporium*. The delignified straw, after further processing, has been used for the production of chemicals, such as hydrogen and lactic acids (Mass et al. 2006). Also, the nutritional value of wheat straw for ruminants increased after lignin degradation (Raghuwanshi et al. 2014).

Several studies have shown that enzymes produced by white rot fungi, such as laccase, xylanase and MnP, are useful in fermentation of wastes from food industries and beneficial in other applications (Selinheimo et al. 2006; Songulashvili et al. 2007). The evaluation of the activities of laccase and MnP from eighteen strains of basidiomycetes in mandarin peelings and ethanol production waste indicated that the expression of enzyme activity is species- and strain-dependent (Selinheimo et al. 2006). In this study, all species of the genus *Trametes* expressed comparatively high laccase activity; however, laccase and MnP production was much dependent on the lignocellulosic growth substrate.

Enzymatic polymerisation in organic synthesis using laccases as a biocatalyst has several applications, including: (i) oxidation of functional groups and (ii) coupling of phenols and steroids. This attribute led to increased use of this enzyme in the production (synthesis) of anesthetics, anti-inlammatroy, antibiotic and sedatives (Aktas and Tanyolac 2003; Mikolasch and Schauer 2009).

9.5.5 Biofuel Industry[1]

Biofuel is produced from cellulosic and lignocellulosic biomass through biological processes. Currently bioethanol is produced by fermentation of sugars from plant carbohydrates, and crops such as sugarcane and corn are the most commonly used. Cellulosic biomass derived from nonfood sources such as trees and wood wastes is also being investigated as a suitable feedstock for ethanol production.

Conversion of polysaccharides from lignocellulosic biomass, such as wood and wood residues, into ethanol requires several processing steps, including enzymatic hydrolysis of the substrate and fermentation of sugars derived from polysaccharides. However, enzyme accessibility of polysaccharides in the native lignocellulosic cell walls is restricted and an effective pretreatment is required to enhance enzymatic accessibility (Zheng et al. 2009). This is because in wood cell walls,

[1]Additional information on biofuel production is presented in Part IV—*Biotehcnology to reduce reliance on fossil fuels: from biomass to biofuel* of this book.

polysaccharides and lignin are interwoven forming a complex association. Pretreatment of lignocellulosic substrates loosens the cell wall microstructure and increases cell wall porosity, which facilitates enzymatic accessibility of cellulose (Singh et al. 2016). The use of wood rotting fungi and their enzymes is showing much promise (Vaidya and Singh 2012; Knežević et al. 2013).

White rot basidiomycetes have been widely used for pretreatment (Lee et al. 2007; Zhang et al. 2007; Kneževica et al. 2013; Li and Chen 2014), and in recent years the use of brown rot fungi and their enzymes has also been explored (Ray et al. 2010; Vaidya and Singh 2012). Ray et al. (2010) found that exposure of *radiata* pine wood to brown rot fungi *Coniophora puteana* and *Postia placenta* resulted in significant increases in glucose yield after enzymatic saccharification. Similar trend was observed in a more recent study where the effect of fungal pretreatment on enzymatic digestibility of *radiata* pine substrates including wood blocks, wood chips and steam exploded wood (SEW), was evaluated using a white rot fungus and three brown rot fungi (Vaidya and Singh 2012). Fungal pretreatment in this study demonstrated enhanced sugar yield, the degree of increase being substrate dependant. The fungal treatment of thermo-chemically treated substrate gave ten times higher sugar yield compared to wood block or chips. A subsequent study using transmission electron microscopy showed that the brown rot fungus *Oligoporus placenta* caused more widespread and greater loosening of cell wall microstructure compared to the white rot fungus *Trametes versicolor* (Singh et al. 2016). In the initial stages of cell wall degradation brown rot fungi are more effective in loosening the structure of lignocellulosic cell walls, as they deploy reactive oxygen species, such as hydroxyl radicals (Enoki et al. 1988; Kneževica et al. 2013) that can readily penetrate into the lignocellosic cell walls.

More recently the potential of other basidiomycete fungi, such as *Flammulina velutipes* (golden needle mushroom), has been explored with promising results in the bioprocessing of lignocellulosic biomass to bioethanol (Mahara et al. 2013).

Biodiesel is another common biofuel and some researchers have shown the potential of selected fungi, such as *Gliocladium roseum* in the production of so called 'myco-diesel' from cellulose (Sergeeva et al. 2008). These fungi have the unique ability of converting cellulose into medium length hydrocarbons typically found in diesel fuels (Strobel et al. 2008).

9.6 Conclusions

Human activities are the main cause of the rising environmental pollution confronting us. Sources of pollution are many, with heavy industries topping the list. Efforts on developing environmentally friendly methods to tackle these challenges are growing, and one such attractive way to minimise waste and environmental pollution is to make use of wood rotting fungi and their enzymes. This chapter comprehensively reviews the information available to date on the role of white and brown rot fungi, the most efficient degraders of lignocellulosic materials, and their enzymes in the control of: (i) organic materials that make up the bulk of organic wastes from varied industrial and nonindustrial sources, and (ii) harmful chemicals used in the processing of industrial products that causes contamination in the ecosystem during their service and the post-service disposal. It is hoped that the information presented will further raise awareness of the pressing need to protect our environment from polluting agents and stimulate activities aimed at developing environmentally compatible technologies based on the use of white and brown rot fungi that are generally regarded as destructive agents.

References

Abuhasan J, Pellinen J, Joyce TW, Chang H-M, 1988. Delignification of chemi-thermomechanical pulp by a fungal treatment. In: Proceedings of the AICHE annual meeting. New Developments in Pulp and Bleaching, Washington, DC, pp 1–7

Akhtar M, Blanchette RA, Myers G, Kirk TK (1998) An overview of biomechanical pulping research. In: Young R, Akhtar M (eds) Environmentally friendly technologies for the pulp and paper industry. Wiley, New York, pp 309–339

Akhtar M, Scott GM, Sweney RE, Shipley DF (2000) Biochemical pulping: a mill-scale evaluation. Resour Conserv Recycl 28:241–252

Aktas N, Tanyolac A (2003) Reaction conditions for laccase catalysed polymerization of catechol. Bioresour Technol 87:209–214

Arantes V, Jellison J, Goodell B (2012) Peculiarities of brown rot fungi and biochemical Fenton reaction with regard to their potential as a model for bioprocessing biomass. Appl Microbiol Biotechnol 94:323–338

Arantes V, Goodell B (2014) Current understanding of brown rot fungal biodegradation mechanisms: a review. In: Deterioration and protection of sustainable biomaterials (eds. Schultz et al.), ACS symposium series. American Chemical Society, Washington DC, pp 3–21

Bajpai P, Bajpai PK, Akhtar M, Jauhari MB (2001) Biokraft pulping of eucalyptus with selected lignin-degrading fungi. J Pulp Pap Sci 27:235–242

Bar-Lev SS, Kirk TK, Chang H-M (1982) Fungal treatment can reduce energy requirement for secondary refining of TMP. Tappi J 65:111–113

Barbosa ES, Perrone D, Vendramini ALA, Leite SGF (2006) Vanillin production by Phanerochaete chrysosporium grown on green coconut agro-industrial husk in solid state fermentation. BioResources 3:1042–1050

BBC Research Report (2011) BIO030 F enzymes in industrial applications: global markets. BBC Research, Wellesley, MA, USA

Beatson RP, Zhang X, Stebbing D, Saddler JN (1999) The dissolved and colloid fractions of white water: impact on paper quality and degradation by enzymes. ISWPC, 7–10 June 1999, Yokohama, Japan

Blanchette RA, Nilsson T, Daniel G, Abad A (1990) Biological degradation of wood. In: Archaeological wood: properties, chemistry and preservation (eds. Rowell RM, Barbour RJ), advances in chemistry 225. American Chemical Society, Washington DC, pp 141–174

Blanchette RA (1995) Degradation of the lignocellulose complex in wood. Can J Bot 73:S999–1010

Buchert J, Mustranta A, Tamminen T, Spetz P, Holmbom B (2002) Modification of spruce lignans with *Trametes hirsute* lacasse. Holzforschung 56:579–584

Catcheside DEA, Mallet KL (1991) Solubilization of Australian lignites by fungi and other microorganisms. Energy Fuels 5:141–145

Chandra RP, Ragauskas AJ (2005) Modidfication of high lignin kraft pulps by laccase. Part 2. Xylanase-enhancedstrength benefits. Biotechnol Prog 21:1302–1306
Clausen CA, Smith RL (1998) Removal of CCA from treated wood by oxalic acid extraction, steam explosion, and bacterial fermentation. J Indust Microbiol Biotechnol 20:251–257
Cowling EB (1961) Comparative biochemistry of the decay of sweetgum sapwood by white rot and brown rot fungi. USDA Tech Bull 1258
Daniel G (2014) Fungal and bacterial biodegradation: white rots, brown rots, soft rots and bacteria. In: Deterioration and protection of sustainable biomaterials (eds. Schultz et al.), ACS symposium series. American Chemical Society, Washington DC, pp 23–58
Daniel G (2015) Fungal degradation of wood cell walls. In: Secondary xylem biology: origins, functions and applications (eds. Kim YS, Funada R, Singh AP), Elsevier, Waltham, MA, USA. In press
Decker P, Cohen B, Butala JH, Gordon T (2002) Exposure to wood dust and heavy metals in workers using CCA pressure-treated wood. AIHA J 63:166–171
Del Rio JC, Gutiérrez A, González-Vila FJ, Martin F (1999) Application of pyrolysis-gas chromatography-mass spectrometry to the analysis of pitch deposits and synthetic polymers in pulp and paper mills. J Anal Appl Pyrol 49:165–177
Dubé E, Shareck F, Hurtubise Y, Beauregard M, Daneault C (2008) Enzyme-based approaches for pitch control in thermomechanical pulping of softwood and pitch removal in process water. J Chem Technol Biotechnol 83:1261–1266
Eberhardt TL, Han JS, Micales JA, Young RA (1994) Decay resistance in conifer seed cones—role of resin acids as inhibitors of decomposition by white rot fungi. Holzforschung 48:278–284
Enoki A, Tanaka H, Fuse G (1988) Degradation of lignin-related compounds, pure cellulose and wood components by white rot and brown rot fungi. Holzforschung 42:85–93
Eriksson KE, Blanchette RA, Ander P (1990) Microbial and enzymatic degradation of wood and wood components. Springer, Berlin
Faison BD (1991) Microbial conversions of low rank coals. Bio/Technology 9:951–956
Farrell RL, Blanchette RA, Brush TS, Hadar Y, Iverson S, Krisa K et al (1993) Cartpip^{TM}, a biopulping product for control of pitch and resin acid problems in pulp mills. J Biotechnol 1993(30):115–122
Gilbertson RL (1980) Wood rotting fungi of North America. Mycologia 72:1–49
Gokcay C, Kolankaya N, Dilek F (2001) Microbial solubilization of lignites. Fuel 80:1421–1433
Gutiérrez A, del Rio JC, Rencoret J, Ibarra D, Martinez AT (2006) Main lipophilic extractives in different paper pulp types can be removed using the laccase-mediater system. Appl Microbiol Biotechnol 72:845–851
Gutiérrez A, del Rio JC, Martinez AT (2009) Microbial and enzymatic control of pitch in the pulp and paper industry. Appl Microbiol Biotechnol 82:1005–1018
Haider R, Ghauri MA, Jones EJ, Orem WH, SamFilipo JR (2015) Structural degradation of Thar lignite using MWI fungal isolate: optimization studies. Int Biodeterior Biodegrad 100:149–154
Hatakka A, Hammel KE (2010) Fungal biodegradation of lignocellulosics. In: Hofrichter M (ed) Industrial applications: the mycota X. Springer-Verlag, Berlin, pp 319–340
Hillis WE, Sumimoto M (1989) Effect of extractives on pulping. In: Rowe JW (ed) Natural products of woody plants: chemical extraneous to the lignocellulosic cell wall, vol 2. Springer-Verlag, Berlin, pp 880–920
Hofrichter M, Fritsche W (1997) Depolymerisation of low-rank coal by extracellular fungal enzyme systems. The ligninolytic enzymes of the coal-humic-acid-depolymerizing fungus *Nematoloma frowardii B19*. Appl Microbiol Biotechnol 47:419–424
Jamberck JR, Townsend T, Solo-Gabriele H (2006) Leaching of chromated copper arsenate (CCA)—treated wood in a simulated monofil and its potential impacts to landfill leachate. J Hazard Mater 135:21–31
Karlsson S, Holmbom B, Spetz P, Mustranta A, Buchert J (2001) Reactivity of *Trametes* laccases with fatty and resin acids. Appl Microbiol Biotechnol 55:317–320

Kartal SN, Kose C (2003) Remediation of CCA-C treated wood using chelating agents. Holz Roh Werst 61:382–387

Kartal SN, Munir E, Kakitani T, Imamura Y (2004) Bioremediation of CCA-treated wood by brown rot fungi *Fomitopsis palustris*, *Coniophora puteana*, and *Laetiporus sulphureus*. J Wood Sci 50:182–188

Kim YS, Singh AP (2000) Micromorphological characteristics of wood biodegradation in wet environments: a review. IAWA J 21:135–155

Koenigs JW (1974) Production of hydrogen peroxide by wood rotting fungi in wood and its correlation with weight loss, depolymerisation and pH changes. Arch Microbiol 99:129–145

Kneževića A, Milovanovića I, Stajića M, Lončarb N, Brčeskib I, Vukojevića J et al (2013) Lignin degradation by selected fungal species. Bioresource Technol. 138:117–123

Leatham GF, Myers GC, Wegner TH (1990) Biomechanical pulping of aspen chips: energy savings resulting from different fungal treatments. Tappi J 73:197–200

Lee J-W, Gwak K-S, Park J-Y, Park M-J, Choi D-H, Kwon M et al (2007) Biological pretreatment of softwood *Pinus densiflora* by three white rot fungi. J Microbiol 45:485–491

Li G, Chen H (2014) Synergistic mechanism of steam explosion combined with fungal treatment by *Phellinus baumii* for the pretreatment of corn stalk. Biomass Bioenergy 67:1–7

Liu N, Qin M, Gao Y, Li Z, Fu Y, Xu Q (2012) Pulp properties and fiber characteristics of xylanase-treated aspen APMP. BioResources 7:3367–3377

Mahara T, Hitomi I, Ichinose H, Furukawa T, Ogasawara W, Takabatake K et al (2013) Ethanol production from high cellulose concentration by the basidiomycete fungus *Flammulina velutipes*. Fungal Biol 117:220–226

Mansfield SD (2002) Laccase impregnation during mechanical pulp processing—improved refining efficiency and sheet strength. Appita J 55:49–53

Mardones L, Gomide JL, Freer J, Ferraz A, Rodríguez J (2006) Kraft pulping of Eucalyptus nitens wood chips biotreated by *Ceriporiopsis subvermispora*. J Chem Technol Biotechnol 81:608–613

Mass RHW, Bakker RR, Eggink G, Weusthuis RA (2006) Lactic acid production from xylose by the fungus Rhizopus oryzae. Appl Microbiul Biotechnol 72(831–868):2006

Martinez-Inigo MJ, Immerzeal P, Gutiérrez A, del Rio JC, Sierra-Alvarez R (1999) Biodegradability of extractives in sapwood and heartwood from Scots pine by sapstain and white rot fungi. Holzforschung 53:247–252

Mikolasch A, Schauer F (2009) Fungal laccases as tools for the synthesis of new hybrid molecules and biomaterials. Appl Microbiol Biotechnol 82:605–624

Minussi RC, Pastore GM, Duran N (2002) Potential applications of laccase in the food industry. Trends Food Sci Technol 13:205–216

Munkittrick KR, Sandstrom O (2003) Eological assessment of pulp mill impacts: issues, concerns, myths, and research needs. In: Stuthridge TR, van den Heuvel MR, Marvin NA, Slade AH, Gifford J (eds) Environmental impacts of pulp and paper waste streams. SETAC Press, Pensacola, Florida, pp 352–362

Ralph JP, Catcheside DE (1997) Transformations of low rank coal by Phanerochaete chrysosporium and other wood-rot fungi. Fuel Process Techn 52:79–93

Raghuwanshi S, Misra S, Saxena RK (2014) Treatment of wheat straw using tannase and white rot fungus to improve feed utilization by rumiants. J Anim Sci Biotechnol 5:1–18

Ray MJ, Leak DJ, Spanu PD, Murphy RJ (2010) Brown rot fungal early stage decay mechanism as a biological pretreatment for softwood biomass in biofuel production. Biomass Bioenergy 34:1257–1262

Rodriguez S, Toca JL (2006) Industrial and biotechnological applications of laccases: A review. Biotechnol Adv 24:500–513

Schubert M, Volkmer T, Lehringer C, Schwarze FWMR (2011) Resistance of bioincised wood treated with wood preservatives to blue-stain and wood-decay fungi. Int Biodeterorat Biodegrad 65:108–115

Sekhola ML, Igbinigie EE (2013) Cowan AK (2013) Biological degradation and solubilisation of coal. Biodegrad 24:305–318

Selinheimo E, Kruus K, Buchert J, Hopia A, Autio K (2006) Effect of laccase, xylanase and their combination on the rheological properties of wheat doughs. J Cereal Sci 42:152–159

Sergeeva YE, Galanina LA, Andrianova DA, Feofilova EP (2008) Lipids of filamentous fungi as a material for producing biodiesel fuel. Appl Biochem Microbiol 44:523

Sierra-Alvarez R (2009) Removal of copper, chromium and arsenic from preservative-treated wood by chemical extraction-fungal bioleaching. Waste Manag 29:1885–1891

Singh AP (2012) A review of microbial decay types found in wooden objects of cultural heritage recovered from buried and waterlogged environments. J Cult Herit 13:S16–S20

Singh AP, Daniel G, Nilsson T (2002) High variability in the thickness of the S3 layer in Pinus radiata tracheids. Holzforschung 56:111–116

Singh AP, Kim YS, Singh T (2015) Bacterial degradation of wood cell walls. In: Kim YS, Funada R, Singh AP (eds) Secondary xylem biology: origins, functions and applications. Elsevier, Waltham, MA, USA, pp 169–190

Singh T, Vaidya AA, Singh AP (2016) Improvement in the enzymatic hydrolysis of biofuel substrate by a combined thermochemical and fungal pretreatment. Wood Sci Technol. (In press)

Sjöström E (1993) Wood chemistry, fundamentals and applications. Academic, San Diego

Songulashvili G, Elisashvili V, Wasser SP, Nevo E, Hadar Y (2007) Basidiomycetes laccases and manganese peroxidase activity in submerged fermentation of food industry wastes. Enzym Micobial Technol 41:57–61

Strobel G, Knighton B, Kluck K, Ren Y, Livinghouse T, Griffin M et al (2008) The production of myco-diesel hydrocarbons and their derivatives by the endophytic fungus Gliocladium roseum (NRRL 50072). Microbiology 154:3319–3328

Vaidya AA, Singh T (2012) Pre-treatment of P. radiata substrate by basidiomycetes fungi to enhance enzymatic hydrolysis. Biotechnol Lett 34:1263–1267

Vanholme R, Demedts B, Morreel K, Ralph J, Boerjan W (2010) Lignin biosynthesis and structure. Plant Phys 153:895–905

Viswanath B, Rajesh B, Janardhan A, Kumar AP, Narasimha G (2014) Fungal laccases and their applications in bioremediation. Enzym Res Article ID 163242, 21 p

Wolfe-Simon F, Blum JS, Kulp TR, Gordon GW, Hoeft SE, Pett-Ridge J et al (2011) A bacterium that can grow by using arsenic instead of phosphorus. Science 332:1163–1166

Yang Q, Zhan H, Wang S, Fu S, Li K (2008) Modification of eucalyptus CTMP fibres with white rot fungus trametes hirsute—Effects on fibre morphology and paper physical strengths. Bioresour Technol 99:8118–8124

Zhang X, Renaud S, Paice M (2005) The potential of laccase to remove extractives present in pulp and white water from TMP newsprint mills. J. Pulp Paper Sci. 31:175–180

Zhang XY, Yu HB, Huang HY, Liu YX (2007) Evaluation of biological pretreatment with white rot fungi for the enzymatic hydrolysis of bamboo culms. Int Biodeter Biodegrad 60:159–164

Zheng Y, Pan Z, Zhang R (2009) Overview of biomass pretreatment for cellulosic ethanol production. Int J Agric Biol Eng 2:51–68

Zhi Z, Wang H (2014) White-rot fungal pretreatment of wheat straw with Phanerochaete chrysosporium for biohydrogen production: simultaneous saccharification and fermentation. Bioprocess Biosyst Eng 37:1447–1458

Chapter 10
Mycoremediation of Heavy Metal/Metalloid-Contaminated Soil: Current Understanding and Future Prospects

Wai Kit Chan, Drik Wildeboer, Hemda Garelick and Diane Purchase

10.1 Introduction

Heavy metals are members of an ill-defined subset of elements that exhibit metallic properties. For many years, the term 'heavy metal' has been broadly accepted as referring to any metals and metalloids that are of relatively high density (>5 g cm^{-3}) (Alloway 2013). The term is thus applicable to the transition metals, some metalloids, lanthanides and actinides. For the purpose of this article, the term 'heavy metals' refers to the common transitional metals such as copper (Cu), cadmium (Cd), lead (Pb), mercury (Hg) and zinc (Zn); whereas the term 'metalloids' denotes a group of chemical elements that exhibit characteristics in-between metals and non-metals—they occur naturally as poly-hydroxylated species. Usually grouped under this classification are the chemical elements boron (B), silicon (Si), germanium (Ge), arsenic (As), antimony (Sb) and tellurium (Te). The rare elements such as polonium (Po) and astatine (At) are also sometimes included (Britannica Academic 2015).

The presence of heavy metals and metalloids in the environment either in water or soil could have significant adverse effects on human and ecosystems health. According to Ochiai (1997), the general mode of action for toxic metal ions is via the blocking of proteins including enzymes either by displacing an essential metal ion in the biomolecules or modifying the active conformation of the biomolecules resulting in the loss of specific activity.

The ability of metals and metalloids to interrupt the function of crucial biological molecules, such as proteins and DNA, in living organisms can be seen as the effect of exposure to toxic metals (Jaishankar et al. 2014). The exposure to Pb, As and Hg can result in various human neurological problems which have been

W.K. Chan · D. Wildeboer · H. Garelick · D. Purchase (✉)
Department of Natural Sciences, Faculty of Science and Technology, Middlesex University, The Burroughs, London NW4 4BT, United Kingdom
e-mail: d.purchase@mdx.ac.uk

D. Purchase (ed.), *Fungal Applications in Sustainable Environmental Biotechnology*, Fungal Biology, DOI 10.1007/978-3-319-42852-9_10

well-documented (e.g. Factor-Litvak et al. 1999; Wasserman et al. 2004; Rothenberg and Rothenberg 2005; Trasande et al. 2005). Nearly 200 enzymes can be inhibited in the presence of As (Ratnaike 2003); for example, glutathione reductase (Muller et al. 1995; Styblo et al. 1997), glutathione S-transferase (Chouchane and Snow 2001), DNA ligase (Snow et al. 1999) and pyruvate kinase galectin (Lin et al. 2006). Cellular damage can be caused by the presence of heavy metal/metalloids through the formation of free radicals and other reactive oxygen species that can induce oxidative stress (Dietz et al. 1999).

Fungi have had a major contribution to the functioning of the biosphere, industry, medicine and research throughout human history (Schwantes 1996; Scholtmeijer et al. 2001). Many fungi are of the microscopic varieties living in soils, playing an important role in recycling by releasing enzymes to break down complex organic compounds to simpler chemical constituents, which are subsequently used by themselves or other organisms (Stajich et al. 2009). In addition to their functions as decomposers of plant polymers such as cellulose and lignin in the ecosystem (Romaní et al. 2006), fungi also received attentions for their potentials to bioremediate inorganic pollutants as they are highly versatile and able to tolerate extreme levels of metal concentration, nutrient availability, pH or temperature (Dhankhar and Hooda 2011). Mycoremediation offers a number of advantages such as high efficiency, low operating cost, reduce usage of chemicals and produce fewer toxic chemical by-products (Vijayaraghavan et al. 2006; Gadd 2009; Harms et al. 2011). The current knowledge and mechanisms involved in mycoremediation of metals/metalloids is being discussed in the chapter, aiming to provide a better understanding of the role of fungi in the cleaning up of the polluted environments.

10.2 Mycoremediation of Metals/Metalloids: A Brief Overview[1]

Environmental contaminations by metals and metalloids from previous mining industries, wastes from commercial products as well as natural source such as volcanic activity have generated interest all over the world because of their effects to the health of humans, animals and plants (Tchounwou et al. 2012; Vodyanitskii 2013). In view of the high cost of treating and remediating contaminated sites, a cost-effective, rapid and environmental friendly method would be advantageous as an alternative. Bioremediation, a process of using or introducing living organisms such as bacteria, fungi, plants or worms to the polluted sites, has become an increasingly popular solution for the decontamination of toxic chemicals due to its low operating cost (Gadd 2001; Sylvia et al. 2005). The diverse metabolism capabilities of fungi play an important role in the detoxification of organic

[1]Additional information on metal and metalloid mycoremediation is available in Chap. 15—*Geomycology*.

pollutants and are involved in remediating heavy metals in the environments (Gadd 2009). Unlike bacteria, fungi are robust organisms that have the ability to tolerate high levels of toxic chemicals including heavy metal/metalloid(s) (Joshi 2014; Dixit et al. 2015). Fungi have been extensively studied by researchers since the 1980(s) for their bioremediation potential; metal-tolerant filamentous fungi have been isolated from heavy metal-contaminated soils and sediments, for instance, *Fusarium* (e.g. Saxena et al. 2006; Ezzouhri et al. 2009), *Penicillium* (e.g. Xu et al. 2012; Bhargavi and Savitha 2014), *Rhizopus* (e.g. Zafar et al. 2007; Srivastava et al. 2011), *Aspergillus* (e.g. Maheswari and Murugesan 2009; Iskandar et al. 2011; Srivastava et al. 2011), *Beauveria* (e.g. Purchase et al. 2009; Joseph et al. 2012), *Cladosporium* (Urik et al. 2014) and *Trichoderma* (e.g. Anand et al. 2006; Tripathi et al. 2013) species. Many endophytic fungi that colonised the roots of plants growing in extremely heavy metal-contaminated soil are also found to show increased metal tolerance (e.g. Zhang et al. 2008; Li et al. 2012; Deng et al. 2014).

With the presence of mycelia, fungi can diversify expeditiously in soil because of their large filaments surface area (Kulshreshtha et al. 2014). The toleration of elevated concentration of toxic heavy metals is also supported by the ability of fungi to produce extracellular degradative enzymes that reduce the toxicity of the heavy metals when taken inside the cell (Danesh et al. 2013). The secretion of enzymes coupled with the mechanical support through the penetration of fungi hypha into the soil help to facilitate and maximise the transformation of contaminants in contaminated soils (Singh 2006). The unique composition of the fungal cell wall offers the advantage of providing excellent metal-binding properties (Gupta et al. 2000). Fungal cells can also use entrapment in extracellular capsules and precipitation of metals as means to bind the pollutants to their cell wall. In addition, intracellular accumulation and sequestration can be a part of the strategies to immobilise metals/metalloids. The membrane transport systems can play a part in suppressing cellular influx or active efflux of metals and metalloids. When metals and metalloids are present in the cells, they can be sequestered via intracellular precipitation, chelation, localisation in organelles; or be biotransformed to less toxic form (Rajendran et al. 2003; Malik 2004; Gadd 2009, 2010). Current research mainly focuses on the development of microbial materials that have the capability to increase affinity, capacity and selectivity for the target metals. Figure 10.1 provides a diagrammatic overview of the detoxification mechanisms observed in fungi.

10.3 Extracellular Metal Sorption in Fungi

Removal of heavy metals by sorption using fungi has been studied extensively using various species of live fungal cells and inactivated fungal biomass (Diels et al. 2002). Fungi species are well-known for their heavy metal sorption capacity and have been considered as hyper-accumulators of heavy metals compared to other organisms such as bacteria and algae (Purvis and Halls 1996; Zhou 1999). The cell surface of fungi consists of mostly polysaccharides, peptides, melanin, phenolic

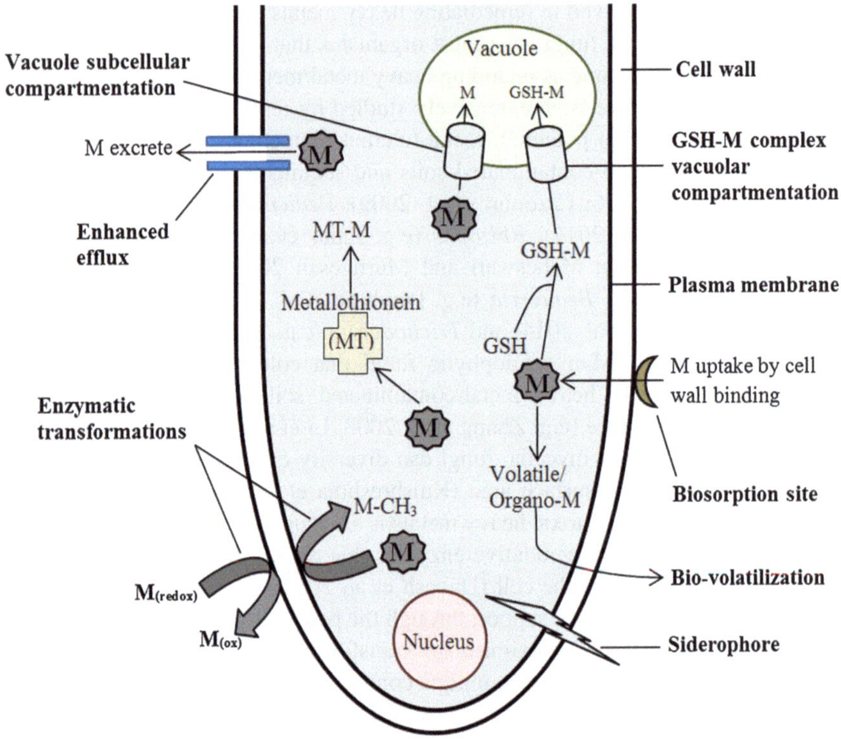

Fig. 10.1 Diagrammatic overview of cellular detoxification mechanisms in metal/metalloid tolerance in observed fungi. *M* metal ion; *GSH* Glutathione

molecules and chitins. They provide excellent metal-binding sites such as the carboxyl, hydroxyls and amides (Wang and Chen 2009, 2014) and are crucial in affording fungi first-line defence against metal/metalloid (Tamás et al. 2005; González-Guerrero et al. 2009; Wysocki and Tamás 2010; Pócsi 2011). Chitins and chitosans are integral parts of fungal cell walls and have been shown to bind metals effectively in a number of fungal species (Zhou 1999; Gadd 2004; Wang and Chen 2014). The acetylglucosamine or glucosamine of chitosan can act as effective binding sites for metals (Guibal et al. 1995). Tereshina et al. (1999) demonstrated that metal sorption activity in fungi is dependent on structural cell wall polysaccharides and chitin:glucan ratio.

Cell surface sorption of metals and metalloids generally occurs as a result of physicochemical interaction between the metal ions and the functional groups such as cell surface proteins. This reaction is rapid, reversible and metabolism independent (Dhankhar and Hooda 2011). A number of fungi have been investigated as potential biosorbents to remove heavy metals from aqueous solution. A thorough review by Wang and Chen (2009) provides detailed information on the biosorption capacity of a wide range of fungi including *Saccharomyces cerevisiae, Penicillium* sp., *Aspergillus* and other filamentous fungi. Biosorption by macrofungus such as *Trametes versicolor* (Bayramoğlu et al. 2003), *Amanita rubescens* (Sari and Tuzen 2009a) and *Inonotus hispidus* (Sari and Tuzen 2009b) have also been reported. Rizzo et al. (1992) demonstrated that the outer melanised parts of the mycelia of *Armillaria* spp. were able to concentrate Al, Zn, Fe and Cu ions to 50–100 times the level found in surrounding soil (ranged up to 3440, 1930, 1890, 15 and 680 $\mu g\ g^{-1}$, respectively). Similar results were found with the melanized pseudosclerotial plates of *Phellinus weirii* that accumulated heavy metals (Al, Ca, Cu, Fe, K, Mg, Mn, Na, Ni, Pb and Zn) from soil (McDougall and Blanchette 1996).

In addition to the cell surface sorption process, metals and metalloids may be bound within fungal cell walls or form precipitates as oxalates, sulfates and phosphates (Fomina et al. 2005; Purchase et al. 2009; Gadd 2010; Joseph et al. 2012; Wei et al. 2013). Whilst cell surface sorption is metabolically independent, metal precipitation depends on metabolic reactions; the fungal cells produce compounds in response to the presence of metal/metalloids that aide the precipitation process (Dhankhar and Hooda 2011). The production of organic acids with strong metal-chelating properties (e.g. oxalic, dicarboxylic, tricarboxylic and citric acids) lead to the dissolution of metals such as Ca, Cd, Co, Cu, Mg, Mn, Ni, Pb, Sr and Zn from the surrounding matrices (White et al. 1997; Fomina et al. 2005; Gadd 2007, 2011). The subsequent formation of insoluble metal–salt complexes via precipitation reduces metal bioavailability, allowing many fungi to survive and grow in highly polluted environments (Baldrian 2003; Jarosz-Wilkołazka and Gadd 2003; Gadd 2011).

10.4 The Role of Transport Mechanisms in Fungal Resistance to Metals and Metalloids

The intracellular uptake of metals can be as efficient as extracellular sorption in many fungi. In *Tichoderma asperellum* and *Fusarium oxysporum* exposed to As (V), the intracellular As accumulation accounted for 82.2 and 63.4 % of the total accumulated As; and in *Pencillium janthinellum*, approximately 50 % (Su et al. 2010). The routes used by heavy metals to enter or exit the cells are normally through the transporters and channels that facilitate the uptake of essential metal micronutrients, sugars and sugar derivatives (Tamás et al. 2005; Wysocki and Tamás 2010). Table 10.1 summarises some of the reported transport mechanisms in yeasts and fungi that are resistant to metals and metalloids.

The elimination of Zrt1p, a Zn(II) transporter; Smf1p and Smf2p, Mn(II) transporters have been found to confer Cd (II) resistance in *Saccharomyces cerevisiae* (Gomes et al. 2002; Gitan et al. 2003; Ruotolo et al. 2008). Similarly, the removal of Sul1p and Sul2p sulfate transporters improved Cr(VI) and Se(VI) tolerance in *S. cerevisiae* (Cherest et al. 1997).

Table 10.1 Reports on the role of transport mechanisms in yeasts and fungi that are resistant to metals and/or metalloids

Organism	Metal/metalloid	Transporter/channel	Comments	Citation
Saccharomyces cerevisiae	Cr(VI), Se (VI)	Sul1p, Sul2p	Sulfate transports	Cherest et al. (1997)
Saccharomyces cerevisiae	As(V)	Pho84p, Pho87p	Phosphate transporters	Bun-ya et al. (1992)
Saccharomyces cerevisiae	Cd(II)	Zrt1p, Smf1p; Smf2p	Zn (II) transporter Mn(II) transporter	Gomes et al. (2002) Gitan et al. (2003), Ruotolo et al. (2008)
Saccharomyces cerevisiae	As(III), Sb(III)	Acr3p	Plasma membrane metalloid-specific antiporter	Wysocki et al. (1997)
Saccharomyces cerevisiae	As(III), Sb(III)	Ycf1p	Vacuolar ABC transporter	Ghosh et al. (1999), Wysocki et al. (2004)
Saccharomyces cerevisiae	As(III)	Fps1p	Aquaglyceroporin channel	Maciaszczyk-Dziubinska et al. (2010)
Saccharomyces cerevisiae	As(III)	Hx1-Hx16	Hexose transporters	Liu et al. (2004)
Glomus mosseae; *Gigaspora margarita*	As(III)	PgPOR29	Porin transport; As resistance on the host plant *Pteris vittata*	Bona et al. (2010)
Glomus intraradices and *Hymenoscyphus ericae*	As(V)	GiPT	Phosphate transporter; As resistance on the host plant *Medicago truncatula* Enter via AMF hyphae and reduced to As (III) by arsenate reductase	González-Chávez et al. (2011) Sharples et al. (2000)
Pisolithus albus	Ni	P-type-ATPase; ABC transporter; MFS	Overexpressed in the presence of high concentration of Ni	Majorel et al. (2014)

Previous research observed that in *S. cerevisiae* exposed to metal/metalloid, As (V) has high affinity to Pho84p and low affinity to Pho87p phosphate transporters (Bun-Ya et al. 1992), whereas the Acr3p plasma membrane transporter was found to be overexpressed in a number of resistant fungal strains in order to remove toxic metals/metalloids such as As(III) (Wysocki et al. 1997). For *S. cerevisiae* cells that lack Ycf1, they were shown to be moderately sensitive to As(III), while the double *acr3*Δ *ycf1*Δ mutant was the most sensitive to metalloids, especially As(III) and Sb (III) (Ghosh et al. 1999; Wysocki et al. 2004). Thus, the plasma membrane metalloid-specific antiporter Acr3 and the vacuolar ABC transporter Ycf1 constitute two distinct but complementary metalloid detoxification pathways in yeast.

Aquaglyceroporin channel, Fps1p was shown to be involved in transporting As (III) and Sb(III) into yeast cells (Maciaszczyk-Dziubinska et al. 2010) as the *fps1*Δ mutant exhibited decreased As(III) excretion out of the cell and also increased As (V) sensitivity. Hexose transporters play an important role in the metalloid permeability in eukaryotes, especially for As(III). It has been found that the accumulation of As(III) was very high in the yeast *fps1*Δ mutant in the absence of glucose (Liu et al. 2004). When glucose levels were restored, no accumulation of As(III) was observed in wild type yeast cells.

In plant root-colonising symbiotic arbuscular mycorrizal fungi (AMF) such as *Glomus mosseae* and *Gigaspora margarita*, the porin transport PgPOR29 was found to be upregulated when exposed to As and confer As resistance to their host plant *Pteris vittata* by increased As uptake or by As(III) efflux into the vacuole in the frond of this fern (Bona et al. 2010). In *Glomus intraradices*, the overexpression of the high-affinity phosphate transporter GiPT is thought to be induced by the presence of As(V), this As exclusion mechanism is thought to contribute to the As tolerance of the host plant *Medicago truncatula* (González-Chávez et al. 2011). In another study on As(V) resistance of *Holcus lanatus*, González-Chávez et al. (2002) suggested that As resistance was conferred by AMF *Glomus mosseae* and *Glomus caledonium*, via the suppression of high-affinity arsenate/phosphate transporters.

Arsenate resistance is exhibited by the ericoid mycorrhizal fungus *Hymenoscyphus ericae* collected from As-contaminated mine soils. The fungus was found to improve As resistance to its host plant *Calluna vulgaris* through an As exclusion mechanism where As(V) was reduced to As(III) and As(III) was transported out of the fungal cells (Sharples et al. 2000). A similar mechanism was proposed by González-Chávez et al. (2011) for arbuscular mycorrhiza *Glomus intraradices*, where As(V) enters the AMF hypha via the high-affinity phosphate transporter (*GiPT*) and once inside the hypha, As(V) is reduced to As(III) by a putative arsenate reductase. As(III) is then pumped out of the cell and back to the soil via an arsenite efflux pump (*GiArsA*).

In the presence of high concentration of Ni, a Ni-tolerant ectomycorrhizal *Pisolithus albus* demonstrated Ni efflux in its fungal tissue; three genes were identified in the efflux mechanisms: a P-type ATPase, an ABC transporter and a major facilitator superfamily permease (MFS), which were overexpressed when Ni was present (Majorel et al. 2014).

Transport of metals from fungal cells can also be promoted via siderophores (Ahmed and Holmström 2014). Siderophores are metal-chelating agents with low molecular masses (200–2000 Da) produced by microorganisms and plants, especially under Fe-limiting conditions (Schwyn and Neilands 1987). Most fungi produce a variety of different extracellular and intracellular siderophores that cover a wide range of physicochemical properties in order to improve iron solubility and uptake (Winkelmann 2007). Depending on the functionalities of their ligands, siderophores could have a strong affinity or selectivity for a particular metal other than Fe (Hernlem et al. 1999). In general, the stability constants for the formation of ferric siderophore complexes range from $ß = 10^{25}$ to $ß = 10^{50}$ (Crumbliss and Harrington 2009). For example, siderophores have been found to be extremely effective in solubilizing and increasing the mobility of heavy metals such as Cd, Cu, Ni, Pb, Zn and the actinides Th(IV), U(IV) and Pu(IV) (Schalk et al. 2011). Regulation of siderophore biosynthesis is essential in fungi and allows an economic use of siderophores and metabolic resources. Extracellular and intracellular reduction of siderophores may occur depending on the fungal strain (Winkelmann 2007).

10.5 Intracellular Detoxification Mechanisms

Fungi are capable of affecting heavy metal concentrations in the environment by transforming metal/metalloid from a toxic to a less toxic species via various enzymatic actions. Organomercury compounds can be detoxified by conversion to Hg(II) by fungal organomercury lyase, the resulting Hg(II) can be further reduced to the more volatile elemental Hg(0) by mercuric reductase (Gadd 1993).

After the uptake of As(V) into the fungal cells via the phosphate transporters, a number of common detoxification pathways within the fungal cell could be involved: these include the reduction to As(III) by arsenate reductases, followed by exclusion or sequestration of As(III) in AMF (Sharples et al. 2000; González-Chávez et al. 2011).

In *S. cerevisiae* three contiguous genes, ACR1 (encoding a transcriptional regulator), ACR2 (an arsenate reductase) and ACR3 (a plasma membrane arsenite efflux pump) are found to be responsible for its As resistance (Rosen 2002).

Another mechanism for metal/metalloid resistance in fungi involves the methylation of inorganic arsenic to produce volatile derivatives (Tamaki and Frankenberger 1992; Čerňanský et al. 2009). Methylation of metals (e.g. Hg, Sn and Pb) and metalloids (e.g. As, Se and Te) are mediated by an enzymatic reaction where the metal is transferred to the methyl group; the methylated metals compounds often has different solubility, volatility and toxicity to that of the parent compounds (Barkay and Wagner-Döbler 2005). *Candida humicola* (Cullen and Reimer 1989), *Glicladium roseum* and *Penicillium* sp. (Cox and Alenxander 1973) are capable of converting non-volatile monomethylarsonic acid and dimethylarsenic acid to volatile trimethylarsine oxide. Effective biovolatalization (up to 23 % of total As) was observed in the heat-resistant *Neosartorya fischeri* (Čerňanský et al.

2007). Biomethylation of As and Sb have also been observed in *Scopulariopsis brevicaulis* (Andrewes et al. 2000) and *Cryptococcus humicolus* (Hartmann et al. 2003). Both studies stipulated that As and Sb methylation occur in very similar if not identical mechanisms in these fungi.

Heavy metals such as iron, copper, cadmium, chromium, lead and mercury exhibited the ability to produce reactive oxygen species (ROS), resulting in oxidative stress, alteration of calcium homeostasis and DNA damage (Klaunig et al. 1998). The production of ROS could induce toxicity in fungi and cause damages to a number of essential biomolecules such as proteins, nucleic acids and lipids (Fridovich 1998; Bai et al. 2003). The ability of fungi to eliminate ROS has been associated with metal/metalloid tolerance in fungi (Fujs et al. 2005). Small proteins (around 2–7 kDa) such as metallothioneins can bind metal ions for storage and/or detoxification in both eukaryotes and prokaryotes because of their high thiolate sulfur contents. Cystein-rich peptides such as phytochelatins and other thiol compounds, including non-protein sulfhydryl groups (NP-SH), protein-bound sulfhydryl groups (PB-SH), glutathione (GSH) and glutathione disulfide or oxidised glutathione (GSSG) have been reported to bind metal ions (Reese et al. 1988; Cobbett and Goldsbrough 2002; Romero-Isart and Vašák 2002) and scavenge ROS (Quesada et al. 1996; Tsuji et al. 2002; Bai et al. 2003).

Antioxidant enzymes also play a considerable role in responses to metal/metalloid exposure or their detoxification in fungi (Cobbett 2000; Ma et al. 2001; Zhang and Cai 2003; Raab et al. 2004). A number of antioxidant enzymes have been isolated from fungi that can remove ROS and their derivatives, or repair the damages caused by these compounds (Bai et al. 2003; Jamieson 1998). Superoxide dismutase (SOD), catalase (CAT), glutathione peroxidase (POD), glutathione reductase (GR) and glutathione S-transferases (GSTs) have all been identified to be capable of protecting cells from metal/metalloid induced stress (Jamieson 1998; Todorova et al. 2010; Adamo et al. 2012; Lazarova et al. 2014; Jiang et al. 2015; Shen et al. 2015). Jiang et al. (2015) demonstrated the biosynthesis of NP-SH, PB-SH, GSH and GSSG and the activation of antioxidant enzymes (including SOD, POD, CAT and GR) in *Oudemansiella radicata* played a considerable role in responses to Cu exposure or Cu detoxification.

10.6 Arsenic and Antimony Removal by Extremophilic Fungi

Although the principal natural source of As is volcanic activity of which the majority of which ends up in the soil and the ocean (Matschullat 2000), arsenic also enters the soil via a variety of anthropogenic routes such as the mining industry, commercial waste and the steel industry. The greatest source of Sb emissions into the environment is also from mining and smelting operation (Adriano 1986). Toxicological effects of As and Sb are mediated by their trivalent species (Gebel

1997). As and Sb can permeate cells using molecular mimicry via transporters that evolved for accumulation of fundamental ions and nutrients. Both trivalent arsenite and antimonite exist in the tri-hydroxylated uncharged forms [$As(OH)_3$ and $Sb(OH)_3$] which structurally resemble glycerol at the physiological pH in aqueous solution (Ramirez-Solis et al. 2004; Porquet and Filella 2007). The aquaglyceroporins (membrane proteins) that are permeable to water and glycerol also permits accumulation of As(III) and Sb(III) (Bhattacharjee et al. 2008). Arsenate is taken up via phosphate transporters due to its chemical similarity to inorganic phosphate (Wysocki and Tamás 2010). In contrast, antimonate exists as $Sb(OH)_6$ and does not compete with phosphate uptake (Tschan et al. 2008). Whilst Sb(III) appears to share similar transport system as As(III), the pathway of Sb(V) entry into the cells has yet to be fully elucidated (Maciaszczyk-Dziubinska et al. 2012).

According to Wang et al. (2011), the sensitivity of soil microbial populations under Sb and As combined and simplex pollution were: bacteria > fungi > actinomycetes. Sb(III) and As(III) had a stronger effect on all the biochemical parameters tested than their pentavalent counterparts. The authors also observed that combined pollution of Sb(III) and As(III) has an antagonistic inhibiting effect on urease and acid phosphatase. However, the combination of Sb(V) and As(V) has a synergistic inhibiting effect on urease and protease.

Conditions such as extremely acidic soil and water ($pH < 3$) and high concentrations of toxic heavy metals and metalloids are some examples of extremity that extremophilic microorganisms can tolerate to live and survive. Extremophilic fungi have been isolated from highly polluted or extreme environments such as mining wastewaters (pH 1.1) (Aguilera et al. 2007; Aguilera 2013; Oggerin et al. 2013), a uranium mine (Vázquez-Campos et al. 2015) and a volcanic geothermal system (Chiacchiarini et al. 2010).

Hitherto, the data of extremophilic fungi diversity in extremely acidic conditions is rather limited; however, melanised and meristematic fungi have been found to be the prevailing groups in such habitats (Baker et al. 2004; López-Archilla et al. 2004; Selbmann et al. 2008; Hujslová et al. 2013). For example, *Acidomyces acidophilus* strain has been isolated from a natural gas purification plant adjacent to a sulfur pilefield (Sigler and Carmichael 1974); *Hortaea acidophila* was isolated from brown coal that contains humic and fulvic acid at pH around 0.6 (Hölker et al. 2004); two acidophilic strains *Hortaea werneckii* and *Acidomyces acidophilum* were isolated from Rio Tinto in Spain, where the mean soil pH value was 2.3 and the ground contained high concentrations of Fe, Cu, Zn, As, Mn and Cr (Amaral et al. 2002). Another black fungus, *Exophiala sideris,* was isolated from an arsenic mine polluted with akylbenzenes in Złoty Stok, Poland, for which the authors claimed potential use for the purpose of bioremediation of metalloids (Seyedmousavi et al. 2011).

Acidomyces acidophilus is of particular interests. Ivarsson and Morita (1982) showed that acidity is a crucial factor in the ecology of this fungus where good growth was observed when adjusting the pH of the medium to 0.5 with HCl. It was first isolated by Starkey and Waksman (1943) in extremely acidic, sulfate-containing industrial water. According to Gould et al. (1974) the species was the only organism isolated from a sulfur-containing soil at pH 1.1, where it occurred at high viable counts. The melanin-containing cell wall protects *A. acidophilus* and enables it to adapt to adverse environmental conditions (Jacobson et al. 1995; Tetsch et al. 2006). The hydrophobicity and negative charges of melanin offer the fungus protection from oxidative stress (Jung et al. 2006). Another interesting property of melanin is that it can shield organisms from ionising radiation. Since melanin has a stable free radical population, it is thought that the radio-protective properties are due to scavenging of free radicals generated by radiation (Eisenman and Casadevall 2012); in addition, the controlled dissipation of high-energy recoil electrons by melanin prevents secondary ionizations and the generation of damaging free radical species (Schweitzer et al. 2009). *Acidomyces acidophilus* has been successfully isolated from uranium mine drain water (Harrison et al. 1966), a Canadian uranium mine in acidic water (pH 2) (Ivarson and Morita 1982) and radioactive contaminated soils at the Nevada Test Site and areas around the damaged Chernobyl nuclear reactors (Dighton et al. 2008). The enzymes secreted by this fungus can function at low pH and high temperature and could have potential applications for variety of industries (Polizeli et al. 2005; Hess 2008).

10.6.1 A Case Study

An arsenic-resistant fungal strain was isolated in our laboratory from a waste roaster pile in a disused mine in Cornwall, UK (Site 3, Fig. 10.2) where arsenic is a by-product of tin-mining activities, primarily in the arsenopyrite (FeAsS) form. The concentration of As in soil adjacent to the mining sites could reach as high as 20,000 mg kg^{-1} (Bai et al. 2008). Roaster piles waste, in particular, contained high levels of arsenic, iron, antimony and zinc (dos Santos et al. 2013). The isolated fungal strain, designated WKC-1, was identified as *Acidomyces acidophilus* sequenced using 16s rRNA sequencing and matrix-assisted laser desorption/ionisation time of flight mass spectrometry (MALDI-TOF MS). The morphological features of strain WKC-1 is presented in Fig. 10.3.

This isolate showed significant resistance to As(V) and Sb(V) where the minimum inhibitory concentration (MIC) of these two metalloids were found to be 22,500 mg L^{-1} and 100 mg L^{-1}, respectively, on CDA medium; it was significantly more resistant to As(V) than the *A. acidophilus* type strains CBS 335.97 and CCF 4251 (Table 10.2). Since the presence of phosphate significantly affected the MIC of As(V), removal experiments were carried out in CDA with minimum phosphate concentration (0.02 mg L^{-1}) and it was found that strain WKC-1 was able to remove 70.30 % of As(V) (100 mg L^{-1}) present (Fig. 10.4). Biosorption experiments using 100 mg L^{-1} As(V) showed that the maximum capacity of As uptake was 170.82 mg g^{-1} dry biomass as predicted by a Langmuir model (R^2 = 0.989), which is greater than the uptake by previously studied fungal strains such as *Inonotus hispidus* (59.6 mg g^{-1}) (Sari and Tuzen 2009b) and *Picea abies* (1.369 mg g^{-1})

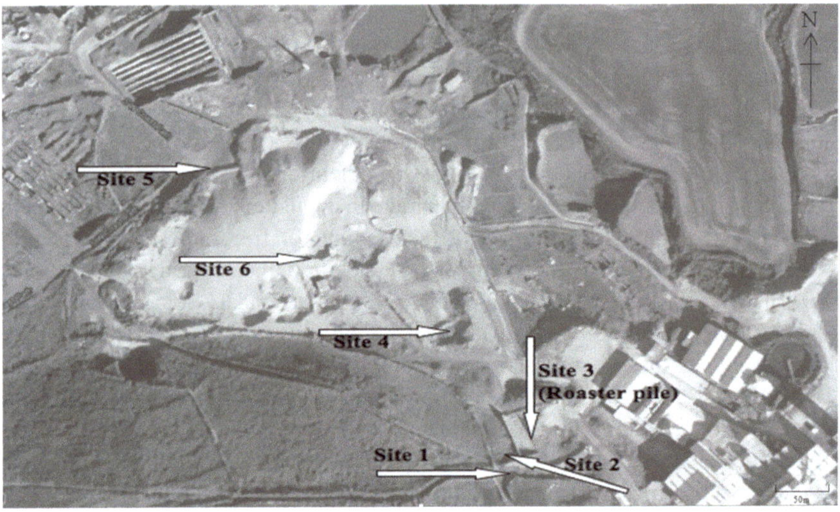

Fig. 10.2 An aerial photograph of the Geevor Tin Mine in Pendene, Penzance, Cornwall, UK and locations of soil sampling sites (Geevor Tin Mine as viewed on 17 July 2013, https://www.google.co.uk/maps/place/Geevor+Tin+Mine/@50.1519033,-5.6744307,1404m). *Arrows* indicate the sampling sites of this study

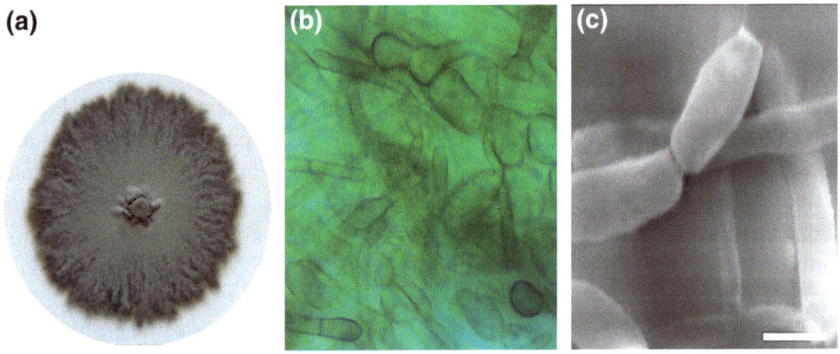

Fig. 10.3 Morphological features of *Acidomyces acidophilus* WKC-1. **a** colony of the strain WKC-1grown on CDA medium; **b** hyphae of the strain observed by light microscope at a magnification of 1000x; **c** scanning electron microscope (SEM) at a magnification 2200x. Scale bar = 2 μm

(Urik et al. 2009). The key functional groups on the cell surface were –OH, –NH, –CH, $-SO_3$ and PO_4 as identified by Fourier Transform Infrared Spectroscopy (FT-IR). As (V) and Sb(V) appeared to compete for the same binding sites in these functional groups as the presence of Sb(V) reduced the As(V) uptake by nearly 40 %. The Pho84p and Pho87p transporters that are involved in *S. cerevisiae* cells exposed to As(V) were also detected in *A. acidophilus* WKC-1 as determined by

Table 10.2 The average diameter measurement ± standard error of *Acidomyces acidophilus* WKC-1 and two *A. acidophilus* type strains grown on CDA medium containing different concentration of As(V). The experiment was carried out in triplicates

As(V) concentration (mg L^{-1})	*A. acidophilus* Strain WCK-1 (cm)	Reference strains	
		A. acidophilus CBS 335.07 (cm)	*A. acidophilus* CBS 4251 (cm)
Control	4.7 ± 0.2	3.9 ± 0.4	4.2 ± 0.2
1000	4.5 ± 0.1	3.8 ± 0.1	3.9 ± 0.4
7500	4.3 ± 0.2	2.7 ± 0.2	2.4 ± 0.1
10,000	4.1 ± 0.2	1.4 ± 0.1	NG
12,500	3.9 ± 0.1	NG	NG
15,000	3.7 ± 0.3	NG	NG
17,500	3.4 ± 0.1	NG	NG
20,000	2.7 ± 0.2	NG	NG
22,500	2.2 ± 0.2	NG	NG
25,000	NG	NG	NG

NG indicates no growth

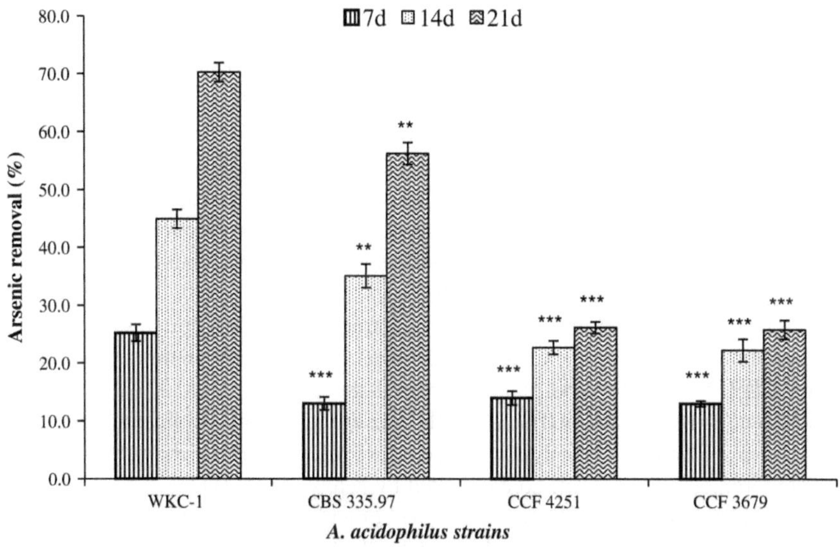

Fig. 10.4 Percentage of arsenate removal by *Acidomyces acidophilus* WKC-1 and *A. acidophilus* type strains after 7, 14 and 21 days on modified CDA medium containing initial arsenate concentration of 100 mg L^{-1}. *Vertical bars* indicate the standard error of the mean of three replicates. *Asterisks* indicate statistical significance of differences tested by ANOVA where $^{**}p < 0.01$, $^{***}p < 0.001$ compared to *A. acidophilus* WKC-1

matrix-assisted laser desorption/ionisation (MALDI) and liquid chromatography–mass spectrometry (LC-MS). They are found to be overexpressed in the presence of As(V), suggesting the high biosorption capability of this fungus is mediated via these permeases. Increased activities of a number of antioxidative enzymes including SOD, CAT and GR were also detected in strain WKC-1 and an *A. acidophilus* type strain (CBS335.97) at the exposed As concentration (100 mg L^{-1}) (results not shown). Interestingly, a decrease of GST activity was observed in the exposure experiment, suggesting GST does not play a major role in the As detoxification mechanism in *A. acidophilus*.

A. acidophilus WKC-1 appears to tolerate high As(V) concentration through increased biosorption of As(V), together with its ability to tolerate high levels of Sb (a co-contaminant in tin mine) and low pH made it a potential candidate for bioremediation of As-contaminated mining sites.

10.7 Future Prospects of Fungal Bioremediation of Metal/Metalloid-Contaminated Land

10.7.1 The Biotechnological Prospective[2]

Genetic engineering offers a number of ways to improve the tolerance of metal/metalloid-resistant fungi to these contaminants. Using *S. cerevisiae* as a model, Pócsi (2011) suggested the most promising targets for genetic manipulations are: (i) increase production of extracellular and intracellular metal chelators, (ii) elimination of metal transporters to reduce metal influx or induce their expression to encourage metal efflux from cells, (iii) overproduction of elements of antioxidative defence system, (iv) modification of the regulatory network and (v) interfering with apoptotic cell death.

Engineered fungal strains have been constructed by protoplast fusion using an endophytic fungi *Mucor* sp. CBRF59 (Deng et al. 2013). Inoculation of the mutant strain CBRF59T3 into rape plants increased the concentration of Cd(II) in rape shoots by 35–189 % in Cd(II) and Pb(II) contaminated soils, indicating that it is a feasible and efficient method to improve stress tolerance of uncharacterized fungi for phytoremediation of soils contaminated by heavy metals (Deng et al. 2013). Qiu et al. (2015) were able to increase the GSH contents by threefold and the tolerance to H_2O_2, temperature, furfural, hydroxymethylfurfural and Cd(II) in the engineered *S. cerevisiae* BY-G strain compared to the reference strain. Although promising, the use of genetically engineered (GM) microorganisms to improve bioremediation need to take into consideration public acceptance of such technology; as such, the

[2]Additional information on isolation and cloning of genes from resistant fungi is presented in Chap. 2—*The genetic basis of abiotic stress resistance in extermophilic fungi: the genes cloning and application*.

potential to apply GM fungi to treat heavy metal/metalloids is still limited. Nevertheless, the availability of sequence information on the genome of metal/metalloid fungi allowed for postgenomic approaches to obtain comprehensive information on their gene functions and regulatory mechanisms, the results obtained with these 'omic' approaches can result in a better understanding of the complexity of cell responses to metals/metalloid toxicity in fungi.

10.7.2 The Nano-Biotechnological Prospective[3]

Metal reduction resulting in immobilisation has applications in bioremediation as well as production of novel biomaterials and catalysts (Gadd 2011). Biosorption-based biosynthesis of nanoparticles is proposed as a means of removing heavy metals from wastewaters and soils and it also aids the development of heavy metal nanoparticles that may have an application within the technology industry (Karman et al. 2015). The application of the highly structured physical and biosynthetic activities of microbial cells to the synthesis of nanosized materials has emerged as a novel approach to the synthesis of metal nanoparticles (NPs) (Gericke and Pinches 2006). Fungi have several advantages over other microorganisms for NPs synthesis as they are easy to handle, require simple nutrients and possess a high cell wall-binding capacity and high intracellular metal uptake capabilities (Dias et al. 2002; Sanghi and Verma 2009).

The synergy of biosynthesis of nanomaterials and the potential of fungi to remove toxic metals from polluted sites have been explored by a number of researchers. Velmurugan et al. (2010) demonstrated that *Fusarium* sp. obtained from a Zn-contaminated mine in South Korea were able to bioabsorb up to 320 mg L^{-1} of Zn and possessed the ability of producing ZnO NPs. Dead biomass of *Hypocrea lixii* has also been successfully used to convert Cu and Ni ions into CuO and NiO NPs in aqueous solution (Salvadori et al. 2014a, 2015). Similarly, *Rhodotorula mucilaginosa* (Salvadori et al. 2014b) and *Trichoderma koningiopsis* (Salvadori et al. 2014c) have been successfully used to biosynthesise Cu NPs.

Nanoparticles have also been found to complement the bioremediation activity of fungi. Zuo et al. (2015) demonstrated a tenfold increase in the removal capacity of Cd(II) in *Phanerochaete chrysosporium* when complemented with the optimal concentration of 1 mg L^{-1} Ag-NPs. Immobilised *P. chrysosporium* loaded with TiO_2 nanoparticles (PTNs) have been shown to be novel high-value bioremediation materials for adsorbing Cd and for degrading 2,4-dichlorophenol where the antioxidative defence system and the physiological fluxes effectively protected the PTN to oxidative stress (Tan et al. 2015).

[3]Further information fungal biosynthesis of nanoparticles can be found in Chap. 13—*Fungal biosynthesis of nanoparticles, a cleaner alternative.*

Interestingly, fungi are also found to be effective in removing NPs from an aqueous medium. Jakubiak et al. (2014) demonstrated that *Pleurotus eryngii* and *Trametes versicolor* can efficiently remove Al_2O_3 NPs, up to 86 and 61 % of total amount of NPs, respectively. *P. eryngii* also capable in removing 58 % of the Pt NPs but is less effective towards Co NPs.

10.8 Conclusion

A number of key sustainability indicators for evaluating remediated systems have been identified by Tripathi et al. (2015) including: (i) reduction in pollutant level/residual concentration after the remediation process, (ii) key variables depicting the improvement of the physicochemical properties of soil, (iii) the enrichment of microbial biomass and their functional diversity in soil, (iv) the positive changes in biodiversity component including the sensitive and key indicator species after the remediation process, (v) the carbon emission/accounting during each and every step of the remediation process, (vi) potential for bio-based economy and entrepreneurial activities and (vii) social acceptability of the remediation process. Fungi bioremediation provides a plausible and sustainable solution for heavy metal/metalloid-contaminated land as it fulfils a number of the key criteria listed above in rectifying and re-establishing the natural condition of soil.

Despite the challenges facing the use of molecular approaches to enhance the remediation prospects, fungal bioremediation remains an attractive and promising alternative to the conventional physicochemical methods such as chemical precipitation, oxidation or reduction, filtration, ion-exchange, reverse osmosis, membrane technology, evaporation and electrochemical treatment. With the advent of nanotechnology and improved understanding of the nanoparticle's life cycle, the application of metal-resistant fungi offers a promising solution for metal/metalloid pollution.

References

Adamo GM, Brocca S, Passolunghi S, Salvato B, Lotti M (2012) Laboratory evolution of copper tolerant yeast strains. Microb Cell Fact. doi:10.1186/1475-2859-11-1

Adriano DC (1986) Trace elements in the terrestrial environment. Springer, Heidelberg

Aguilera A (2013) Eukaryotic organisms in extreme acidic environments, the Río Tinto case. Life (Basel) 3(3):363–74. doi:10.3390/life3030363

Aguilera A, Zettler E, Gómez F, Amaral-Zettler L, Rodríguez N, Amils R (2007) Distribution and seasonal variability in the benthic eukaryotic community of Río Tinto (SW, Spain), an acidic, high metal extreme environment. Syst Appl Microbiol 30(7):531–546 Epub 2007 Jul 17

Ahmed E, Holmström SJM (2014) Siderophores in environmental research: roles and applications. Microb Biotechnol 7:196–208. doi:10.1111/1751-7915.12117

Alloway BJ (2013) Heavy metals in soils: trace metals and metalloids in soils and their bioavailability. Springer, The Netherlands. doi:10.1007/978-94-007-4470-7

Amaral LA, Gómez F, Zettler E, Keenan BG, Amils R, Sogin ML (2002) Eukaryotic diversity in Spain's river of fire. Nature 417:137. doi:10.1038/417137a

Anand P, Isar J, Saran S, Saxena RK (2006) Bioaccumulation of copper by *Trichoderma viride*. Bioresour Technol 97(8):1018–1025. doi:10.1016/j.biortech.2005.04.046

Andrewes P, Cullen WR, Polishchuk E (2000) Antimony biomethylation by *Scopulariopsis brevicaulis*: characterization of intermediates and the methyl donor. Chemosphere 41 (11):1717–1725. doi:10.1016/S0045-6535(00)00063-1

Bai J, Lin X, Yin R, Zhang H, Junhua W, Xueming C, Yongming L (2008) The influence of arbuscularmycorrhizal fungi on As and P uptake by maize (*Zea mays* L.) from As-contaminated soils. Appl Soil Ecol 38(2):137–145

Bai Z, Harvey LM, McNeil B (2003) Oxidative stress in submerged cultures of fungi. Crit Rev Biotechnol 23:267–302

Baker BJ, Lutz MA, Dawson SC, Bond PL, Banfield JF (2004) Metabolically active eukaryotic communities in extremely acidic mine drainage. Appl Environ Microbiol 70(10):6264–6271

Barkay T, Wagner-Döbler I (2005) Microbial transformations of mercury: potentials, challenges, and achievements in controlling mercury toxicity in the environment. Adv Appl Microbiol 57:1–52. doi:10.1016/S0065-2164(05)57001-1

Baldrian P (2003) Interactions of heavy metals with white-rot fungi. Enzyme Microb Technol 32:78–91. doi:10.1016/S0141-0229(02)00245-4

Bayramoğlu G, Bektaş S, Arica MY (2003) Biosorption of heavy metal ions on immobilized white-rot fungus *Trametes* versicolor. J Hazard Mater 101(3):285–300

Bhargavi SD, Savitha J (2014) Arsenate resistant *Penicillium coffeae*: a potential fungus for soil bioremediation. B Environ Contam Toxicol 92(3):369–373. doi:10.1007/s00128-014-1212-y

Bhattacharjee H, Mukhopadhyay R, Thiyagarajan S, Rosen BP (2008) Aquaglyceroporins: ancient channels for metalloids. J Biol 7. doi:10.1186/jbiol91

Bona E, Cattaneo C, Cesaro P, Marsano F, Lingua G, Cavaletto M et al (2010) Proteomic analysis of *Pteris vittata* fronds: two arbuscular mycorrhizal fungi differentially modulate protein expression under arsenic contamination. Proteomics 10:3811–3834

Britannica Academic. Metalloid (2015) http://www.britannica.com/science/metalloid. Accessed 13 Oct 2015

Bun-Ya M, Harashima S, Oshima Y (1992) Putative GTP-binding protein, Gtr1, associated with the function of the Pho84 inorganic phosphate transporter in *Saccharomyces cerevisiae*. Mol Cell Biol 12(7):2958–2966

Čerňanský S, Kolenčik M, Ševc J, Urík M, Hiller E (2009) Fungal volatilization of trivalent and pentavalent arsenic under laboratory conditions. Bioresour Technol 100:1037–1040

Čerňanský S, Urik M, Sevc J, Khun M (2007) Biosorption and biovolatilization of arsenic by heat-resistant fungi. Env Sci Poll Res 14:31–35. doi:10.1065/espr2006.11.361

Cherest H, Davidian JC, Thomas D, Benes V, Ansorge W, Surdin-Kerjan Y (1997) Molecular characterization of two high affinity sulfate transporters in *Saccharomyces cerevisiae*. Genetics 3:627–635

Chiacchiarini P, Lavalle L, Giaveno A, Donati E (2010) First assessment of acidophilic microorganisms from geothermal Copahue-Cavahue system. Hydrometallurgy 104(3–4): 334–341. doi:10.1016/j.hydromet.2010.02.020

Chouchane S, Snow ET (2001) In vitro effect of arsenical compounds on glutathione-related enzymes. Chem Res Toxicol 14:517–522. doi:10.1021/tx000123x

Cobbett CS (2000) Phytochelatins and their roles in heavy metal detoxification. Plant Physiol 123 (3):825–832

Cobbett C, Goldsbrough P (2002) Phytochelatins and metallothioneins: roles in heavy metal detoxification and homeostasis. Ann Rev Plant Biol 53:159–182

Crumbliss AL, Harrington JM (2009) Iron sequestration by small molecules: thermodynamic and kinetic studies of natural siderophores and synthetic model compounds. In: van Eldik R, Hubbard CD (eds) Advanced Inorganic Chemistry, vol 61. Academic Press, London, pp 179–249

Cox DP, Alexander M (1973) Production of trimethylarsine gas from various arsenic compounds by three sewage fungi. Bull Environ Contam Toxicol 9(2):84–88

Cullen WR, Reimer KJ (1989) Arsenic speciation in the environment. Chem Rev 89(4):713–764. doi:10.1021/cr00094a002
Danesh YR, Tajbakhsh M, Goltapeh EM, Varma A (2013) Mycoremediation of heavy metals. In: Goltapeh EM, Danesh YR, Varma A (eds) Fungi as Bioremediators. Soil Biology, vol 32. Springer, Berlin, Germany, pp 245–67
Deng Z, Zhang R, Shi Y, Hu L, Tan H, Cao L (2014) Characterization of Cd-, Pb-, Zn-resistant endophytic *Lasiodiplodia* sp. MXSF31 from metal accumulating *Portulaca oleracea* and its potential in promoting the growth of rape in metal-contaminated soils. Environ Sci Pollut Res Int 21(3):2346–2357. doi:10.1007/s11356-013-2163-2
Deng Z, Zhang R, Shi Y, Hu L, Tan H, Cao L (2013) Enhancement of phytoremediation of Cd- and Pb-contaminated soils by self-fusion of protoplasts from endophytic fungus *Mucor* sp. CBRF59. Chemosphere 91(1):41–47. doi:10.1016/j.chemosphere.2012.11.065
Dhankhar R, Hooda A (2011) Fungal biosorption—an alternative to meet the challenges of heavy metal pollution in aqueous solutions. Environ Technol 32(5):467–491. doi:10.1080/09593330.2011.572922
Dias MA, Lacerda ICA, Pimentel PF, De Castro HF, Rosa CA (2002) Removal of heavy metals by an *Aspergillus terreus* strain immobilized in a polyurethane matrix. Lett Appl Microbiol 34:46–50
Diels L, Van der Lelie N, Bastiaens L (2002) New developments in treatment of heavy metal contaminated soils. Rev Environ Sci Biotechnol 1(1):75–82
Dietz KJ, Baier M, Krämer U (1999) Free radicals and reactive oxygen species as mediators of heavy metal toxicity in plants. In: Prasad MNV, Hagaemeyer J (eds) Heavy metal stress in plants. Springer, Berlin Heidelberg, pp 73–97
Dighton J, Tugay T, Zhdanova N (2008) Fungi and ionizing radiation from radionuclides. FEMS Microbiol Lett 281:109–120
Dixit R, Malaviya D, Pandiyan K, Singh UB, Sahu A, Shukla R et al (2015) Bioremediation of heavy metals from soil and aquatic environment: an overview of principles and criteria of fundamental processes. Sustainability 7(2):2189–2212
dos Santos JV, de Melo Rangel W, Guimaraes AA, Jaramillo PMD, Rufini M, Marra LM, de Souza Moreira FM (2013) Soil biological attributes in arsenic-contaminated gold mining sites after revegetation. Ecotoxicology 22(10):1526–1537
Eisenman HC, Casadevall A (2012) Synthesis and assembly of fungal melanin. Appl Microbiol Biotechnol 93(3):931–940. doi:10.1007/s00253-011-3777-2
Ezzouhri L, Castro E, Moya M, Espinola F, Lairini K (2009) Heavy metal tolerance of filamentous fungi isolated from polluted sites in Tangier. Morocco. Afr J Microbiol Res. 3:25–48
Factor-Litvak P, Wasserman G, Kline JK, Graziano P (1999) The Yugoslavia prospective study of environmental lead exposure. Environ Health Perspect 107:9–15
Fomina M, Hillier S, Charnock JM, Melville K, Alexander IJ, Gadd GM (2005) Role of oxalic acid overexcretion in transformation of toxic metal minerals by *Beauveria caledonica*. Appl Env Microbiol. 71:371–381. doi:10.1128/AEM.71.1.371-381.2005.1
Fridovich I (1998) Oxygen toxicity: a radical explanation. J Exp Biol 201:1203–1209
Fujs S, Gazdag Z, Poljšak B, Stibilj V, Milačič R, Pesti M (2005) The oxidative stress response of the yeast *Candida intermedia* to copper, zinc, and selenium exposure. J Basic Microbiol 45:125–135
Gadd GM (1993) Interactions of fungi with toxic metals. New Phytologist 124(1):25–60
Gadd GM (ed) (2001) Fungi in bioremediation (No. 23). Cambridge University Press
Gadd GM (2004) Mycotransformation of organic and inorganic substrates. Mycologist 18:60–70
Gadd GM (2007) Geomycology: biogeochemical transformations of rocks, minerals, metals and radionuclides by fungi, bioweathering and bioremediation. Mycol Res 111:3–49
Gadd GM (2009) Biosorption: critical review of scientific rationale, environmental importance and significance for pollution treatment. J Chem Technol Biotechnol 84(1):13–28
Gadd GM (2010) Metals, minerals and microbes: geomicrobiolgy and bioremediation. Microbiology 156:609–643. doi:10.1099/mic0.037143.0
Gadd GM (2011) Geomycology. In: Reitner J, Thiel V (eds) Encyclopedia of geobiology, part 7. Springer, Heidelberg, pp 416–432

Gebel T (1997) Arsenic and antimony: comparative approach on mechanistic toxicology. Chem Biol Interact 107(3):131–144
Gericke M, Pinches A (2006) Biological synthesis of metal nanoparticles. Hydrometallurgy 83:132–140
Ghosh M, Shen J, Rosen BP (1999) Pathways of As(III) detoxification in *Saccharomyces cerevisiae*. Proc Natl Acad Sci USA 96(9):5001–5006
Gimmler H, de Jesus J, Greiser A (2001) Heavy metal resistance of the extreme acidotolerant filamentous fungus *Bispora* sp. Microbial Ecol 42:87–98
Gitan RS, Shababi M, Kramer M, Eide DJ (2003) A cytosolic domain of the yeast Zrt1 zinc transporter is required for its post-translational inactivation in response to zinc and cadmium. J Biol Chem 278:3955–3964
Gomes DS, Fragoso LC, Riger CJ, Panek AD, Eleutherio EC (2002) Regulation of cadmium uptake by *Saccharomyces cerevisiae*. Biochim Biophys Acta 1573:21–25
González-Chávez MC, Ortega-Larrocea Mdel P, Carrillo-González R, López-Meyer M, Xoconostle-Cázares B, Gomez SK et al (2011) Arsenate induces the expression of fungal genes involved in as transport in arbuscular mycorrhiza. Fungal Biol 115(12):1197–1209. doi:10.1016/j.funbio.2011.08.005
González-Guerrero M, Benabdellah K, Ferrol N, Azcón-Aguilar C (2009) Mechanisms underlying heavy metal tolerance in arbuscular mycorrhizas. In: Azcón-Aguilar C, Barea JM, Gianinazzi S, Gianinazzi-Pearson V (eds) Mycorrhizas—functional processes and ecological impacts. Springer, Berlin , pp 107–22
González-Chávez C, Harris PJ, Dodd J, Meharg AA (2002) Arbuscular mycorrhizal fungi confer enhanced arsenate resistance on *Holcus lanatus*. New Phytol 155:163–171
Gould WD, Fujikawa JI, Cook FD (1974) A soil fungus tolerant to extreme acidity and high salt concentrations. Can J Microbiol 20:1023–1027
Guibal E, Roulph C, Le Cloirec P (1995) Infrared spectroscopic study of uranyl biosorption by fungal biomass and materials of biological origin. Environ Sci Technol 29(10):2496–2503. doi:10.1021/es00010a007
Gupta R, Ahuja P, Khan S, Saxena RK, Mohapatra H (2000) Microbial biosorbents: meeting challenges of heavy metal pollution in aqueous solutions. Curr Sci 78(8):967–973
Harms H, Schlosser D, Wick LY (2011) Untapped potential: exploiting fungi in bioremediation of hazardous chemicals. Nat Rev Microbiol 9:177–192. doi:10.1038/nrmicro2519
Harrison VF, Gow WA, Ivarsson KC (1966) Leaching of uranium from Elliott Lake ore in the presence of bacteria. Can Min J 87:64–67
Hartmann LM, Craig PJ, Jenkins RO (2003) Influence of arsenic on antimony methylation by the aerobic yeast *Cryptococcus humicolus*. Arch Microbiol 180(5):347–352. doi.10.1007/s00203-003-0600-1
Hernlem BJ, Vane LM, Sayles GD (1999) The application of siderophores for metal recovery and waste remediation: examination of correlations for prediction of metal affinities. Water Res 33:951–960
Hess M (2008) Thermoacidophilic proteins for biofuel production. Trends Microbiol 16(9): 414–419
Hölker U, Bend J, Pracht R, Tetsch L, Müller T, Höfer M (2004) *Hortaea acidophila*, a new acid-tolerant black yeast from lignite. Antonie Van Leeuwenhoek 86(4):287–294
Hujslová M, Kubátová A, Kostovčík M, Kolařík M (2013) *Acidiella bohemica* gen. et sp. nov. and *Acidomyces* spp. (*Teratosphaeriaceae*), the indigenous inhabitants of extremely acidic soils in Europe. Fungal Div 58(1):33–45
Iskandar NL, Zainudin NA, Tan SG (2011) Tolerance and biosorption of copper (Cu) and lead (Pb) by filamentous fungi isolated from a freshwater ecosystem. J Environ Sci (China) 23(5): 824–830
Ivarsson KC, Morita H (1982) Single-cell protein production by the acid-tolerant fungus *Scytalidium acidophilum* from acid hydrolysates of waste paper. Appl Environ Microbiol 43: 643–647

Jacobson ES, Hove E, Emery HS (1995) Antioxidant function of melanin in black fungi. Infect Immun 63(12):4944–4945

Jaishankar M, Tseten T, Anbalagan N, Mathew BB, Beeregowda KN (2014) Toxicity, mechanism and health effects of some heavy metals. Interdisc Toxicol 7(2):60–72. doi:10.2478/intox-2014-0009

Jakubiak M, Giska I, Asztemborska M, Bystrzejewska-Piotrowska G (2014) Bioaccumulation and biosorption of inorganic nanoparticles: factors affecting the efficiency of nanoparticle mycoextraction by liquid-grown mycelia of *Pleurotus eryngii* and *Trametes versicolor*. Mycol Prog 13:525–532. doi:10.1007/s11557-013-0933-3

Jamieson DJ (1998) Oxidative stress responses of the yeast *Saccharomyces cerevisiae*. Yeast 14:1511–1527

Jarosz-Wilkołazka A, Gadd GM (2003) Oxalate production by wood-rotting fungi growing in toxic-metal-amended medium. Chemospere 52:541–547. doi:10.1016/S0045-6535(03)00235-2

Jiang J, Qin C, Shu X, Chen R, Song H, Li Q et al (2015) Effects of copper on induction of thiol-compounds and antioxidant enzymes by the fruiting body of *Oudemansiella radicata*. Ecotoxicol Environ Saf 111:60–65. doi:10.1016/j.ecoenv.2014.09.014

Joseph E, Cario S, Simon A, Wörle M, Mazzeo R, Junier P, Job D (2012) Protection of metal artifacts with the formation of metal-oxalates complexes by *Beauveria bassiana*. Front Microbiol 2:270. doi:10.3389/fmicb.2011.00270. eCollection 2011

Joshi BH (2014) Evaluation and characterization of heavy metal resistant fungi for their prospects in bioremediation. J Environ Res Dev 8(04):876–882

Jung WH, Sham A, White R, Kronsta JW (2006) Iron regulation of the major virulence factors in the AIDS-associated pathogen *Cryptococcus neoformans*. PLoS Biol. doi:10.1371/journal.pbio.0040410

Karman SB, Zaleha S, Diah M, Gebeshuber IC (2015) Raw materials synthesis from heavy metal industry effluents with bioremediation and phytomining: a biomimetic resource management approach. Adv Mater Sci Eng. doi:10.1155/2015/185071

Klaunig JE, Xu Y, Isenberg JS, Bachowski S, Kolaja KL, Jiang J et al (1998) The role of oxidative stress in chemical carcinogenesis. Environ. Health Perspect 106:289–295

Kulshreshtha S, Mathur N, Bhatnagar P (2014) Mushroom as a product and their role in mycoremediation. AMB Express. 4:29. doi:10.1186/s13568-014-0029-8

Lazarova N, Krumova E, Stefanova T, Georgieva N, Angelova M (2014) The oxidative stress response of the filamentous yeast *Trichosporon cutaneum* R57 to copper, cadmium and chromium exposure. Biotechnol Biotechnol Equip 28(5):855–862

Li HY, Li DW, He CM, Zhou ZP, Mei T, Xu HM (2012) Diversity and heavy metal tolerance of endophytic fungi from six dominant plant species in a Pb-Zn wasteland in China. Fungal Ecol 5:309–315. doi:10.1016/j.funeco.2011.06.002

Lin CH, Huang CF, Chen WY, Chang YY, Ding WH, Lin MS et al (2006) Characterization of the interaction of galectin-1 with sodium arsenite. Chem Res Toxicol 19:469–474. doi:10.1021/tx0503348

Liu Z, Boles E, Rosen BP (2004) Arsenic trioxide uptake by hexose permeases in *Saccharomyces cerevisiae*. J Biol Chem 279:17312–17318

López-Archilla AI, González AE, Terrón MC, Amils R (2004). Ecological study of the fungal populations of the acidic Tinto River in southwestern Spain. Can J Microbiol 50(11):923–934

Ma LQ, Komar KM, Tu C, Zhang W, Cai Y, Kennelley ED (2001) A fern that hyperaccumulates arsenic: a hardy, versatile, fast-growing plant helps to remove arsenic from contaminated soils. Nature 409:579. doi:10.1038/35054664

Maciaszczyk-Dziubinska E, Wawrzycka D, Wysocki R (2012) Arsenic and antimony transporters in eukaryotes. Int J Mol Sci 13(3):3527–3548. doi:10.3390/ijms13033527

Maciaszczyk-Dziubinska E, Migdal I, Migocka M, Bocer T, Wysocki R (2010) The yeast aquaglyceroporin Fps1p is a bidirectional arsenite channel. FEBS Lett 584:726–732

Maheswari S, Murugesan AG (2009) Remediation of arsenic in soil by Aspergillus nidulans isolated from an arsenic-contaminated site. Environ Technol 30(9):921–926. doi:10.1080/09593330902971279

Majorel C, Hannibal L, Ducousso M, Lebrun M, Jourand P (2014) Evidence of nickel (Ni) efflux in Ni-tolerant ectomycorrhizal *Pisolithus albus* isolated from ultramafic soil. Environ Microbiol Rep 6:510–518. doi:10.1111/1758-2229.12176

Malik A (2004) Metal bioremediation through growing cells. Environ Int 30:261–278

Matschullat J (2000) Arsenic in the geosphere—a review. Sci Total Environ 249(1–3):297–312

McDougall DN, Blanchette RA (1996) Metal ion adsorption by pseudoschlerotial plates of *Phelliunus weirii*. Mycologia 88:98–103

Muller S, Walter RD, Fairlamb AH (1995) Differential susceptibility of filarial and human erythrocyte glutathione reductase to inhibition by the trivalent organic arsenical melarsen oxide. Mol Biochem Parasitol 71:211–219. doi:10.1016/0166-6851(94)00053-P

Ochiai E (1997) Bio-inorganic chemistry: an Inroduction. Align and Bawn, Boston

Oggerin M, Tornos F, Rodríguez N, del Moral C, Sánchez-Román M, Amils R (2013) Specific jarosite biomineralization by *Purpureocillium lilacinum*, an acidophilic fungi isolated from Río Tinto. Environ Microbiol 15(8):2228–2237. doi:10.1111/1462-2920.12094

Pócsi I (2011) Toxic metal/metalloid tolerance in fungi—a biotechnological-oriented approach. In: Bánfalvi G (ed) Cellular effects of heavy metals. Springer, Berlin, pp 31–58. doi:10.1007/978-94-0097-0428-2_2

Polizeli M, Rizzatti L, Monti T, Terenzi A, Jorge C, Amorim S (2005) Xylanases from fungi: properties and industrial applications. Appl Microbiol Biotechnol 67(5):577–591

Porquet A, Filella M (2007) Structural evidence of the similarity of $Sb(OH)_3$ and $As(OH)_3$ with glycerol: implications for their uptake. Chem Res Toxicol 20:1269–1276

Purchase D, Scholes LNL, Revitt DM, Shutes RBE (2009) Effects of temperature on metal tolerance and the accumulation of Zn and Pb by metal-tolerant fungi isolated from urban runoff treatment wetlands. J Appl Microbiol 106:1163–1174

Purvis OW, Halls C (1996) A review of lichens in metal-enriched environments. Lichenologist 28:571–601

Qiu Z, Deng Z, Tan H, Zhou S, Cao L (2015) Engineering the robustness of *Saccharomyces cerevisiae* by introducing bifunctional glutathione synthase gene. J Ind Microbiol Biotechnol 42(4):537–542. doi:10.1007/s10295-014-1573-6

Quesada AR, Byrnes RW, Krezoski SO, Petering DH (1996) Direct reaction of H_2O_2 with sulfhydryl groups in HL-60 cells: zinc-metallothionein and other sites. Arch Biochem Biophys 334:241–250

Raab A, Feldmann J, Meharg AA (2004) The nature of arsenic phytochelatin complexes in *Holcus lanatus* and *Pteris cretica*. Plant Physiol 134:1113–1122. doi:10.1104/pp.103.033506

Rajendran P, Muthukrishnan J, Gunasekaran P (2003) Microbes in heavy metal remediation. Indian J Exp Biol 41(9):935–944

Ramirez-Solis A, Mukopadhyay R, Rosen BP, Stemmler TL (2004) Experimental and theoretical characterization of arsenite in water: insights into the coordination environment of As-O. Inorg Chem 43(9):2954–2959

Ratnaike RN (2003) Acute and chronic arsenic toxicity. Postgrad Med J 79:391–396. doi:10.1136/pmj.79.933.391

Reese RN, Mehra RK, Tarbet EB, Winge DR (1988) Studies on the γ-glutamyl Cu-binding peptide from *Schizosaccharomyces pombe*. J Biol Chem 263:4186–4192

Rizzo DM, Blanchette RA, Palmer MA (1992) Biosorption of metal ions by *Armillaria rhizomorphs*. Can J Bot 1992(70):1515–1520

Romero-Isart N, Vašák M (2002) Advances in the structure and chemistry of metallothioneins. J Inorg Biochem 88:388–396

Romaní AM, Fischer H, Mille-Lindblom C, Tranvik LJ (2006) Interactions of bacteria and fungi on decomposing litter: differential extracellular enzyme activities. Ecology 87(10):2559–2569

Rosen BP (2002) Biochemistry of arsenic detoxification. FEBS Lett 529:86–92

Rothenberg SJ, Rothenberg JC (2005) Testing the dose-response specification in epidemiology: public health policy consequences for lead. Environ Health Perspect 113:1190–1195

Ruotolo R, Marchini G, Ottonello S (2008) Membrane transporters and protein traffic networks differentially affecting metal tolerance: a genomic phenotyping study in yeast. Genome Biol 9: R67. doi:10.1186/gb-2008-9-4-r67

Salvadori MR, Ando RA, Oller do Nascimento CA, Corrêa B (2014a) Intracellular biosynthesis and removal of copper nanoparticles by dead biomass of yeast isolated from the wastewater of a mine in the Brazilian Amazonia. PLoS ONE 9(1):e87968. doi:10.1371/journal.pone.0087968. eCollection 2014

Salvadori MR, Nascimento CA, Corrêa B (2014b) Nickel oxide nanoparticles film produced by dead biomass of filamentous fungus. Sci Rep. doi:10.1038/srep06404

Salvadori MR, Ando RA, Do Oller Nascimento CA, Corrêa B (2014c) Bioremediation from wastewater and extracellular synthesis of copper nanoparticles by the fungus *Trichoderma koningiopsis*. J Environ Sci Health A Tox Hazard Subst Environ Eng 49(11):1286–1295. doi:10.1080/10934529.2014.910067

Salvadori MR, Ando RA, Nascimento CA, Corrêa B (2015) Extra and intracellular synthesis of nickel oxide nanoparticles mediated by dead fungal biomass. PLoS One 10(6):e0129799. doi:10.1371/journal.pone.0129799

Sanghi R, Verma P (2009) Biomimetic synthesis and characterisation of protein capped silver nanoparticles. Bioresour Technol 100:501–504. doi:10.1016/j.biortech.2008.05.048

Sari A, Tuzen M (2009a) Kinetic and equilibrium studies of biosorption of Pb(II) aned Cd(ii) from aqueous solution by macrofungus (*Amanita rubescens*) biomass. J Hazard Mater 164: 1004–1011. doi:10.1016/j.jhazmat.2008.09.002

Sari A, Tuzen M (2009b) Biosorption of As(III) and As(V) form aqueous solution by macrofungus (*Inonotus hispidus*) biomass: equilibrium and kinetic studies. J Hazard Mater 164(2–3): 1372–1378. doi:10.1016/j.jhazmat.2008.09.047

Saxena P, Bhattacharyya AK, Mathur N (2006) Nickel tolerance and accumulation by filamentous fungi from sludge of metal finishing industry. Geomicrobiol J 23:333–340

Schalk IJ, Hannauer M, Braud A (2011) Minireview new roles for bacterial siderophores in metal transport and tolerance. Environ Microbiol 13:2844–2854

Schweitzer AD, Howell RC, Jiang Z, Bryan RA, Gerfen G, Chen CC et al (2009) Physico-chemical evaluation of rationally designed melanins as novel nature-inspired radioprotectors. PLoS ONE 4(9):e7229. doi:10.1371/journal.pone.0007229

Scholtmeijer K, Wessels J, Wösten H (2001) Fungal hydrophobins in medical and technical applications. Appl Microbiol Biotechnol 56(1–2):1–8

Schwantes HO (1996) The biology of fungi: an introduction to applied mycology. Verlag Eugen Ulmer GmbH

Schwyn B, Neilands JB (1987) Universal chemical assay for the detection and determination of siderophores. Anal Biochem 160:47–56

Selbmann L, De Hoog GS, Zucconi L, Isola D, Ruisi S, van den Ende AG et al (2008) Drought meets acid: three new genera in a dothidealean clade of extremotolerant fungi. Stud Mycol 61:1–20

Seyedmousavi S, Badali H, Chlebicki A, Zhao J, Prenafeta-Boldu FX, De Hoog GS (2011) *Exophiala sideris*, a novel black yeast isolated from environments polluted with toxic alkyl benzenes and arsenic. Fungal Biol 115(10):1030–1037

Sharples JM, Meharg AA, Chambers SM, Cairney JWG (2000) Mechanism of arsenate resistance in the ericoid mycorrhizal fungus *Hymenoscyphus ericae*. Plant Physiol 124:1327–1334

Shen M, Zhao D-K, Qiao Q, Liu L, Wang J-L, Cao G-H et al (2015) Identification of glutathione S-transferase (GST) genes from a dark septate endophytic fungus (*Exophiala pisciphila*) and their expression patterns under varied metal stress. PLoS ONE 10(4):e0123418. doi:10.1371/journal.pone.0123418

Sigler L, Carmichael JW (1974) A new acidophilic *Scytalidium*. Can J Microbiol 20(2):267–268

Singh H (2006) Mycoremediation: fungal bioremediation. Wiley

Snow ET, Hu Y, Yan CC, Chouchane S (1999) Modulation of DNA repair and glutathione levels in human keratinocytes by micromolar arsenite. In: Chappell WR, Abernathy CO, Calderon RL (eds) Arsenic exposure and health effects. Elsevier, Oxford, pp 243–251. doi:10.1016/b978-008043648-7/50028-5

Srivastava PK, Vaish A, Dwivedi S, Chakrabarty D, Singh N, Tripathi RD (2011) Biological removal of arsenic pollution by soil fungi. Sci Total Environ 409(12):2430–2442. doi:10.1016/j.scitotenv.2011.03.002

Stajich JE, Berbee ML, Blackwell M, Hibbett DS, James TY, Spatafora, JW et al (2009) The fungi. Curr Biol 19(18):R840–5

Starkey RL, Waksman SA (1943) Fungi tolerant to extreme acidity and high concentrations of copper sulfate. J Bacteriol 45(5):509–691

Styblo M, Serves SV, Cullen WR, Thomas DJ (1997) Comparative inhibition of yeast glutathione reductase by arsenicals and arsenthiols. Chem Res Toxicol 10:27–33. doi:10.1021/tx960139g

Su S, Zeng X, Bai L, Jiang X, Li L (2010) Bioaccumulation and biovolatilisation of pentavalent arsenic by *Penicillin janthinellum*, *Fusarium oxysporum* and *Trichoderma asperellum* under laboratory conditions. Curr Microbiol 61(4):261–266. doi:10.1007/s00284-010-9605-6

Sylvia DM, Fuhrmann JJ, Hartel PG, Zuberer DA (2005) Principles and applications of soil microbiology. Pearson Prentice Hall, Upper Saddle River, NJ

Tamaki S, Frankenberger WT Jr (1992) Environmental biochemistry of arsenic. Rev Environ Contam Toxicol 124:79–110

Tamás MJ, Labarre J, Toledano MB, Wysocki R (2005) Mechanisms of toxic metal tolerance in yeast. In: Tamás MJ, Martinoia E (eds) Molecular biology of metal homeostasis and detoxification: from microbes to man. Springer, Heidelberg, pp 395–454

Tan Q, Chen G, Zeng G, Chen A, Guan S, Li Z et al (2015) Physiological fluxes and antioxidative enzymes activities of immobilized *Phanerochaete chrysosporium* loaded with TiO_2 nanoparticles after exposure to toxic pollutants in solution. Chemosphere 2015(128):21–27. doi:10.1016/j.chemosphere.2014.12.088

Tchounwou PB, Yedjou CG, Patlolla AK, Sutton DJ (2012) Heavy metal toxicity and the environment. EXS 101:133–164. doi:10.1007/978-3-7643-8340-4_6

Tereshina VM, Mar'in AP, Kosyakov VN, Kozlov VP, Feofilova EP (1999) Different metal sorption capacities of cell wall polysaccharides of *Aspergillus niger*. Appl Biochem Micorbiol 35:389–392

Tetsch L, Bend J, Hölker U (2006) Molecular and enzymatic characterisation of extra- and intracellular laccases from the acidophilic ascomycete *Hortaea acidophila*. Antonie Van Leeuwenhoek 90(2):183–194

Todorova TT, Kujumdzieva AV, Vuilleumier S (2010) Non-enzymatic roles for the URE2 glutathione S-transferase in the response of Saccharomyces cerevisiae to arsenic. Arch Microbiol 192(11):909–918. doi:10.1007/s00203-010-0614-4

Trasande L, Landrigan PJ, Schechter C (2005) Public health and economic consequences of methyl mercury toxicity to the developing brain. Environ Health Perspect 113:590–596

Tripathi P, Singh PC, Mishra A, Chauhan PS, Dwivedi S, Bais RT et al (2013) *Tricoderma*: a potential bioremediator for environmental clean up. Clean Technol Environ Policy 15: 541–550. doi:10.1007/s10098-012-0553-7

Tripathi V, Fraceto LF, Abhilash PC (2015) Sustainable clean-up technologies for soils contaminated with multiple pollutants: plant-micorb-pollutant and climate nexus. Ecol Eng 82:330–335. doi:10.1016/j.ecoleng.2015.05.027

Tschan M, Robinson B, Schulin R (2008) Antimony uptake by *Zea mays* (L.) and *Helianthus annuus* (L.) from nutrient solution. Environ Geochem Health 30:187–191

Tsuji N, Hirayanagi N, Okada M, Miyasaka H, Hirata K, Zenk MH et al (2002) Enhancement of tolerance to heavy metals and oxidative stress in *Dunaliella tertiolecta* by Zn-induced phytochelatin synthesis. Biochem Biophys Res Commun 293:653–659

Urik M, Littera P, Kolen M (2009) Removal of arsenic (V) from aqueous solutions using chemically modified sawdust of spruce (*Picea abies*): kinetics and isotherm studies. Int J Environ Sci Technol 6(3):451–456

Urik M, Hlodak M, Mikusova P, Matus P (2014) Potential of microscopic fungi isolated from mercury contaminated soils to accumulate and volatilize mercury(II). Water Air Soil Pollut 225:2219–2229. doi:10.1007/s11270-014-2219-z

Vázquez-Campos X, Kinsela AS, Collins RN, Neilan BA, Aoyagi N, Waite TD (2015) Uranium binding mechanisms of the acid-tolerant fungus *Coniochaeta fodinicola*. Environ Sci Technol 49(14):8487–8496. doi:10.1021/acs.est.5b01342

Velmurugan P, Shim J, You Y, Choi S, Kamala-Kannan S, Lee KJ, Kim HJ, Oh BT (2010) Removal of zinc by live, dead, and dried biomass of *Fusarium* spp. isolated from the abandoned-metal mine in South Korea and its perspective of producing nanocrystals. J Hazard Mater 182(1–3):317–324. doi:10.1016/j.jhazmat.2010.06.032

Vijayaraghavan K, Padmesh TVN, Palanivelu K, Velan M (2006) Biosorption of nickel(II) ions onto Sargassum wightii: application of two-parameter and three-parameter isotherm models. J Hazard Mater 133(1–3):304–308

Vodyanitskii YN (2013) Contamination of soils with heavy metals and metalloids and its ecological hazard (analytic review). Eurasian Soil Sci 46(7):793–801

Wang J, Chen C (2014) Chitosan-based biosorbents: modification and application for biosorption of heavy metals and radionuclides. Bioresour Technol 160:129–141. doi:10.1016/j.biortech.2013.12.110

Wang J, Chen C (2009) Biosorbents for heavy metals removal and their future. Biotech Adv 27:195–226. doi:10.1016/j.biotechadv.2008.11.002

Wang Q, He M, Wang Y (2011) Influence of combined pollution of antimony and arsenic on culturable soil microbial populations and enzyme activities. Ecotoxicology 20(1):9–19

Wasserman GA, Liu X, Parvez F, Ahsan H, Factor-Litvak P, van Geen A (2004) Water arsenic exposure and children's intellectual function in Araihazar. Bangladesh Environ Health Perspect. 112:1329–1333

Wei Z, Liang X, Pendlowski H, Hillier S, Suntornvongsagul K, Sihanonth P et al (2013) Fungal biotransformation of zinc silicate and sulfide mineral ores. Environ Microbiol 15(8):2173–2186. doi:10.1111/1462-2920.12089

White C, Sayer JA, Gadd GM (1997) Microbial solubilization and immobilization of toxic metals: key biogeochemical processes for treatment of contamination. FEMS Microbiol Rev 20(3–4):503–516

Winkelmann G (2007) Ecologyof siderophores with special reference to the fungi. Biometals 20:379–392. doi:10.1007/s10534-006-9076-1

Wysocki R, Bobrowicz P, Ułaszewski S (1997) The *Saccharomyces cerevisiae* ACR3 gene encodes a putative membrane protein involved in arsenite transport. J Biol Chem 272 (48):30061–30066

Wysocki R, Fortier PK, Maciaszczyk E, Thorsen M, Leduc A, Odhagen A, Owsianik G, Ulaszewski S, Ramotar D, Tamás MJ (2004) Transcriptional activation of metalloid tolerance genes in Saccharomyces cerevisiae requires the AP-1-like proteins Yap1p and Yap8p. Mol Biol Cell 15:2049–2060

Wysocki R, Tamás MJ (2010) How *Saccharomyces cerevisiae* copes with toxic metals and metalloids. FEMS Microbiol Rev 34:925–951

Xu X, Xia L, Huang Q, Gu JD, Chen W (2012) Biosorption of cadmium by a metal-resistant filamentous fungus isolated from chicken manure compost. Environ Technol 33(14):1661–1670

Zafar S, Aqil F, Ahmad I (2007) Metal tolerance and biosorption potential of filamentous fungi isolated from metal contaminated agricultural soil. Bioresour Technol 98(13):2557–2561

Zhang W, Cai Y (2003) Purification and characterization of thiols in an arsenic hyperaccumulator under arsenic exposure. Anal Chem 15, 75(24):7030–7035

Zhang YJ, Zhang Y, Liu MJ, Shi XD, Zhao ZW (2008) Dark septate endophyte (DSE) fungi isolated from metal polluted soils: their taxonomic position, tolerance, and accumulation of heavy metals in vitro. J Microbiol 46(6):624–632. doi:10.1007/s12275-008-0163-6

Zhou JL (1999) Zn biosorption by *Rhizopus arrhizus* and other fungi. Appl Microbiol Biotechnol 51:686–693

Zuo Y, Chen G, Zeng G, Li Z, Yan M, Chen A, Guo Z, Huang Z, Tan Q (2015) Transport, fate, and stimulating impact of silver nanoparticles on the removal of Cd(II) by *Phanerochaete chrysosporium* in aqueous solutions. J Hazard Mater 285:236–244. doi:10.1016/j.jhazmat.2014.12.003

Part IV
Biotechnology to Reduce Reliance on Fossil Fuels: From Biomass to Biofuel

Chapter 11
Fungi in Consolidated Bioprocessing of Lignocellulosic Materials

Anastasia P. Galanopoulou and Dimitris G. Hatzinikolaou

11.1 Industrial Approaches in Biomass Conversion—An Introduction

Over the last 30 years, lignocellulose biotechnology has made solid steps toward its transformation from a laboratory research topic to an industrial process of favorable financial perspective. The major driving force for this was the growing concerns over the emergence of the first generation biofuel technology. Since this approach mainly involved the exploitation of sucrose and starch derived sugars as its main substrates, it rapidly created a competition for land and water that were traditionally used for food production (Sims et al. 2010; Ho et al. 2014). Lignocellulosic biomass (LCB), on the other hand, is a renewable carbon source that predominantly comprises of agricultural and forestry residues and by-products, uniformly produced in large quantities throughout the biosphere. As a result, it appears as an ideal alternative for exploitation, through novel and efficient technologies cumulatively incorporated into the "2nd generation biorefinery" concept (Rowlands et al. 2008; Adsul et al. 2011; Dugar and Stephanopoulos 2011).

The carbon stored in lignocellulosic biomass is distributed among three general types of polymers, namely: cellulose, hemicellulose, and lignin. These polymers are strongly interconnected with practically all types of chemical bonds, to yield a three-dimensional superstructure of high chemical recalcitrance (http://www.nature.com/scitable/content/structure-of-lignocellulose-14464273). In order to use lignocellulose as raw material for microbial transformations, it has to be broken down into soluble sugars and low molecular weight compounds. The β-1,4-glycosidic bond that is dominant in cellulose and hemicellulose structures, is among the most stable bonds in nature, and its chemical decomposition requires very harsh conditions

A.P. Galanopoulou · D.G. Hatzinikolaou (✉)
Microbial Biotechnology Group (MBG), Department of Biology,
National and Kapodistrian University of Athens, 15784 Attica, Athens, Greece
e-mail: dhatzini@biol.uoa.gr

D. Purchase (ed.), *Fungal Applications in Sustainable Environmental Biotechnology*, Fungal Biology, DOI 10.1007/978-3-319-42852-9_11

of high temperature and extreme pH values. Controlled enzymatic hydrolysis following mild chemical pretreatment, appears as the only sustainable and environmentally benign alternative for biomass degradation (Alvira et al. 2010; Naik et al. 2010; Van Dyk and Pletschke 2012). The enzymes required for biomass degradation have been isolated from a variety of species that expand throughout the tree of life and range from simple prokaryotes to insects and crustaceans (Cragg et al. 2015). A large variety of these enzymes has been isolated, characterized, and systematically categorized (www.cazy.org). From a practical point of view, exploitable quantities of biomass degrading enzymes are produced by certain microbial species, distinguished as natural biomass degraders. Fungi dominate this group since many of them produce literally all the required enzyme activities (Guerriero et al. 2015).

11.1.1 Process Considerations in Biorefinery

A lignocellulose biorefinery may adopt various process configurations depending on the type of biomass used, the nature of enzyme and microorganism, the target product, etc. In any case, the minimum necessary stages are given below:

(1) Biomass pretreatment
(2) Hydrolytic enzyme production (cellulases, hemicellulases, and auxiliary activities)
(3) Hydrolysis of pretreated biomass
(4) Microbial transformation (fermentation) of hexoses and pentoses into fuels and/or chemicals.

From a process standpoint, three clearly distinguishable general approaches exist, as shown in Fig. 11.1. They differ in the degree of integration among these four stages. The first one, denoted as **Separate Hydrolysis and Fermentation—SHF**—totally separates all the stages offering the advantage of simplicity, and the opportunity to perform hydrolysis and fermentation at *the specific optimum conditions for each one*. As a drawback, it appears as the choice with the highest installation costs and lowest overall productivity. Since the majority of biorefinery products are *low unit price—high production volume* chemicals, installation costs greatly affect economic viability, a fact that calls for higher integration among the various stages (Menon and Rao 2012).

The second approach is **Simultaneous Saccharification and Fermentation—SSF**—a process configuration developed toward higher integration, by combining biomass hydrolysis and fermentation in the same vessel and, usually, at the same time (Fig. 11.1). In SSF systems, the fermenting microorganism is added together (or with a few hours delay) with the hydrolytic enzymes in the pretreated biomass vessel, resulting in a process with significant overall productivity gains. Productivity is also enhanced by the continuous removal of monomeric sugars through fermentation, since hexose and pentose oligosaccharides are strong inhibitors of their corresponding hydrolases. The major drawback of an SSF biorefinery

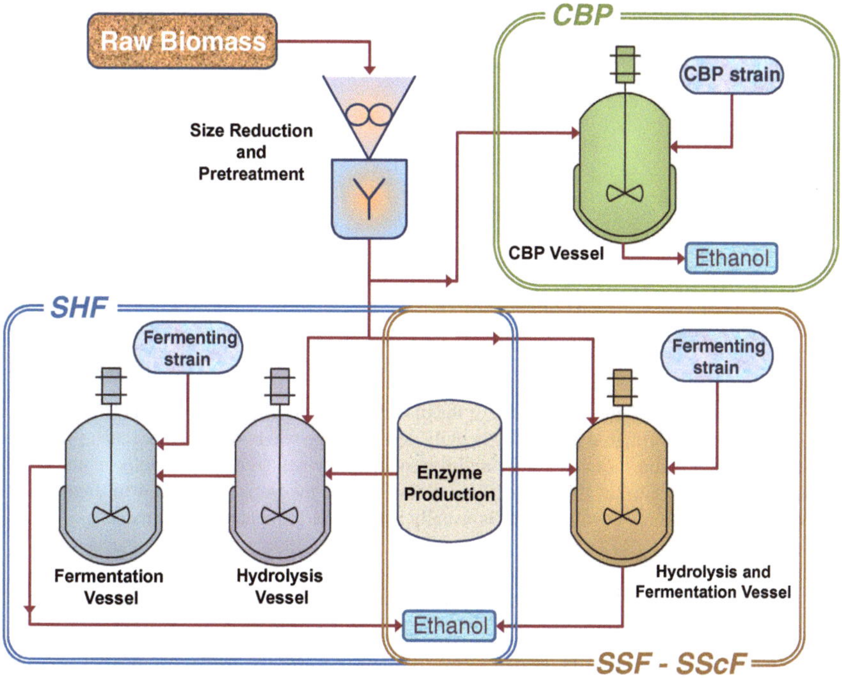

Fig. 11.1 Schematic flowchart of the different process configurations in ethanol biorefineries. Following size reduction and pretreatment of the lignocellulosic substrate, a separate Enzyme Production step is necessary for both Separate Hydrolysis and Fermentation (SHF) and Simultaneous Saccharification and Fermentation or co-fermentation (SSF/SScF) processes. Consolidated Bioprocessing (CBP) integrates all steps in a single vessel

approach arises from the fact that the optimum operating conditions (pH, temperature, etc.) for the action of hydrolytic enzymes usually do not coincide with those of the microbial growth and fermentation. Consequently, a compromise between hydrolysis and fermentation conditions is necessary (Lynd et al. 2005). A variation of SSF is often encountered under the term **Simultaneous Saccharification and co-Fermentation—SScF**. The term is used for processes that take also into account the hemicellulosic part of the substrate and target toward microbial species capable of co-fermenting both glucose and pentoses.

The third approach, encompassing the maximum integration of all biorefinery steps, is referred as **Consolidated Bioprocessing—CBP**. It implies the use of a single microorganism that is capable of producing most—if not all—enzymes necessary for biomass hydrolysis and transforming the resulting hexoses and pentoses to the desired product (i.e., ethanol) simultaneously. Since there is no wild-type microbial strain capable of performing all stages at an industrially exploitable capacity, the success of CBP technology heavily relies on genetic manipulations on a microorganism toward either:

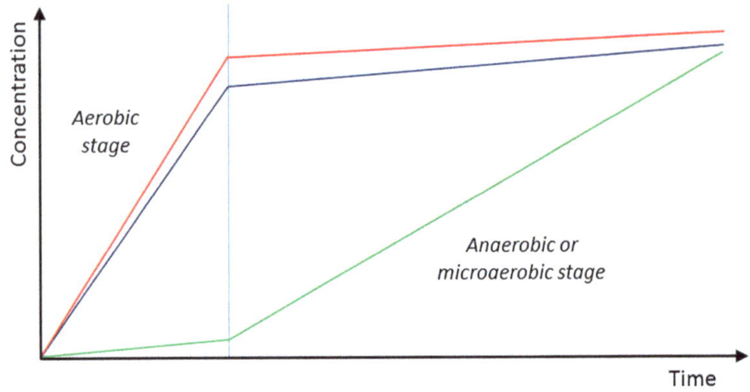

Fig. 11.2 Simplified schematic diagram of the distinct phases encountered in CBP. The process is usually progressing through a first initial heavily aerobic stage where the microorganism grows on the substrate producing biomass (*blue*) and hydrolytic enzymes (*red*). Negligible or no product (*green*) formation occurs at this stage. At a certain point, the vessel is shifted to anaerobic or microaerobic conditions, new substrate is usually added and fermentation commences

(1) engineering efficient natural biomass degraders to produce ethanol or other desirable compounds, or
(2) engineering organisms with exceptional biosynthetic traits for a specific product, to produce biomass degrading hydrolases during growth on pretreated lignocellulosic materials.

The engineered microorganisms developed for these two options are often referred as Type I and Type II CBP transformants, respectively. At this point, it is important to clarify that although a CBP process takes place in a single vessel and by the same microorganism, the three biorefinery stages mentioned above (enzyme production, biomass hydrolysis, and fermentation of the oligosaccharides) usually are not expected to take place simultaneously, especially when fungal species are considered (Fig. 11.2).

11.1.2 Fungi in Lignocellulose Pretreatment

Regardless the selected process approach employed in any biorefinery, an essential pretreatment step is required, in order to loosen-up the lignocellulose structure and facilitate enzyme access. Several types of pretreatments have been considered, which involve the use of mild acid or alkali, steam, hot water, various solvents, or combinations of them (Menon and Rao 2012; Narron et al. 2016; Rabemanolontsoa and Saka 2016). Of particular relevance to this book is the recently introduced term of "biological pretreatment" (Sindhu et al. 2016). This approach involves the use of microorganisms, and especially fungi as the actual pretreatment agents. So far, the most successful efforts have been focused on the use of white-rot fungi that preferentially attack lignin, a fact that is expected to increase hydrolysis rates of the remaining biomass polysaccharides through the action of cellulases and hemicellulases (Larran et al. 2015; Saha et al. 2016).[1] Biological pretreatment is supported as the only truly eco-friendly pretreatment, since it resembles the natural process of biomass degradation and as a consequence it yields minimum or zero accumulation of potential inhibitors against the enzymes and microorganisms that will be subsequently used. As a result, it appears particularly attractive for use in single vessel or consolidated bioprocessing approaches. In addition, since the process employs fungal species, a significant amount of hydrolases could be produced simultaneously, enhancing the subsequent hydrolysis steps (Isroi et al. 2011; Narayanaswamy et al. 2013). Currently, none of the second generation industrial processes involves biological pretreatments due to its most significant limitation—the long incubation times required for effective delignification. This can be minimized to an extent, using suitable microbial consortia combined with fine tuning of the process parameters (Larran et al. 2015; Sindhu et al. 2016).

11.2 Develop of Lignocellulose Degrading Traits in Natural Ethanol Producers

This first approach for the production of a transformant suitable for bioethanol production in CBP systems, concerns the genetic manipulations of natural ethanologenic microorganisms. In principle, this methodology involves the heterologous expression in the ethanol producer of enzymes necessary for the hydrolysis of cellulose and/or hemicellulose, to yield a **Type II CBP** transformant. Significant efforts have been attempted with ethanologenic bacteria such as *Zymomonas mobilis* (la Grange et al. 2010; Linger et al. 2010; Luo and Bao 2015) or *Esherichia coli* (Mazzoli et al. 2012; Vinuselvi and Lee 2012) but with limited practical success. More promising results, even though with their own limitations, have been achieved with various yeast genera, such as *Saccharomyces* and *Kluyveromyces*.

11.2.1 Saccharomyces *spp. in Type II CBP Processes*

Saccharomyces cerevisiae is one of the most efficient natural ethanol producers and has long been considered as the industry standard in the area (Della-Bianca

[1]For additional information on decomposition of lignin by white-rot fungi, please refer to Chap. 9 *White and brown rot fungi as decomposers of lignocellulosic materials and their role in waste and pollution control.*

et al. 2012). Among its advantages are: (i) high ethanol productivities (over 3 g/L/h), (ii) ability to work at low pH values, (iii) high ethanol tolerance, and (iv) it is a well-known member of the Generally Recognized As Safe (GRAS) family of microorganisms (Hasunuma et al. 2015; Albergaria and Arneborg 2016). However, wild-type *S. cerevisiae* strains lack two basic traits essential to CBP applications; the ability to grow and ferment pentoses and, mainly, the production of cellulolytic and hemicellulolytic enzymes. As a result, genetic engineering efforts for *Saccharomyces* species have focused, either toward the incorporation of genes related to xylose transport and catabolism or the introduction of cellulolytic and hemicellulolytic abilities. Since the former perspective is mainly directed toward co-fermentation approaches and not for pure CBP applications, the numerous efforts for introducing pentose utilization traits in yeasts will not be discussed in this chapter. Interested readers may consult some of the available reviews on the subject (Matsushika et al. 2009; Cai et al. 2012; Moysés et al. 2016).

S. cerevisiae has been used for over 20 years as a production platform for the heterologous expression of eukaryotic enzymes. Since the most efficient and well studied cellulases and hemicellulases come from fungal species such as *Trichoderma*, *Neurospora* and *Aspergillus*, their simultaneous and controlled expression in *Saccharomyces* initially seemed as a plausible target (Elkins et al. 2010). This perception was also supported by the fact that *S. cerevisiae* is one of the model microbial systems in molecular biology studies and as a result there are many well-established methodologies for its genetic manipulations and transformation (Hasunuma et al. 2015).

Over the last decade, all these characteristics fueled a large number of efforts to incorporate the three main cellulase classes (endoglucanase, cellobiohydrolase and β-glucosidase) into *S. cerevisiae*. Several approaches were employed covering many different aspects and combinations for the expression strategy. For instance:

1. The use of inducible or constitutive promoters.
2. The use of plasmids of various copy numbers or the integration of cellulase gene cassettes directly into the chromosome.
3. Haploid or diploid *Saccharomyces* hosts.
4. The molar ratio of the three cellulase classes.
5. The source of cellulase—fungal (*Aspergillus*, *Trichoderma*, *Thermoascus*, etc.) or bacterial (*Clostridium*).
6. The enzyme excretion system—free, tethered, or cellulosomal.

All the above efforts, resulted in a significant amount of publications that are thoroughly discussed in some recently published reviews on the subject (van Zyl et al. 2007; la Grange et al. 2010; Kricka et al. 2014; den Haan et al. 2015; Hasunuma et al. 2015). Despite the successful incorporation and expression of several cellulases, until today there has been no Type II CBP *S. cerevisiae* strain of practical value. Several reasons have been proposed for the limited advance in these efforts that are mostly related to: (i) the detrimental effect of co-expression of multiple heterologous genes on yeast metabolism, (ii) the deregulation of the transcriptional mechanism of the cells by the simultaneous co-expression of multiple genes, and (iii) the improper folding and/or secretion of some of the proteins (Xu et al. 2009). Thus, even with the most successful attempts, the reported final ethanol concentrations do not exceed 20 g/L at a time scale that extends from 4 to 10 days. Moreover, these values were obtained using only model cellulosic substrates such as phosphoric acid swollen cellulose (PASC), microcrystalline, or amorphous cellulose. Only very recently, final ethanol concentrations at the level of 35 g/L were obtained by *S. cerevisiae* transformants co-expressing—at an optimized activity ratio—a β-glucosidase from *Trichoderma reesei* and an endoglucanase from *Aspergillus aculeatus* using pretreated corn cob. However, exogenous commercial cellulase was necessary to be added in order to achieve the above ethanol titers (Feng et al. 2016).

11.2.2 Potential of **Kluyveromyces** *spp. in Type II CBP*

Kluyveromyces species are considered more appropriate candidates for SSF and CBP applications, compared to *Saccharomyces cerevisiae*. This is because they are able to utilize and ferment a wider spectrum of monosaccharides, including pentoses. One additional advantage anticipated was the fact that, in contrast to *S. cerevisiae*, *Kluyveromyces* species are slightly thermophilic, with optimum fermentation temperatures between 40 and 45 °C (Ballesteros et al. 1991). This temperature is closer to the optimum temperature range for the activity of most commercial cellulases and hemicellulases, favoring the use of *Kluyveromyces* species in single-vessel SSF or CBP systems.

In the late 1980s, *Kluyveromyces fragilis* has been employed in the SSF of pretreated wheat straw in synergy with *T. reesei* cellulases (Szczodrak 1988, 1989) resulting in ethanol yields higher than 30 % (w/w) on a dry feedstock basis. The trend continued during the 1990s with significant success in terms of productivity and substrate conversion, despite the fact that the initially reported yields were never achieved in SSF systems as shown in the results summary of Table 11.1. These studies also revealed that *Kluyveromyces* species are more sensitive to inhibitors produced during the various substrate pretreatments and less ethanol tolerant than *Saccharomyces* species, a fact that resulted in lower ethanol titers compared to *S. cerevisiae*.

Despite the drawback of less well-established transformation and genetic handling protocols, some progress has been accomplished for the biotechnological exploitation of *Kluyveromyces* species through molecular approaches, including the efficient production of cellulases and hemicellulases (Fonseca et al. 2008). The first attempts on the subject involved the successful introduction of a cellulase cassette of three cellulolytic enzymes, namely an endo-β-1,4-glucanase, a cellobiohydrolase, and a β-glucosidase from the thermophilic fungus *Thermoascus aurantiacus* into the chromosome of *K. marxianus* NBRC1777. Although all three enzymes were functionally expressed in the recombinant strain, the latter could not produce

Table 11.1 Representative studies for ethanol production from various lignocellulosic substrates by *K. marxianus* in SSF systems

Strain	Enzyme added	Feedstock and pretreatment	Maximum ethanol (g/L)	Substrate conversion (% dry basis)	Productivity at max. ethanol (g/L/h)	Reference
K. marxianus IMB3	Commercial *T. reesei* cellulase	Cellulose	10	20	0.15	Barron et al. (1995)
K. marxianus ATCC 36907	Commercial *T. reesei* cellulase	Recycled paper sludge	35	18	0.49	Lark et al. (1997)
K. marxianus IMB3	Commercial *T. reesei* cellulase	Alkali pretreated barley straw	12	20	0.20	Boyle et al. (1997)
K. marxianus CECT 10875	Celluclast (NOVO Nordisk)	Steam explosion pretreated substrates	16–19	0.16–0.19	0.22–0.28	Ballesteros et al. (2004)
K. marxianus IMB4	Commercial cellulase (Iogen, Canada)	Hydrothermally pretreated switchgrass	17	23	0.23	Suryawati et al. (2008)
K. marxianus CECT 10875	Cellulase NS50013 supplemented with β-glucosidase NS50010 (Novozymes}	Steam-exploded wheat straw	36.2	26	0.50	Tomás-Pejó et al. (2009)
K. marxianus IMB3	Accellerase 1500 (Genencor)	Kanlow switchgrass subjected to hydrothermolysis	32	40	0.67	Pessani et al. (2011)
K. marxianus CECT 10875	Commercial cellulase/xylanase mixture (Novozymes)	Steam pretreated barley straw	29.4	40	0.41	García-Aparicio et al. (2011)

ethanol when grown on cellulose, but it effectively fermented cellobiose at a yield of 43 % w/w (Hong et al. 2007). In a similar study, the same *K. marxianus* strain harboring only the β-glucosidase gene produced 28 g/L of ethanol from 100 g/L cellobiose in a ventilation-mediated CBP system (Matsuzaki et al. 2012).

Until today, the most successful attempt for developing a *Kluyveromyces* sp. suitable for CBP applications, concerns a synthetic biology technique called "promoter-based gene assembly and simultaneous overexpression" (PGASO) (Chang et al. 2012). The method was applied in *K. marxianus* KY3 for its simultaneous transformation and expression of seven genes—five cellulases, one cellodextrin transporter and a selection marker—all of various fungal origins. The recombinant strain KR7 could express the five heterologous cellulase genes and more importantly, to convert crystalline cellulose into ethanol (Chang et al. 2013). Although the reported ethanol yield (approx. 14 % of max. theoretical) and productivity (0.14 g/L/day) are very low to consider any practical application, the proposed methodology represents a significant step in developing an efficient yeast strain for CBP applications.

11.2.3 Other Yeast Genera

Several additional yeast genera such as *Pichia* (currently *Scheffersomyces*), *Candida*, *Pachysolen*, *Geotrichum* and others, have been used in bioethanol production processes showing significant potential, mainly due to their ability to co-ferment pentoses (Zhang et al. 2015a). However, these yeasts ferment xylose less efficiently and at significantly lower yields than glucose, while they suffer from glucose repression and low ethanol tolerance (Harner et al. 2014). Efforts to improve their properties using combinations of classical and molecular methods have not yet resulted in the development of an industrially competent strain (Harner et al. 2014; Sànchez Nogué and Karhumaa 2015). In addition, since lignocellulose degrading enzymes production in these genera is either absent or very inefficient, their application has been limited only to SHF or SSF systems (Kricka et al. 2014).

Efforts to develop true Type II CBP transformants have been restricted to *Sheffersomyces stipitis*. Although there are reports on low-level xylanase production by *Sheffersomyces* isolates (Lara et al. 2014) and even an early reference concerning direct ethanol production (up to 2 g/L) from xylan (Lee et al. 1986), the low production levels and lack of cellulases prompted genetic modification efforts on *S. stipitis* for the overexpression of heterologous lignocellulolytic enzymes. In an early attempt toward this direction, the endoxylanase gene from the basidiomycetes *Cryptococcus albidus* was expressed in *S. stipitis* resulting in a transformant strain that could produce small amounts of ethanol (40 mM) when grown on xylan (Morosoli et al. 1993). At a step forward, the simultaneous expression of β-xylanase and β-xylosidase from two ascomycetes in *S. stipitis,* enhanced its xylanolytic ability and growth on xylan substrates, although ethanol production was not examined in this work (Den Haan and Van Zyl 2003). Very recently, the genes of an *A. niger* endoxylanase and an *Aspergillus aculeatus* endoglucanase were effectively integrated into the genome of *S. stipitis* following codon optimization. The transformed yeast was able to grow and produce ethanol (3 g/L) from a mixture of glucan and xylan and also from pretreated corn cob (Puseenam et al. 2015).

11.2.4 Type II CBP Processes on Jerusalem Artichoke

The most effective Type II CBP-like process that involves yeast species, concerns the production of ethanol from Jerusalem artichoke tubers (Jat) and as a result it deserves to separately be discussed. The tubers of this plant are rich in inulin (~60 %), a nondietary polymeric carbohydrate, comprising chain-terminating glucosyl moieties, and a repetitive fructosyl moiety (up to 60 fructose monomers) interconnected through β-1,2 glycosidic bonds (Kim and Kim 2014). Although inulin can be used as food ingredient or for fructose production, the tubers of Jerusalem artichoke grown under stressful conditions, such as drought and saline soils, produce much lower yield and are not suitable for extracting high quality inulin; instead they represent an excellent feedstock for biorefinery (Long et al. 2016). Inulin is not as recalcitrant as cellulose or hemicellulose and its use in CBP systems usually requires minimal pretreatment limited only to size reduction and hot water extraction (Yuan et al. 2012). In addition, inulinase, the enzyme responsible for the hydrolysis of inulin can be produced at significant quantities from *Saccharomyces* and, mainly, *Klyuveromyces* species that can also easily ferment the resulting fructose (Hu et al. 2012). As a result, direct ethanol production from artichoke tubers appears as the only true CPB approach that involves wild-type Type II strains. In fact the process has been extensively studied for strain selection and process parameter optimization (Table 11.2), yielding final ethanol concentrations well above 40 g/L, which is the minimum economically acceptable production level for lignocellulose biorefineries (Hahn-Hägerdal et al. 2006; Galbe et al. 2007).

Genetic engineering of yeast strains has also been attempted, either toward enhancement of inulinase production by *Kluyveromyces* sp. or toward the development of *S. cerevisiae* transformants that heterologously produce inulinase. A number of CBP bioreactor studies using Jat tubers showed that homologous overexpression of inulinase in *K. marxianus* resulted in only a small increase (11 %) of ethanol yield and productivity compared to the wild-type strain (Yuan et al. 2013b), and that the limiting factor of the process is yeast ethanol tolerance rather than the substrate hydrolysis rates. Thus, enhancement of inulinase expression in *Saccharomyces* was also examined. Integration of *K. marxianus* inulinase gene under the control of phosphoglycerate kinase promoter into *S. cerevisiae* resulted in a 2.5 fold increase in enzyme production and the transformant was able to produce 73 g/L ethanol from 300 g/L Jat tubers at a productivity of 1.51 g/L/h

Table 11.2 CBP approaches for the production of ethanol from Jerusalem artichoke by wild-type yeast species

Strain	Substrate pretreatment	Maximum ethanol, g/L	Maximum productivity, g/L/h	Reported conversions	Reference
K. marxianus ATCC8554	Untreated pulverized raw tubers from halophilic Jat	60.9	0.73	46.7 % of the dry raw substrate (91 % of theoretical)	Yuan et al. (2008)
K. marxianus PT-1	Untreated pulverized raw tubers	73.6	1.53	36.8 % of the dry raw substrate (90 % of theoretical)	Hu et al. (2012)
S. cerevisiae JZ1C	Untreated pulverized raw tubers	65.2	0.91	32.6 % of the dry raw substrate (79.7 % theoretical)	Hu et al. (2012)
S.cerevisiae DQ1	Untreated pulverized raw tubers	128.7	1.78	Helical ribbon stirring bioreactor, 36.7 % of the dry raw substrate (74 % of theoretical)	Guo et al. (2013a)
K. marxianus CBS1555	Acid pretreated stalks	45.3	1.51	0.252 g ethanol per g dry biomass or 0.32 g ethanol/g fermentable sugars	Kim and Kim (2014)
K. marxianus DBKKU Y-102	Untreated, grinded into mash, raw tubers	104.8	4.3	41.9 % of total sugars in the substrate	Charoensopharat et al. (2015)

(Yuan et al. 2013a). In a similar study, selection of the appropriate promoter and secretory signal sequence for the *K. marxianus* inulinase, resulted in a recombinant *Saccharomyces* strain that was capable in transforming inulin efficiently to ethanol in a bioreactor at a yield of 0.43 g ethanol per gram inulin and a productivity of 1.22 g/L/h (Hong et al. 2014).

11.3 Natural Biomass Degraders as Biofuel Producers

A Type I microorganism represents the second category of candidate biocatalysts for CBP applications. In principle, this approach appears as a more promising perspective; ethanologenesis in fungi requires the efficient function of only pyruvate decarboxylase and acetaldehyde dehydrogenase (Fig. 11.3), which compares favorably to the much larger number of cellulases and hemicellulases required for effective biomass hydrolysis in Type II microbial systems.

The ability to extensively degrade and assimilate residual plant biomass is among the most profound traits of filamentous fungi and certain aerobic and anaerobic bacteria. Each one of these microbial groups employs a different system for enzyme excretion that affects their potential in CBP applications. Some anaerobic bacteria and fungi produce an elaborate cellulosome structure that retains enzymes in close proximity to the cell surface (Bayer et al. 2004), in a strategy that provides significant benefits for survival in natural environments. Fungal strains producing cellulosomes have not yet studied in CBP processes (Olson et al. 2012) and till today, cellulosomal CBP applications involve mainly anaerobic bacteria of *Clostridium* species (Xu et al. 2010a; Jin et al. 2011; Yee et al. 2014; Papanek et al. 2015). Aerobic and facultative anaerobic fungi on the other hand, and especially ascomycetes and basidiomycetes, excrete a wide variety of free enzymes that completely decompose lignocellulose to its constituent monomers (Brink and Vries 2011), and present the most efficient natural enzyme secretion system. In certain cases such as *T. reesei* the enzyme secretion level can reach 100 g/L; at least an order of magnitude higher than the best performing bacteria (Nevalainen and Peterson 2014). In addition, a number of related fungal genera such as *Aspergillus*, *Fusarium*, *Monilia*, *Mucor*, *Rhizopus*, *Neurospora, Paecilomyces*, etc., are also facultative anaerobes capable of fermenting hexoses and pentoses to ethanol (Wu et al. 1986). As a result, several of these wild-type strains have been used in the direct microbial conversion of cellulosic substrates to ethanol since the 1980s (Singh et al. 1992).

11.3.1 Wild-Type Fungi as Type I CBP Microorganisms

The first single-vessel approach for ethanol production from lignocellulosic by-products dates back to 1975 when *Trichoderma viride* was grown on alkali pretreated barley straw, when co-cultured with either *Saccharomyces cerevisiae* or *Candida utilis* (Peitersen 1975). The first true CBP study involving a fungal species, appeared in 1981 when the saprophytic filamentous fungus *Monillia* sp. was shown to produce 16 g/L of ethanol when grown in a medium containing 5 % w/v microcrystalline cellulose as a sole carbon source (Gong et al. 1981). Since then, a substantial number of studies have appeared concerning the use wild-type fungal species for the production of ethanol in CBP systems using various artificial and natural lignocellulosic substrates. Most of these studies involve the genera

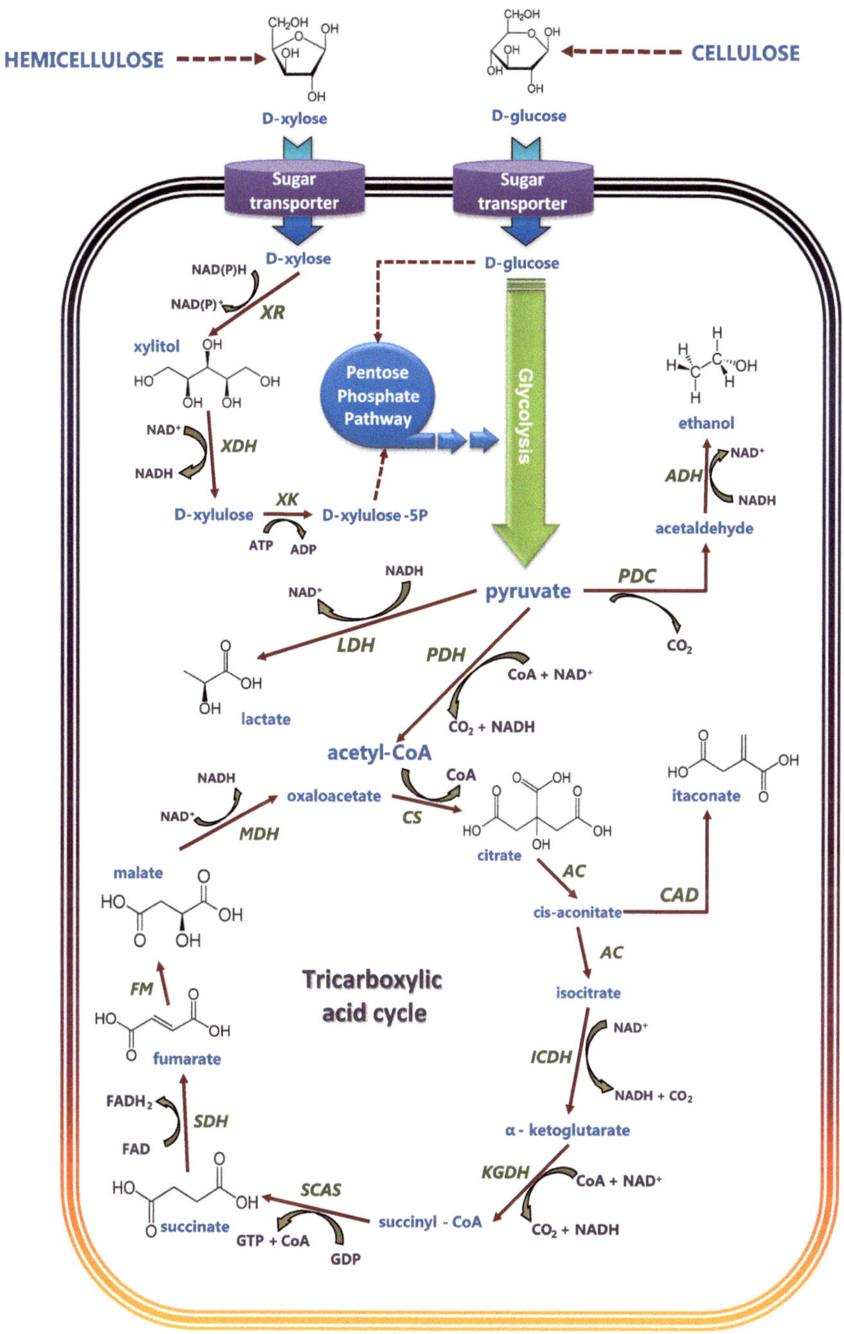
HEMICELLULOSE
CELLULOSE
D-xylose
D-glucose
Sugar transporter
Sugar transporter
D-xylose
D-glucose
NAD(P)H
NAD(P)+
XR
xylitol
Pentose Phosphate Pathway
Glycolysis
NAD+
XDH
NADH
D-xylulose
XK
D-xylulose-5P
ATP
ADP
ethanol
ADH
NAD+
NADH
acetaldehyde
NADH
NAD+
pyruvate
PDC
CO2
LDH
PDH
CoA + NAD+
CO2 + NADH
lactate
acetyl-CoA
NADH
CoA
NAD+
oxaloacetate
CS
MDH
citrate
itaconate
malate
AC
CAD
cis-aconitate
AC
isocitrate
Tricarboxylic acid cycle
FM
fumarate
NAD+
ICDH
NADH + CO2
FADH2
SDH
FAD
α - ketoglutarate
KGDH
CoA + NAD+
SCAS
succinyl - CoA
succinate
GTP + CoA
GDP
CO2 + NADH

◀ **Fig. 11.3** Distribution of potential CBP products covered in this chapter, within the general fungal metabolism (depicted by their structural formulas). Glucose and xylose are the main sugar monomers produced from the enzymatic hydrolysis of cellulose and hemicellulose, respectively. Following active entrance into the cell, glucose is mainly metabolized through glycolysis to pyruvate. The existence of *XR*, *XDH* and *XK* results in xylose catabolism to pyruvate through the pentose phosphate pathway. In certain genera, part of metabolized xylose is excreted following conversion to xylitol. Pyruvate fuels ethanol and/or lactic acid production under oxygen limitation, or the TCA cycle under aerobic conditions yielding organic acid products. Biosynthetic pathways for fatty acid esters and mycodiesel constituents are not depicted, but originate from Acetyl-CoA. Enzyme names: *AC* aconitase; *ADH* alcohol dehydrogenase; *CS* citrate synthase; *FM* fumarase; *ICDH* isocitrate dehydrogenase; *KGDH* α-ketoglutarate dehydrogenase, *LDH* lactate dehydrogenase; *MDH* malate dehydrogenase; PDC pyruvate decarboxylase; *PDH* pyruvate dehydrogenase; *SCAS* succinyl-CoA synthetase; *SDH* succinate dehydrogenase; *XDH* xylitol dehydrogenase; *XK* xylulose kinase; *XR* xylose reductase

Trichoderma, *Aspergillus*, *Fusarium,* and *Rhizopus* and are mainly focused on the optimization of process parameters, such as pH, temperature, substrate concentration, pretreatment, inoculum size, medium composition, etc. A number of reviews have been carried out on fungal CBP (Xu et al. 2009; Amore and Faraco 2012; Ali et al. 2016) and their conclusions can be summarized in the following general points:

1. Most fungal species used in these experiments produce a full array of biomass degrading enzymes at adequate quantities for hydrolysis when grown aerobically on various lignocellulosic substrates. However, some strains produce low levels of β-glucosidase and cellobiohydrolase.
2. The majority of these enzymes remain active and continue to function toward oligosaccharide production during the anaerobic ethanologenic part of the process.
3. Sugar transport does not seem to be a rate-limiting step for the overall process. In contrast, the various sugar transporters function at different sugar concentrations allowing the organism to assimilate sugars with a wide range of availability.
4. Although ethanol related genes are present and active, ethanol production rates are low and the fermentative part usually lasts for days.
5. Final ethanol titres—for wild-type Type I CBP fungi rarely exceed the 40 g/L barrier indicating poor ethanol tolerance of the corresponding strains.
6. Certain fungal species direct a significant amount of their carbon flow toward unwanted by-product formation (i.e., acetate).

The last three points constitute significant limitations for the implementation of an economically efficient Type I fungal CPB strain. To overcome these limitations some of the following approaches could be selected:

(1) Overexpression of certain hydrolases
(2) Knockouts of genes that lead to by-product formation
(3) Alteration of ethanol tolerance
(4) Increase inherent metabolic rates

The first two strategies are less demanding for genetic manipulations than the last two, since the latter involve major interventions in the fungal physiology without clearly identified gene targets (Stephanopoulos 2007). In addition, with the exception of *Aspergillus* spp., well-established molecular tools (e.g., efficient transformation protocols, lack of auxotrophies, well annotated genomes) are not yet widely available among Type I species.

11.3.2 Examples of Genetic Engineering Efforts to Improve Fungal Type I CBP Performance

11.3.2.1 *Trichoderma* and *Aspergillus* spp.

While being the most efficient cellulase and hemicellulase producers (Seidl and Seiboth 2010), *Trichoderma,* and *Aspergillus* species are also among the most problematic as far as ethanol tolerance and yields are concerned (Xu et al. 2009; Amore and Faraco 2012). As a result, genetic engineering efforts in these genera have mainly been focused in the homologous or heterologous overexpression of ligno-cellulose degrading enzymes, in order to optimize their performance as enzyme production platforms for SHF or SSF (Kück and Hoff 2010; Nevalainen and Peterson 2014). There is only one study on *T. reesei* CICC 40360 attempting to improve Type I CBP capabilities. In this case, genome shuffling resulted in improved ethanol production and a simultaneous enhancement of ethanol resistance. The shuffled strain efficiently converted lignocellulosic sugars to ethanol under aerobic conditions. Furthermore, it was able to produce up to 4 g/L ethanol directly from sugar-cane bagasse only after 120 h of fermentation (Huang et al. 2014).

11.3.2.2 *Fusarium* spp.

Fusarium spp has been the subject of extensive studies for the identification of metabolic bottlenecks that affect ethanol productivity in Type I CBP. The ability of wild-type members of this genus to directly produce ethanol when grown on cellulose or lignocellulosic substrates has been recognized more than 25 years ago (Christakopoulos et al. 1989; Singh and Kumar 1991). Since then, several studies targeting especially to *Fusarium oxysporum* strains, have verified that the diversity, catalytic performance and production levels of the hydrolases from wild-type *F. oxysporum* are similar to those of *Aspergillus* and *Trichoderma* species (Amore and Faraco 2012; Ali et al. 2016). Another advantage of *Fusarium* sp. compared to other fungi, is its relatively high tolerance to sugars and fermentation inhibitors such as acetate (Singh and Kumar 1991; Hennessy et al. 2013). Finally, glucose fermentation by *Fusarium* sp. is usually not inhibited by ethanol concentrations up to 40–50 g/L (Enari and Suihko 1983; Singh and Kumar 1991).

Almost all the work conducted so far for the development of efficient Type I CBP transformants of *F. oxysporum* has been based on two wild-type strains, namely F3 (Christakopoulos et al. 1989) and 11C (Ali et al. 2012), both selected for their relatively high ethanol production yields from lignocellulosic substrates. Early work on these and other wild-type Fusaria had already identified slow growth and fermentation rates, sugar transport limitations, as well as acetate accumulation as the major barriers in their exploitation in CBP processes (Ali et al. 2016).

Studies on intracellular metabolite profiles combined with principal component analysis in *F. oxysporum* F3, identified two metabolic bottlenecks related to the expression levels of phosphoglucomutase and transaldolase (Panagiotou et al. 2005a, c). The first was based on the observation that high levels of glucose-1,6-diphosphate (G-1,6-2P) were accumulated in *F. oxysporum* F3 cultures with glucose as carbon source. This was attributed to reduced activities of phosphoglucomutase (Kourtoglou et al. 2011), that prevented channeling of glucose toward cell wall biosynthesis, glycolysis and ethanol production (Panagiotou et al. 2005a). The second bottleneck was identified when the profile of the phosphorylated intermediates was studied under oxygen-limited conditions during *F. oxysporum* F3 fermentation of glucose–xylose mixtures (Panagiotou et al. 2005b). Due to the reduced transaldolase activity, accumulation of sedoheptulose-7-P (S7P) was observed indicating a block in the Pentose Phosphate Pathway (PPP). This resulted in an overall negative effect on the production of NADPH which finally caused an increased acetate production and reduced xylose consumption (Panagiotou et al. 2005b).

In an attempt to bypass the above limitations, both phosphoglucomutase and transaldolase were constitutively overexpressed in *F. oxysporum* F3. Under aerobic conditions, the transformant displayed on average 20 % higher specific growth rate upon growth on glucose and xylose and 30 % higher biomass yield on xylose (Anasontzis et al. 2014). When the cultures were shifted to the anaerobic stage, the transformant strain presented a 40 % increase in ethanol yield with a more than75 % reduction of acetate (Anasontzis et al. 2016). In a similar approach, the transaldolase gene from *Pichia stipitis* and *S. cerevisiae* were also constitutively overexpressed in a wild-type *F. oxysporum.* The transformant harboring the *P. stipitis* gene showed only 10 % improvement in ethanol yields from glucose and xylose (Fan et al. 2010) while the transformant carrying the *S. cerevisiae* gene showed an over 25 % improvement compared to the wild-type (Fan et al. 2011). The latter was also used in CBP cultures with pretreated rice straw, producing up to 5 g/L ethanol with a 25 % improvement against the wild-type.

Engineering of *F. oxysporum* F3 for higher CBP productivities on lignocellulosic substrates was also investigated in the direction of biomass degrading enzymes overproduction. For instance, endo-β-1,4-xylanase was incorporated into the *F. oxysporum* genome under constitutive regulation (Anasontzis et al. 2011). The obtained transformants presented high extracellular xylanase activities in cultures with glucose which is normally a repressor for the production of this enzyme. Additionally, these transformants produced approximately 60 % more ethanol compared to the wild-type on simple CBP systems using corn cob or wheat straw as carbon source (Anasontzis et al. 2011).

Other studies focused on the impact of sugar transporters in ethanol production and tolerance in *F. oxysporum* (Ali et al. 2013; Hennessy et al. 2013). Indeed, Ali et al. (2013), overexpressed a high affinity glucose transporter (Hxt) in *F. oxysporum* 11C and the resulted transformants showed very favorable profiles in lab-scale CBP cultures with alkali pretreated wheat straw. The overexpression also increased the xylose uptake rate resulting in enhanced overall fermentation rates.

Using suppression subtractive hybridisation (SSH), the performance of strain 11C was compared with *F. oxysporum* strain 7E—a strain characterized with comparatively lower CBP efficacy. In wheat straw/bran CBP cultures, over 200 transcripts were identified as being overexpressed in strain 11C, encoding for proteins involved in various cellular functions such as carbohydrate metabolism, energy, protein and sugar transport and detoxification (Ali et al. 2014). Post-translational silencing of three of these genes in *F. oxysporum* 11C significantly reduced the capacity of the fungus to produce ethanol from a straw/bran mix (Ali et al. 2014). These results pointed out the capacity of this experimental approach to identify target genes for genetic manipulation of *F. oxysporum* to enhance CBP performance.

11.3.2.3 Other Fungi

As mentioned above (Sect. 11.3.1), additional fungal genera with wild-type members of CBP potential have been proposed. These include facultative anaerobic fungi such as *Rhizopus*, *Paecilomyces,* and *Mucor* spp., several anaerobic fungal genera and some thermophilic fungi including *Thermoascus aurantiacus*, *Sporotrichum thermophile* and *Talaromyces emersonii*. However, their study almost exclusively involved the biochemistry and diversity of their biomass degrading enzymes (Amore and Faraco 2012; Herrera Bravo de Laguna et al. 2015). Although several of the above fungal genera are reported to produce ethanol from simple lignocellulosic sugars and hydrolysates (Millati et al. 2005; Wikandari et al. 2012) there are no reports on the direct ethanol production from biomass. A notable exception is *Paecilomyces variotii* ATHUM 8891. This strain was able to efficiently ferment glucose and xylose mixtures to ethanol without the distinct diauxic behavior of glucose preference. Furthermore, it possessed the necessary enzyme factory for the degradation of lignocellulosic biomass, as it was able to grow and produce ethanol from untreated agro-industrial derivatives such as corn cob and brewers spent grain (Zerva et al. 2012).

11.4 Consolidated Bioprocessing for Chemicals Other Than Ethanol

Although the vast majority of fungal applications in CBP systems have been targeted toward the production of ethanol, over the last decade the prospect of using biomass derived sugars for the production of other chemicals has also gained significant attention. This is quite reasonable, since ethanol is among the high volume commodity chemicals with low market prices while the diverse metabolic capability of CBP microorganisms, especially in fungi and yeasts, can certainly support the efficient biosynthesis of several compounds of higher economical value (Jäger and Büchs 2012). Indeed, the chemicals that can be derived from lignocellulosic biomass with the most favorable market prospects have already been identified by the US-DOE since 2004 and many of them can be produced through SSF or CBP approaches using fungi and yeasts (Aden et al. 2004). These include, among others, C3–C6—organic acids, amino acids, and sugar alcohols (Chen and Dou 2015; Mondala 2015) as shown in Fig. 11.3.

11.4.1 Organic Acids

11.4.1.1 Itaconic Acid

Itaconic acid, 2-methylidenebutanedioic acid, (IA) is a C5 dicarboxylic acid with an internal double bond, used in the industrial synthesis of polyesters, plastics, and artificial glass, as well as in the production of bioactive compounds for agricultural and medical applications (Okabe et al. 2009). It is produced naturally by a limited number of fungal species such as *Aspergillus terreus*, *Ustilago*, *Rhodotorula, and Candida* spp. (Okabe et al. 2009). Its current commercial biobased production centers on *A. terreus* that produces exploitable IA quantities (>50 g/L) in the presence of high initial sugar concentration, high dissolved oxygen concentrations and limiting concentrations of metal ions and nitrogen (Klement and Büchs 2013). Hexoses such as glucose and sucrose are the preferred substrates that are converted through glycolysis and the tricarboxylic acid cycle to cis-aconitate and finally to itaconic acid by the enzyme cis-aconitate decarboxylase (Mondala 2015) (Fig. 11.3). Since *A. terreus* is capable of producing a variety of biomass degrading enzymes upon growth on lignocellulosic substrates (Gao et al. 2008; de Siqueira et al. 2010; Narra et al. 2012), the feasibility of IA production from such feedstock in SSF or CPB modes was investigated. In this direction, substrates such as hardwood and softwood hydrolysates (or their hemicellulosic part) were utilized for the production of IA itaconic acid but the yields and productivities were not satisfactory due to the presence of inhibitory compounds from wood that negatively influenced the growth and production from *A. terreus* (Klement and Büchs 2013). Grass silage and its juices were also tested as substrates in IA production by *A. terreus* but the fungus appeared to be sensitive to certain components of the substrate that had to be removed during pretreatment (Ulber et al. 2010).

Ustilago maydis was also evaluated for IA production from enzymatically hydrolysed cellulose in seawater, salt-assisted organic acid hydrolyzed cellulose, or hemicellulose hydrolysed from fractionated beechwood (Klement et al. 2012). It was proven that *U. maydis* was able to produce significant amounts of IA in high osmotic media and real seawater. Despite the tolerance of the fungus to most impurities from pretreatment, high amounts of salts or residues of organic acids was shown to slow its growth and decreased IA production (Klement et al. 2012).

11.4.1.2 Fumaric Acid

Fumaric acid, (E)-2-butenedioic acid, (FA) is a C4 dicarboxylic acid that has been applied as a chemical intermediate for other platform chemicals such as succinic and maleic acids, and as a building block for the manufacture of polymers to be used as plasticizers, paper resins, alkyl resins, and other unsaturated polyester resins (Mondala 2015). Although all commercial FA is currently derived from the petrochemical industry, several zygomycetes (mainly *Rhizopus* sp.) and ascomycetes (*Aspergillus* and *Penicillium* spp.) have been reported to produce FA from simple hexoses in titers that sometimes exceed 100 g/L (Gangl et al. 1990; Riscaldati et al. 2000; Zhou et al. 2000). Several enzymatic hydrolysates of lignocellulosic substrates have been evaluated for FA production over the last 15 years. The hydrolysate of cassava bagasse was used as the sole carbon source to produce FA by submerged fermentation using several *Rhizopus* strains. Following extensive optimization, a strain of *R. formosa* was selected yielding 21.3 g/L of FA at the end of the fermentation (Carta et al. 1999). Somewhat higher final concentrations of FA (27.8 g/L) were obtained from enzymatically hydrolyzed acid pretreated corn straw by *Rhizopus oryzae*, previously grown in the xylose-rich acid treatment hydrolysates (Xu et al. 2010b). Recently, an approach based on metabolic profiling was employed for the *direct* microbial production of FA from acid pretreated wheat bran by *R. oryzae*. Supplementation of the growth medium with metabolites related to certain metabolic deficiencies, in combination with seed medium optimization, resulted in an almost 500 % increase in FA production (20.2 g/L) relative to the initial medium (Wang et al. 2015).

11.4.1.3 Malic Acid

Malic acid, hydroxybutanedioic acid, (MA) is another C4 dicarboxylic acid that can potentially be produced through second generation biorefinery approaches. It is a product with many applications in food and beverage industry. The standard chemical process for MA production takes place at high temperatures and pressures, yielding a racemic mixture of D- and L-malic acid, resulting in high production costs that hinder a more widespread use of L-malic acid (Chi et al. 2016). The current biochemical routes for MA production involve microbial fermentation using glucose and also starch hydrolysates by several *Aspergillus* and yeast species (Brown et al. 2013; Chi et al. 2016). There is only one reference for direct MA production from agricultural by-products concerning the successful evaluation of *A. niger* strains in MA production from thin stillage (West 2011). Recently, a *Rhizopus delemar* malate producing strain was isolated. Following a metabolic flux regulation approach, the strain was able to transform the C5 and C6 sugars from corn straw hydrolysate to MA at high productivity, reaching a final concentration of 120 g/L (Li et al. 2014).

11.4.1.4 Lactic Acid

Lactic acid, 2-hydroxypropanoic acid, (LA) is a C3 hydroxy-acid that finds many applications as an additive in food, cosmetic and pharmaceutical industries and as a building block for the production of biodegradable polymers (polylactate, PLA) (Vijayakumar et al. 2008). LA is produced in complex media by lactic acid bacteria through the homolactic fermentation, but the high medium costs has led to the exploitation of low cost substrates, such as lignocellulosic residues (Eiteman and Ramalingam 2015). Although several yeasts have been engineered to produce LA from simple sugars derived from starch or sugarcane, *Rhizopus oryzae* is the only fungal species reported to be able to convert renewable materials such as raw starch, waste paper, woody residues and corn cobs to LA in SSF-CBP resembling modes (Zhang et al. 2007; John et al. 2009). In the same framework, *R. oryzae* UMIP4.77 was shown to produce 24 g/L LA from partially hydrolyzed unbleached paper (Vially et al. 2009) while *R. oryzae* NLX-M-1 reached production levels of 60.3 g/L LA from corncob waste residue (Zhang et al. 2015b), both in SSF bioreactors.

Since Aspergilli are considered more versatile microbial cell factories than *Rhizopus*, the feasibility of lactic acid production with *Aspergillus* was also recently investigated. The enzyme NAD-dependent lactate dehydrogenase (LDH) that is responsible for lactic acid production by *Rhizopus oryzae* was overexpressed in *Aspergillus brasiliensis* BRFM103 (Liaud et al. 2015). The best lactic acid transformant from this study was able to produce up to 32.2 g/L lactic acid (conversion yield: 44 %, w/w) from glucose. More importantly though, was its ability to produce LA directly from starch and especially hemicellulose, although at a less than 20 % yield.

11.4.2 Other Compounds

Several additional compounds can potentially be produced through CBP by fungi and yeasts. Fatty Acid Esters (FAE), such as those used in biodiesel, are among those being proposed recently. Currently FAE are produced by the transesterification of plant or residual oils with an alcohol (methanol or ethanol) through acid, base or enzymatic catalysis. However, the high cost of this process combined with the use of plant oils for nonfood purposes made the search for alternative routes both economically and strategically attractive. In this direction the oleaginous yeast *Yarrowia lipolytica* shows a very promising perspective since it has recently been engineered to accumulate lipids over 90 % of its dry way at very high volumetric titers (Blazeck et al. 2014). In addition, β-glucosidase and cellodextrin transporter genes from *Neurospora crassa* where heterologously overexpressed in *Y. lipolytica* and the transformant was able to efficiently grow on cellobiose producing 0.37 g citrate per g substrate (Lane et al. 2015). Homologous overexpression of two endogenous β-glucosidases also yielded similar results (Guo et al. 2015), paving the way for further development in the application of *Y. lipolytica* in SSF of CBP schemes for FAE production from lignocellulosic substrates.

Sugar alcohols such as mannitol, sorbitol and xylitol are compounds of high market potential as food supplements and starting materials for the production of commodity chemicals (Park et al. 2016). Xylitol is of special interest as it can be produced from wild-type fungal species that naturally metabolize xylose, such as the yeast genera of *Candida* and *Scheffersomyces* (Fig. 11.3). Current commercial production of xylitol involves the catalytic hydrogenation of xylose at high temperatures and pressures (Park et al. 2016). These conditions increase production costs and have a negative environmental impact. As a result, microbial production of xylitol using lignocellulosic materials has been considered over the last decade, and a significant number of related studies have been comprehensively reviewed (de Albuquerque et al. 2014). In the majority of these studies no SSF or CBP approaches were employed, and xylitol production was always performed in SHF systems using enzymatically or chemically hydrolysed substrates. The first direct transformation of a lignocellulosic substrate to xylitol was accomplished by constructing a novel xylan conversion pathway in *Candida tropicalis.* A xylanase gene and a β-xylosidase gene from *A. terreus* were cloned and expressed in the xylose assimilating *C. tropicalis* BIT-Xol-1 strain (Guo et al. 2013b). The engineered yeast strain could simultaneously saccharify and ferment xylan to xylitol at yields of 77.1 and 66.9 % using xylan and corncob as substrates, respectively. In a very recent study, xylitol was directly produced from rice straw hemicellulosic oligosaccharides using a recombinant *S. cerevisiae* strain expressing cytosolic xylose reductase from *S. stipitis*, along with *A. aculeatus* β-glucosidase, *A. oryzae* β-xylosidase, and *T. reesei* xylanase (co-)displayed on the cell surface (Guirimand et al. 2016). The concerted action of these enzymes contributed to the consolidated bioprocessing (CBP) of the lignocellulosic hydrolysate to xylitol to produce 5.8 g/L xylitol with 79.5 % of theoretical yield. In addition, nanofiltration of the substrate provided removal of fermentation inhibitors increasing final xylitol concentration up to 38 g/L in the CBP system (Guirimand et al. 2016).

Recently, endophytic fungi have been identified as producers of hydrocarbons with promising fuel potential (Mycodiesel).[2] The compounds found in the volatile phases of the corresponding cultures include linear and branched alkanes, cyclic hydrocarbonds, alkyl alcohols and ketones, benzenes and polyaromatic hydrocarbons (Strobel 2015). The endophytic fungi involved in the biosynthesis are mostly ascomycete species such as *Muscodor albus* (Strobel 2011), *Ascocoryne sarcoides* (Gianoulis et al. 2012) and *Nodulisporium* sp. (Riyaz-Ul-Hassan et al. 2013). Of particular importance is that production of these compounds appears to be feasible also from lignocellulosic substrates (Mallette et al. 2014), thus representing a tremendous potential for further exploitation of lignocellulosic materials to completely new directions (Deneyer et al. 2015).

11.5 Conclusions and Future Prospects

Fueled only by solar energy, photosynthetically derived biomass is the only *truly renewable* source of reduced carbon available to our society. Geopolitical balances may currently keep oil prices low, but oil reserves will eventually be exhausted. More importantly, the deep concerns over the increased green house gas emissions from our "petrochemical economy" are pushing for modest growth rates based on renewable resources characterized by minimal carbon fingerprints (Kircher 2015). As a result, residual lignocellulosic biomass will inevitably become one of the future key raw materials for exploitation toward the production of basic platform chemicals and fuels.

In this direction, biobased processes are playing a pivotal role, as they provide for a diverse product range delivered in an environmental friendly and sustainable way. However, in a competitive and globalized economy, any novel production process has to compete with existing ones in productivity and profitability terms, in order to penetrate the corresponding market (Galbe et al. 2007). This is also true for any biotechnological approach and particularly for lignocellulose biorefineries, where substrate recalcitrance increases the production costs and reduces expected yields and productivities (Menon and Rao 2012). Among the available biorefinery technologies, consolidated bioprocessing (CBP) is considered as the one with the maximum cost reducing effect (Lynd et al. 2008) and extensive research has been conducted for its implementation, mainly toward ethanol production.

From the analysis described in the previous paragraphs it is clear that, despite the tremendous progress accomplished over the last two decades, the realization of an economically viable CBP process for lignocellulosic ethanol based on a single wild-type or engineered species is still under question. This finding applies to all

[2]Additional information on endophytic fungi as biofactories is presented in Chap. 14 *Unravelling the chemical interactions of fungal endophytes for exploitation as microbial factories.*

microbial species evaluated so far for CBP and not only to fungi that are the subject of this book (Olson et al. 2012; Kawaguchi et al. 2016).

In nature, biomass degradation to simple compounds is performed by microorganisms in order to cover their growth needs and not for the biosynthesis of specific products. In addition, in most environments, degradation and production are split among different organisms. Thus, it is very difficult to identify a wild-type strain tailor-made for CBP. Consequently, the construction of Type I or Type II CBP transformants through genetic engineering is and will continue to be the focus of current and future research in the field. Especially for Type II CBP, the combination of synthetic biology with classic molecular biology techniques is expected to expand our spectrum of genome editing tools for yeasts other than *S. cerevisiae*; for instance, the nonconventional yeast production systems of *Hansenula polymorpha*, *K. lactis*, *P. pastoris*, and *Y. lipolytica* (Wagner and Alper 2015). The recent innovations for targeted editing of microbial genomes through CRISPR-Cas systems are expected to have a key impact (Sander and Joung 2014). In our opinion, the large number of enzymes required for efficient biomass hydrolysis, the difficulties in regulation of their expression by the host strain and the attainment of trouble-free excretion from the cell envelope, makes the construction of a Type II CBP transformant a relatively distant target. On the other hand, Type I CBP transformants still lack the productivities and ethanol tolerance that are needed for their exploitation in biofuel production from cellulosic materials.

All the above considerations may lead to the conclusion that an efficient CBP process for bioethanol production from lignocellulose, based on a single fungal or bacterial transformant is perhaps a difficult goal for the near future. Mixed-cultures could represent a more feasible target as long as the single-vessel approach is maintained in order to achieve parallel hydrolysis and fermentation. At the moment there are a number of available fungal and bacterial strains tailored to optimally perform in each stage of an ethanol biorefinery. On the other hand, the preparation of a couple of additional seed cultures is not expected to increase costs to any substantial level. The compatibility of operating conditions among the different species is still an issue that has to be tackled, but there are several promising compatible degrader-fermenter combinations (Chen and Fu 2016; Choudhary et al. 2016).

Another important aspect that has to be taken into consideration when selecting CBP organisms is the specific characteristics of each potential substrate. Although plant biomass of various origin contains the same structural polymers (cellulose—hemicellulose—lignin), each lignocellulosic product is characterized by differences in their relative abundance and recalcitrance. In addition, the potential inhibitory compounds released from the various pretreatments also vary significantly among lignocellulosic substrates. Consequently, a substrate-oriented instead of the more commonly used product-oriented approach toward the selection of the microbial producers could at times be chosen to avoid the requirement for extensive metabolic engineering (Rumbold et al. 2009).

Finally and especially for fungal species, it is quite clear that there are several additional products in addition to ethanol, that could be produced from lignocellulosic materials through SSF or CBP (Kawaguchi et al. 2016). Targeting for those products, is likely more important from a practical stand point, as market ethanol prices are comparatively low. The examples presented in this chapter are only the first research attempts toward this direction and the continuous emergence of fungal metabolic wealth suggests that new candidate products are very likely to appear.

References

Aden AJ, Bozell HJ, White J, Manheim A (2004) Top value added chemicals from biomass. Volume I: results of screening for potential candidates from sugars and synthesis gas. In: Werpy T, Petersen G (eds) Pacific Northwest National Laboratory (PNNL), National Renewable Energy Laboratory (NREL) and Office of Biomass Program (EERE)

Adsul M, Singhvi M, Gaikaiwari S, Gokhale D (2011) Development of biocatalysts for production of commodity chemicals from lignocellulosic biomass. Bioresour Technol 102(6):4304–4312

Albergaria H, Arneborg N (2016) Dominance of *Saccharomyces cerevisiae* in alcoholic fermentation processes: role of physiological fitness and microbial interactions. Appl Microbiol Biotechnol 100(5):2035–2046

Ali SS, Khan M, Fagan B, Mullins E, Doohan FM (2012) Exploiting the inter-strain divergence of *Fusarium oxysporum* for microbial bioprocessing of lignocellulose to bioethanol. AMB Express 2(1):1–9

Ali SS, Nugent B, Mullins E, Doohan FM (2013) Insights from the fungus *Fusarium oxysporum* point to high affinity glucose transporters as targets for enhancing ethanol production from lignocellulose. PLoS ONE 8(1):e54701

Ali SS, Khan M, Mullins E, Doohan FM (2014) Identification of *Fusarium oxysporum* genes associated with lignocellulose bioconversion competency. Bioenergy Res 7(1):110–119

Ali SS, Nugent B, Mullins E, Doohan FM (2016) Fungal-mediated consolidated bioprocessing: the potential of *Fusarium oxysporum* for the lignocellulosic ethanol industry. AMB Express 6 (1):1–13

Alvira P, Tomás-Pejó E, Ballesteros M, Negro MJ (2010) Pretreatment technologies for an efficient bioethanol production process based on enzymatic hydrolysis: a review. Bioresour Technol 101(13):4851–4861

Amore A, Faraco V (2012) Potential of fungi as category I Consolidated BioProcessing organisms for cellulosic ethanol production. Renew Sustain Energy Rev 16(5):3286–3301

Anasontzis G, Zerva A, Stathopoulou P, Haralampidis K, Diallinas G, Karagouni A, Hatzinikolaou D (2011) Homologous overexpression of xylanase in *Fusarium oxysporum* increases ethanol productivity during consolidated bioprocessing (CBP) of lignocellulosics. J Biotechnol 152(1–2):16–23

Anasontzis GE, Kourtoglou E, Mamma D, Villas-Boâs SG, Hatzinikolaou DG, Christakopoulos P (2014) Constitutive homologous expression of phosphoglucomutase and transaldolase increases the metabolic flux of *Fusarium oxysporum*. Microb Cell Fact 13(1):1–13

Anasontzis GE, Kourtoglou E, Villas-Boas SG, Hatzinikolaou DG, Christakopoulos P (2016) Metabolic engineering of *Fusarium oxysporum* to improve its ethanol-producing capability. Front Microbiol 7:632

Ballesteros I, Ballesteros M, CabaÑas A, Carrasco J, MartÍn C, Negro MJ, Saez F, Saez R (1991) Selection of thermotolerant yeasts for simultaneous saccharification and fermentation (SSF) of cellulose to ethanol. Appl Biochem Biotechnol 28–29(1):307–315

Ballesteros M, Oliva JM, Negro MJ, Manzanares P, Ballesteros I (2004) Ethanol from lignocellulosic materials by a simultaneous saccharification and fermentation process (SFS) with *Kluyveromyces marxianus* CECT 10875. Process Biochem 39(12):1843–1848

Barron N, Marchant R, McHale L, McHale AP (1995) Studies on the use of a thermotolerant strain of *Kluyveromyces marxianus* in simultaneous saccharification and ethanol formation from cellulose. Appl Microbiol Biotechnol 43(3):518–520

Bayer EA, Belaich JP, Shoham Y, Lamed R (2004) The cellulosomes: multienzyme machines for degradation of plant cell wall polysaccharides. Ann Rev Microbiol 58:521–554

Blazeck J, Hill A, Liu L, Knight R, Miller J, Pan A, Otoupal P, Alper HS (2014) Harnessing *Yarrowia lipolytica* lipogenesis to create a platform for lipid and biofuel production. Nat Commun 5

Boyle M, Barron N, McHale AP (1997) Simultaneous saccharification and fermentation of straw to ethanol using the thermotolerant yeast strain *Kluyveromyces marxianus* imb3. Biotechnol Lett 19(1):49–51

Brink J, Vries RP (2011) Fungal enzyme sets for plant polysaccharide degradation. Appl Microbiol Biotechnol 91(6):1477–1492

Brown SH, Bashkirova L, Berka R, Chandler T, Doty T, McCall K, McCulloch M, McFarland S, Thompson S, Yaver D, Berry A (2013) Metabolic engineering of Aspergillus oryzae NRRL 3488 for increased production of l-malic acid. Appl Microbiol Biotechnol 97(20):8903–8912

Cai Z, Zhang B, Li Y (2012) Engineering *Saccharomyces cerevisiae* for efficient anaerobic xylose fermentation: reflections and perspectives. Biotechnol J 7(1):34–46

Carta FS, Soccol CR, Ramos LP, Fontana JD (1999) Production of fumaric acid by fermentation of enzymatic hydrolysates derived from cassava bagasse. Bioresour Technol 68(1):23–28

Chang J-J, Ho C-Y, Ho F-J, Tsai T-Y, Ke H-M, Wang C, Chen H-L, Shih M-C, Huang C-C, Li W-H (2012) PGASO: a synthetic biology tool for engineering a cellulolytic yeast. Biotechnol Biofuels 5(1):53

Chang J-J, Ho F-J, Ho C-Y, Wu Y-C, Hou Y-H, Huang C-C, Shih M-C, Li W-H (2013) Assembling a cellulase cocktail and a cellodextrin transporter into a yeast host for CBP ethanol production. Biotechnol Biofuels 6(1):19

Charoensopharat K, Thanonkeo P, Thanonkeo S, Yamada M (2015) Ethanol production from Jerusalem artichoke tubers at high temperature by newly isolated thermotolerant inulin-utilizing yeast *Kluyveromyces marxianus* using consolidated bioprocessing. Antonie Van Leeuwenhoek 108(1):173–190

Chen R, Dou J (2015) Biofuels and bio-based chemicals from lignocellulose: metabolic engineering strategies in strain development. Biotechnol Lett 38(2):213–221

Chen H, Fu X (2016) Industrial technologies for bioethanol production from lignocellulosic biomass. Renew Sustain Energy Rev 57:468–478

Chi Z, Wang Z-P, Wang G-Y, Khan I, Chi Z-M (2016) Microbial biosynthesis and secretion of l-malic acid and its applications. Crit Rev Biotechnol 36(1):99–107

Choudhary J, Singh S, Nain L (2016) Thermotolerant fermenting yeasts for simultaneous saccharification fermentation of lignocellulosic biomass. Electron J Biotechnol 21:82–92

Christakopoulos P, Macris BJ, Kekos D (1989) Direct fermentation of cellulose to ethanol by *Fusarium oxysporum*. Enzyme Microb Technol 11(4):236–239

Cragg SM, Beckham GT, Bruce NC, Bugg TDH, Distel DL, Dupree P, Etxabe AG, Goodell BS, Jellison J, McGeehan JE, McQueen-Mason SJ, Schnorr K, Walton PH, Watts JEM, Zimmer M (2015) Lignocellulose degradation mechanisms across the Tree of Life. Curr Opin Chem Biol 29:108–119

de Albuquerque TL, da Silva IJ Jr, de Macedo GR, Rocha MVP (2014) Biotechnological production of xylitol from lignocellulosic wastes: a review. Process Biochem 49(11): 1779–1789

de Siqueira FG, de Siqueira EG, Jaramillo PMD, Silveira MHL, Andreaus J, Couto FA, Batista LR, Filho EXF (2010) The potential of agro-industrial residues for production of holocellulase from filamentous fungi. Int Biodeterior Biodegradation 64(1):20–26

Della-Bianca BE, Basso TO, Stambuk BU, Basso LC, Gombert AK (2012) What do we know about the yeast strains from the Brazilian fuel ethanol industry? Appl Microbiol Biotechnol 97 (3):979–991

Den Haan R, Van Zyl WH (2003) Enhanced xylan degradation and utilisation by *Pichia stipitis* overproducing fungal xylanolytic enzymes. Enzyme Microb Technol 33(5):620–628

den Haan R, van Rensburg E, Rose SH, Görgens JF, van Zyl WH (2015) Progress and challenges in the engineering of non-cellulolytic microorganisms for consolidated bioprocessing. Curr Opin Biotechnol 33:32–38

Deneyer A, Renders T, Van Aelst J, Van den Bosch S, Gabriëls D, Sels BF (2015) Alkane production from biomass: chemo-, bio- and integrated catalytic approaches. Curr Opin Chem Biol 29:40–48

Dugar D, Stephanopoulos G (2011) Relative potential of biosynthetic pathways for biofuels and bio-based products. Nat Biotechnol 29(12):1074–1078

Eiteman MA, Ramalingam S (2015) Microbial production of lactic acid. Biotechnol Lett 37 (5):955–972

Elkins JG, Raman B, Keller M (2010) Engineered microbial systems for enhanced conversion of lignocellulosic biomass. Curr Opin Biotechnol 21(5):657–662

Enari TM, Suihko ML (1983) Ethanol production by fermentation of pentoses and hexoses from cellulosic materials. Crit Rev Biotechnol 1(3):229–240

Fan J-X, Yang Q, Liu Z-H, Huang X-M, Song J-Z, Chen Z-X, Sun Y, Liang Q, Wang S (2010) The characterization of transaldolase gene tal from *Pichia stipitis* and its heterologous expression in *Fusarium oxysporum*. Mol Biol Rep 38(3):1831–1840

Fan J-X, Yang X-X, Song J-Z, Huang X-M, Cheng Z-X, Yao L, Juba OS, Liang Q, Yang Q, Odeph M, Sun Y, Wang Y (2011) Heterologous expression of transaldolase gene Tal from *Saccharomyces cerevisiae* in *Fusarium oxysporum* for enhanced bioethanol production. Appl Biochem Biotechnol 164(7):1023–1036

Feng C, Zou S, Liu C, Yang H, Zhang K, Ma Y, Hong J, Zhang M (2016) Ethanol production from acid- and alkali-pretreated corncob by endoglucanase and β-glucosidase co-expressing *Saccharomyces cerevisiae* subject to the expression of heterologous genes and nutrition added. World J Microbiol Biotechnol 32(5):1–7

Fonseca GG, Heinzle E, Wittmann C, Gombert AK (2008) The yeast *Kluyveromyces marxianus* and its biotechnological potential. Appl Microbiol Biotechnol 79(3):339–354

Galbe M, Sassner P, Wingren A, Zacchi G (2007) Process engineering economics of bioethanol production. In: Olsson L (ed) Biofuels. Springer, Berlin Heidelberg, pp 303–327

Gangl IC, Weigand WA, Keller FA (1990) Economic comparison of calcium fumarate and sodium fumarate production by *Rhizopus arrhizus*. Appl Biochem Biotechnol 24–25(1):663–677

Gao J, Weng H, Zhu D, Yuan M, Guan F, Xi Y (2008) Production and characterization of cellulolytic enzymes from the thermoacidophilic fungal *Aspergillus terreus* M11 under solid-state cultivation of corn stover. Bioresour Technol 99(16):7623–7629

García-Aparicio MP, Oliva JM, Manzanares P, Ballesteros M, Ballesteros I, González A, Negro MJ (2011) Second-generation ethanol production from steam exploded barley straw by *Kluyveromyces marxianus* CECT 10875. Fuel 90(4):1624–1630

Gianoulis TA, Griffin MA, Spakowicz DJ, Dunican BF, Alpha CJ, Sboner A, Sismour AM, Kodira C, Egholm M, Church GM, Gerstein MB, Strobel SA (2012) Genomic analysis of the hydrocarbon-producing, cellulolytic, endophytic fungus ascocoryne sarcoides. PLoS Genet 8 (3):e1002558

Gong C-S, Maun CM, Tsao GT (1981) Direct fermentation of cellulose to ethanol by a cellulolytic filamentous fungus, *Monilia sp*. Biotechnol Lett 3(2):77–82

Guerriero G, Hausman J-F, Strauss J, Ertan H, Siddiqui KS (2015) Destructuring plant biomass: Focus on fungal and extremophilic cell wall hydrolases. Plant Sci 234:180–193

Guirimand G, Sasaki K, Inokuma K, Bamba T, Hasunuma T, Kondo A (2016) Cell surface engineering of *Saccharomyces cerevisiae* combined with membrane separation technology for xylitol production from rice straw hydrolysate. Appl Microbiol Biotechnol 100(8):3477–3487

Guo L, Zhang J, Hu F, Dy Ryu D, Bao J (2013a) Consolidated bioprocessing of highly concentrated jerusalem artichoke tubers for simultaneous saccharification and ethanol fermentation. Biotechnol Bioeng 110(10):2606–2615

Guo X, Zhang R, Li Z, Dai D, Li C, Zhou X (2013b) A novel pathway construction in *Candida tropicalis* for direct xylitol conversion from corncob xylan. Bioresour Technol 128:547–552

Guo Z, Duquesne S, Bozonnet S, Cioci G, Nicaud J-M, Marty A, O'Donohue MJ (2015) Development of cellobiose-degrading ability in *Yarrowia lipolytica* strain by overexpression of endogenous genes. Biotechnol Biofuels 8(1):1–16

Hahn-Hägerdal B, Galbe M, Gorwa-Grauslund MF, Lidén G, Zacchi G (2006) Bio-ethanol—the fuel of tomorrow from the residues of today. Trends Biotechnol 24(12):549–556

Harner NK, Wen X, Bajwa PK, Austin GD, Ho C-Y, Habash MB, Trevors JT, Lee H (2014) Genetic improvement of native xylose-fermenting yeasts for ethanol production. J Ind Microbiol Biotechnol 42(1):1–20

Hasunuma T, Ishii J, Kondo A (2015) Rational design and evolutional fine tuning of *Saccharomyces cerevisiae* for biomass breakdown. Curr Opin Chem Biol 29:1–9

Hennessy RC, Doohan F, Mullins E (2013) Generating phenotypic diversity in a fungal biocatalyst to investigate alcohol stress tolerance encountered during microbial cellulosic biofuel production. PLoS ONE 8(10):e77501

Herrera Bravo de Laguna I, Toledo Marante FJ, Mioso R (2015). Enzymes and bioproducts produced by the ascomycete fungus *Paecilomyces variotii*. J Appl Microbiol 119:1455–1466

Ho DP, Ngo HH, Guo W (2014) A mini review on renewable sources for biofuel. Bioresour Technol 169:742–749

Hong J, Wang Y, Kumagai H, Tamaki H (2007) Construction of thermotolerant yeast expressing thermostable cellulase genes. J Biotechnol 130(2):114–123

Hong S-J, Kim HJ, Kim J-W, Lee D-H, Seo J-H (2014) Optimizing promoters and secretory signal sequences for producing ethanol from inulin by recombinant *Saccharomyces cerevisiae* carrying *Kluyveromyces marxianus* inulinase. Bioprocess Biosyst Eng 38(2):263–272

Hu N, Yuan B, Sun J, Wang S-A, Li F-L (2012) Thermotolerant *Kluyveromyces marxianus* and *Saccharomyces cerevisiae* strains representing potentials for bioethanol production from Jerusalem artichoke by consolidated bioprocessing. Appl Microbiol Biotechnol 95(5): 1359–1368

Huang J, Chen D, Wei Y, Wang Q, Li Z, Chen Y, Huang R (2014) Direct Ethanol production from lignocellulosic sugars and sugarcane bagasse by a recombinant *Trichoderma reesei* strain HJ48. Sci World J 2014:8

Isroi R, Millati S, Syamsiah C, Niklasson MN, Cahyanto K Lundquist, Taherzadeh MJ (2011) Biological pretreatment of lignocelluloses with white-rot fungi and its applications: a review. BioResources 6(4):5224–5259

Jäger G, Büchs J (2012) Biocatalytic conversion of lignocellulose to platform chemicals. Biotechnol J 7(9):1122–1136

Jin M, Balan V, Gunawan C, Dale BE (2011) Consolidated bioprocessing (CBP) performance of *Clostridium phytofermentans* on AFEX-treated corn stover for ethanol production. Biotechnol Bioeng 108(6):1290–1297

John RP, Anisha GS, Nampoothiri KM, Pandey A (2009) Direct lactic acid fermentation: Focus on simultaneous saccharification and lactic acid production. Biotechnol Adv 27(2):145–152

Kawaguchi H, Hasunuma T, Ogino C, Kondo A (2016) Bioprocessing of bio-based chemicals produced from lignocellulosic feedstocks. Curr Opin Biotechnol 42:30–39

Kim S, Kim CH (2014) Evaluation of whole Jerusalem artichoke (*Helianthus tuberosus* L.) for consolidated bioprocessing ethanol production. Renew Energy 65:83–91

Kircher M (2015) Sustainability of biofuels and renewable chemicals production from biomass. Curr Opin Chem Biol 29:26–31

Klement T, Büchs J (2013) Itaconic acid—a biotechnological process in change. Bioresour Technol 135:422–431

Klement T, Milker S, Jäger G, Grande PM, Domínguez de María P, Büchs J (2012) Biomass pretreatment affects *Ustilago maydis* in producing itaconic acid. Microb Cell Fact 11(1):1–13

Kourtoglou E, Anasontzis GE, Mamma D, Topakas E, Hatzinikolaou DG, Christakopoulos P (2011) Constitutive expression, purification and characterization of a phosphoglucomutase from *Fusarium oxysporum*. Enzyme Microb Technol 48(3):217–224

Kricka W, Fitzpatrick J, Bond U (2014) Metabolic engineering of yeasts by heterologous enzyme production for degradation of cellulose and hemicellulose from biomass: a perspective. Front Microbiol 5

Kück U, Hoff B (2010) New tools for the genetic manipulation of filamentous fungi. Appl Microbiol Biotechnol 86(1):51–62

la Grange D, den Haan R, van Zyl W (2010) Engineering cellulolytic ability into bioprocessing organisms. Appl Microbiol Biotechnol 87(4):1195–1208

Lane S, Zhang S, Wei N, Rao C, Jin Y-S (2015) Development and physiological characterization of cellobiose-consuming *Yarrowia lipolytica*. Biotechnol Bioeng 112(5):1012–1022

Lara CA, Santos RO, Cadete RM, Ferreira C, Marques S, Gírio F, Oliveira ES, Rosa CA, Fonseca C (2014) Identification and characterisation of xylanolytic yeasts isolated from decaying wood and sugarcane bagasse in Brazil. Antonie van Leeuwenhoek Int J Gen Mol Microbiol 105(6):1107–1119

Lark N, Xia Y, Qin C-G, Gong CS, Tsao GT (1997) Production of ethanol from recycled paper sludge using cellulase and yeast, *Kluveromyces marxianus*. Biomass Bioenergy 12(2):135–143

Larran A, Jozami E, Vicario L, Feldman SR, Podestá FE, Permingeat HR (2015) Evaluation of biological pretreatments to increase the efficiency of the saccharification process using *Spartina argentinensis* as a biomass resource. Bioresour Technol 194:320–325

Lee H, Biely P, Latta RK, Barbosa MFS, Schneider H (1986) Utilization of xylan by yeasts and its conversion to ethanol by *Pichia stipitis* strains. Appl Environ Microbiol 52(2):320–324

Li X, Liu Y, Yang Y, Zhang H, Wang H, Wu Y, Zhang M, Sun T, Cheng J, Wu X, Pan L, Jiang S, Wu H (2014) High levels of malic acid production by the bioconversion of corn straw hydrolyte using an isolated *Rhizopus delemar* strain. Biotechnol Bioprocess Eng 19(3): 478–492

Liaud N, Rosso M-N, Fabre N, Crapart S, Herpoël-Gimbert I, Sigoillot J-C, Raouche S, Levasseur A (2015) L-lactic acid production by *Aspergillus brasiliensis* overexpressing the heterologous ldha gene from *Rhizopus oryzae*. Microb Cell Fact 14(1):1–9

Linger JG, Adney WS, Darzins A (2010) Heterologous expression and extracellular secretion of cellulolytic enzymes by *Zymomonas mobilis*. Appl Environ Microbiol 76(19):6360–6369

Long X-H, Shao H-B, Liu L, Liu L-P, Liu Z-P (2016) Jerusalem artichoke: a sustainable biomass feedstock for biorefinery. Renew Sustain Energy Rev 54:1382–1388

Luo Z, Bao J (2015) Secretive expression of heterologous β-glucosidase in *Zymomonas mobilis*. Bioresour Bioprocess 2(1):1–6

Lynd LR, van Zyl WH, McBride JE, Laser M (2005) Consolidated bioprocessing of cellulosic biomass: an update. Curr Opin Biotechnol 16(5): 577–583

Lynd LR, Laser MS, Bransby D, Dale BE, Davison B, Hamilton R, Himmel M, Keller M, McMillan JD, Sheehan J, Wyman CE (2008) How biotech can transform biofuels. Nat Biotechnol 26(2):169–172

Mallette N, Pankratz EM, Parker AE, Strobel GA, Busse SC, Carlson RP, Peyton BM (2014) Evaluation of cellulose as a substrate for hydrocarbon fuel production by *Ascocoryne sarcoides* (NRRL 50072). J Sustain Bioenergy Syst 4:33–49

Matsushika A, Inoue H, Kodaki T, Sawayama S (2009) Ethanol production from xylose in engineered *Saccharomyces cerevisiae* strains: current state and perspectives. Appl Microbiol Biotechnol 84(1):37–53

Matsuzaki C, Nakagawa A, Koyanagi T, Tanaka K, Minami H, Tamaki H, Katayama T, Yamamoto K, Kumagai H (2012) *Kluyveromyces marxianus*-based platform for direct ethanol fermentation and recovery from cellulosic materials under air-ventilated conditions. J Biosci Bioeng 113(5):604–607

Mazzoli R, Lamberti C, Pessione E (2012) Engineering new metabolic capabilities in bacteria: lessons from recombinant cellulolytic strategies. Trends Biotechnol 30(2):111–119

Menon V, Rao M (2012) Trends in bioconversion of lignocellulose: Biofuels, platform chemicals & biorefinery concept. Prog Energy Combust Sci 38(4):522–550

Millati R, Edebo L, Taherzadeh MJ (2005) Performance of *Rhizopus*, *Rhizomucor*, and *Mucor* in ethanol production from glucose, xylose, and wood hydrolyzates. Enzyme Microb Technol 36 (2–3):294–300

Mondala AH (2015) Direct fungal fermentation of lignocellulosic biomass into itaconic, fumaric, and malic acids: current and future prospects. J Ind Microbiol Biotechnol 42(4):487–506

Morosoli R, Zalce E, Durand S (1993) Secretion of a *Cryptococcus albidus* xylanase in *Pichia stipitis* resulting in a xylan fermenting transformant. Curr Genet 24(1):94–99

Moysés DN, Reis VCB, de Almeida JRM, de Moraes LMP, Torres FAG (2016) Xylose fermentation by *Saccharomyces cerevisiae*: Challenges and prospects. Int J Mol Sci 17(3)

Naik SN, Goud VV, Rout PK, Dalai AK (2010) Production of first and second generation biofuels: a comprehensive review. Renew Sustain Energy Rev 14(2):578–597

Narayanaswamy N, Dheeran P, Verma S, Kumar S (2013) Biological pretreatment of lignocellulosic biomass for enzymatic saccharification. In: Fang Z (ed) Pretreatment techniques for biofuels and biorefineries. Springer, Berlin Heidelberg, pp 3–34

Narra M, Dixit G, Divecha J, Madamwar D, Shah AR (2012) Production of cellulases by solid state fermentation with *Aspergillus terreus* and enzymatic hydrolysis of mild alkali-treated rice straw. Bioresour Technol 121:355–361

Narron RH, Kim H, Chang H-M, Jameel H, Park S (2016) Biomass pretreatments capable of enabling lignin valorization in a biorefinery process. Curr Opin Biotechnol 38:39–46

Nevalainen H, Peterson R (2014) Making recombinant proteins in filamentous fungi- are we expecting too much? Front Microbiol 5

Okabe M, Lies D, Kanamasa S, Park E (2009) Biotechnological production of itaconic acid and its biosynthesis in *Aspergillus terreus*. Appl Microbiol Biotechnol 84(4):597–606

Olson DG, McBride JE, Joe Shaw A, Lynd LR (2012) Recent progress in consolidated bioprocessing. Curr Opin Biotechnol 23(3):396–405

Panagiotou G, Christakopoulos P, Olsson L (2005a) The influence of different cultivation conditions on the metabolome of *Fusarium oxysporum*. J Biotechnol 118(3):304–315

Panagiotou G, Christakopoulos P, Villas-Boas SG, Olsson L (2005b) Fermentation performance and intracellular metabolite profiling of *Fusarium oxysporum* cultivated on a glucose–xylose mixture. Enzyme Microb Technol 36(1):100–106

Panagiotou G, Villas-Bôas SG, Christakopoulos P, Nielsen J, Olsson L (2005c) Intracellular metabolite profiling of *Fusarium oxysporum* converting glucose to ethanol. J Biotechnol 115 (4):425–434

Papanek B, Biswas R, Rydzak T, Guss AM (2015) Elimination of metabolic pathways to all traditional fermentation products increases ethanol yields in *Clostridium thermocellum*. Metab Eng 32:49–54

Park Y-C, Oh EJ, Jo J-H, Jin Y-S, Seo J-H (2016) Recent advances in biological production of sugar alcohols. Curr Opin Biotechnol 37:105–113

Peitersen N (1975) Cellulase and protein production from mixed cultures of *Trichoderma viride* and a yeast. Biotechnol Bioeng 17(9):1291–1299

Pessani NK, Atiyeh HK, Wilkins MR, Bellmer DD, Banat IM (2011) Simultaneous saccharification and fermentation of Kanlow switchgrass by thermotolerant *Kluyveromyces marxianus* IMB3: the effect of enzyme loading, temperature and higher solid loadings. Bioresour Technol 102(22):10618–10624

Puseenam A, Tanapongpipat S, Roongsawang N (2015) Co-expression of endoxylanase and endoglucanase in *Scheffersomyces stipitis* and its application in ethanol production. Appl Biochem Biotechnol 177(8):1690–1700

Rabemanolontsoa H, Saka S (2016) Various pretreatments of lignocellulosics. Bioresour Technol 199:83–91

Riscaldati E, Moresi M, Federici F, Petruccioli M (2000) Direct ammonium fumarate production by *Rhizopus arrhizus* under phosphorous limitation. Biotechnol Lett 22(13):1043–1047

Riyaz-Ul-Hassan S, Strobel G, Geary B, Sears J (2013) An endophytic *Nodulisporium* sp. from central America producing volatile organic compounds with both biological and fuel potential. J Microbiol Biotechnol 23(1):29–35

Rowlands WN, Masters A, Maschmeyer T (2008) The biorefinery—challenges, opportunities, and an Australian perspective. Bull Sci Technol Soc 28(2):149–158

Rumbold K, van Buijsen H, Overkamp K, van Groenestijn J, Punt P, Werf M (2009) Microbial production host selection for converting second-generation feedstocks into bioproducts. Microb Cell Fact 8(1):64

Saha BC, Qureshi N, Kennedy GJ, Cotta MA (2016) Biological pretreatment of corn stover with white-rot fungus for improved enzymatic hydrolysis. Int Biodeterior Biodegradation 109: 29–35

Sànchez Nogué V, Karhumaa K (2015) Xylose fermentation as a challenge for commercialization of lignocellulosic fuels and chemicals. Biotechnol Lett 37(4):761–772

Sander JD, Joung JK (2014) CRISPR-Cas systems for editing, regulating and targeting genomes. Nat Biotechnol 32(4):347–355

Seidl V, Seiboth B (2010) *Trichoderma reesei*: genetic approaches to improving strain efficiency. Biofuels 1(2):343–354

Sims RE, Mabee W, Saddler JN, Taylor M (2010) An overview of second generation biofuel technologies. Bioresour Technol 101(6):1570–1580

Sindhu R, Binod P, Pandey A (2016) Biological pretreatment of lignocellulosic biomass—an overview. Bioresour Technol 199:76–82

Singh A, Kumar P (1991) *Fusarium oxysporum*: status in bioethanol production. Crit Rev Biotechnol 11(2):129–147

Singh A, Kumar P, Schügerl K (1992). Bioconversion of cellulosic materials to ethanol by filamentous fungi. Enzymes and products from bacteria fungi and plant cells, vol 45. Springer, Berlin Heidelberg, pp 29–55

Stephanopoulos G (2007) Challenges in engineering microbes for biofuels production. Science 315(5813):801–804

Strobel G (2011) *Muscodor* species- endophytes with biological promise. Phytochem Rev 10 (2):165–172

Strobel GA (2015) Bioprospecting—fuels from fungi. Biotechnol Lett 37(5):973–982

Suryawati L, Wilkins MR, Bellmer DD, Huhnke RL, Maness NO, Banat IM (2008) Simultaneous saccharification and fermentation of Kanlow switchgrass pretreated by hydrothermolysis using *Kluyveromyces marxianus* IMB4. Biotechnol Bioeng 101(5):894–902

Szczodrak J (1988) Enzymatic hydrolysis and fermentation of pretreated wheat straw to ethanol. Biotechnol Bioeng 32(6):771–776

Szczodrak J (1989) Use of cellulases from a β-glucosidase-hyperproducing mutant of *Trichoderma reesei* in simultaneous saccharification and fermentation of wheat straw. Biotechnol Bioeng 33 (9):1112–1116

Tomás-Pejó E, Oliva JM, González A, Ballesteros I, Ballesteros M (2009) Bioethanol production from wheat straw by the thermotolerant yeast *Kluyveromyces marxianus* CECT 10875 in a simultaneous saccharification and fermentation fed-batch process. Fuel 88(11):2142–2147

Ulber R, Sieker T, Tippkötter N, Bart HJ, Dimitrova D, Heinzle E, Neuner A (2010) Grassilage als Rohstoff für die chemische Industrie. Chem Ing Tech 82(8):1153–1159

Van Dyk JS, Pletschke BI (2012) A review of lignocellulose bioconversion using enzymatic hydrolysis and synergistic cooperation between enzymes—factors affecting enzymes, conversion and synergy. Biotechnol Adv 30(6):1458–1480

van Zyl W, Lynd L, den Haan R, McBride J (2007) Consolidated bioprocessing for bioethanol production using *Saccharomyces cerevisiae*. In: Olsson L (ed) Biofuels, vol 108. Springer, Berlin Heidelberg, pp 205–235

Vially G, Marchal R, Guilbert N (2009) L(+) Lactate production from carbohydrates and lignocellulosic materials by *Rhizopus oryzae* UMIP 4.77. World J Microbiol Biotechnol 26 (4):607–614

Vijayakumar J, Aravindan R, Viruthagiri T (2008) Recent trends in the production, purification and application of lactic acid. Chem Biochem Eng Q 22(2):245–264

Vinuselvi P, Lee SK (2012) Engineered *Escherichia coli* capable of co-utilization of cellobiose and xylose. Enzyme Microb Technol 50(1):1–4

Wagner JM, Alper HS (2015) Synthetic biology and molecular genetics in non-conventional yeasts: current tools and future advances. Fungal Genet Biol

Wang G, Huang D, Li Y, Wen J, Jia X (2015) A metabolic-based approach to improve xylose utilization for fumaric acid production from acid pretreated wheat bran by *Rhizopus oryzae*. Bioresour Technol 180:119–127

West TP (2011) Malic acid production from thin stillage by *Aspergillus* species. Biotechnol Lett 33(12):2463–2467

Wikandari R, Millati R, Lennartsson PR, Harmayani E, Taherzadeh MJ (2012) Isolation and characterization of zygomycetes fungi from tempe for ethanol production and biomass applications. Appl Biochem Biotechnol 167(6):1501–1512

Wu JF, Lastick SM, Updegraff DM (1986) Ethanol production from sugars derived from plant biomass by a novel fungus. Nature 321:887–888

Xu Q, Singh A, Himmel ME (2009) Perspectives and new directions for the production of bioethanol using consolidated bioprocessing of lignocellulose. Curr Opin Biotechnol 20 (3):364–371

Xu C, Qin Y, Li Y, Ji Y, Huang J, Song H, Xu J (2010a) Factors influencing cellulosome activity in consolidated bioprocessing of cellulosic ethanol. Bioresour Technol 101(24):9560–9569

Xu Q, Li S, Fu Y, Tai C, Huang H (2010b) Two-stage utilization of corn straw by *Rhizopus oryzae* for fumaric acid production. Bioresour Technol 101(15):6262–6264

Yee KL, Rodriguez M Jr, Thompson OA, Fu C, Wang Z-Y, Davison BH, Mielenz JR (2014) Consolidated bioprocessing of transgenic switchgrass by an engineered and evolved *Clostridium thermocellum* strain. Biotechnol Biofuels 7(1):1–6

Yuan WJ, Zhao XQ, Ge XM, Bai FW (2008) Ethanol fermentation with *Kluyveromyces marxianus* from Jerusalem artichoke grown in salina and irrigated with a mixture of seawater and freshwater. J Appl Microbiol 105(6):2076–2083

Yuan WJ, Chang BL, Ren JG, Liu JP, Bai FW, Li YY (2012) Consolidated bioprocessing strategy for ethanol production from Jerusalem artichoke tubers by *Kluyveromyces marxianus* under high gravity conditions. J Appl Microbiol 112(1):38–44

Yuan W-J, Li N-N, Zhao X-Q, Chen L-J, Kong L, Bai F-W (2013a) Engineering an industrial *Saccharomyces cerevisiae* strain with the inulinase gene for more efficient ethanol production from Jerusalem artichoke tubers. Eng Life Sci 13(5):472–478

Yuan W, Zhao X, Chen L, Bai F (2013b) Improved ethanol production in Jerusalem artichoke tubers by overexpression of inulinase gene in *Kluyveromyces marxianus*. Biotechnol Bioprocess Eng 18(4):721–727

Zerva AP, Stathopoulou PM, Katsifas EA, Karagouni AD, Hatzinikolaou DG (2012) The filamentous fungus *Paecilomyces variotii* as a potential candidate for bioethanol production via consolidated bioprocessing of lignocellulosics. New Biotechnol 29S:S5

Zhang ZY, Jin B, Kelly JM (2007) Production of lactic acid from renewable materials by *Rhizopus* fungi. Biochem Eng J 35(3):251–263

Zhang GC, Liu JJ, Kong II, Kwak S, Jin YS (2015a) Combining C6 and C5 sugar metabolism for enhancing microbial bioconversion. Curr Opin Chem Biol 29:49–57

Zhang L, Li X, Yong Q, Yang S-T, Ouyang J, Yu S (2015b) Simultaneous saccharification and fermentation of xylo-oligosaccharides manufacturing waste residue for l-lactic acid production by *Rhizopus oryzae*. Biochem Eng J 94:92–99

Zhou Y, Du J, Tsao GT (2000) Mycelial pellet formation by *Rhizopus oryzae* ATCC 20344. Appl Biochem and Biotechnol Part A Enzyme Eng Biotechnol 84–86:779–789

Chapter 12
Fungal Enzymes and Their Role in Bioenergy

Cristiane Sanchez Farinas and Fernanda Marisa da Cunha

12.1 Introduction

The use of enzymes in different industrial sectors as an alternative to conventional chemical processes is currently increasing as a result of the technico-economic and environmental advantages inherent to such biocatalysts. The use of enzymes can contribute to energy reduction because enzymes act under mild temperature and pressure conditions. Reduction of the downstream processing steps is expected as well, since the formation of undesirable byproducts is reduced due to the high selectivity and specificity of enzymes. Moreover, when the enzymes can be immobilized onto solid supports it is possible to reuse them, which can significantly contribute to the cost reduction of the overall process. These characteristics make the use of enzymes in different sectors very favorable in comparison to the use of chemical catalysts, including the bioenergy sector.

Developments in the bioenergy sector have been focusing on the use of biotechnological routes employing enzymes such as amylases, cellulases, and lipases in the production process of bioethanol and biodiesel in an attempt to meet the growing demand of the biofuels industry in a sustainable way. The industrial use of amylolytic enzymes in the process for bioethanol production from amylaceous sources, such as corn, is a very well-established technology. However, to date the industrial use of cellulases and lipases for biofuels production is still under development, with only a few large-scale industrial plants in operation.

C.S. Farinas (✉) · F.M. da Cunha
Embrapa Instrumentation, Rua XV de Novembro 1452, São Carlos, SP 13560-970, Brazil
e-mail: cristiane.farinas@embrapa.br

C.S. Farinas
Graduate Program of Chemical Engineering, Federal University of São Carlos, São Carlos, SP 13565-905, Brazil

D. Purchase (ed.), *Fungal Applications in Sustainable Environmental Biotechnology*, Fungal Biology, DOI 10.1007/978-3-319-42852-9_12

Cellulolytic enzymes are used to hydrolyze the cellulose present in lignocellulosic materials, such as agroindustrial and forest residues, into glucose that can be converted via fermentation into bioethanol, the so-called cellulosic ethanol, or second generation (2G) ethanol. In the enzymatic conversion of lignocellulosic biomass, the carbohydrate structures are advantageously hydrolyzed into monomeric sugars and such process is being considered as a key technology in future biorefineries. On the other hand, lipases are the enzymes that catalyze the transesterification reaction of triacylglycerides into fatty acid alkyl esters, which constitute the biodiesel. The main advantage of the use of enzymatic route for biodiesel production is related to the wider variety of oils or fats, either vegetable or animal origin, virgin or recycled, that can be processed into biodiesel (Hasan et al. 2006; Singh and Mukhopadhyay 2012; Hwang et al. 2014).

The main obstacle preventing the wider application of cellulase and lipase enzymes in the industrial production of biofuels is the high cost of these enzymes. Technical challenges related to the characteristics of the enzymes such as high specificity and stability also need to be improved. Besides, for cellulolytic enzymes, the quality of enzymatic complexes, in terms of their composition, is also an important issue, since cocktails containing cellulases, hemicellulases, pectinases, and other accessory enzymes, acting in synergy in the degradation process, are often necessary due to the high recalcitrance of plant biomass. Although some enzymes can be obtained from vegetal sources, the main sources of industrial enzymes are microbial, including bacteria and fungi. Aerobic fungi, especially, are recognized for their high rates of growth and protein secretion (de Vries 2003; Polizeli et al. 2005; Sukumaran et al. 2005; Guimaraes et al. 2006; van den Brink and de Vries 2011; Andersen et al. 2012; Singh and Mukhopadhyay 2012; Correa et al. 2014; Guerriero et al. 2015).

Microbial cultivation processes for the industrial production of enzymes can be conducted using a liquid medium (submerged fermentation, SmF) or a solid medium (solid-state fermentation, SSF). In the latter case, the cultivation employs a solid substrate with sufficient moisture only for maintenance of growth and metabolism of the microorganism (in other words, there is no free water). In SmF, on the other hand, the medium essentially consists of water containing dissolved nutrients. Submerged cultivation techniques have advantages related to instrumentation and process control, and are widely used for the production of industrial enzymes and other bioproducts (Singhania et al. 2010; Delabona et al. 2013a; Cunha et al. 2015). SSF can be particularly advantageous for the cultivation of filamentous fungi, because it simulates the natural habitat of the microorganisms leading to higher enzymatic productivity, compared to SmF processes. From the environmental perspective, an important advantage of SSF is the ability to use solid substrates consisting of agroindustrial residues that serve as sources of carbon and energy for microorganism growth and enzyme production. Nevertheless, certain operational limitations of SSF, such as difficulty in controlling the moisture level of the substrate and avoiding heat build-up, have held back its wider implementation in large-scale industrial processes (Pandey et al. 1999; Holker et al. 2004; Farinas et al. 2011; Thomas et al. 2013; Farinas 2015).

Therefore, there are several technical and economic issues regarding the implementation of the use of enzymes in the bioenergy sector that must be overcome for the biotechnological route to become feasible in large-scale industrial process. This chapter will address this issue by presenting an overview about cellulases and the application of these fungal enzymes in the bioenergy sector. Recent developments in the production process of cellulases and application of fungal enzymes in bioenergy are also presented.

12.2 Enzymes in Bioenergy: Cellulases

Cellulase enzymes comprise a set of glycoside hydrolases whose action involves hydrolysis of the β-1,4-glycosidic bonds of cellulose, the major polymer present in biomass. These enzymes show synergistic action during degradation of the polymeric cellulose chain. The most widely accepted mechanism of action of cellulases involves three classes of enzymes: endoglucanases, exoglucanases, and β-glucosidases. Endoglucanases hydrolyze accessible intramolecular β-1,4-glycosidic bonds of the cellulose chains randomly, producing new chain ends; exoglucanases progressively cleave cellulose chains at the ends to release soluble cellobiose or glucose; and β-glucosidases hydrolyze cellobiose to glucose (Zhang et al. 2006).

In addition to the breakdown of cellulose, deconstruction of the hemicellulose fraction to the constituent sugars is essential for the efficient production of biofuels and other chemicals from plant biomass (Dodd and Cann 2009). The use of hemicellulases and other auxiliary enzymes, in conjunction with cellulolytic enzymes, can improve cellulose conversion by removing hemicellulose and increasing the access of cellulases to the substrate (Gao et al. 2011). The endo-1,4-β-xylanase (xylanase) enzymes cleave the β-1,4-glycosidic linkage between xylose residues in the backbone of xylans. This is one of the most important enzymatic activities required for depolymerization of the hemicellulosic constituent of plant cell walls (Dodd and Cann 2009), because xylan is the most abundant hemicellulose (Saha 2003). The resulting oligosaccharides can be further degraded to xylose by the action of β-xylosidases.

Furthermore, the recent discovery of the role of lytic polysaccharide monooxygenases (LPMO) and other accessory proteins in increasing the degradation of cellulose has resulted in the inclusion of a new category in the CAZy database, called "auxiliary activities" (AA), which includes a group of catalytic modules involved in the degradation of plant cell walls (Levasseur et al. 2013). The discovery of these enzymes represents a revolution in the enzymatic processing of biomass and suggests a new paradigm for the enzymatic degradation of cellulose, in which the action of the classical hydrolytic cellulases is facilitated by the lytic action of the polysaccharide monooxygenases (Horn et al. 2012).

Several filamentous fungi have been used for the industrial production of cellulolytic cocktails. *Trichoderma* and *Aspergillus* strains are considered the workhorses, presenting good fermentation characteristics, such as high protein secretion rates and ability to produce a wide range of extracellular enzymes (Singhania et al. 2010; Farinas 2015). Although most industrial enzymes are produced by microbial cultivation under SmF, often using genetically modified microorganisms, most of these enzymes could be produced by SSF using wild-type microorganisms (Holker et al. 2004). Some of the most important aspects of SmF and SSF bioprocess engineering for biomass-degrading enzymes production by filamentous fungi are discussed in the following sections.

12.3 Recent Advances on the Production Process of Fungal Enzymes

Microbial cultivation processes for the industrial production of fungal enzymes can be conducted using a liquid medium (submerged fermentation, SmF) or a solid medium (solid-state fermentation, SSF). There have been several recent reports on the production of cellulolytic enzymes using SmF and SSF under different process conditions (Delabona et al. 2013b; Rodriguez-Zuniga et al. 2013; Vasconcellos et al. 2015). Microbial cultivations for cellulolytic enzymes production under SmF in a stirred tank bioreactor and in SSF in a column-type bioreactor are shown in Fig. 12.1.

The cultivation of microorganisms under submerged fermentation (SmF) involves the inoculation and growth directly into a liquid nutrient medium. In the case of filamentous fungi, the cultivation can be initiated by either inoculating the spores or the mycelium produced in a pre-cultivation step. This type of fermentation is used industrially for the production of enzymes, being a well-consolidated process in terms of operation and control. The production of enzymes by filamentous fungi under SmF is affected by several operational process parameters such as the pH and temperature of the medium, oxygen supply, cultivation medium, type of bioreactor, and mode of operation whether batch, fed-batch or continuous (Ahamed and Vermette 2009, 2010; Michelin et al. 2011; Cunha et al. 2015).

In solid-state fermentation (SSF), the microbial cultivation takes place on a solid substrate with sufficient moisture only to maintain growth and metabolism of the microorganism (in the absence of free water). SSF can be particularly advantageous for the cultivation of filamentous fungi, because it simulates the natural habitat of the microorganisms, leading to higher enzymatic productivity, compared to SmF processes (Pandey et al. 1999; Holker and Lenz 2005; Farinas 2015). The efficiency of SSF processes for obtaining the desired products is dependent on the microorganism used as well as the operational and environmental conditions. Factors such as temperature, pH, moisture content, aeration, nutrient concentrations, and the nature of the substrate are among the key variables that influence SSF processes. It is also important to carefully consider the choice of bioreactor design.

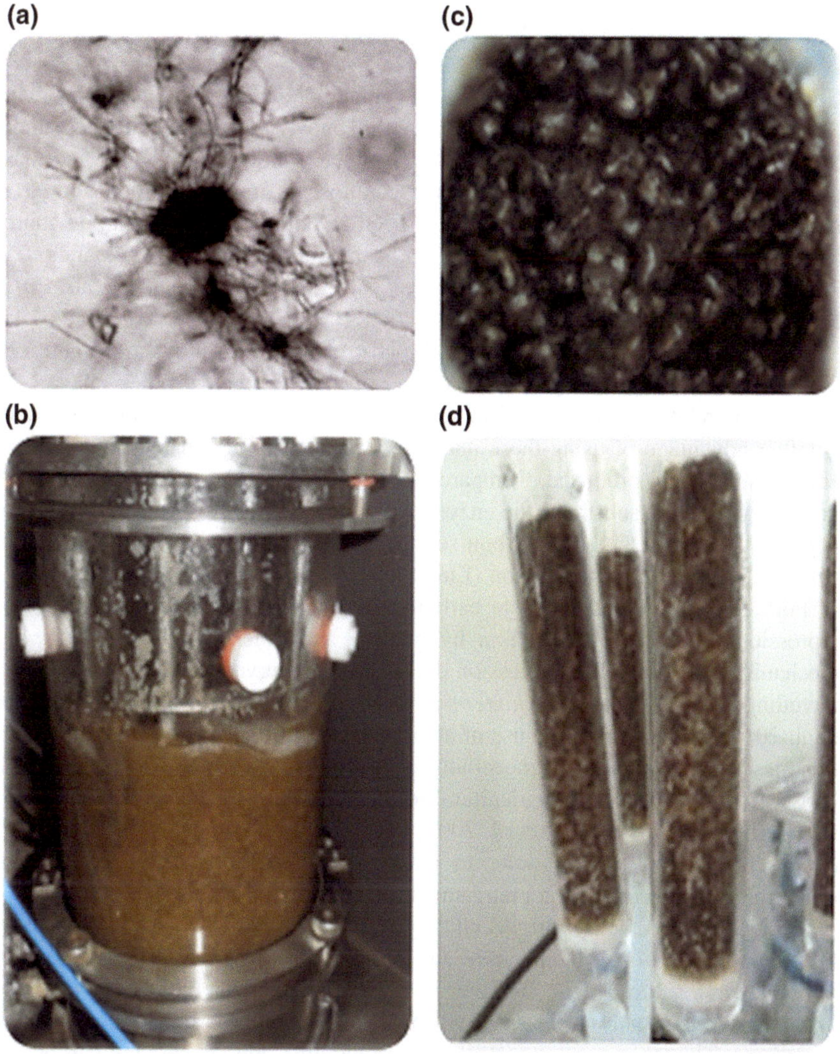

Fig. 12.1 Pictures of *Aspergillus niger* (A12) when cultivated for enzyme production in submerged fermentation (SmF) and solid-state fermentation (SSF). **a** Predominant morphology of *Aspergillus niger* after 12 h of cultivation in a stirred tank bioreactor using 1 % w/w of sugarcane bagasse as inducer substrate (400-fold magnification). **b** Cultivation broth after 12 h of SmF in a 4 L stirred tank bioreactor. **c** Spores of *Aspergillus niger* observed after 72 h of fermentation in wheat bran as solid inducer substrate. **d** Colum-type bioreactor used for enzyme production in SSF

Recent developments concerning the effects of operational and environmental conditions on cellulolytic enzymes production by different fungal strains cultivated under SmF and SSF are discussed below.

12.3.1 Production of Cellulases by SmF

Currently, most commercial cellulolytic enzymatic preparations are produced using filamentous fungi of the genera *Trichoderma* and *Aspergillus,* cultivated under SmF. Microbial cellulolytic enzymes are considered inducible enzymes and maximum yields are usually obtained when cellulose is used as carbon source or inducer. In fact, cellulosic materials have been widely used as both carbon source and as an inducer for cellulase production by filamentous fungi (Sukumaran et al. 2009; Singhania et al. 2010; Cunha et al. 2012; Vitcosque et al. 2012; Rodriguez-Zuniga et al. 2013).

As an effective approach to reduce the cost of cellulases production, pure cellulosic substrates commonly used as inducer of the enzymatic synthesis have been replaced by relatively cheaper substrates, such as residual lignocellulosic materials. Another advantage of using these materials is related to the characteristics of the enzymatic cocktail produced. Compared to mono or disaccharides, when fungi are grown on complex substrates the enzymes produced are more specifically divert to degrade that biomass, as the carbon source may affect not only the production of cellulase, but also its composition (Liu et al. 2013; van den Brink et al. 2014).

Furthermore, although soluble carbohydrates may produce faster growth, the repression of cellulase production by glucose and other soluble sugars by the mechanism of catabolite repression have been observed in cellulolytic fungi including *Aspergillus* and *Trichoderma* (Ilmen et al. 1996; Sohail et al. 2009; Liu et al. 2015), which makes the use of alternative inducer substrates more attractive. For these reasons, various lignocellulosic substrates have been evaluated for cellulase production including agricultural residues such as sugarcane bagasse, wheat straw, among others (Sohail et al. 2009; Cunha et al. 2012, 2015; van den Brink et al. 2014). However, in order to make such lignocellulosic materials more accessible to fungi growth, a pretreatment step is usually required.

Besides medium composition, cellulolytic enzyme production can be influenced by manipulation of the temperature and pH conditions of the fermentation. The optimal temperature for cellulolytic enzyme production by *Aspergillus* and *Trichoderma* is usually in the range of 28–35 °C, since these fungi are mesophilic organisms. Though a very important process variable, the temperature can be relatively easily monitored and controlled in SmF cultivations. On the other hand, the pH is more subject to variability during cultivation. The pH of the medium affects the germination of fungal spores as well as their growth morphology (Papagianni 2004). Current research in the fungal production of cellulases usually use the initial pH between 4.0 and 7.0, in some cases with control and monitoring over time and other cases with free pH (Cunha et al. 2012, 2015; Delabona et al. 2012, 2013a).

The agitation and oxygen supply also play an important role in filamentous fungi growth and cellulolytic enzyme production under SmF. Nevertheless, only few studies have been done to evaluate the influence of operating conditions (e.g., agitation and aeration rates) and oxygen transfer rates on cellulolytic enzyme production by filamentous fungi (Chipeta et al. 2008; Shahriarinour et al. 2011;

Cunha et al. 2015). Agitation and aeration in SmF cultivations are directly related to the morphology of the filamentous fungi, which can vary from freely dispersed mycelium to highly dense clumps or pellets. The optimal morphology for enzyme production is usually related to each specific microorganism. However, changes in filamentous fungi morphology can impact the rheology of cultivation media, which consequently affects the flow patterns in the bioreactor and the overall performance of the cultivation. In terms of process requirements, a dispersed hyphal morphology may result in a medium with high viscosity, which will be more complicated to mix and aerate than a cultivation medium where pellets are the predominant morphology.

Changes in fungal growth related to the agitation speed are expected, since it is well known that the agitation speed affects (i) mixing, heat and mass transfer, (ii) the shear stress to which the fungus is exposed, (iii) the growth patterns of filamentous fungi, and (iv) the production of metabolites (Papagianni 2004; Lin et al. 2010; Gabelle et al. 2012). During SmF cultivations for cellulase production by filamentous fungi, the usual approach is to maintain the dissolved oxygen (DO) concentration above 20 % air saturation by manipulating the agitation and aeration rates or by increasing the oxygen concentration in the sparge air (Schell et al. 2001). Therefore, finding an optimum condition for agitation and aeration is highly important to improve cellulolytic enzyme production.

The types of bioreactors used for the production of enzymes under SmF can be classified into (i) conventional stirred tank bioreactors (STB) and (ii) nonconventional pneumatic type, such as the airlift and bubble column bioreactors. Both types of SmF bioreactors have been evaluated for (hemi)cellulases production by filamentous fungi such as *Aspergillus* and *Trichoderma* (Kim et al. 1997; Siedenberg et al. 1997; Michelin et al. 2011, 2013; Cunha et al. 2012, 2015).

The stirred tank bioreactors (STB) are widely used in industry for cellulolytic enzyme production. The main advantage of these bioreactors is the versatility of operation and ease of control and monitoring of the process variables, being a very well-consolidated system from the engineering standpoint. The disadvantage of this system is the high energy demand required to maintain the homogeneity of the culture broth.

Pneumatic type bioreactors have been applied in the production of cellulolytic enzyme mostly in laboratory scale. The advantages of pneumatic bioreactors include: (i) ease of construction, operation and maintenance due to the absence of a mechanical seal, (ii) low power consumption, and (iii) high oxygen transfer rates (Campesi et al. 2009; Cerri and Badino 2010). In pneumatically agitated bioreactors the air flow oxygenates the biomass and induces a global circulation of the solid particles. Such type of agitation usually leads to lower shear stress, indicating a strong potential for the growth of filamentous fungi. However, the main disadvantage of the pneumatic bioreactors is the fact that both aeration and mixing as are promoted by the injection of air into the reactor and, therefore, these systems are less versatile than the conventional STB.

SmF bioreactors can be operated under batch, fed-batch, and continuous modes. Most of the studies for cellulase production have used the batch processes.

However, with the use of fed-batch cultivation the agitation and aeration difficulties are minimized, as well as possible catabolic repression effects caused by accumulation of reducing sugar in the medium. Mixed cultures of *Trichoderma* and *Aspergillus* grown as fed-batch mode in a 3L STB has been used for enhanced cellulase production (Ahamed and Vermette 2008). Therefore, the understanding of the fungal physiology along with the bioprocess engineering aspects needs to be considered for the selection of cultivation conditions to improve the production efficiency of cellulolytic enzymes by fungi in SmF cultivations.

12.3.2 Production of Cellulases by SSF

Despite the previously mentioned advantages of SSF over SmF, the implementation of enzymes production by SSF at an industrial scale has been held back, mainly due to difficulties in monitoring and controlling the different process variables (Mitchell et al. 2000; Nagel et al. 2001; Chen and He 2012). As a consequence, even though there have been several reports of the production of fungal cellulolytic enzymes using SSF (Singhania et al. 2010), most investigations were carried out at the laboratory scale and were focused on the cultivation of an individual microorganism, employing specific operational conditions or varying one condition at a time (Jecu 2000; Mamma et al. 2008; Sohail et al. 2009). Identification of the optimum conditions for enzyme production under SSF requires evaluation of parameters such as temperature, moisture content, and air flow rate, as well as the interactions between parameters.

Temperature is one of the most important process variables affecting SSF, because microbial growth under aerobic conditions results in the release of metabolic heat. At extreme levels, this can cause denaturation of the enzymes produced, as well as other deleterious effects on microorganism growth and metabolite production (Raghavarao et al. 2003; Singhania et al. 2009; Farinas 2015). Since the SSF process occurs in the absence of free water, it is difficult to remove the metabolic heat produced during microbial growth, due to the limited thermal conductivity of the solid substrate and the low thermal capacity of air (Raghavarao et al. 2003). The influence of temperature on the production of cellulolytic enzymes has been reported for a variety of fungal strains cultivated under SSF (Jecu 2000; Pirota et al. 2013b; Castro et al. 2015; El-Shishtawy et al. 2015). Most studies have focused on finding an optimal temperature for fungal growth and enzyme production. This type of characterization can be used to predict the effects of changes in temperature on enzymatic productivity in SSF bioreactors.

An ideal humidity level must be also maintained during SSF cultivation, since the water content is one of the most important operating parameters affecting the efficiency of the process. If the moisture content is too high, the void spaces of the solids are filled with water, resulting in restricted gaseous diffusion. At the other extreme, if the moisture content is low, the growth of the microorganism will be harmed (Raghavarao et al. 2003). The optimal moisture content value depends on both the solid substrate and the microorganism used (Raimbault 1998), and a critically important consideration is that it varies during the course of a fermentation process.

Besides temperature and moisture content, aeration also plays an important role in SSF for the production of fungal cellulolytic enzymes. In SSF, the function of aeration is to maintain aerobic conditions and remove the carbon dioxide generated; it also assists on regulating the temperature and moisture level of the substrate (Raimbault 1998). The gas environment in SSF can significantly affect the levels of cell growth and enzyme production. Most of the studies concerning the effect of aeration on cellulolytic enzymes production under SSF have only been conducted at the laboratory scale (Farinas et al. 2011; Vitcosque et al. 2012; Castro et al. 2015).

As mentioned before, microbial cellulolytic enzymes are considered inducible enzymes and maximum yields are usually obtained when cellulose is used as the carbon source or inducer. The selection of suitable substrates for SSF should take into consideration their costs and availability (Pandey et al. 2000). The possibility of using agroindustrial wastes as substrates in SSF is a major advantage of this technique, because these residues are usually byproducts of other industrial processes. In addition, these residues are generated in large amounts and have low commercial value. Furthermore, their disposal or subsequent use, often without any kind of treatment, can lead to environmental problems.

A variety of agroindustrial wastes can be used as solid substrates for SSF, including sugarcane, cassava, and orange bagasse, as well as cereal bran and coffee pulp, amongst others. The basic structures of these materials (consisting of cellulose, hemicellulose, lignin, starch, pectin, and fibers) determine the properties of the solid substrate and provide sources of carbon and energy for microorganism growth. In addition, physical characteristics of the substrate, such as crystallinity, accessible area, surface area, porosity, and particle size have an important influence on the SSF process (Pandey et al. 1999; Farinas 2015).

The factors that affect SSF bioreactor design are much more complex compared to SmF. Besides oxygen transfer, factors that need to be considered include temperature and the water content of the solid medium, as well as the morphology of the fungus, its resistance to mechanical agitation, and the requirement for sterilization (Durand 2003). In general, the bioreactors commonly used for SSF can be distinguished by the type of aeration (forced or unforced) and the mixing system employed (with or without mixing) (Ali and Zulkali 2011). The SSF bioreactor configurations that have been used for cellulolytic enzymes production include tray-type, packed-bed, and horizontal rotating drum bioreactors, among others.

12.4 Application of Fungal Enzymes on Biomass Conversion

The cellulosic or second generation (2G) ethanol produced from lignocellulosic biomass has been considered to be the biofuel with the greatest potential to replace fossil fuels. However, 2G technology is not as mature as that of conventional first generation (1G) ethanol that uses starch or sucrose-based feedstocks, and the process is still less economically feasible. The enzymatic conversion of lignocellulosic materials to obtain fermentable sugars has been identified as a promising and of great industrial interest to increase the ethanol productivity in a sustainable manner (see Chap. 11—*Fungi in consolidated bioprocessing*). However, the enzymatic conversion of biomass, although an alternative of lower environmental impact, still requires the development of technologies that can reduce the costs of the enzymes. The cost of cellulases is considered to be one of the major constraints for the commercialization of 2G ethanol (Klein-Marcuschamer et al. 2012).

Plant cell walls consist primarily of cellulose (20–50 % on a dry weight basis), hemicellulose (15–35 %), and lignin (10–30 %) (Chundawat et al. 2011). In addition to the breakdown of cellulose, deconstruction of the hemicellulose fraction to the constituent sugars is essential for the efficient production of biofuels and other chemicals from plant biomass (Dodd and Cann 2009). The use of hemicellulases and other auxiliary enzymes, in conjunction with cellulolytic enzymes, can improve cellulose conversion by removing hemicellulose and increasing the access of cellulases to the substrate (Gao et al. 2011).

The feasibility of using the entire solid-state fermentation medium, containing the enzymes, mycelia, and residual solid substrate, for the saccharification of a lignocellulosic biomass such as sugarcane bagasse for 2G ethanol production was demonstrated by Pirota et al. (2013a, 2014). A major advantage of this configuration is that it enables the use of a single reactor system, avoiding any need for the additional separation steps required in traditional SSF.

In traditional SSF processes, the enzymes synthesized by the microorganisms are extracted from the solid substrate by conventional solid-liquid extraction after the cultivation period, with the final products after filtration being a liquid supernatant containing the enzymes of interest, which can be further concentrated, together with a solid residue. Use of the whole SSF medium offers a potential way of avoiding production of the solid residue, because the enzyme extraction/filtration steps can be eliminated from the process. This could be highly advantageous in terms of reduced enzyme and process costs, as well as the avoidance of unnecessary effluent streams. The enzymatic conversion of biomass using whole SSF media is therefore a potential alternative process configuration that conforms to the biorefinery concept.

12.5 Concluding Remarks

Biotechnological advances concerning the production of enzymatic complexes by filamentous fungi play a key role for the economy of biofuels in industrial scale. The use of lignocellulosic materials as both inducer substrate to enzyme production and as raw materials for bioconversion into fermentable sugars can offer many advantages to the bioenergy sector, such as the reduction of the overall process cost and the production of cocktails with composition tailored by fungi to application on lignocellulose bioconversion. Furthermore, filamentous fungi are able to produce a wide range of enzymes that can be applied to biofuels industry in bioprocess performed with different bioreactors and operational conditions, which may be optimized for the production of each particular enzyme.

References

Ahamed A, Vermette P (2008) Enhanced enzyme production from mixed cultures of *Trichoderma reesei* RUT-C30 and *Aspergillus niger* LMA grown as fed batch in a stirred tank bioreactor. Biochem Eng J 42(1):41–46

Ahamed A, Vermette P (2009) Effect of culture medium composition on *Trichoderma reesei's* morphology and cellulase production. Bioresour Technol 100(23):5979–5987

Ahamed A, Vermette P (2010) Effect of mechanical agitation on the production of cellulases by *Trichoderma reesei* RUT-C30 in a draft-tube airlift bioreactor. Biochem EngJ 49(3):379–387

Ali HKQ, Zulkali MMD (2011) Design aspects of bioreactors for solid-state fermentation: a review. Chem Biochem Eng Q 25(2):255–266

Andersen MR, Giese M, de Vries RP, Nielsen J (2012) Mapping the polysaccharide degradation potential of *Aspergillus niger*. BMC Genom 13:17

Campesi A, Cerri MO, Hokka CO, Badino AC (2009) Determination of the average shear rate in a stirred and aerated tank bioreactor. Bioprocess Biosyst Eng 32(2):241–248

Castro AM, Castilho LR, Freire DMG (2015) Performance of a fixed-bed solid-state fermentation bioreactor with forced aeration for the production of hydrolases by *Aspergillus awamori*. Biochem Eng J 93:303–308

Cerri MO, Badino AC (2010) Oxygen transfer in three scales of concentric tube airlift bioreactors. Biochem Eng J 51(1–2):40–47

Chen HZ, He Q (2012) Value-added bioconversion of biomass by solid-state fermentation. J Chem Technol Biotechnol 87(12):1619–1625

Chipeta ZA, du Preez JC, Christopher L (2008) Effect of cultivation pH and agitation rate on growth and xylanase production by *Aspergillus oryzae* in spent sulphite liquor. J Ind Microbiol Biotechnol 35(6):587–594

Chundawat S, Beckham G, Himmel M, Dale B, Prausnitz J (2011) Deconstruction of lignocellulosic biomass to fuels and chemicals. Ann Rev Chem Biomol Eng 2(2):121–145

Correa RCG, Rhoden SA, Mota TR, Azevedo JL, Pamphile JA, de Souza CGM et al (2014) Endophytic fungi: expanding the arsenal of industrial enzyme producers. J Ind Microbiol Biotechnol 41(10):1467–1478

Cunha FM, Esperanca MN, Zangirolami TC, Badino AC, Farinas CS (2012) Sequential solid-state and submerged cultivation of *Aspergillus niger* on sugarcane bagasse for the production of cellulase. Bioresour Technol 112:270–274

Cunha FM, Esperanca MN, Florencio C, Vasconcellos VM, Farinas CS, Badino AC (2015) Three-phasic fermentation systems for enzyme production with sugarcane bagasse in stirred tank bioreactors: effects of operational variables and cultivation method. Biochem Eng J 97:32–39

de Vries RP (2003) Regulation of *Aspergillus* genes encoding plant cell wall polysaccharide-degrading enzymes; relevance for industrial production. Appl Microbiol Biotechnol 61(1):10–20

Delabona P, Farinas C, da Silva M, Azzoni S, Pradella J (2012) Use of a new *Trichoderma harzianum* strain isolated from the Amazon rainforest with pretreated sugar cane bagasse for on-site cellulase production. Bioresour Technol 107:517–521

Delabona PD, Farinas CS, Lima DJD, Pradella JGD (2013a) Experimental mixture design as a tool to enhance glycosyl hydrolases production by a new *Trichoderma harzianum* P49P11 strain cultivated under controlled bioreactor submerged fermentation. Bioresour Technol 132: 401–405

Delabona PD, Pirota R, Codima CA, Tremacoldi CR, Rodrigues A, Farinas CS (2013b) Effect of initial moisture content on two Amazon rainforest Aspergillus strains cultivated on agro-industrial residues: biomass-degrading enzymes production and characterization. Ind Crops Prod 42:236–242

Dodd D, Cann I (2009) Enzymatic deconstruction of xylan for biofuel production. Global Change Biol Bioenergy 1(1):2–17

Durand A (2003) Bioreactor designs for solid state fermentation. Biochem Eng J 13(2–3):113–125

El-Shishtawy RM, Mohamed SA, Asiri AM, Gomaa ABM, Ibrahim IH, Al-Talhi HA (2015) Saccharification and hydrolytic enzyme production of alkali pre-treated wheat bran by *Trichoderma virens* under solid state fermentation. BMC Biotechnol 15:13

Farinas C, Vitcosque G, Fonseca R, Neto V, Couri S (2011) Modeling the effects of solid state fermentation operating conditions on endoglucanase production using an instrumented bioreactor. Ind Crops Prod 34(1):1186–1192

Farinas CS (2015) Developments in solid-state fermentation for the production of biomass-degrading enzymes for the bioenergy sector. Renew Sustain Energy Rev 9

Gabelle JC, Jourdier E, Licht RB, Ben Chaabane F, Henaut I, Morchain J et al (2012) Impact of rheology on the mass transfer coefficient during the growth phase of *Trichoderma reesei* in stirred bioreactors. Chem Eng Sci 75:408–417

Gao DH, Uppugundla N, Chundawat SPS, Yu XR, Hermanson S, Gowda K et al (2011) Hemicellulases and auxiliary enzymes for improved conversion of lignocellulosic biomass to monosaccharides. Biotechnol Biofuels 4:11

Guerriero G, Hausman JF, Strauss J, Ertan H, Siddiqui KS (2015) Destructuring plant biomass: focus on fungal and extremophilic cell wall hydrolases. Plant Sci 234:180–193

Guimaraes LHS, Peixoto-Nogueira SC, Michelin M, Rizzatti ACS, Sandrim VC, Zanoelo FF et al (2006) Screening of filamentous fungi for production of enzymes of biotechnological interest. Braz J Microbiol 37(4):474–480

Hasan F, Shah AA, Hameed A (2006) Industrial applications of microbial lipases. Enzyme Microb Technol 39(2):235–251

Holker U, Hofer M, Lenz J (2004) Biotechnological advantages of laboratory-scale solid-state fermentation with fungi. Appl Microbiol Biotechnol 64(2):175–186

Holker U, Lenz J (2005) Solid-state fermentation—are there any biotechnological advantages? Curr Opin Microbiol 8(3):301–306

Horn SJ, Vaaje-Kolstad G, Westereng B, Eijsink VGH (2012) Novel enzymes for the degradation of cellulose. Biotechnol Biofuels 5

Hwang HT, Qi F, Yuan CL, Zhao XB, Ramkrishna D, Liu DH, Varma A (2014) Lipase-catalyzed process for biodiesel production: protein engineering and lipase production. Biotechnol Bioeng 111(4):639–653

Ilmen M, Thrane C, Penttila M (1996) The glucose repressor gene cre1 of *Trichoderma*: isolation and expression of a full-length and a truncated mutant form. Mol Gen Genet 251(4):451–460

Jecu L (2000) Solid state fermentation of agricultural wastes for endoglucanase production. Ind Crops Prod 11(1):1–5

Kim SW, Kang SW, Lee JS (1997) Cellulase and xylanase production by *Aspergillus niger* KKS in various bioreactors. Bioresour Technol 59(1):63–67

Klein-Marcuschamer D, Oleskowicz-Popiel P, Simmons BA, Blanch HW (2012) The challenge of enzyme cost in the production of lignocellulosic biofuels. Biotechnol Bioeng 109(4): 1083–1087

Levasseur A, Drula E, Lombard V, Coutinho PM, Henrissat B (2013) Expansion of the enzymatic repertoire of the CAZy database to integrate auxiliary redox enzymes. Biotechnol Biofuels 6:14

Lin P-J, Scholz A, Krull R (2010) Effect of volumetric power input by aeration and agitation on pellet morphology and product formation of *Aspergillus niger*. Biochem Eng J 49(2):213–220

Liu D, Li J, Zhao S, Zhang R, Wang M, Miao Y et al (2013) Secretome diversity and quantitative analysis of cellulolytic *Aspergillus fumigatus* Z5 in the presence of different carbon sources. Biotechnol Biofuels 6

Liu X, Ma Y, Zhang M (2015) Advances in expansins and expansion-like proteins involved in lignocellulose degradation. Biotechnol Lett

Mamma D, Kourtoglou E, Christakopoulos P (2008) Fungal multienzyme production on industrial by-products of the citrus-processing industry. Bioresour Technol 99(7):2373–2383

Michelin M, Mota AMD, Polizeli M, da Silva DP, Vicente AA, Teixeira JA (2013) Influence of volumetric oxygen transfer coefficient (k(L)a) on xylanases batch production by *Aspergillus niger* van Tieghem in stirred tank and internal-loop airlift bioreactors. Biochem Eng J 80: 19–26

Michelin M, Polizeli M, da Silva DP, Ruzene DS, Vicente AA, Jorge JA et al (2011) Production of xylanolytic enzymes by *Aspergillus terricola* in stirred tank and airlift tower loop bioreactors. J Ind Microbiol Biotechnol 38(12):1979–1984

Mitchell DA, Krieger N, Stuart DM, Pandey A (2000) New developments in solid-state fermentation II. Rational approaches to the design, operation and scale-up of bioreactors. Process Biochem 35(10):1211–1225

Nagel F, Tramper J, Bakker MSN, Rinzema A (2001) Temperature control in a continuously mixed bioreactor for solid-state fermentation. Biotechnol Bioeng 72(2):219–230

Pandey A, Selvakumar P, Soccol CR, Nigam P (1999) Solid state fermentation for the production of industrial enzymes. Curr Sci 77(1):149–162

Pandey A, Soccol CR, Mitchell D (2000) New developments in solid state fermentation: I-bioprocesses and products. Process Biochem 35(10):1153–1169

Papagianni M (2004) Fungal morphology and metabolite production in submerged mycelial processes. Biotechnol Adv 22(3):189–259

Pirota R, Baleeiro FCF, Farinas CS (2013a) Saccharification of biomass using whole solid-state fermentation medium to avoid additional separation steps. Biotechnol Prog 29(6):1430–1440

Pirota R, Tonelotto M, Delabona PD, Fonseca RF, Paixao DAA, Baleeiro FCF et al (2013b) Enhancing xylanases production by a new Amazon Forest strain of *Aspergillus oryzae* using solid-state fermentation under controlled operation conditions. Ind Crops Prod 45:465–471

Pirota R, Delabona PS, Farinas CS (2014) Simplification of the biomass to ethanol conversion process by using the whole medium of filamentous fungi cultivated under solid-state fermentation. Bioenergy Res 7(2):744–752

Polizeli M, Rizzatti ACS, Monti R, Terenzi HF, Jorge JA, Amorim DS (2005) Xylanases from fungi: properties and industrial applications. Appl Microbiol Biotechnol 67(5):577–591

Raghavarao K, Ranganathan T, Karanth N (2003) Some engineering aspects of solid-state fermentation. Biochem Eng J 13(2–3):127–135

Raimbault R (1998) General and microbiological aspects of solid substrate fermentation. Electron J Biotechnol 3–45

Rodriguez-Zuniga UF, Couri S, Neto VB, Crestana S, Farinas CS (2013) Integrated strategies to enhance cellulolytic enzyme production using an instrumented bioreactor for solid-state fermentation of sugarcane bagasse. Bioenergy Res 6(1):142–152

Saha BC (2003) Hemicellulose bioconversion. J Ind Microbiol Biotechnol 30(5):279–291

Schell DJ, Farmer J, Hamilton J, Lyons B, McMillan JD, Saez JC, Tholudar A (2001) Influence of operating conditions and vessel size on oxygen transfer during cellulase production. Appl Biochem Biotechnol 91–3:627–642

Shahriarinour M, Ramanan RN, Wahab MNA, Mohamad R, Mustafa S, Ariff AB (2011) Improved cellulase production by *Aspergillus terreus* using oil palm empty fruit bunch fibre as substrate in a stirred tank bioreactor through optimization of the fermentation conditions. Bioresources 6 (3):2663–2675

Siedenberg D, Gerlach SR, Czwalinna A, Schugerl K, Giuseppin MLF, Hunik J (1997) Production of xylanase by *Aspergillus awamori* on complex medium in stirred tank and airlift tower loop reactors. J Biotechnol 56(3):205–216

Singh AK, Mukhopadhyay M (2012) Overview of fungal lipase: a review. Appl Biochem Biotechnol 166(2):486–520

Singhania RR, Patel AK, Soccol CR, Pandey A (2009) Recent advances in solid-state fermentation. Biochem Eng J 44(1):13–18

Singhania RR, Sukumaran RK, Patel AK, Larroche C, Pandey A (2010) Advancement and comparative profiles in the production technologies using solid-state and submerged fermentation for microbial cellulases. Enzyme Microb Technol 46(7):541–549

Sohail M, Siddiqi R, Ahmad A, Khan SA (2009) Cellulase production from *Aspergillus niger* MS82: effect of temperature and pH. New Biotechnol 25(6):437–441

Sukumaran RK, Singhania RR, Pandey A (2005) Microbial cellulases—production, applications and challenges. J Sci Ind Res 64(11):832–844

Sukumaran RK, Singhania RR, Mathew GM, Pandey A (2009) Cellulase production using biomass feed stock and its application in lignocellulose saccharification for bio-ethanol production. Renew Energy 34(2):421–424

Thomas L, Larroche C, Pandey A (2013) Current developments in solid-state fermentation. Biochem Eng J 81:146–161

van den Brink J, de Vries RP (2011) Fungal enzyme sets for plant polysaccharide degradation. Appl Microbiol Biotechnol 91(6):1477–1492

van den Brink J, Maitan-Alfenas GP, Zou G, Wang CS, Zhou ZH, Guimaraes VM et al (2014) Synergistic effect of *Aspergillus niger* and *Trichoderma reesei* enzyme sets on the saccharification of wheat straw and sugarcane bagasse. Biotechnol J 9(10):1329–1338

Vasconcellos VM, Tardioli PM, Giordano RLC, Farinas CS (2015) Production efficiency *versus* thermostability of (hemi)cellulolytic enzymatic cocktails from different cultivation systems. Process Biochem

Vitcosque GL, Fonseca RF, Rodríguez-Zúniga UF, Bertucci Neto V, Couri V, Farinas CS (2012) Production of biomass-degrading multienzyme complexes under solid-state fermentation of soybeanmeal using a bioreactor. Enzyme Res 2012:8

Zhang Y, Himmel M, Mielenz J (2006) Outlook for cellulase improvement: screening and selection strategies. Biotechnol Adv 24(5):452–481

Part V
Fungi as Biofactories and Other Biotechnological Applications

Chapter 13
Fungal Biosynthesis of Nanoparticles, a Cleaner Alternative

Ernestina Castro-Longoria

13.1 Introduction

Nanotechnology involves physical, chemical, biological, and engineering sciences and has emerged as a cutting-edge technology to investigate novel techniques capable to manipulate single atoms and molecules for the production and application of materials at the nanoscale. Such material in the form of nanoparticles (NPs), range from 0.1 to 100 nm in size, is synthesized in a wide range of shapes, presenting different organizational levels. When particles are created in such small dimensions, their magnetic, electronic, and optical properties change significantly from those at larger scales.

Due to their unique properties, NPs have several important applications in various fields, including: (i) sensor technology (Zhang et al. 2012), (ii) optical devices (Maier 2006), (iii) catalysis (Kim et al. 2003), (iv) biological labeling (Lin et al. 2009), (v) personal care products (Nohynek and Dufour 2012), (vi) development of new technologies for neuroimaging (Re et al. 2012), and (vii) agents for cancer therapy (Jain et al. 2012) among others. There are a wide variety of nanomaterials that can be employed: nanospheres, nanotubes, liposomes, nanowires, nanosheets, nanoclusters, nanomicelles are just a few examples of nanostructures under current evaluation. For instance, nanospheres have promising applications for the treatment of some cancers (O'Neal et al. 2004).

Typically, NPs are produced by physical and chemical methods, but the majority of them are regarded as having a relatively high environmental cost since they are often energy intensive, employ toxic chemicals, and require high temperatures. For that reason, there is an increasing interest in developing clean, nontoxic, and environmentally friendly processes for the synthesis and assembly of nanostructured

E. Castro-Longoria (✉)
Department of Microbiology, Center for Scientific Research and Higher Education of Ensenada (CICESE), Ensenada, Baja California, Mexico
e-mail: ecastro@cicese.mx

D. Purchase (ed.), *Fungal Applications in Sustainable Environmental Biotechnology*, Fungal Biology, DOI 10.1007/978-3-319-42852-9_13

materials. These eco-friendly methods represent a cleaner alternative since the use of molecules found in biological material replace chemicals used as reducing, capping, and stabilizer agents.

In finding eco-friendly processes for the synthesis of NPs, researchers found inspiration from nature, since some organisms are able to live and grow at high metal ion concentrations (Haefeli et al. 1984) and can synthesize inorganic materials, like the magnetotactic bacteria which internally synthesize a chain of magnetic nanoparticles (Spring and Schleifer 1995; Chen et al. 2010). Hence, one of the first attempts to produce metallic NPs was carried out using the silver-resistant bacterial strain *Pseudomonas stutzeri* (Klaus et al. 1999). This organism was cultured in high concentrations of silver salts, and the intracellular formation of silver-based single crystals, such as equilateral triangles and hexagons, was reported (Klaus et al. 1999). Since then, many other organisms like actinomycetes, yeasts, fungi, plants, and algae, have been evaluated for the synthesis of nanoparticles (Quester et al. 2013a). In particular, fungi are excellent candidates to explore new eco-friendly methods for NPs production since many species are easy to culture and maintain in the laboratory. More than 70,000 species of fungi have been described and many more remain to be discovered; recent data estimates, based on high-throughput sequencing methods, that as many as 5.1 million fungal species exist (Blackwell 2011).

Nanoparticles can be broadly grouped into organic nanoparticles (biopolymer-based nanoparticles) (Nitta and Numata 2013) and inorganic nanoparticles (metallic nanoparticles and those of some minerals). Since most of the literature on fungal biosynthesis is related to metallic bioreduction, this chapter focuses in the production of metallic nanoparticles synthesized using fungal species.

13.2 Formation of Metallic Nanoparticles Using Fungal Biomass

Biosynthesis of metallic nanoparticles using fungi was first demonstrated using the biomass of *Verticillium* sp, the authors challenged the fungus with aqueous Ag^+ ions which resulted in the intracellular formation of silver nanoparticles (Mukherjee et al. 2001). Since then, several fungal species including yeasts have been utilized to evaluate the production of NPs (Table 13.1).

A general protocol when using fungal biomass involves the incubation of the fungus in liquid cultures and the recovery and extensive washing of biomass to eliminate all traces of the liquid media; then biomass is placed in a solution of metallic ions at known concentrations. Size and shape of the nanoparticles will depend on the incubation conditions such as pH, temperature, metal concentration, and exposure time to ionic solution (Gericke and Pinches 2006). Thus, the resulting NPs will depend, to some extent, on the parameters used during the reduction process. After the incubation period, biomass is recovered and analyzed for the

Table 13.1 Fungal species used to produce nanoparticles via their biomass

FUNGI	Nanoparticles	Shape*	Size (nm)	Source or reference
Aspergillus flavus	Ag	Spherical	8.92	Vigneshwaran et al. (2007)
	TiO_2	Spherical	62–74	Rajakumar et al. (2012)
	Au	Spherical, trigonal, cubic, tetragonal, hexagonal	18–22	Gupta and Bector (2013)
	Ag	Spherical	2–22	Vala et al. (2014)
Aspergillus fumigatus	Mg	Quasi-spherical**	40–80	Kaul et al. (2012)
	Au	NR	42.4	Kaul et al. (2012)
	Au	Spherical, trigonal, cubic, tetragonal, and hexagonal	18–26	Gupta and Bector (2013)
Aspergillus nidulans	Au	Minute dots, large triangular plates	NR	Prusinkiewicz et al. (2012)
Aspergillus niger	Au	Nonoplates	50–500	Xie et al. (2007)
	Ag	Circular, oval	20–100	Ganbarov et al. (2014)
Aspergillus oryzae var. *viridis*	Au	Spheres	10–60	Binupriya et al. (2010a)
		Triangles, pentagons, hexagons	30–400	
Aspergillus sydowii	Au	Spherical	10	Vala (2015)
Aspergillus versicolor	Hg	NR	NR	Das et al. (2009a)
Aspergillus wentii	Fe	NR	64.5	Kaul et al. (2012)
Chaetomium globosum	Fe	Quasi-spherical**	25.3	Kaul et al. (2012)
Cochliobolus lunatus	Ag	Spherical	3–21	Salunkhe et al. (2011)
Colletotrichum sp.	Au	Spherical	8–40	Shankar et al. (2003)
Coriolis versicolor	Ag	Spherical	444–491	Sanghi and Verma (2009a)
	CdS	Spherical	5–9	Sanghi and Verma (2009b)
	Au	Spherical and ellipsoidal	100–300	Sanghi and Verma (2010)

(continued)

Table 13.1 (continued)

FUNGI	Nanoparticles	Shape*	Size (nm)	Source or reference
Curvularia lunata	Fe	NR	20.8	Kaul et al. (2012)
Cylindrocladium floridanum	Ag	Spherical	25	Narayanan et al. (2013)
Epicoccum nigrum	Au	Spherical, rods	5–50	Sheikhloo et al. (2011)
Fusarium semitectum	Au, Au–Ag	Spherical	25, 18	Sawle et al. (2008)
Fusarium spp.	Zn	Irregular, some spherical	100–200	Velmurugan et al. (2010)
F. oxysporum *F. oxysporum* f. sp. *lycopersici*	Ag	Highly variable	5–15	Ahmad et al. (2002)
	CdS	Hexagonal	5–20	Ahmad et al. (2003)
	Au	Spherical, triangular	46, 21^	Shankar et al. (2004)
	Ag	Spherical	20–50	Durán et al. (2005)
	Au–Ag	Quasi-spherical**	8–14	Senapati et al. (2005)
	Ba	Quasi-spherical	4–5	Bansal et al. (2006)
	Fe_3O_4	Irregular, quasi-spherical	20–50	Bharde et al. (2006)
	Cd	Spherical	9–15	Kumar et al. (2007a)
	Pt	Hexagonal, pentagonal, circular, squares, rectangles	10–100	Riddin et al. (2006)
Geotricum sp.	Ag	Quasi-spherical**	30–50	Jebali et al. (2011)
Hormoconis resinae	Ag	Spherical, triangles	20–80, 10–20	Varshney et al. (2009)
Neurospora crassa	Ag, Au, Ag–Au	Mainly spherical or ellipsoidal	3–50, 3–100, 3–100	Castro-Longoria et al. (2011)
	Pt	Quasi-spherical, nanoaggregates	4–35, 20–110	Castro-Longoria et al. (2012)
Penicillium chrysogenum	Ag	Spherical	30–50	Mohammadi and Salouti (2015)
Penicillium sp.	Au	Spherical	50	Du et al. (2011)
Phaenerochaete chrysosporium	Ag	Pyramidal	50–200	Vigneshwaran et al. (2006)

(continued)

Table 13.1 (continued)

FUNGI	Nanoparticles	Shape*	Size (nm)	Source or reference
Pochonia chlamydosporium	Mg	NR	20–60	Kaul et al. (2012)
Pycnoporus sanguineus	Ag	Spherical, ellipsoidal	52.8 ± 0.06	San Chan and Don (2013)
Rhizopus oryzae	Au	Quasi-spherical**	10	Das et al. (2009b)
Schizophyllum commune	Ag	Spherical, ellipsoidal	53.9 ± 0.01	San Chan and Don (2013)
	Ag	Spherical	54–99	Arun et al. (2014)
Streptomyces hygroscopicus	Au	Spherical, pentagonal, hexagonal	2–20, 10–20, 30	Sadhasivam et al. (2012)
Trichosporon beigelii	Ag	Spherical	50–100	Ghodake et al. (2011)
Trichoderma harzianum	Au	Spherical	26–34	Tripathi et al. (2014)
Trichoderma ressei	Ag	Quasi-spherical**	5–50	Vahabi et al. (2011)
Verticillium luteoalbum	Au	Spherical, triangular, hexagonal and others	Various	Gericke and Pinches (2006)
Verticillium sp.	Ag	Spherical	25	Mukherjee et al. (2001)
	Fe_3O_4	Cubo-octahedral	100–400	Bharde et al. (2006)
YEAST				
Candida glabrata	Cd	NR	20 Å, 29 Å	Dameron et al. (1989)
	CdS	NR	NR	Krumov et al. (2007)
Pichia jadinii (*Candida utilis*)	Au	Various	NR	Gericke and Pinches (2006)
Shizosaccharomyces pombe	Cd	NR	18 Å, 29 Å 1–1.5	Dameron et al. (1989)
	CdS	Hexagonal	NR	Kowshik et al. (2002)
	CdS	NR		Krumov et al. (2007)

(continued)

Table 13.1 (continued)

FUNGI	Nanoparticles	Shape*	Size (nm)	Source or reference
Yarrowia lipolytica	Au	Particles and plates	Varying	Pimprikar et al. (2009)
Yeast strain MKY3	Ag	Twinned or multitwinned, some hexagonal	2–5	Kowshik et al. (2003)

Abbreviations NR not reported
*Shape is named as originally reported, **Not reported but visible in image, ^When immobilized in stearic acid and octadecylamine films, respectively

presence of NPs. The supernatant is also analyzed in order to determine if the fungus produce extracellular NPs via its excreted compounds.

There are various analyses that must be implemented in order to confirm NPs formation. The first, easily performed, includes removing aliquots of the supernatant to monitor the bioreduction process using a UV-vis spectrophotometer. Metallic NPs scatter and absorb light at certain wavelengths due to the resonant collective excitations of charge density at the interface between a conductor and an insulator, phenomena known as surface plasmon resonance. The optical response of metallic NPs depends on their size, shape, and environment (Noguez 2007). This first analysis gives an indication of size and shape of the obtained NPs, but further analyses are necessary to fully characterize them.

In some cases, depending on the metal, the formation of NPs is apparent from the gradual change in color of the incubated biomass. One of the clearest examples for this is the intracellular formation of gold NPs, where the appearance of the mycelium exhibits a red to purple color (Fig. 13.1) upon completion of the reaction with the auric ion. A range of colors can be observed on aqueous solutions containing gold NPs, depending on size and shape. Aqueous solutions of gold nanospheres less than 100 nm are red to purple, and yellowish for larger particles (Mody et al. 2010).

Other type of analyses frequently used to report the presence of NPs on the fungal biomass is by scanning electron microscopy (SEM), where biomass must be

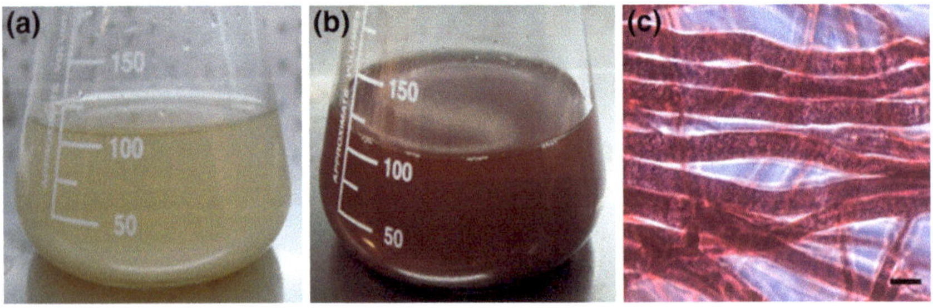

Fig. 13.1 Visual inference of gold nanoparticles synthesis by *Neurospora crassa* biomass. **a** culture at time zero, **b** culture after 24 h in $HAuCl_4$, **c** bright field microscopy of *N. crassa* hyphae after 24 h in $HAuCl_4$. Scale bar = 10 μm

desiccated before examined under the microscope. The microscope must be equipped with an energy dispersive spectrometer (EDS), in which case can provide rapid qualitative, or even with adequate standards, quantitative analysis of elemental composition, and thus corroborate the presence of the metal and other elements. However, SEM analysis only reveal the presence of metallic NPs on the external fungal cell wall, requiring also that samples be prepared for sectioning with an ultramicrotome and analyzed under transmission electron microscopy (TEM) in order to determine the intracellular formation of NPs.

Most studies using fungal biomass report only the formation of NPs in the supernatant, for instance using the biomass of *Cylindrocladium floridanum* very good results were obtained (Fig. 13.2). As previously mentioned, NPs are also present intracellularly as it has been demonstrated by TEM analysis of fine sections

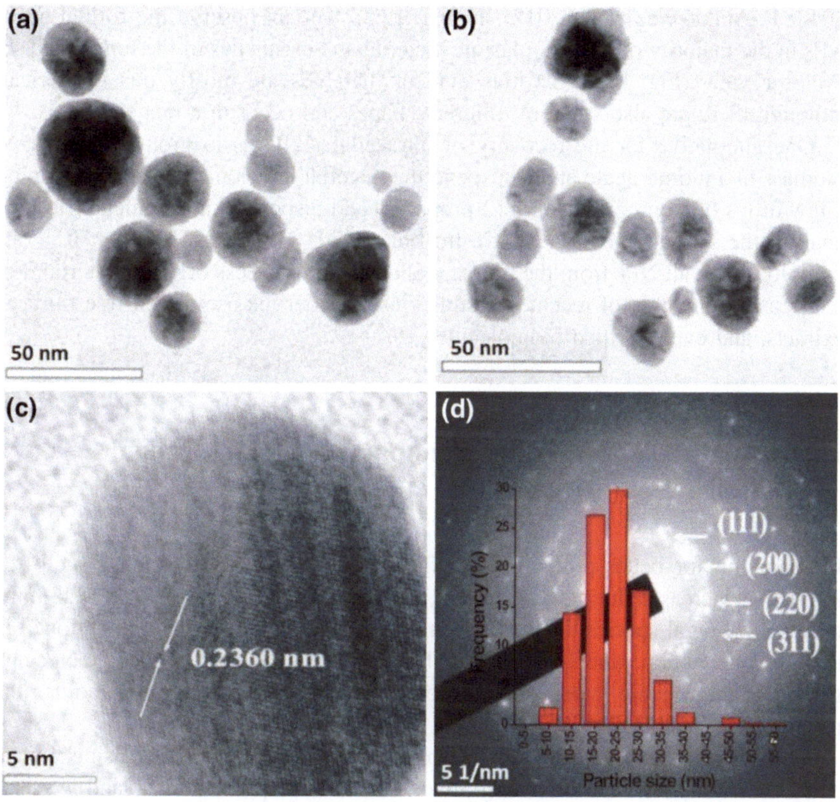

Fig. 13.2 TEM micrographs of extracellular mycogenic silver nanoparticles. **a** and **b** Representative TEM images of mycogenic silver nanoparticles by *Cylindrocladium floridanum*. **c** Spherical particle as viewed perpendicular to the pore axis of nanoparticle at a magnification of x800 k. **d** SAED pattern showing nanoparticle diffraction rings arisen due to the reflections from (111), (200), (220), and (311). The inset shows histogram of particle size distribution of mycogenic silver nanoparticles. Reproduced from Narayanan et al. (2013) with permission from Elsevier B.V., the Netherlands

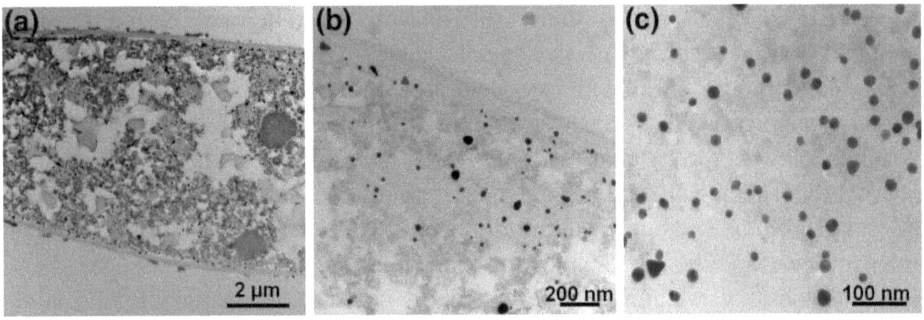

Fig. 13.3 TEM images of *Neurospora crassa* showing intracellular accumulation of Au nanoparticles at different magnifications after 24 h exposure in $HAuCl_4$

(Mukherjee et al. 2001; Gericke and Pinches 2006; Castro-Longoria et al. 2011, 2012; Prusinkiewicz et al. 2012). In Fig. 13.3a we can observe the formation of NPs in the majority of the cytoplasmic material and organelles of the fungal cell. In a closer view (Fig. 13.3b, c) it is evident that NPs are mostly quasi-spherical, although there are also some of different shapes and occur in a range of sizes.

One alternative for the recovery of intracellular NPs is to ground lyophilized biomass in a mortar agate and resuspend the material in deionized water containing a few drops of isopropyl alcohol (2-propanol) and disperse it in a sonicator to then analyze the sample under TEM (Castro-Longoria et al. 2011). However, it is difficult to obtain all NPs from the biomass and get rid of excess organic material. For that reason a number of recent investigations consider the use of cell-free filtrates, extracts, and even purified biomolecules.

13.3 Synthesis of Metallic Nanoparticles Using Fungal Cell-Free Filtrate

In searching for better protocols for NPs production, the use of fungal cell-free filtrate has increased during recent years, with the consequent increase as well in the number of fungal species that have been considered (Table 13.2). As mentioned in the previous section, when using fungal biomass it is difficult to extract the NPs and get rid of the organic material. However, using only the excreted metabolites the problem can be solved.

General protocols for the production of metallic nanoparticles using fungal cell-free filtrate, is first obtaining mycelium from liquid cultures and wash it thoroughly with distilled water to eliminate any trace of culture media. Washed biomass (10–20 g wet weight) is placed in 100 mL sterile distilled water and incubated for typically 48–72 h under agitation at 120 rpm, and then the aqueous solution is separated by filtration. Cell filtrate is then challenged with metallic ions (usually 10^{-3}M) and incubated in dark conditions until NPs have formed. Until

Table 13.2 Fungal species used to produce nanoparticles via cell-free filtrate

FUNGI	Nanoparticles	Shape*	Size (nm)	Source or reference
Alternaria alternata[§]	Au	Spherical, triangular	12 ± 5	Sarkar et al. (2012)
A. alternata	Fe	Cubic	9 ± 3	Mohamed et al. (2015)
Alternaria solani	Ag	Spherical	5–20	Devi et al. (2014)
Amylomyces rouxii KSU-09	Ag	Spherical	5–27	Musarrat et al. (2010)
Aspergillus aeneus	ZnO	Spherical	100–140	Jain et al. (2013)
Aspergillus clavatus	Ag	NR	550–650	Saravanan and Nanda (2010)
	Ag	Spherical, hexagonal	10–25	Verma et al. (2010)
Aspergillus flavus	Ag	Spherical	17	Jain et al. (2011)
	Ag	Spherical	20–60	Anand et al. (2015)
Aspergillus foetidus	Ag	Spherical	20–40	Roy et al. (2013)
Aspergillus fumigatus	Ag	Mostly spherical, some triangular	5–25	Bhainsa and D'Souza (2006)
	Ag	Mostly spherical	15–45	Alani et al. (2012)
Aspergillus japonicus	Fe_2O_3	Cubic	60–70	Bhargava et al. (2013)
Aspergillus niger	Au	Plates, aggregates, spherical	Various	Xie et al. (2007)
	Au	Spherical and elliptical	12.79 ± 5.61	Bhambure et al. (2009)
	Au	Spherical	3–30	Jaidev and Narasimha (2010)
	Ag	Spherical	5–35	Kathiresan et al. (2010)
	Ag	Spherical	1–20	Sagar and Ashok (2012)
Aspergillus oryzae	Ag	Spherical	20–60	Anand et al. (2015)
Aspergillus oryzae var. *viridis*	Au	Spheres	10–60	Binupriya et al. (2010a)
		Triangles, pentagons, hexagons	30–400	
Aspergillus parasiticus	Ag	Spherical	2–50	Moazeni et al. (2014)
Aspegillus tamari	Ag	Spherical	25–50	Kumar et al. (2012)

(continued)

Table 13.2 (continued)

FUNGI	Nanoparticles	Shape*	Size (nm)	Source or reference
Aspergillus terreus	Ag	Spherical	1–20	Li et al. (2012)
	Ag	Spherical	5–50	Devi and Joshi (2012)
Aspergillus tubingensis	Ag	Spherical	35 ± 10	Rodrigues et al. (2013)
Beauveria bassiana	Ag	Spherical	36.88–60.93	Banu and Balasubramanian (2014a)
Bionectria ochroleuca	Ag	Spherical	35 ± 10	Rodrigues et al. (2013)
Bipolaris nodulosa§	Ag	Spherical, hexahedral, pentagonal, semi-pentagonal	10–60	Saha et al. (2010)
Botrytis cinerea§	Au	Triangular, hexagonal, spherical, decahedral, and pyramidal	1–100	Castro et al. (2014)
Chrysosporium tropicum	Ag	Spherical	20–50	Soni and Prakash (2012)
Cladosporium cladosporioides	Ag	Mostly spherical	10–100	Balaji et al. (2009)
Cryphonectria sp.	Ag	NR	30–70	Dar et al. (2013)
Colletotrichum sp.	Ag	Spherical, near to spherical, triangular, and hexagonal	20–60	Azmath et al. (2015)
Coriolis versicolor§	Ag	Spherical	25–75	Sanghi and Verma (2009a)
	Au	Spherical and ellipsoidal	20–100	Sanghi and Verma (2010)
Epicoccum nigrum	Ag	Spherical, quasi-spherical	1–22	Qian et al. (2013)
Fusarium acuminatum	Ag	Spherical	5–40	Ingle et al. (2008)
Fusarium oxysporum f. sp. *lycopersici*	Pt	Hexagonal, pentagonal, circular, squares, rectangles	10–100	Riddin et al. (2006)
	Ag	Spherical	20–50	Soni and Prakash (2012)
F. oxysporum	Ag	Spherical	10–45	Joshi et al. (2013)
	Ag	Spherical	5–60	Selvi and Sivakumar (2012)
F. oxysporum§	Ag	Spherical	12–30	Ravindra and Rajasab (2015)

(continued)

Table 13.2 (continued)

FUNGI	Nanoparticles	Shape*	Size (nm)	Source or reference
Fusarium sp.	Ag	Spherical	5–50	Devi and Joshi (2012)
F. oxysporum f sp. *vasinfectum*	Ag	Spherical	10–30	Joshi et al. (2013)
F. oxysporum f. sp. *ciceris*§	Ag	Spherical	12–75	Ravindra and Rajasab (2015)
Fusarium solani	Ag	Spherical	5–35	Ingle et al. (2009)
	Ag	NR	3–8	El-Rafie et al. (2010)
Fusarium udum§	Ag	Spherical	12–75	Ravindra and Rajasab (2015)
Isaria fumosorosea	Ag	Spherical	51.31 and 111.02	Banu and Balasubramanian (2014b)
Macrophomina phaseolina	Ag	Spherical	8–25	Joshi et al. (2013)
Neurospora intermedia	Ag	Spherical	30	Hamedi et al. (2014)
N. intermedia§	Ag	Spherical	19	Hamedi et al. (2014)
Nigrospora oryzae§	Ag	Spherical	30–80	Saha et al. (2011)
Nigrospora sphaerica	Ag	Spherical	20–70	Muhsin and Hachim (2014)
Paecilomyces lilacinus	Ag	Spherical	5–50	Devi and Joshi (2012)
Penicillium aurantiogriseum§	Cu_2O	Spherical	89–250	Honary et al. (2012)
Penicillium brevicompactum	Ag	Irregular**	23–105	Shaligram et al. (2009)
Penicillium citrinum§	Cu_2O	Spherical	85–295	Honary et al. (2012)
	Ag	Spherical	109	Honary et al. (2013)
Penicillium diversum§	Ag	Spherical	5–45	Ganachari et al. (2012)
Penicillium fellutanum	Ag	Mostly spherical	5–25	Kathiresan et al. (2009)
Penicillium funiculosum	Ag	Spherical	5–10	Devi et al. (2014)
Penicillium nalgiovense	Ag	Spherical	15.2 ± 2.6	Maliszewska et al. (2014)
Penicillium sp.	Ag	Mostly spherical	10–100	Maliszewska and Sadowski (2009)
	Ag	Spherical	25	Singh et al. (2014)

(continued)

Table 13.2 (continued)

FUNGI	Nanoparticles	Shape*	Size (nm)	Source or reference
Penicillium waksmanii§	Cu_2O	Spherical	79–179	Honary et al. (2012)
Pestalotia sp.	Ag	Spherical	10–40	Raheman et al. (2011)
Phoma glomerata	Ag	Spherical	60–80	Birla et al. (2009)
	Ag	Spherical	19	Gade et al. (2014)
Pleurotus sapidus§	Au	Spherical, triangular, hexagonal	65 ± 5	Sarkar et al. (2013)
Pleurotus sajor caju	Ag	NR	5–50	Nithya and Ragunathan (2009)
Puccinia graminis	Ag	Spherical with aggregates	30–120	Kirthi et al. (2012)
Rhizopus oryzae	Au	Triangular, hexagonal, pentagonal, spheroidal, urchinlike, 2D nanowires, nanorods	Various	Das et al. (2010)
Rhizopus sp.	Ag	Spherical	27	Hiremath et al. (2014)
Rhizopus stolonifer	Ag	Quasi-spherical**	25–30	Binupriya et al. (2010b)
	Au	NR	1–5	Binupriya et al. (2010b)
	Ag	Spherical	5–50	Banu and Rathod (2011)
	Ag	Spherical	5–50	Afreen and Ranganath (2011)
Schizophyllum commune§	Ag	Spherical	51–93	Arun et al. (2014)
Schizophyllum radiatum§	Ag	Different shapes and aggregates	10–40	Krishna et al. (2015)
Stachybotrys chartarum	Ag	Spherical	65–108	Mohamed (2014)
Stemphylium vericans§	Ag	Spherical	25–60	Ravindra and Rajasab (2015)
Stereum hirsutum	Cu	Spherical	5–20	Cuevas et al. (2015)
Streptomyces albidoflavus	Ag	Spherical	10–40	Prakasham et al. (2012)
Streptomyces hygroscopicus§	Ag	Spherical	20–30	Sadhasivam et al. (2010)
Streptomyces sp.	Ag	Mostly spherical	15–25	Alani et al. (2012)

(continued)

Table 13.2 (continued)

FUNGI	Nanoparticles	Shape*	Size (nm)	Source or reference
Talaromyces flavus	Ag	Spherical	20–60	Anand et al. (2015)
Trichoderma asperellum	Ag	Quasi-spherical**	13–18	Mukherjee et al. (2008)
Tricholoma crissum	Ag	Mostly spherical	5–50	Ray et al. (2011)
Trichoderma gamsii	Ag	Spherical	20–60	Anand et al. (2015)
Trichoderma harzianum	Ag	Spherical, rods	30–50	Singh and Raja (2011)
	Ag	Spherical, irregular	51.1	Ahluwalia et al. (2014)
Trichoderma viride	Ag	Spherical, rodlike	5–40	Fayaz et al. (2009a)
	Ag	Spherical	2–4, 10–40, 80–100	Fayaz et al. (2009b)
	Ag	Mostly spherical	2–4	Fayaz et al. (2010a)
	Ag	Spherical, rodlike	5–40	Fayaz et al. (2010b)

Abbreviations NR not reported
*Shape is named as originally reported, **Not reported but visible in image, §Mycelium-free culture filtrate

now, this is the most widely explored protocol to biosynthesize NPs using fungal species.

Biosynthesized NPs are proven to be stable for several months (Mukherjee et al. 2008; Quester et al. 2013b); this is due to the excreted metabolites that serve as reducing as well as capping/stabilizer agents. Stabilization of NPs is essential since physical and chemical properties depend not only on their composition, but also on their size and shape (Henglein 1989; Burda et al. 2005). For instance, AgNPs are highly reactive and tend to agglomerate and precipitate if they are uncoated (Radziuk et al. 2007); therefore stabilization is crucial in order to preserve their size and thus their physical properties. For instance, some processes such as the aggregation of NPs can modify their physical and chemical properties; even if those aggregations contain NPs of the same size; also their biological effects are altered by affecting ion release from the surface (Hotze et al. 2010).

By the use of cell-free filtrate of *Trichoderma asperellum* (Mukherjee et al. 2008), a controlled and upscalable biosynthetic route to produce AgNPs with well-defined morphology and size (13–18 nm) (Fig. 13.4) was reported. The authors found the particles to be stable even after prolonged storage for over 6 months, having the additional advantage of using a strain that is nonpathogenic and a commercially viable biocontrol agent.

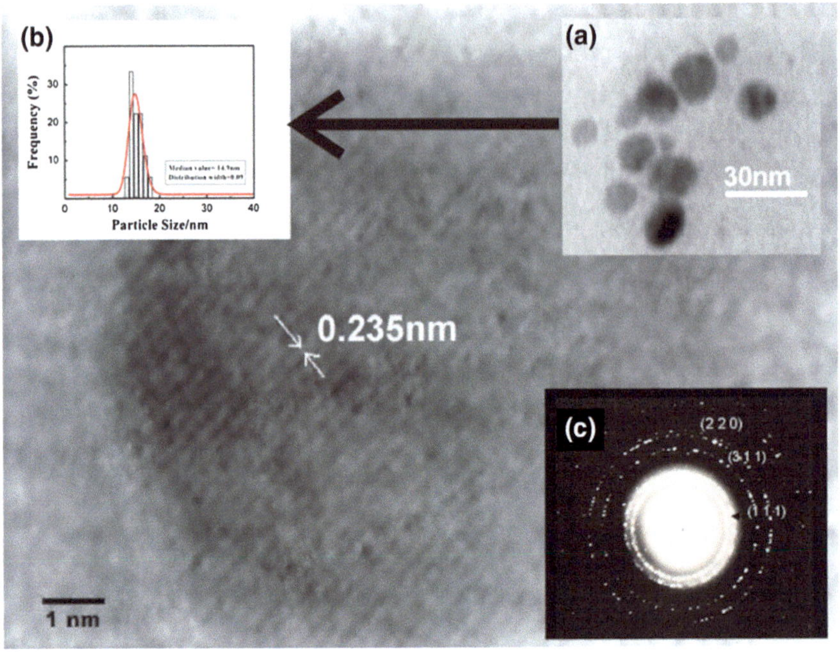

Fig. 13.4 High-resolution transmission electron micrograph of drop-cast silver nanoparticles produced using the cell-free filtrate of *Trichoderma asperellum*. *Inset*: **a** low-resolution micrographs showing size of the particulates, **b** histogram of particle size distribution as obtained from TEM, and **c** SAED pattern recorded on the same sample. Reproduced from Mukherjee et al. (2008) with permission from IOP Publishing, U.K

Some reports documented the use of mycelium-free culture filtrate as bioreducing agent (Table 13.2); this liquid contains fungal metabolites as well as the nutrients in which the fungus was cultivated. By applying this mixture, NPs can also be obtained, as seen in Fig. 13.5, in which the authors reported AuNPs of different shapes, ranging between 1–100 nm and 15–100 nm using *Botrytis cinerea* (Castro et al. 2014) and *Pleurotus sapidus* (Sarkar et al. 2013), respectively.

However, even if using the same fungal species, it is important to note that the use of cell filtrate or culture filtrate will result in differences in NPs production. This was demonstrated by Hamedi et al. (2014) in which the production of AgNPs was compared using cell-free filtrate and the culture supernatant of *Neurospora intermedia*. The authors report the smallest average size and the highest productivity of AgNPs when using the culture supernatant, and the formation of particles with higher stability and monodispersity when using cell-free filtrate. This difference can be due to the higher amounts of proteins, enzymes, and also the presence of metabolites in the culture supernatant, which may act rapidly in the AgNPs synthesis compared to the cell-free filtrate that only contains fungal metabolites.

It is assumed that extracellular putative NADH-dependent reductase participates in the reduction of metallic ions and the subsequent formation of NPs. However, it

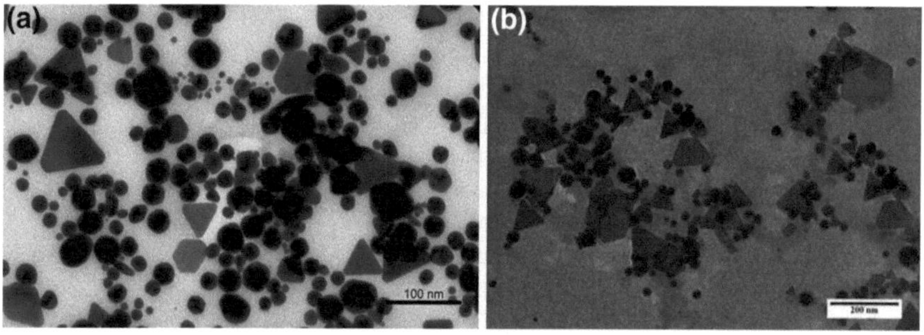

Fig. 13.5 Representative transmission electron micrographs of AuNPs synthesized using culture filtrate of *Botrytis cinerea* (**a**) and *Pleurotus sapidus* (**b**). Reproduced and modified from Castro et al. (2014) and Sarkar et al. (2013), respectively, with permission from Elsevier B.V., the Netherlands

is believed that this enzyme is not the only active component that participates in this process, other molecules such as extracellular laccases also contribute to the formation of NPs (Faramarzi and Forootanfar 2011; Castro et al. 2014). What has been proved in several studies is that NPs are stabilized by surface bound protein molecules that also prevent aggregation.

13.4 Production of Metallic Nanoparticles Using Fungal Extract

The control of size and shape of biosynthesized nanomaterial still remains a challenge. Searching for alternative methods, a few studies have used fungal extract to explore an improved bioreduction protocol (Table 13.3). A general procedure is to obtain fungal mycelium from liquid cultures and after thoroughly wash with distilled water, the cells must be broken to release cytoplasmic content, either with an agate mortar (Castro-Longoria et al. 2012) or using glass beads (Gericke and Pinches 2006). Freshly harvested biomass is ground in deionized water, and then this mixture is centrifuged. The solution obtained is further filtered through a fine filter to discard solid material, and finally this extract is challenged with metallic ions and incubated for a preestablished time. Usually, shape and size of NPs depend on the incubation conditions, therefore most of the time it is necessary to perform experiments at different conditions of pH, temperature, proportion of salts/reducing agent, in order to optimize the conditions where the desired NPs are obtained (Das et al. 2010; Sanghi and Verma 2010; Quester 2014). For instance, different sizes and shapes of AuNPs were obtained when using *N. crassa* extract at different pH values (Fig. 13.6). NPs obtained at pH 3, are significantly bigger than those obtained at pH 5.5 or pH 10, also less variability in shape was obtained when using pH 10 (Fig. 13.6c). Additionally, it was also demonstrated that AuNPs maintained their sizes and shapes for over 10 months at room temperature (Quester et al. 2013b).

Table 13.3 Fungal species used to produce nanoparticles via their extract

FUNGI	Nanoparticles	Shape*	Size (nm)	Source or reference
Aspergillus niger	Au	Nanowalls, spiral plates, spherical	Various	Xie et al. (2007)
Fusarium oxysporum	Au	Spherical and triangular	8–40	Mukherjee et al. (2002)
Neurospora crassa	Pt	Quasi-spherical, nanoaggregates	4–35, 20–110	Castro-Longoria et al. (2012)
	Au	Triangles, hexagons, pentagons	10–200	Quester et al. (2013b)
	Au	Quasi-spheres	6–23	Quester et al. (2013b)
	Au	Quasi-spheres	3–12	Quester et al. (2013b)
Verticillium luteoalbum	Au	Triangular and hexagonal plates	100–500	Gericke and Pinches (2006)
Volvariella volvaceae	Ag	Spherical	15	Philip (2009)
	Au	Triangular, spherical, hexagonal	20–150	Philip (2009)
	Au–Ag	NR	NR	Philip (2009)

Abbreviations NR not reported
*Shape is named as originally reported

When the reducing agent (same fungal species) is present either as: biomass, mycelia-free spent medium or extract, different types of nanoparticles are produced even if there is no variation in the time and incubation conditions. In the first case, biomass include all molecules as electron donors, while in the mycelia-free spent media there are only excreted molecules by the fungus and in the extract there are only soluble molecules. There are not many examples of studies in which the same fungal species is used as the reduced agent in the three different forms mentioned above. As far as I am aware, only one clear example was published by Xie et al.

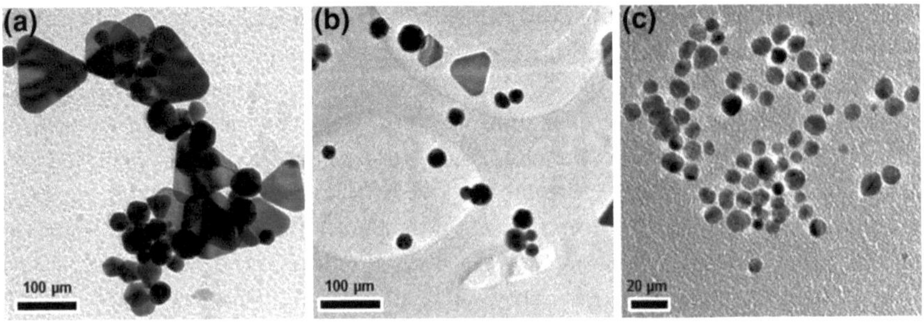

Fig. 13.6 TEM images of gold nanoparticles synthesized by *N. crassa* extract at 60 °C and different pH values. pH 3 (**a**), pH 5.5 (**b**), and pH 10 (**c**). Reproduced and modified from Quester (2014)

(2007) with *Aspergillus niger*; the resulting gold nanoparticles (AuNPs) were of different size and shape. The authors found that using biomass resulted in the formation of AuNPs only in the mycelium, since the aqueous supernatant was colorless and no absorption in the 400–800 nm region by UV-vis spectroscopy was detected (Fig. 13.7a, spectrum 1) and purple mass separated from the aqueous solution contained fungal biomass with gold nanoparticles. Analysis by field emission scanning electron microscopy (FESEM) revealed that the hyphal surface was decorated by spherical and triangular gold nanoparticles (Fig. 13.7b). Using the fungal cell-free filtrate, planar nanoparticles were the most abundant, accounting for $\sim 60\ \%$ of the total nanoparticles synthesized, the remaining $\sim 40\ \%$ was gold nanospheres (Fig. 13.7c). The application of fungal extract produced mainly hexagonal nanoplates, most of them had three or more stacked layers (Fig. 13.7d). This demonstrates that a wide variety of methodologies can be developed to find the best option to obtain a particular type of nanostructure.

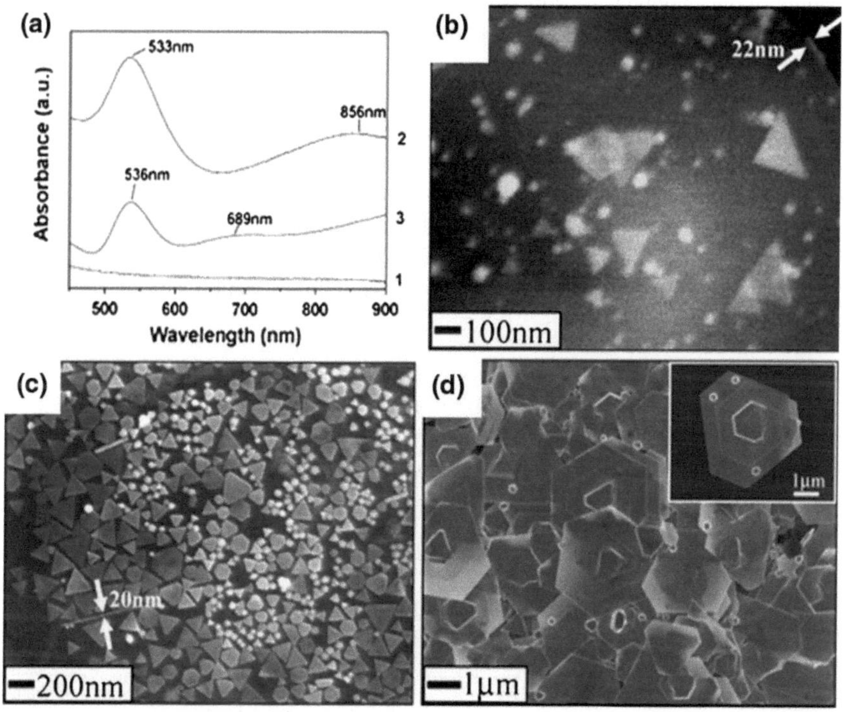

Fig. 13.7 Representative FESEM images of gold nanoparticles synthesized using *Aspergillus niger*. **a** UV-vis spectra of gold nanoparticles synthesized by fungal biomass for 24 h (spectrum 1); the spent medium (spectrum 2); and by the fungal extract (spectrum 3). **b** Resulting AuNPs synthesized by the fungal biomass; **c** by the mycelia-free spent medium, and **d** the fungal extract (inset shows nanoplates, with three stacked layers). *Arrows* point out to thickness of gold nanoplates, which were standing perpendicularly. Reproduced and modified from Xie et al. (2007) with permission by the American Chemical Society., Washington, DC

13.5 Production of Metallic Nanoparticles Using Fungal Purified Biomolecules

The use of fungal purified molecules for the synthesis of NPs is starting to be employed in order to optimize biocompatibility; in this process only one type of molecule serves as bioreducing agent and could represent promising applications for biological/medical procedures (see Sect. 13.6). However, currently there are very few studies in this respect and all of them utilizing purified enzymes (Table 13.4). The protocol to obtain NPs using purified biomolecules is basically the same as using fungal extract, with the exception that before exposing the ionic solutions to the reducing agent (biomolecule), this must be purified from the fungal cultures. This represents additional efforts since usually it takes time and skills to obtain the required molecule. Recently, the enzyme nitrate reductase from *Fusarium oxysporum* was purified by a combination of ultrafiltration and ion exchange chromatography on DEAE Sephadex (Gholami-Shabani et al. 2014). Then the molecular weight of the enzyme was estimated by gel filtration on Sephacryl S-300. Thereafter, it was used for the production of AgNPs, which according with the authors, yielded stable nonaggregating nanoparticles, spherical in shape with an average size of 50 nm (Fig. 13.8a) (Gholami-Shabani et al. 2014). Similar methodology for enzyme purification was used previously by Kumar et al. (2007b) to obtain AgNPs with nitrate reductase purified from the same fungus. Durán et al. (2014) also applied nitrate reductase to obtain silver chloride nanoparticles (Ag@AgCl) (Fig. 13.8b) with a semi-purified laccase from *Trametes versicolor*.

Despite the additional steps for enzyme purification, this type of research has good potential for industrial applications, since the purification step of NPs can be omitted.

Table 13.4 Fungal species used to produce nanoparticles via purified biomolecules

FUNGI	Reducing agent	Nanoparticles	Shape*	Size (nm)	Source or reference
Fusarium oxysporum	Nitrate reductase	Ag	Aggregates	10–25	Kumar et al. (2007b)
	Hydrogenase	Pt	Rectangular, triangular, spherical, aggregates	70–180	Govender et al. (2009)
	Nitrate reductase	Ag	Spherical	50	Gholami-Shabani et al. (2014)
Paraconiothyrium variabile	Purified laccase	Au	Quasi-spherical	71–266	Faramarzi and Forootanfar (2011)
Trametes versicolor	Semi-purified laccase	Ag@AgCl	Spherical	90–370	Durán et al. (2014)

*Shape is named as originally reported

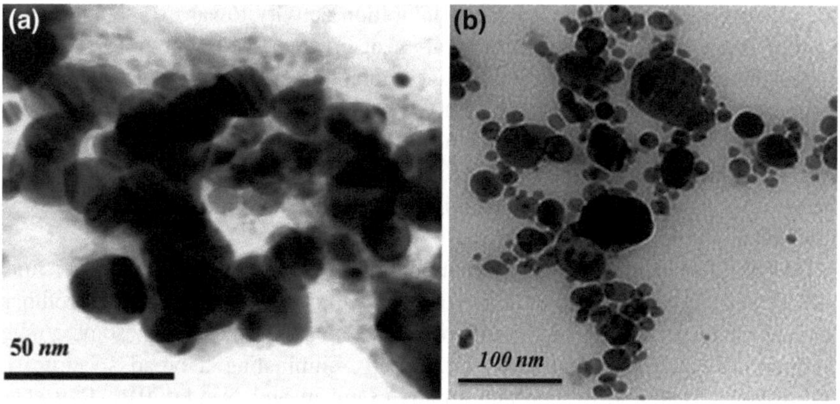

Fig. 13.8 Representative transmission electron micrographs of AgNPs synthesized using nitrate reductase purified from *Fusarium oxysporum* (**a**) and semi-purified laccase from *Trametes versicolor* (**b**). Reproduced and modified from Gholami-Shabani et al. (2014) and Durán et al. (2014), respectively, with permission from Springer Science+Business Media B.V., Germany

13.6 Potential Uses of Fungal-Mediated Metallic Nanoparticles

As mentioned previously, metallic NPs have a wide range of potential applications; therefore, the development of clean, nontoxic, and environmentally friendly procedures will greatly reduce the contaminants release and the production costs. In fact, some of the most recent literature proposes a number of potential uses of the fungal-mediated NPs. The antimicrobial properties of silver has been long recognized, therefore, several studies have carried out inhibition assays and proved the excellent antimicrobial properties of the fungal-mediated AgNPs. In most cases, AgNPs inhibited an extended number of microorganisms including different bacterial and fungal pathogens.

Successful inhibition has been demonstrated against bacterial strains such as *Staphylococcus epidermidis*, *Klebsiella pneumoniae, Bacillus cereus*, *Proteus vulgarius*, *Micrococcus luteus*, *Staphylococcus aureus*, *Pseudomonas aeruginosa*, *Bacillus subtilis*, *Enterococcus faecalis*, *Escherichia coli*, *Salmonella typhimurium* (Saha et al. 2010; Sadhasivam et al. 2010; Li et al. 2012; Rodrigues et al. 2013; San Chan and Don 2013; Joshi et al. 2013; Singh et al. 2014).

AgNPs also proven to be a good antifungal agent against fungal pathogens including *Candida* spp. (Musarrat et al. 2010; Li et al. 2012; Sagar and Ashok 2012; Qian et al. 2013; Rodrigues et al. 2013), *Aspergillus* spp. (Li et al. 2012; Qian et al. 2013; Roy et al. 2013), *Fusarium oxysporum* (Musarrat et al. 2010; Ravindra and Rajasab 2015), the dermatophytic fungal pathogens *Trichophyton* spp. (Arun et al. 2014) and the main agent causing dandruff *Malassezia furfur* (Joshi et al. 2013).

Furthermore, AgNPs showed good inhibition activity toward some strains that were resistant to the common antifungal agent Fluconazole (Li et al. 2012). Some studies have also demonstrated that a combination of AgNPs with commercial antimicrobial drugs enhances their activity, showing maximum inhibition against bacteria (Birla et al. 2009; Fayaz et al. 2010b; Devi and Joshi 2012) and fungal strains (Gajbhiye et al. 2009). These results show that fungal-mediated AgNPs could be considered as a potent antimicrobial agent, implicating their possible biomedical application.

On the other hand, the necessary amount of silver to eliminate bacteria or fungi will depend on the size and/or shape of NPs. Although is not possible to compare between studies because of the differences in biosynthesized NPs, some studies reported their excellent antimicrobial activity, eliminating a broad spectrum of microbes, even multidrug resistant strains (Saravan and Nanda 2010; Ray et al. 2011; Rathod and Ranganath 2011; Hiremath et al. 2014; Nithya and Ragunathan 2014). In a recent study, spherical AgNPs (chemically synthesized) of small size (5 nm) demonstrated the best results and mediated the fastest bactericidal activity against the tested strains (*E. coli*, *B. subtilis*, *S. aureus*) compared to AgNPs having 7 nm and 10 nm sizes at similar bacterial concentrations (Agnihotri et al. 2014).

AgNPs are among the nanomaterials that are most commercialized (Ahmed et al. 2008), they have been included in medical devices such as catheters and surgical cloths to avoid microbial proliferation. Also, it has been demonstrated that AgNPs may have potential for use as a preservative in cosmetics since they showed good preservation efficacy against mixed bacteria and mixed fungi, and did not penetrate normal human skin (Kokura et al. 2010). Thus, the development of an eco-friendly method to produce AgNPs could greatly improve biocompatibility and its potential use in medical treatments. Purifying and using single biomolecules to produce AgNPs has made some progress in this respect, for instance the enzyme nitrate reductase is found to be responsible for the extracellular synthesis of nanoparticles (Durán et al. 2005; Kumar et al. 2007b; Gade et al. 2008; Ingle et al. 2008). A number of attempts have been made to obtain AgNPs using this enzyme. Recently, the antimicrobial activity of AgNPs produced by nitrate reductase purified from *F. oxysporum* was evaluated; a strong inhibitory activity against all tested human pathogenic fungi and bacteria was found, representing a promising alternative for microbial infections (Gholami-Shabani et al. 2014).

Antimicrobial properties of other metallic NPs have been also evaluated; for instance the nanogold-bioconjugate prepared using *R. oryzae* showed high antimicrobial activity against pathogenic bacteria such as *P. aeruginosa*, *E. coli*, *B. subtilis*, *S. aureus*, *Salmonella* sp., and against the yeasts *S. cerevisiae* and *C. albicans* (Das et al. 2009b). TiO_2 NPs also showed to be a good novel antibacterial agent (Rajakumar et al. 2012) and although iron nanoparticles were also tested for antimicrobial activity against some bacterial species, they did not show the same effectiveness as AgNPs (Abdeen et al. 2013).

Other evaluations for fungal-mediated metallic NPs include their potential use for mosquito control. Silver and gold nanoparticles have been successfully tested against the dengue vector, *Aedes aegypti* (Salunkhe et al. 2011; Banu and

Balasubramanian 2014a, b; Soni and Prakash 2012; Sundaravadivelan and Padmanabhan 2014) and the mosquito vector of malaria *Anopheles stephensi* (Salunkhe et al. 2011). AgNPs have also been considered as food preservatives, particles were incorporated into sodium alginate films and suggested as vegetable and fruit preservative (Fayaz et al. 2009a). Anticancer activity was also tested on human cell lines on epidermoid larynx carcinoma (HEP-2), obtaining 27.2 and 64 % mortality at concentrations of 10 and 100 μg/mL, respectively (Arun et al. 2014). Also the anticancer properties were evaluated against MCF-7 cell line (breast carcinoma) causing a significant decrease in cell viability (Yehia and Al-Sheikh 2014). Also it was shown for the first time that biosynthesized AuNPs using *N. crassa* extract display good Surface-enhanced Raman scattering (SERS) properties, presenting promising potential applications in sensor technology (Quester et al. 2013b).

13.7 Summary and Conclusions

This chapter has placed emphasis on the different approaches employed for the formation of metallic nanoparticles using fungal species. Successful production of NPs has been obtained for all the tested fungal strains, then the use of these organisms in nanotechnology seems promising; however the control of size and shape still remains a challenge in most cases. Therefore, further research is necessary to find more efficient protocols for the production of monodisperse particles. Also, the applications of the produced NPs using these cleaner methods remain to be fully explored.

Acknowledgments Part of this work was funded by a SEP-CONACyT grant (CB2011/169154).

References

Abdeen S, Isaac RS, R, Geo S, Rose A, Praseetha PK (2013) Evaluation of antimicrobial activity of biosynthesized iron and silver nanoparticles using the fungi *Fusarium oxysporum* and Actinomycetes sp. on human pathogens. Nano Biomed Eng 5:39–45

Afreen RV, Ranganath E (2011) Synthesis of monodispersed silver nanoparticles by *Rhizopus Stolonifer* and its antibacterial activity against MDR strains of *Pseudomonas Aeruginosa* from burnt patients. Int J Environ Sci 1(7):1582–1592

Agnihotri S, Mukherji S, Mukherji S (2014) Size-controlled silver nanoparticles synthesized over the range 5–100 nm using the same protocol and their antibacterial efficacy. RSC Advances 4 (8):3974–3983

Ahluwalia V, Kumar J, Sisodia R, Shakil NA, Walia S (2014) Green synthesis of silver nanoparticles by *Trichoderma harzianum* and their bio-efficacy evaluation against *Staphylococcus aureus* and *Klebsiella pneumonia*. Ind Crops Prod 55:202–206

Ahmad A, Mukherjee P, Mandal D, Senapati S, Khan MI, Kumar R et al (2002) Enzyme mediated extracellular synthesis of CdS nanoparticles by the fungus *Fusarium oxysporum*. J Am Chem Soc 124:12108–12109

Ahmad A, Mukherjee P, Senapati S, Mandal D, Khan MI, Kumar R, et al. (2003) Extracellular biosynthesis of silver nanoparticles using the fungus *Fusarium oxysporum*. Colloids Surf B 28:313–318

Ahmed M, Karns M, Goodson M, Rowe J, Hussain S, Schlager J et al (2008) DNA damage response to different surface chemistry of silver nanoparticles in mammalian cells. Toxicol Appl Pharmacol 233:404–410

Alani F, Moo-Young M, Anderson W (2012) Biosynthesis of silver nanoparticles by a new strain of *Streptomyces* sp. compared with *Aspergillus fumigatus*. World J Microbiol Biotechnol 28:1081–1086

Anand BG, Thomas CN, Prakash S, Kumar CS (2015) Biosynthesis of silver nanoparticles by marine sediment fungi for a dose dependent cytotoxicity against HEp2 cell lines. Biocatal Agr Biotechnol 4:150–157

Arun G, Eyini M, Gunasekaran P (2014) Green synthesis of silver nanoparticles using the mushroom fungus *Schizophyllum commune* and its biomedical applications. Biotechnol Bioprocess Eng 19(6):1083–1090

Azmath P, Baker S, Rakshith D, Satish S (2015) Mycosynthesis of silver nanoparticles bearing antibacterial activity. Saudi Pharm J. doi:10.1016/j.jsps.2015.01.008

Balaji DS, Basavaraja S, Deshpande R, Mahesh DB, Prabhakar BK, Venkataraman A (2009) Extracellular biosynthesis of functionalized silver nanoparticles by strains of *Cladosporium cladosporioides* fungus. Colloids Surf B 68:88–92

Bansal V, Poddar P, Ahmad A, Sastry M (2006) Room-temperature biosynthesis of ferroelectric barium titanate nanoparticles. J Am Chem Soc 128:11958–11963

Banu A, Rathod V (2011) Synthesis and characterization of silver nanoparticles by *Rhizopus stolonier*. Int J Biomed Adv Res 2:148–158

Banu AN, Balasubramanian C (2014a) Myco-synthesis of silver nanoparticles using *Beauveria bassiana* against dengue vector, *Aedes aegypti* (Diptera: Culicidae). Parasitol Res 113 (8):2869–2877

Banu AN, Balasubramanian C (2014b) Optimization and synthesis of silver nanoparticles using *Isaria fumosorosea* against human vector mosquitoes. Parasitol Res 113(10):3843–3851

Bhainsa KC, D'Souza SF (2006) Extracellular biosynthesis of silver nanoparticles using the fungus *Aspergillus fumigatus*. Colloids Surf B 47:160–164

Bhambure R, Bule M, Shaligram N, Kamat M, Singhal R (2009) Extracellular biosynthesis of gold nanoparticles using *Aspergillus niger*–its characterization and stability. Chem Eng Technol 32 (7):1036–1041

Bhargava A, Jain N, Barathi M, Akhtar MS, Yun YS, Panwar J (2013) Synthesis, characterization and mechanistic insights of mycogenic iron oxide nanoparticles. In: Nanotechnology for sustainable development (pp 337–348). Springer International Publishing

Bharde A, Rautaray D, Bansal V, Ahmad A, Sarkar I, Yusuf SM et al (2006) Extracellular biosynthesis of magnetite using fungi. Small 2:135–141

Binupriya AR, Sathishkumar M, Vijayaraghavan K, Yun S-I (2010a) Bioreduction of trivalent aurum to nano-crystalline gold particles by active and inactive cells and cell-free extract of *Aspergillus oryzae* var. *viridis*. J Hazard Mater 177:539–545

Binupriya AR, Sathishkumar M, Yun S-I (2010b) Biocrystallization of silver and gold ions by inactive cell filtrate of *Rhizopus stolonifer*. Colloids Surf B 79:531–534

Birla SS, Tiwari VV, Gade AK, Ingle AP, Yadav AP, Rai MK (2009) Fabrication of silver nanoparticles by *Phoma glomerata* and its combined effect against *Escherichia coli*, *Pseudomonas aeruginosa* and *Staphylococcus aureus*. Lett Appl Microbiol 48:173–179

Blackwell M (2011) The fungi: 1, 2, 3. 5.1 million species? Am J Bot 98:426–438

Burda C, Chen X, Narayanan R, El-Sayed MA (2005) Chemistry and properties of nanocrystals of different shapes. Chem Rev 105:1025–1102

Castro ME, Cottet L, Castillo A (2014) Biosynthesis of gold nanoparticles by extracellular molecules produced by the phytopathogenic fungus *Botrytis cinerea*. Mater Lett 115:42–44

Castro-Longoria E, Vilchis-Nestor AR, Avalos-Borja M (2011) Biosynthesis of silver, gold and bimetallic nanoparticles using the filamentous fungus *Neurospora crassa*. Colloids Surf B 83:42–48

Castro-Longoria E, Moreno-Velásquez SD, Vilchis-Nestor AR, Arenas-Berumen E, Avalos-Borja M (2012) Production of platinum nanoparticles and nanoaggregates using *Neurospora crassa*. J Microbiol Biotechnol 22:1000–1004

Chen L, Bazylinksi DA, Lower BH (2010) Bacteria that synthesize nano-sized compasses to navigate using earth's geomagnetic field. Nat Edu Knowl 1:14

Cuevas R, Durán N, Diez MC, Tortella GR, Rubilar O (2015) Extracellular biosynthesis of copper and copper oxide nanoparticles by *Stereum hirsutum*, a native white-rot fungus from Chilean Forests. J Nanomater. doi:10.1155/2015/789089

Dameron CT, Reese RN, Mehra RK, Kortan AR, Carroll PJ, Steigerwald ML et al (1989) Biosynthesis of cadmium sulphide quantum semiconductor crystallites. Nature 338:596–597

Dar MA, Ingle A, Rai M (2013) Enhanced antimicrobial activity of silver nanoparticles synthesized by *Cryphonectria* sp. evaluated singly and in combination with antibiotics. Nanomed Nanotechnol Biol Med 9(1):105–110

Das SK, Das AR, Guha AK (2009a) Adsorption behavior of mercury on functionalized *Aspergillus versicolor* mycelia: Atomic force microscopic study. Langmuir 25:360–366

Das SK, Das AR, Guha AK (2009b) Gold nanoparticles: microbial synthesis and application in water hygiene management. Langmuir 25:8192–8199

Das SK, Das AR, Guha AK (2010) Microbial synthesis of multishaped gold nanostructures. Small 6:1012–1021

Devi LS, Joshi SR (2012) Antimicrobial and synergistic effects of silver nanoparticles synthesized using soil fungi of high altitudes of Eastern Himalaya. Mycobiology 40(1):27–34

Devi LS, Bareh DA, Joshi SR (2014) Studies on biosynthesis of antimicrobial silver nanoparticles using endophytic fungi isolated from the ethno-medicinal plant *Gloriosa superba* L. Proc Natl Acad Sci India Sect B: Biol Sci 84:1091–1099

Du L, Xian L, Feng JX (2011) Rapid extra-/intracellular biosynthesis of gold nanoparticles by the fungus *Penicillium* sp. J Nanopart Res 13(3):921–930

Durán N, Marcato PD, Alves OL, De Souza GIH, Esposito E (2005) Mechanistic aspects of biosynthesis of silver nanoparticles by several *Fusarium oxysporum* strains. J Nanobiotechnol. doi:10.1186/1477-3155-3-8

Durán N, Cuevas R, Cordi L, Rubilar O, Diez MC (2014) Biogenic silver nanoparticles associated with silver chloride nanoparticles (Ag@AgCl) produced by laccase from *Trametes versicolor*. SpringerPlus 3:645

El-Rafie MH, Mohamed AA, Shaheen TI, Hebeish A (2010) Antimicrobial effect of silver nanoparticles produced by fungal process on cotton fabrics. Carbohydr Polym 80(3):779–782

Faramarzi MA, Forootanfar H (2011) Biosynthesis and characterization of gold nanoparticles produced by laccase from *Paraconiothyrium variabile*. Colloids Surf B 87(1):23–27

Fayaz AM, Balaji K, Girilal M, Kalaichelvan PT, Venkatesan R (2009a) Mycobased synthesis of silver nanoparticles and their incorporation into sodium alginate films for vegetable and fruit preservation. J Agric Food Chem 57:6246–6252

Fayaz AM, Balaji K, Kalaichelvan PT, Venkatesan R (2009b) Fungal based synthesis of silver nanoparticles—an effect of temperature on the size of particles. Colloids Surf B 74:123–126

Fayaz AM, Tiwary CS, Kalaichelvan PT, Venkatesan R (2010a) Blue orange light emission from biogenic synthesized silver nanoparticles using *Trichoderma viride*. Colloids Surf B 75:175–178

Fayaz AM, Balaji K, Girilal M, Yadav R, Kalaichelvan PT, Venketesan R (2010b) Biogenic synthesis of silver nanoparticles and their synergistic effect with antibiotics: A study against gram-positive and gram-negative bacteria. Nanomedicine 6:103–109

Gade AK, Bonde P, Ingle AP, Marcato PD, Durán N, Rai MK (2008) Exploitation of *Aspergillus niger* for synthesis of silver nanoparticles. J Biobased Mater Bioenerg 2:123–129

Gade A, Gaikwad S, Durán N, Rai M (2014) Green synthesis of silver nanoparticles by *Phoma glomerata*. Micron 59:52–59
Gajbhiye M, Kesharwani J, Ingle A, Gade A, Rai M (2009) Fungus-mediated synthesis of silver nanoparticles and their activity against pathogenic fungi in combination with fluconazole. Nanomed Nanotechnol Biol Med 5(4):382–386
Ganachari SV, Bhat R, Deshpande R, Venkataraman A (2012) Extracellular biosynthesis of silver nanoparticles using fungi *Penicillium diversum* and their antimicrobial activity studies. BioNanoScience 2(4):316–321
Ganbarov KG, Ahmadov IS, Ramazanov MA, Musayev EM, Eyvazova QI, Aghamaliyev ZA (2014) Silver nanoparticles synthesized by the Azerbaijanian environmental isolates *Aspergillus niger*. J Microbiol Biotechnol Food Sci 4(2):137–141
Gericke M, Pinches A (2006) Biological synthesis of metal nanoparticles. Hydrometallurgy 83:132–140
Ghodake VP, Kininge PT, Magdum SP, Dive AS, Pillai MM (2011) Biosynthesis of silver nanoparticles using *Trichosporon beigelii* NCIM 3326 and evaluation of their antimicrobial activity. J Eng Res Stud E-ISSN 976:7916
Gholami-Shabani M, Akbarzadeh A, Norouzian D, Amini A, Gholami-Shabani Z, Imani A et al (2014) Antimicrobial activity and physical characterization of silver nanoparticles green synthesized using nitrate reductase from *Fusarium oxysporum*. Appl Biochem Biotechnol 172 (8):4084–4098
Govender Y, Riddin T, Gericke M, Whiteley CG (2009) Bioreduction of platinum salts into nanoparticles: a mechanistic perspective. Biotechnol Lett 31:95–100
Gupta S, Bector S (2013) Biosynthesis of extracellular and intracellular gold nanoparticles by *Aspergillus fumigatus* and *A. flavus*. Antonie Van Leeuwenhoek 103(5):1113–1123
Haefeli C, Franklin C, Hardy K (1984) Plasmid-determined silver resistance in *Pseudomonas stutzeri* isolated from a silver mine. J Bacteriol 158:389–392
Hamedi S, Shojaosadati SA, Shokrollahzadeh S, Hashemi-Najafabadi S (2014) Extracellular biosynthesis of silver nanoparticles using a novel and non-pathogenic fungus, *Neurospora intermedia*: controlled synthesis and antibacterial activity. World J Microbiol Biotechnol 30 (2):693–704
Henglein A (1989) Small-particle research: physicochemical properties of extremely small colloidal metal and semiconductor particles. Chem Rev 89:1861–1873
Hiremath J, Rathod V, Ninganagouda S, Singh D, Prema K (2014) Antibacterial activity of silver nanoparticles from *Rhizopus* spp. against Gram negative *E. coli* MDR strains. J Pure Appl Microbiol 8(1):555–562
Honary S, Barabadi H, Gharaei-Fathabad E, Naghibi F (2012) Green synthesis of copper oxide nanoparticles using *Penicillium aurantiogriseum*, *Penicillium citrinum* and *Penicillium waksmanii*. Dig J Nanomater Bios 7:999–1005
Honary S, Barabadi H, Gharaei-Fathabad E, Naghibi F (2013) Green synthesis of silver nanoparticles induced by the fungus *Penicillium citrinum*. Trop J Pharm Res 12(1):7–11
Hotze EM, Phenrat T, Lowry GV (2010) Nanoparticle aggregation: Challenges to understanding transport and reactivity in the environment. J Environ Qual 39:1909–1924
Ingle A, Gade A, Pierrat S, Sonnichsen C, Rai M (2008) Mycosynthesis of silver nanoparticles using the fungus *Fusarium acuminatum* and its activity against some human pathogenic bacteria. Curr Nanosci 4:141–144
Ingle A, Rai M, Gade A, Bawaskar M (2009) *Fusarium solani*: a novel biological agent for the extracellular synthesis of silver nanoparticles. J Nanopart Res 11(8):2079–2085
Jaidev LR, Narasimha G (2010) Fungal mediated biosynthesis of silver nanoparticles, characterization and antimicrobial activity. Colloids Surf B 81:430–433
Jain N, Bhargava A, Majumdar S, Tarafdar JC, Panwar J (2011) Extracellular biosynthesis and characterization of silver nanoparticles using *Aspergillus flavus* NJP08: a mechanism perspective. Nanoscale 3:635–641
Jain N, Bhargava A, Tarafdar JC, Singh SK, Panwar J (2013) A biomimetic approach towards synthesis of zinc oxide nanoparticles. Appl Microbiol Biotechnol 97(2):859–869

Jain S, Hirst DG (2012) O'sullivan JM. Gold nanoparticles as novel agents for cancer therapy. Br J Radiol 85(1010):101–113

Jebali A, Ramezani F, Kazemi B (2011) Biosynthesis of silver nanoparticles by *Geotricum* sp. J Cluster Sci 22(2):225–232

Joshi PA, Bonde SR, Gaikwad SC, Gade AK, Abd-Elsalam K, Rai MK (2013) Comparative studies on synthesis of silver nanoparticles by *Fusarium oxysporum* and *Macrophomina phaseolina* and it's efficacy against bacteria and *Malassezia furfur*. J Bionanosci 7(4):378–385

Kathiresan K, Manivannan S, Nabeel MA, Dhivya B (2009) Studies on silver nanoparticles synthesized by a marine fungus, *Penicillium fellutanum* isolated from coastal mangrove sediment. Colloids Surf B 71:133–137

Kathiresan K, Alikunhi NM, Pathmnaban S, Nabikhan A, Kandasamy S (2010) Analysis of antimicrobial silver nanoparticles synthesized by coastal strains of *Escherichia coli* and *Aspergillus niger*. Can J Microbiol 56:1050–1059

Kaul RK, Kumar P, Burman U, Joshi P, Agrawal A, Raliya R, Tarafdar JC (2012) Magnesium and iron nanoparticles production using microorganisms and various salts. Mater Sci Pol 30 (3):254–258

Kim YC, Park NC, Shin JS, Lee SR, Lee YJ, Moon DJ (2003) Partial oxidation of ethylene to ethylene oxide over nanosized $Ag/\alpha\text{-}Al_2O_3$ catalysts. Catal Today 87(1):153–162

Kirthi AV, Rahuman AA, Jayaseelan C, Karthik L, Marimuthu S, Santhoshkumar T et al (2012) Novel approach to synthesis silver nanoparticles using plant pathogenic fungi *Puccinia graminis*. Mater Lett 81:69–72

Klaus T, Joerger R, Olsson E, Granqvist CG (1999) Silver-based crystalline nanoparticles, microbially fabricated. Proc Natl Acad Sci USA 96:13611–13614

Kokura S, Handa O, Takagi T, Ishikawa T, Naito Y, Yoshikawa T (2010) Silver nanoparticles as a safe preservative for use in cosmetics. Nanomed Nanotechnol Biol Med 6(4):570–574

Kowshik M, Deshmukh N, Vogel W, Urban J, Kulkarni SK, Paknikar KM (2002) Microbial synthesis of semiconductor CdS nanoparticles, their characterization, and their use in the fabrication of an ideal diode. Biotechnol Bioeng 78:583–588

Kowshik M, Ashtaputre S, Kharrazi S, Vogel W, Urban J, Kulkarni SK et al (2003) Extracellular synthesis of silver nanoparticles by a silver-tolerant yeast strain MKY3. Nanotechnology 14:95–100

Krishna G, Prasad MR, Krishna PS, Bindu NSH, Samatha B, Charya MA (2015) Fungus-mediated synthesis of silver nanoparticles and their activity against Gram positive and Gram negative bacteria in combination with antibiotics. Walailak J Sci Technol (WJST), 12(7)

Krumov N, Oder S, Perner-Nocht I, Angelov A, Posten C (2007) Accumulation of CdS nanoparticles by yeasts in a fed-batch bioprocess. J Biotechnol 132:481–486

Kumar SA, Ansary AA, Ahmad A, Khan MI (2007a) Extracellular Biosynthesis of CdSe Quantum Dots by the Fungus *Fusarium oxysporum*. J Biomed Nanotechnol 3:190–194

Kumar SA, Abyabeh MK, Gosavi SW, Kulkarni SK, Pasricha R, Ahmad A et al (2007b) Nitrate reductase-mediated synthesis of silver nanoparticles from $AgNO_3$. Biotechnol Lett 29:439–445

Kumar RR, Priyadharsani KP, Thamaraiselvi K (2012) Mycogenic synthesis of silver nanoparticles by the Japanese environmental isolate *Aspergillus tamarii*. J Nanopart Res 14:1–7

Li G, He D, Qian Y, Guan B, Gao S, Cui Y et al (2012) Fungus-mediated green synthesis of silver nanoparticles using *Aspergillus terreus*. Int J Mol Sci 13:466–476

Lin CAJ, Yang TY, Lee CH, Huang SH, Sperling RA, Zanella M. et al. (2009) Synthesis, characterization, and bioconjugation of fluorescent gold nanoclusters toward biological labeling applications. Acs Nano 3(2):395–401

Maier SA (2006) Plasmonics: The promise of highly integrated optical devices. IEEE J Sel Topics Quantum Electron 12(6):1671–1677

Maliszewska I, Sadowski Z (2009) Synthesis and antibacterial activity of silver nanoparticles. In: Journal of Physics: Conference Series 146(1):012024. IOP Publishing

Maliszewska I, Juraszek A, Bielska K (2014) Green synthesis and characterization of silver nanoparticles using Ascomycota fungi *Penicillium nalgiovense* AJ12. J Cluster Sci 25(4):989–1004

Moazeni M, Shahverdi AR, Nabili M, Noorbakhsh F, Rezaie S (2014) Green synthesis of silver nanoparticles: The reasons for and against *Aspergillus parasiticus*. Nanomed J 1(4):267–275

Mody VV, Siwale R, Singh A, Mody HR (2010) Introduction to metallic nanoparticles. J Pharm Bioallied Sci 2(4):282–289. doi:10.4103/0975-7406.72127

Mohamed AGMT (2014) *Stachybotrys chartarum*: a novel biological agent for the extracellular synthesis of silver nanoparticles and their antimicrobial activity. Indonesian J Biotechnol 18(2)

Mohamed YM, Azzam AM, Amin BH, Safwat NA (2015) Mycosynthesis of iron nanoparticles by *Alternaria alternata* and its antibacterial activity. Afr J Biotechnol 14(14):1234–1241

Mohammadi B, Salouti M (2015) Extracellular bioynthesis of silver nanoparticles by *Penicillium chrysogenum* and *Penicillium expansum*. Synth React Inorg Met-Org Nano-Met Chem 45 (6):844–847

Muhsin TM, Hachim AK (2014) Mycosynthesis and characterization of silver nanoparticles and their activity against some human pathogenic bacteria. World J Microbiol Biotechnol 30 (7):2081–2090

Mukherjee P, Ahmad A, Mandal D, Senapati S, Sainkar SR, Khan MI et al (2001) Fungus-mediated synthesis of silver nanoparticles and their immobilization in the mycelial matrix: a novel biological approach to nanoparticle synthesis. Nano Lett 1:515–519

Mukherjee P, Senapati S, Mandal D, Ahmad A, Khan MI, Kumar R et al (2002) Extracellular synthesis of gold nanoparticles by the fungus *Fusarium oxysporum*. ChemBioChem 3:461–463

Mukherjee P, Roy M, Mandal BP, Dey GK, Mukherjee PK, Ghatak J et al (2008) Green synthesis of highly stabilized nanocrystalline silver particles by a non-pathogenic and agriculturally important fungus *T. asperellum*. Nanotechnology 19:075103

Musarrat J, Dwivedi S, Singh BS, Al-Khedhairy AA, Azam A, Naqvi A (2010) Production of antimicrobial silver nanoparticles in water extracts of the fungus *Amylomyces rouxii* strain KSU-09. Bioresour Technol 101:8772–8776

Narayanan KB, Park HH, Sakthivel N (2013) Extracellular synthesis of mycogenic silver nanoparticles by *Cylindrocladium floridanum* and its homogeneous catalytic degradation of 4-nitrophenol. Spectrochim Acta Part A Mol Biomol Spectrosc 116:485–490

Nithya R, Ragunathan R (2009) Synthesis of silver nanoparticle using *Pleurotus sajor* caju and its antimicrobial study. Digest J Nanomater Biostruct 4(4):623–629

Nithya R, Ragunathan R (2014) In vitro synthesis, characterization and medical application of silver nanoparticle by using a lower fungi. Middle-East J Sci Res 21(6):922–928

Nitta SK, Numata K (2013) Biopolymer-Based Nanoparticles for Drug/Gene Delivery and Tissue Engineering. Int J Mol Sci 14(1):1629–1654

Noguez C (2007) Surface plasmons on metal nanoparticles: the influence of shape and physical environment. J Phys Chem C 111:3806–3819

Nohynek GJ, Dufour EK (2012) Nano-sized cosmetic formulations or solid nanoparticles in sunscreens: A risk to human health? Arch Toxicol 86:1063–1075

O'Neal DP, Hirsch LR, Halas NJ, Payne JD, West JL (2004) Photo-thermal tumor ablation in mice using near infrared-absorbing nanoparticles. Cancer Lett 209(2):171–176

Philip D (2009) Biosynthesis of Au, Ag and Au–Ag nanoparticles using edible mushroom extract. Spectrochim Acta Part A Mol Biomol Spectrosc 73(2):374–381

Pimprikar PS, Joshi SS, Kumar AR, Zinjarde SS, Kulkarni SK (2009) Influence of biomass and gold salt concentration on nanoparticles synthesis by the tropical marine yeast *Yarrowia lipolytica* NCIM 3589. Colloids Surf B 74:309–316

Prakasham RS, Buddana SK, Yannam SK, Guntuku GS (2012) Characterization of silver nanoparticles synthesized by using marine isolate *Streptomyces albidoflavus*. J Microbiol Biotechnol 22:614–621

Prusinkiewicz MA, Farazkhorasani F, Dynes JJ, Wang J, Kaminskyj SG et al (2012) Proof-of-principle for SERS imaging of *Aspergillus nidulans* hyphae using in vivo synthesis of gold nanoparticles. Analyst 137:4934–4942. doi:10.1039/c2an35620a

Quester K (2014) The use of *Neurospora crassa* soluble proteins as bioreducing agent for the production of gold and silver nanostructures. Doctoral thesis. Centro de Investigación Científica y Educación Superior de Ensenada, Baja California, 166 pp

Quester K, Avalos-Borja M, Castro-Longoria E (2013a) Biosynthesis and microscopic study of metallic nanoparticles. Micron 54:1–27

Quester K, Avalos-Borja M, Vilchis-Nestor AR, Camacho-López MA, Castro-Longoria E (2013b) SERS properties of different sized and shaped gold nanoparticles biosynthesized under different environmental conditions by *Neurospora crassa* extract. PLoS ONE 8(10):e77486

Qian Y, Yu H, He D, Yang H, Wang W, Wan X et al (2013) Biosynthesis of silver nanoparticles by the endophytic fungus *Epicoccum nigrum* and their activity against pathogenic fungi. Bioprocess Biosyst Eng 36(11):1613–1619

Radziuk D, Skirtach A, Sukhorukov G, Shchukin D, Möhwald H (2007) Stabilization of silver nanoparticles by polyelectrolytes and poly(ethylene glycol). Macromol Rapid Commun 28:848–855

Raheman F, Deshmukh S, Ingle A, Gade A, Rai M (2011) Silver nanoparticles: novel antimicrobial agent synthesized from an endophytic fungus *Pestalotia* sp. isolated from leaves of *Syzygium cumini* (L). Nano. Biomed Eng 3(3):174–178

Rajakumar G, Rahuman AA, Roopan SM, Khanna VG, Elango G, Kamaraj C et al (2012) Fungus-mediated biosynthesis and characterization of TiO_2 nanoparticles and their activity against pathogenic bacteria. Spectrochim Acta A Mol Biomol Spectrosc 91:23–29

Rathod V, Banu A, Ranganath E (2011) Biosynthesis of highly stabilized silver nanoparticles by *Rhizopus stolonifer* and their Anti-fungal efficacy. Int J Mol Clin Microbiol 1(2):65–70

Ravindra BK, Rajasab AH (2015) Silver Nanoparticles synthesis from different fungal species and their antifungal effect. Int J Pharm Pharm Sci 7(5):165–170

Ray S, Sarkar S, Kundu S (2011) Extracellular biosynthesis of silver nanoparticles using the mycorrhizal mushroom *Tricholoma crassum* (Berk.) Sacc: its antimicrobial activity against pathogenic bacteria and fungus, including multidrug resistant plant and human bacteria. Dig J Nanomater Biostruc 6:1289–1299

Re F, Moresco R, Masserini M (2012) Nanoparticles for neuroimaging. J Phys D Appl Phys 45:073001. doi:10.1088/0022-3727/45/7/073001

Riddin TL, Gericke M, Whiteley CG (2006) Analysis of the inter- and extracellular formation of platinum nanoparticles by *Fusarium oxysporum* f. sp. *lycopersici* using response surface methodology. Nanotechnology 17:3482–3489

Rodrigues AG, Ping LY, Marcato PD, Alves OL, Silva MC, De Souza AO et al (2013) Biogenic antimicrobial silver nanoparticles produced by fungi. Appl Microbiol Biotechnol 97(2):775–782

Roy S, Mukherjee T, Chakraborty S, Das TK (2013) Biosynthesis, characterization and antifungal activity of silver nanoparticles synthesized by the fungus *Aspergillus foetidus* MTCC8876. Dig J Nanomat Bios 8:197–205

Sadhasivam S, Shammugam P, Yun K (2010) Biosynthesis of silver nanoparticles by *Streptomyces hygroscopicus* and antimicrobial activity against medically important pathogenic microorganisms. Colloids Surf B 81:358–362

Sadhasivam S, Shanmugam P, Veerapandian M, Subbiah R, Yun K (2012) Biogenic synthesis of multidimensional gold nanoparticles assisted by *Streptomyces hygroscopicus* and its electrochemical and antibacterial properties. Biometals 25:351–360

Sagar G, Ashok B (2012) Green synthesis of silver nanoparticles using *Aspergillus niger* and its efficacy against human pathogens. Eur J Exp Biol 2:1654–1658

Saha S, Sarkar J, Chattopadhyay D, Patra S, Chakraborty A, Acharya K (2010) Production of silver nanoparticles by a phytopathogenic fungus *Bipolaris nodulosa* and its antimicrobial activity. Dig J Nanomater Biostruct 5:887–895

Saha S, Chattopadhyay D, Acharya K (2011) Preparation of silver nanoparticles by bio-reduction using *Nigrospora oryzae* culture filtrate and its antimicrobial activity. Digest J Nanomater Biostruct (DJNB) 6(4)

Salunkhe RB, Patil SV, Patil CD (2011) Salunke. Larvicidal potential of silver nanoparticles synthesized using fungus *Cochliobolus lunatus* against *Aedes aegypti* (Linnaeus, 1762) and *Anopheles stephensi* Liston (Diptera; Culicidae). Parasitol Res 109(3):823–831

San Chan Y, Don MM (2013) Biosynthesis and structural characterization of Ag nanoparticles from white rot fungi. Mater Sci Eng C 33(1):282–288

Sanghi R, Verma P (2009a) Biomimetic synthesis and characterization of protein capped silver nanoparticles. Bioresour Technol 100:501–504

Sanghi R, Verma P (2009b) A facile green extracellular biosynthesis of CdS nanoparticles by immobilized fungus. Chem Eng J 155:886–891

Sanghi R, Verma P (2010) pH dependant fungal proteins in the 'green' synthesis of gold nanoparticles. Adv Mat Lett 1:193–199

Saravanan M, Nanda A (2010) Extracellular synthesis of silver bionanoparticles from *Aspergillus clavatus* and its antimicrobial activity against MRSA and MRSE. Colloids Surf B 77:214–218

Sarkar J, Ray S, Chattopadhyay D, Laskar A, Acharya K (2012) Mycogenesis of gold nanoparticles using a phytopathogen *Alternaria alternata*. Bioprocess Biosyst Eng 35:637–643

Sarkar J, Roy SK, Laskar A, Chattopadhyay D, Acharya K (2013) Bioreduction of chloroaurate ions to gold nanoparticles by culture filtrate of *Pleurotus sapidus* Quél. Mater Lett 92:313–316

Sawle BD, Salimath B, Deshpande R, Bedre MD, Prabhakar BK, Venkataraman A (2008) Biosynthesis and stabilization of Au and Au–Ag alloy nanoparticles by fungus, *Fusarium semitectum*. Sci Technol Adv Mater 9(3):035012

Selvi KV, Sivakumar T (2012) Isolation and characterization of silver nanoparticles from *Fusarium oxysporum*. Int J Curr Microbiol App Sci 1(1):56–62

Senapati S, Ahmad A, Khan MI, Sastry M, Kumar R (2005) Extracellular biosynthesis of bimetallic Au-Ag alloy nanoparticles. Small 1:517–520

Shaligram NS, Bule M, Bhambure R, Singhal RS, Singh SK, Szakacs G et al (2009) Biosynthesis of silver nanoparticles using aqueous extract from the compactin producing fungal strain. Process Biochem 44(8):939–943

Shankar SS, Ahmad A, Pasricha R, Sastry M (2003) Bioreduction of chloroaurate ions by geranium leaves and its endophytic fungus yields gold nanoparticles of different shapes. J Mater Chem 13:1822–1826

Shankar SS, Ahmad A, Pasricha R, Khan MI, Kumar R, Sastry M (2004) Immobilization of biogenic gold nanoparticles in thermally evaporated fatty acid and amine thin films. J Colloid Interface Sci 274:69–75

Sheikhloo Z, Salouti M, Katiraee F (2011) Biological synthesis of gold nanoparticles by fungus *Epicoccum nigrum*. J Cluster Sci 22(4):661–665

Singh P, Raja RB (2011) Biological synthesis and characterization of silver nanoparticles using the fungus *Trichoderma harzianum*. Asian J Exp Biol Sci 2(4):600–605

Singh D, Rathod V, Ninganagouda S, Hiremath J, Singh AK, Mathew J (2014) Optimization and characterization of silver nanoparticle by endophytic fungi *Penicillium* sp. isolated from *Curcuma longa* (Turmeric) and application studies against MDR *E. coli* and *S. aureus*. Bioinorg Chem Appl. doi:10.1155/2014/408021

Soni N, Prakash S (2012) Efficacy of fungus mediated silver and gold nanoparticles against *Aedes aegypti* larvae. Parasitol Res 110:175–184. doi:10.1007/s00436-011-2467-4

Spring S, Schleifer KH (1995) Diversity of magnetotactic bacteria. Syst Appl Microbiol 18:147–153

Sundaravadivelan C, Padmanabhan MN (2014) Effect of mycosynthesized silver nanoparticles from filtrate of *Trichoderma harzianum* against larvae and pupa of dengue vector *Aedes aegypti* L. Environ Sci Pollut Res 21(6):4624–4633

Tripathi RM, Gupta RK, Singh P, Bhadwal AS, Shrivastav A, Kumar N et al (2014) Ultra-sensitive detection of mercury (II) ions in water sample using gold nanoparticles synthesized by *Trichoderma harzianum* and their mechanistic approach. Sensors Actuators B: Chem 204:637–646

Vahabi K, Mansoori GA, Karimi S (2011) Biosynthesis of silver nanoparticles by fungus *Trichoderma reesei* (a route for large-scale production of AgNPs). Insci J 1(1):65–79

Vala AK (2015) Exploration on green synthesis of gold nanoparticles by a marine-derived fungus Aspergillus sydowii. Environ Progress Sustain Energy 34(1):194–197

Vala AK, Shah S, Patel R (2014) Biogenesis of silver nanoparticles by marine-derived fungus *Aspergillus flavus* from Bhavnagar Coast, Gulf of Khambhat, India. J Mar Biol Oceanogr 3 (1):2

Varshney R, Mishra AN, Bhadauria S, Gaur MS (2009) A novel microbial route to synthesize silver nanoparticles using fungus *Hormoconis resinae*. Digest J Nanomater Biostruct 4(2):349–355

Velmurugan P, Shim J, You Y, Choi S, Kamala-Kannan S, Lee K.-J, et al. (2010) Removal of zinc by live, dead, and dried biomass of *Fusarium* spp. isolated from the abandoned-metal mine in South Korea and its perspective of producing nanocrystals. J Hazard Mater 182:317–324

Verma VC, Kharwar RN, Gange AC (2010) Biosynthesis of antimicrobial silver nanoparticles by the endophytic fungus *Aspergillus clavatus*. Nanomedicine 5:33–40

Vigneshwaran N, Kathe AA, Varadarajan PV, Nachane RP, Balasubramanya RH (2006) Biomimetics of silver nanoparticles by white rot fungus *Phaenerochete chrysosporium*. Colloids Surf B 53:55–59

Vigneshwaran N, Ashtputre NM, Varadarajan PV, Nachane RP, Paralikar KM, Balasubramanya RH (2007) Biological synthesis of silver nanoparticles using the fungus *Aspergillus flavus*. Mater Lett 61:1413–1418

Xie J, Lee JY, Wang DIC, Ting YP (2007) High-yield synthesis of complex gold nanostructures in a fungal system. J Phys Chem C 111:16858–16865

Yehia RS, Al-Sheikh H (2014) Biosynthesis and characterization of silver nanoparticles produced by *Pleurotus ostreatus* and their anticandidal and anticancer activities. World J Microbiol Biotechnol 30(11):2797–2803

Zhang Y, Gao G, Qian Q, Cui D (2012) Chloroplasts-mediated biosynthesis of nanoscale Au-Ag alloy for 2-butanone assay based on electrochemical sensor. Nanoscale Res Lett 7:475

Chapter 14
Unraveling the Chemical Interactions of Fungal Endophytes for Exploitation as Microbial Factories

Wen-Xuan Wang, Souvik Kusari and Michael Spiteller

14.1 Introduction

Since the emergence of plants on the earth, microbes and plants have been coevolving for millions of years and forming a plethora of multifaceted inter-kingdom interaction systems (Partida-Martínez and Heil 2011; Goh et al. 2013; Christian et al. 2015). Besides epiphytic microorganisms, mycorrhizae, rhizobacteria, and other plant-associated microscopic organisms, there is a group of microorganisms (mostly fungi and bacteria) living inside plants without causing any harm or perceptive symptoms. These special groups of organisms are known as endophytic microorganisms, which are also called endophytes (in Greek, *endon* means living inside and *phyton* means plant) (Fig. 14.1) (Wilson 1995). More recent investigations have showed that such a specific microbial lifestyle, generally known as "endophytism" is a common phenomenon, and every plant species investigated for endophytes has shown to harbor at least one endophyte (Kusari et al. 2012a).

W.-X. Wang · S. Kusari (✉) · M. Spiteller (✉)
Department of Chemistry and Chemical Biology, Chair of Environmental Chemistry and Analytical Chemistry, Institute of Environmental Research (INFU), TU Dortmund, Otto-Hahn-Str. 6, 44221 Dortmund, Germany
e-mail: souvik.kusari@tu-dortmund.de

M. Spiteller
e-mail: michael.spiteller@tu-dortmund.de

D. Purchase (ed.), *Fungal Applications in Sustainable Environmental Biotechnology*, Fungal Biology, DOI 10.1007/978-3-319-42852-9_14

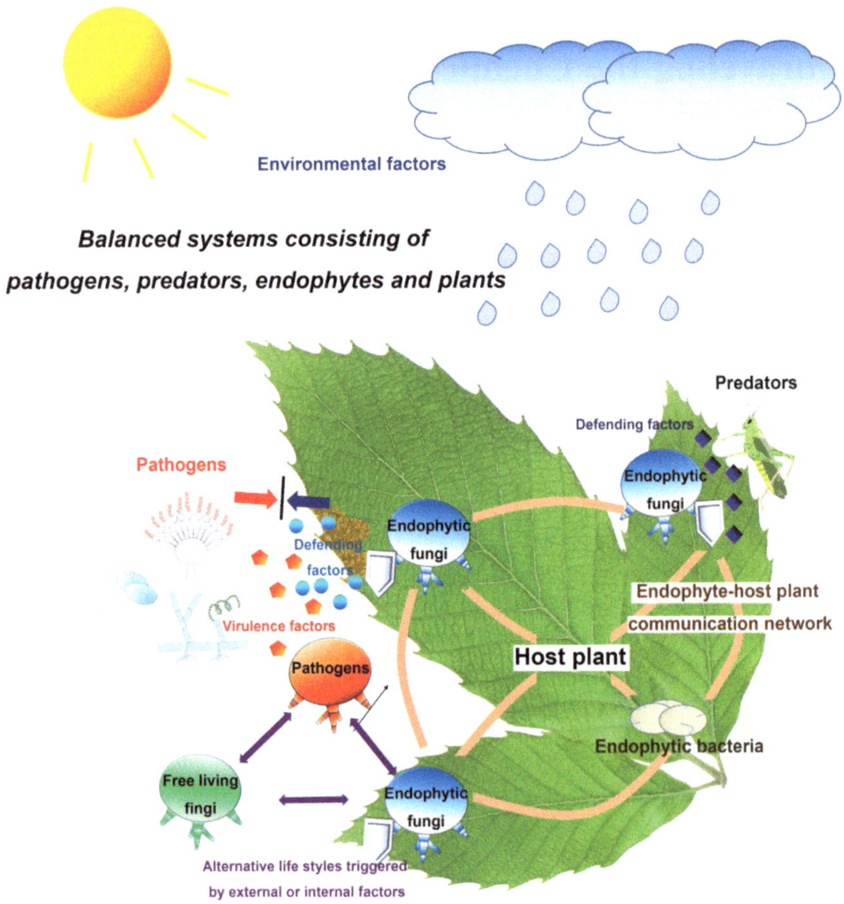

Fig. 14.1 Schematic representation of the network of endophytic microbial community comprising of a plethora of macro and microorganisms

14.2 Potential of Endophytic Fungi for Producing Interesting Natural Products

Although initial insights into the existence of endophytic fungi within plants were developed since the end of the nineteenth century, these fascinating groups of organisms remained unexploited for several decades (Kusari et al. 2012a). The well-known historic investigation of "fescue toxicosis" of livestock caused by the endophytic fungal toxic alkaloids in tall fescue brought the prominence of endophytic fungi to the forefront (Lyons et al. 1986). Further research showed that these endophytic fungi are capable of defending host plants against insects and herbivorous mammals (Clay and Cheplick 1989). This breakthrough suggested that endophytic fungi may not exhibit pathogenic or parasitic lifestyle in their natural ecological niche (inside host plants), but are in fact mutualistic partners providing competitive advantages for their hosts (Kusari et al. 2012a). Numerous examples have now substantiated that by producing bioactive secondary metabolites, endophytic fungi are capable of enhancing the tolerance of host plants toward diverse biotic or abiotic stresses (Tan and Zou 2001; Saikkonen et al. 2013; Nisa et al. 2015). The process of searching for (bio-)functional compounds from endophytic fungi have already led to the discovery of a plethora of compounds encompassing almost all possible chemical scaffolds isolated from the fungi from other habitats (Tan and Zou 2001; Kharwar et al. 2011). It has cemented the notion that the possibility to find unknown bioactive compounds from endophytic fungi is enormous (Kharwar et al. 2011). However, it is noteworthy that such a unique trait of endophytic fungi can also be found in other niches, because investigation on fungi from other habitats including invertebrate animals and extreme conditions (cold or hot temperatures, low pH, salinity, etc.) continue to yield different bioactive compounds (Newman and Cragg 2016).

Interestingly, endophytic fungi have been shown to produce several clinically valuable compounds which were initially isolated and characterized from their host plants. In 1993, paclitaxel (Taxol®) was reported to be produced by an endophytic fungus, *Taxomyces andreanae*, harbored in the inner bark of pacific yew (*Taxus brevifolia*) (Stierle et al. 1993). This milestone discovery of microbial production of a potent anticancer compound (used to treat different forms of cancer including ovarian, breast, lung, bladder, prostate, melanoma, esophageal, etc.) inspired scientists to rethink the biosynthetic origin of bioactive plant compounds in light of plant-endophytic interactions. Over time, several other natural products initially isolated from plants were found to be produced by their associated endophytic fungi, including several well-known compounds (Table 14.1). In 2001, Tan et al. proposed that genetic recombination between the coexisting endophytic fungi and their host plants occurring during their coevolution might be the reason that endophytes are able to produce plant compounds or analogs (Tan and Zou 2001). Given the close relationship between endophytic fungi and host plants, the genetic recombination mechanism might include horizontal gene transfer as reported among endophytic and epiphytic bacteria (Nongkhlaw and Joshi 2016). However, Xiong et al. (2013) suggested that the microbial paclitaxel biosynthetic genes *ts* and *bapt* showed low similarities to the genes from *Taxus* species, which did not support the hypothesis of plant-endophyte genetic recombination.

The advantages of obtaining bioactive compounds from fungi rather than the limited plant resources are obvious (Kharwar et al. 2011), especially for the valuable compounds sourced from threatened plants (for example, paclitaxel from pacific yew) (Huxtable 1992). This initially led to high hopes of producing these compounds for the pharmaceutical industry using endophytic fungi by large-scale fermentation. However, the results so far have been disappointing (Kusari and Spiteller 2011). In almost all cases to date, it has been seen that under in vitro axenic monoculture conditions, endophytic fungi cannot maintain the capacity of producing desired compounds over repeated subcultures (Kusari and Spiteller 2011; Kusari et al. 2014b, c). It appears that certain triggers are lost when a single

Table 14.1 Representative list of important plant natural products with alternative endophytic fungal sources

Natural products	Plant resources	Endophytic fungal producers	References
Azadirachtin A and B	*Azadirachta indica*	*Eupenicillium parvum*	Kusari et al. (2012b)
Camptothecins	*Camptotheca acuminata* *Nothapodytes* spp.	*Trichoderma atroviride* LY357 *Fusarium solani* *Fomitopsis* sp. P. Karst *Alternaria alternata* (Fr.) Keissl *Phomposis* sp. *Neurospora* sp. *Entrophospora infrequens*	Rehman et al. (2008), Puri et al. (2005), Kusari et al. (2009b, 2011), Pu et al. (2013), Shweta et al. (2013)
Ginkgolide B	*Ginkgo biloba*	*Fusarium oxysporum*	Cui et al. (2012)
Gymnemagenin	*Gymnema sylvestre*	*Penicillium oxalicum*	Parthasarathy and Sathiyabama (2014)
Hypericin and emodin	*Hypericum* spp.	*Thielavia subthermophila*	Kusari et al. (2008, 2009a)
Huperzine A	Family Huperziaceae	*Paecilomyces tenuis* YS-13 *Colletotrichum gloeosporioides*	Wang et al. (2011), Zhao et al. (2013), Dong et al. (2014), Shu et al. (2014), Zhang et al. (2015), Su and Yang (2015)
Paclitaxel	*Taxus* spp. *Corylus avellena*	More than 200 fungi belonging to more than 40 fungal genera, see the text	Flores-Bustamante et al. (2010), Hao et al. (2013), Kusari et al. (2014c)
Phillyrin	*Forsythia suspensa*	*Colletotrichum gloeosporioides. (Penz.)*	Zhang et al. (2012)
Piperine	*Piper longum*	*Periconia* sp.	Chithra et al. (2014)
Podophyllotoxins	*Podophyllum* sps.	*Trametes hirsute* *Phialocephala fortinii*	Puri et al. (2006), Eyberger et al. (2006)
Sanguinarine	*Macleaya cordata*	*Fusarium proliferatum* BLH51	Wang et al. (2014)
Silybin A, silybin B, and isosilybin A	*Silybum marianum*	*Aspergillus iizukae*	El-Elimat et al. (2014)
Vinblastine and vincristine	*Catharanthus roseus*	*Fusarium oxysporum*	Kumar et al. (2013)

endophytic fungus is cultivated in artificial conditions, which leads to the gradual shutdown or obliteration of biosynthesis of the desired compounds that are only necessary in the endophytic life cycles (Kusari and Spiteller 2011; Kusari et al. 2014b, c). For example, over 200 fungal strains encompassing 40 fungal genera have so far shown to produce paclitaxel, but no successful industry-scale production has been achieved until now (Kusari et al. 2014c). These failures indicate that the endophytic microbial communities in plants are far more complicated than initially assumed and introducing new strategies to fully understand the multifaceted crosstalk between plants and associated microorganisms is inevitable before endophytes can be used as sustainable microbial factories for industrial production of target compounds.

14.3 Endophytic Fungus-Plant Interactions

During the process of coevolution, plants and endophytes develop strategies to adapt toward coexistence with each other. For example, the endophytic fungus *Piriformospora indica* evades the defense of host plants such as *Arabidopsis* by establishing a biotrophic interaction (Lahrmann and Zuccaro 2012). At the early stage of evasion, *P. indica* interferes with the plant hormone level and secretes fungal lectins and effector proteins to suppress host defenses. Thereafter, *P. indica* preferentially grows in moribund host cells and secretes digestive enzymes to degrade proteins (Lahrmann and Zuccaro 2012). However, this colonization does not cause massive host cell death as other hemibiotrophic or necrotrophic fungi, but benefit the host plant by providing it tolerance to biotic and abiotic adversities (Lahrmann and Zuccaro 2012). Furthermore, endophytic colonization can influence the gene expression of both endophytic fungi and host plants. In the case of biocontrol endophytic *Trichoderma* species, colonization into host plant (*Theobroma cacao*) causes a number of genes, both fungal and plant, to be induced or repressed (Bailey et al. 2006). The switch-on/off of operons in endophytic fungi concomitant to host plants reveals that fungal endophytes and their hosts can establish a genetic crosstalk system when the endophytic association occurs (Bailey et al. 2006). Therefore, it is not surprising that separating endophytic fungi from their habitat (host plant tissue) will alter their genetic as well as phenotypic characteristics, including their ability to produce certain secondary metabolites.

Several recent studies have further revealed that host plants are able to regulate the expression of endophytic secondary metabolite biosynthetic genes (Tanaka et al. 2012). For instance, Young et al. (2006) reported a complex gene cluster (*ltm* cluster) in the grass endophyte *Neotyphodium lolii*, which is responsible for the biosynthesis of lolitrems. Interestingly, all 10 genes composing this gene cluster were highly expressed in the host plant, but only poorly expressed or completely unexpressed in mycelia (Young et al. 2006). Comparison to orthologous clusters from other fungal species suggested that the *ltm* cluster is located in a fast evolving region which renders this gene cluster ready to evolve under the stress of natural selection. Therefore, the lack of stability was considered as a major driving force leading to chemical diversities of compounds produced by the endophyte (Young et al. 2006; Tanaka et al. 2012). In another study, Kusari et al. (2011) expounded the cross-kingdom biosynthesis of camptothecin in an endophytic fungus-host plant system, which is helpful in considering some of the reasons for the observed degeneration of endophytic fungi after separation from their hosts. Detailed examination of an endophytic fungus, *Fusarium solani*, harboring in the Chinese plant *Camptotheca acuminata* revealed that the fungus is able to synthesize precursors of the well-known anticancer compound camptothecin, while the host plant provides the enzyme strictosidine synthase to condense the tryptophan-derived part with the monoterpene part, thereby completing the biogenesis of camptothecin (Kusari et al. 2011). This unique mechanism of cross-species biosynthesis explained why the in vitro monoculture of endophytic *F. solani* could not preserve the capacity of producing camptothecin without the aid of the host plant. Surprisingly, reinfecting the host plant with the endophytic fungus after repeated subculturing did not restore the production of camptothecin. Random nonsynonymous mutations, particularly in the camptothecin biosynthetic genes of the in vitro subcultured endophytic fungus, could be attributed to the observed biosynthetic silencing of the endophyte (Kusari et al. 2011). Instability of selected biosynthetic genes, therefore, is another significant cause of loss of certain biosynthetic functions in endophytes growing under in vitro conditions. The mechanism or selection pressures that enable wild-type endophytic fungal strains to maintain the intact structure and/or function of their biosynthetic genes in host plants are still largely unknown. Nevertheless, it is becoming clear that regulation and apoplastic conditions of the plant's internal environment play an essential role for the endophytes to develop and maintain their functional traits.

14.4 Crosstalk Between Endophytes

In the natural niches of endophytes, there are many coexisting micro and macroorganisms including cells of the host plant, other endophytes, and possibly pathogenic organisms and predators (Kusari et al. 2014b) (Fig. 14.1). There are also varying environmental factors, such as sunshine, moisture, temperature, and nutrients, to name a few. Because of the adaptive traits and natural selection, each organism in these endophytic microbial communities can change their phenotype or even genotype in response to the circumstances of the immediate vicinity, which can also influence other organisms (Soen 2014). These so-called 'awareness' of circumstantial changes and the corresponding gene regulations can be achieved via signal pathways connecting each individual in these mixed communities, which are well-exemplified by the commonly known quorum sensing cascade systems for bacteria (Nickerson et al. 2006; Schaefer et al. 2008; Deng et al. 2011; Eldar 2011; Dandekar et al. 2012) and signal sensing systems for fungi (Albuquerque and Casadevall 2012; Bandara et al. 2012; Grice et al. 2013; Mallick and Bennett 2013;

Feldman et al. 2014). On one hand, signaling pathways are essential to maintain endophytic association (mutualism) that is transposable to pathogenicity upon biotic or abiotic interference (Eaton et al. 2011). On the other hand, mutation or disruption of certain virulent genes (for example, pGMR1 of *Colletotrichum magna*) can also change pathogenic fungi into endophytes (Freeman and Rodriguez 1993; Redman et al. 1999). These kinds of alternative gene expression patterns can also be achieved via natural cascade signal pathways. Therefore, it is compelling that endophytic microbial systems are balanced by the interactions or communications between each individual, which determines the functional roles of the resident microbes. Recently, for example, the antineoplastic so-called plant compound maytansine was ascertained to be produced by the endophytic bacterial community harbored in the roots of *Putterlickia* plants (Kusari et al. 2014a). Given the cytotoxic and antimicrobial efficacies of maytansine, it obviously has impacts on other coexisting organisms. Concomitantly, the endophytic bacterial community of the plant (including the maytansine producers) and host plant cells themselves must have developed strategies to resist the toxicity of maytansine by communicating with each other over an evolutionary period (Kusari et al. 2014a). Following-up this extremely interesting outcome, we have recently discovered another endophytic community harboring stems of a Cameroonian *Maytenus serrata* plant, which are able to biosynthesize maytansine via a plant-endophyte cross-species pathway (Kusari et al. 2016). We proved that the biosynthesis of maytansine in *M. serrata* is shared between the endophytic bacterial community colonizing the stem and the host plant containing non-cultivable cryptic endophytes. Our work demonstrates that maytansine is biosynthesized only when a host plant joins forces with its selected and very eco-specific endophytic bacterial community (Fig. 14.2). Without this aforementioned combination, maytansine cannot be biosynthesized. Similar interactions leading to plant fitness benefits have also been explored and commercially exploited in other plants (see for example, von Maltzahn et al. 2016). As apparent in these examples, such complex communications in endophytic microbial systems might influence the expression of genes, including the biosynthetic genes in endophytic fungi or bacteria that are responsible for the production of pharmaceutically relevant compounds (Suryanarayanan et al. 2009).

The diffusible small signal molecules like *N*-Acyl homoserine lactones (AHLs) have been shown to trigger cryptic natural product biosyntheses (Scherlach and Hertweck 2009), they are probably also the tools of communication between endophytes and plants, or between endophytes and endophytes (Kusari et al. 2012a). Wang et al. (2015) recently reported that a series of novel hexacyclopeptides that act as crosstalk molecules between an endophytic fungus (*F. solani*) harboring *Narcissus tazetta* and an endophytic bacterium *Achromobacter xylosoxidans* isolated from the same host tissue. These crosstalk molecules were secreted by *F. solani* to the surrounding medium, and selectively accumulated by this bacterium over time (Fig. 14.3) (Wang et al. 2015). The capability of this *F. solani* strain to produce these hexacyclopeptides degenerated sharply after isolation. This interesting secretion-uptake behavior supported the corollary of endophytism

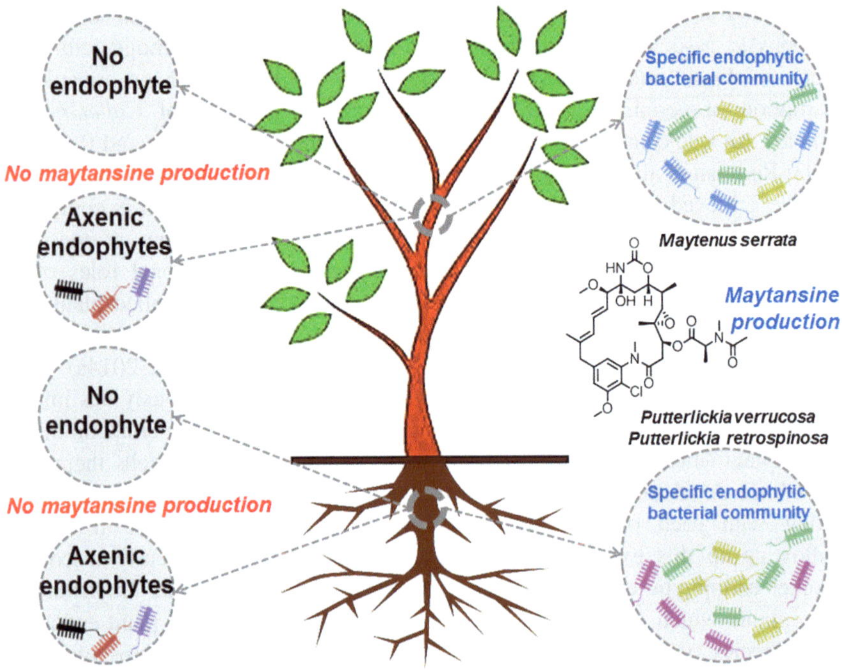

Fig. 14.2 Schematic representation showing that only selected communities of endophytic bacteria (in *Putterlickia* plants), and a combination of endophytic bacterial communities and host plant (in *Maytenus serrata*) lead to production of maytansine. See Kusari et al. (2014a, b, c, 2016)

in the view of coadaption and coevolution (Tan and Zou 2001; Suryanarayanan 2013; Kusari et al. 2015); such microbial neighbor communication tactics further suggested that endophytes generally establish one or more crosstalk strategies in their natural niches for survival (Kusari et al. 2014a; Wang et al. 2015).

14.5 Technologies for Utilizing Endophytic Fungi as Microbial Factories

It is generally held that when Stierle et al. (1993) discovered the fungal strain producing paclitaxel, hopes were high that it could be used as a sustainable source for the paclitaxel manufacture by fermentation. Two decades later, however, success still eludes us not only for the industrial production of paclitaxel using endophytes, but also for other noteworthy compounds (Kusari et al. 2014b).

To solve the difficulties in natural product research, well-established techniques have emerged in the recent years. For example, genome mining and synthetic biology methods enable us to obtain cryptic bioactive compounds which cannot be

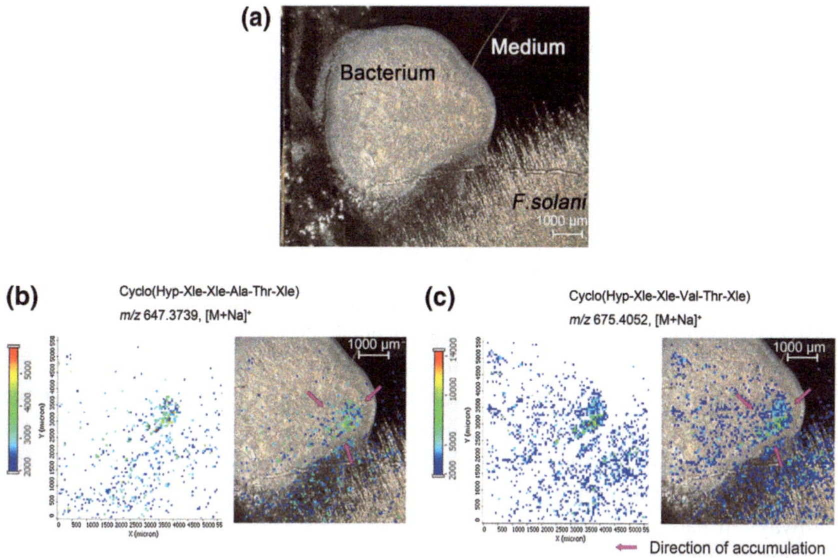

Fig. 14.3 Distribution of two masses detected in Matrix-assisted laser desorption ionization imaging high-resolution mass spectrometry (MALDI-imaging-HRMS) scan of the endophytic fungus *F. solani* and another endophytic bacterium *A. xylosoxidans* harbored in the same tissue of *Narcissus tazetta*. **a** Optical image of the boundary area between *F. solani* and bacterium *A. xylosoxidans*. **b** Ion intensity map of mass *m/z* 647.3739, $C_{30}H_{52}O_8N_6Na$, $[M+Na]^+$, (cyclo (Hyp-Xle-Xle-Ala-Thr-Xle)). **c** Ion intensity map of mass *m/z* 675.4052, $C_{32}H_{56}O_8N_6Na$, $[M+Na]^+$, (cyclo(Hyp-Xle-Xle-Val-Thr-Xle)). Adapted and modified from Wang et al. (2015) with permission

produced in artificial conditions (because the biosynthetic genes are silenced or the organisms cannot be cultivated) (Li et al. 2016). Moreover, engineering biosynthetic pathways in heterologous organisms are promoting the manufacture of valuable natural products (Luo et al. 2015), which was well exemplified by the fermentation of artemisinic acid in heterologous hosts *Escherichia coli* and *Saccharomyces cerevisiae* (Paddon et al. 2013). On the protein level, further, chemoenzymatic approaches directly utilize enzymes in certain step(s) of synthesis (Li 2010), which were applied in the synthesis of several important natural products including heparins (Xu et al. 2011) and thiocillins (Wever et al. 2015). Owing to the fact that these strategies can be potentially used for any biosynthetic pathway, they have been considered for alternative production of natural products sourced from plants (such as paclitaxel). Despite many relevant endeavors, however, there are still several bottlenecks which need to be overcome for using these technologies on an industrial scale, particularly for cascade, compartmentalized pathways at the plant-endophyte interface. As a representative example, the biosynthetic pathway of paclitaxel in yew trees has been demonstrated to consist of 19 enzyme catalyzed steps (Flores-Bustamante et al. 2010). Based on the biosynthetic pathway, a heterologous biosynthetic pathway was first constructed in *E. coli* whereby the key precursor of paclitaxel, taxadiene, was obtained (Huang et al. 2001). However, because of the lack of understanding of several key steps, heterologous biosynthesis of the final product paclitaxel is still impending (Luo et al. 2015). For instance, when the cytochrome P450s (cP450s, key enzymes for the final biosynthesis of paclitaxel) were expressed in heterologous systems, they presented no functionality because of incorrect folding, transportation and insertion to the cell membrane (Howat et al. 2014). Furthermore, the cP450 reductases are also required for the efficient function of cP450s. Despite the achievement of coupling cP450 reductase with first cP450 hydroxylation (Jennewein et al. 2005), the co-expression of all required cP450 reductases is still problematic (Ajikumar et al. 2010). It suggests that delineation of the biosynthetic mechanisms is the prerequisite for manipulating these biosynthetic functions, because the production of natural products may not be only determined by the functionalities of biosynthetic enzymes but also regulated by many other factors including signaling crosstalk and compartmentalization of the pathways in nature.

Thus far, endophytic fungi have become remarkable resources for "discovery" of a plethora of natural products, because they are naturally "engineered" by coevolution as nature's so-called heterologous hosts over time. Unlike free living fungi, it is quite difficult for discerning fermentation methods employing strategies like OSMAC (one strain many compounds) (Bode et al. 2002; Scherlach and Hertweck 2009; Kusari et al. 2012a), or dual and multiple/mixed cultures (Oh et al. 2007; Zuck et al. 2011; Bertrand et al. 2013) to introduce the natural triggers into artificial conditions or mimic the natural environment in which endophytes live (Wang et al. 2015). This is because the endophytic communities consist of many other cryptic microorganisms (including endophytic viruses) that are influenced by numerous environmental factors (Suryanarayanan 2013).

Recently, in situ methods used to investigate uncultured microorganisms have started providing useful clues to reconstruct the natural conditions for endophytes in the laboratory (Ling et al. 2015). Moreover, several innovative biotechnological advances in systems biology (Arkin and Schaffer 2011), including high-throughput and next-generation sequencing, proteomics, metabolomics, secretomics, transcriptomics, and bioinformatics might enable us to comprehensively characterize endophytic metabolism and gene expression, and then analyze them on systems level (Kusari et al. 2012a; Christian et al. 2015). For example, Arabidopsis Interactome Mapping Consortium (2011) worked out a binary protein–protein interaction map of the plant *Arabidopsis thaliana* to describe the interactome network involving 6200 highly reliable interactions between 2700 proteins. This kind of strategy could be applied on the elucidation of the complex interactome network in plant-endophyte systems as well, which may be useful to clarify the relationships between biosynthetic phenotype and genotype of endophytes across space and time.

As opposed to obtaining the overall information about endophytic communities, simplifying the crosstalk system is also critical to unveil the fundamental principles of endophytic interaction network(s). For example, Youk and Lim (2014) constructed a synthetic signal secrete-and-sense circuit motif in yeast, and mathematically demonstrated how microbes achieve versatile social behaviors by altering parameters of the signaling system, including secretion/degradation rate, receptor abundance, and positive feedback linking sensing and secretion. Although interpreting the cross-species crosstalk with multiple signaling systems will be much more challenging, this work exemplified a basic framework to design models to understand the behaviors of endophytic communities.

Remarkably, the development of imaging mass spectrometry (IMS) is also a great tool to visualize, both spatially and temporally, the chemical communication between endophytes (fungi as well as bacteria), and discover important secondary metabolites at physiological concentrations (Wang et al. 2015). IMS also allows us to investigate the minor functional compounds produced by endophytes only in their ecological niches inside host plants, which may be ignored by conventional fermentation techniques and purification strategies (Esquenazi et al. 2009; Yang et al. 2009). With the idea of "distribution represents function," IMS could be developed into a routine method to reveal the ecological functions of natural products from endophytic fungi. For example, in our recent investigation on the bioactive azaphilones, colletotrichones A–C, isolated from the endophytic fungus *Colletotrichum* sp. BS4, colletotrichone A showed significant antibacterial activities against environmental bacteria *E. coli* and *Bacillus subtilis* (Wang et al. 2016). Thereafter, MALDI IMS experiments were performed on its colonies to reveal their ecological function(s) (Fig. 14.4), with the established method (Wang et al. 2015). On rice agar (rice flour 20 g, agar 15 g in 1 L water), *Colletotrichum* sp. BS4 revealed some typical morphological features of the genus (Cano et al. 2004; Photita et al. 2005), such as the dark-colored, thick-walled hyphal clumps (setae) (Fig. 14.4). In the IMS imaging (Fig. 14.4), it was interesting to note that colletotrichone A was secreted into agar after production, while colletotrichone B and/or chermesinone B was localized at the site of production at or around the hyphal clumps (Fig. 14.4c, d). From the ecological point of view, some interesting connotations could be made. Plant-associated *Colletotrichum* species are known to utilize setae (or associated appressoria) for attachment to host surface for penetration into the host tissue (Cano et al. 2004). Concomitantly, it was noteworthy that the bioactive colletotrichone A was dispersed away from the site of production (hyphal clump) by the endophyte while retaining the less active colletotrichone B and/or chermesinone B at the hyphal clumps where they were produced. It might be plausible that colletotrichone A is removed from the site of production by the endophyte, which is also the possible site of its association with the host plant, in order to avoid activation of the plant immune signals and/or to prevent it from interfering with the endophyte-mediated ecological balance in plant tissues (Kogel et al. 2006).

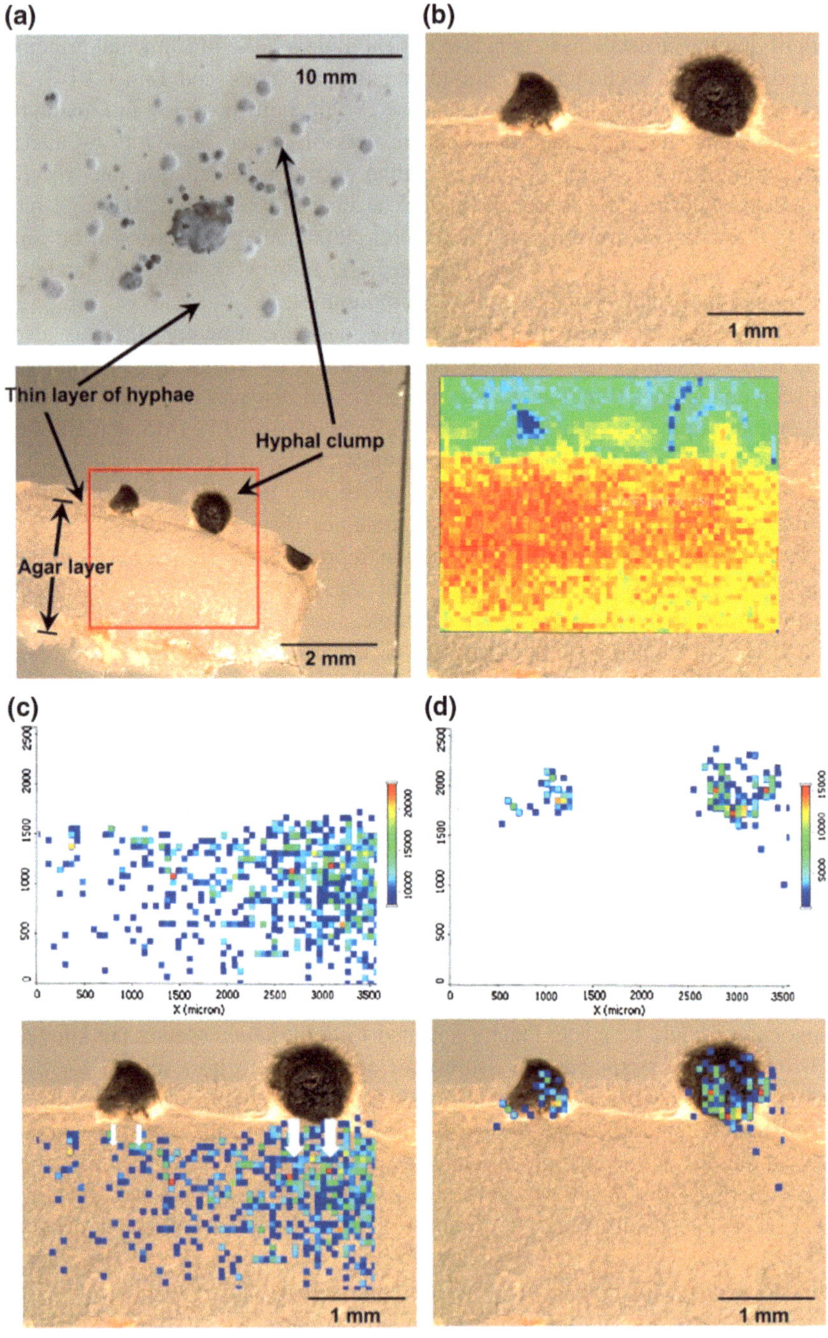
(a)
10 mm
Thin layer of hyphae
Hyphal clump
Agar layer
2 mm
(b)
1 mm
(c)
X (micron)
1 mm
(d)
X (micron)
1 mm

◂ **Fig. 14.4** Spatial distribution of colletotrichone A and colletotrichone B and/or chermesinone B produced by endophytic *Colletotrichum* sp. BS4 on rice agar after 16 days. **a** Optical image of colony and the cross section of agar layer on glass slide. **b** Total ion current (TIC) of the scanned area (*m/z* 100–800). **c** Spatial distribution of colletotrichone A ($[M+K]^+$, *m/z* 387.0841). **d** Spatial distribution of colletotrichone B and/or chermesinone B ($[M+K]^+$, *m/z* 355.0942)

14.6 Conclusion

It is worth to note that the kind of complexity as seen in endophytic microbial communities is also common in other ecological systems. In a recent report, for instance, the gut microbial community was shown to be responsible for the production of aggregation pheromones for the host German cockroach (Wada-Katsumata et al. 2015). Another recent study, Baldwin and colleagues demonstrated that root-associated bacterial community of tobacco plants (*Nicotiana attenuata*) in their native habitat significantly increased the resistance and survivability of host plant toward sudden tissue collapse and black roots (Santhanam et al. 2015). The bacterial consortium rather than any single member in the bacterial community provided protection for the host plant, which unveiled that prokaryotes and plants develop opportunistic mutualisms (Santhanam et al. 2015). Moreover, between parasitic plant *Cuscuta pentagona* and its host plants, the mRNA was exchanged on genomic scale with large proportions (Kim et al. 2014). Therefore, we deem that most ecological niches comprised of multiple members are not a place where organisms are simply living together but communities having complex relationships encompassing each member. The conventional thoughts and strategies for endophytes research, which base on purifying organisms and investigating them in monoculture, should be reconsidered and improved in the future.

We all foresee the great potential of endophytic fungi as natural product factories. Before the promising future of this field is turned into reality, however, we have to investigate the chemical ecology of endophytes in detail. In order to utilize the industrial potential of endophytic fungi, a large number of technical difficulties still remain to be solved and many open questions need to be answered. The silver lining at the present time, interestingly, is that future investigation on the relationships among the endophytic communities will inevitably lead to comprehensive revelations about endophytic fungi and their lifestyles vis-à-vis sustained production of secondary metabolites worth pharmaceutical exploitation.

References

Ajikumar PK, Xiao W-H, Tyo KEJ, Wang Y, Simeon F, Leonard E et al (2010) Isoprenoid pathway optimization for taxol precursor overproduction in *Escherichia coli*. Science 330:70–74

Albuquerque P, Casadevall A (2012) Quorum sensing in fungi—a review. Med Mycol 50:337–345

Arabidopsis Interactome Mapping Consortium (2011) Evidence for network evolution in an *Arbidopsis* interactome map. Science 333:601–607

Arkin AP, Schaffer DV (2011) Network news: innovations in 21st century systems biology. Cell 144:844–849

Bailey BA, Bae H, Strem MD, Roberts DP, Thomas SE, Crozier J et al (2006) Fungal and plant gene expression during the colonization of cacao seedlings by endophytic isolates of four *Trichoderma* species. Planta 224:1449–1464

Bandara HMHN, Lam OLT, Jin LJ, Samaranayake L (2012) Microbial chemical signaling: a current perspective. Crit Rev Microbiol 38:217–249

Bertrand S, Schumpp O, Bohni N, Monod M, Gindro K, Wolfender JL (2013) *De novo* production of metabolites by fungal co-culture of *Trichophyton rubrum* and *Bionectria ochroleuca*. J Nat Prod 76:1157–1165

Bode HB, Bethe B, Höfs R, Zeeck A (2002) Big effects from small changes: possible ways to explore nature's chemical diversity. ChemBioChem 3:619–627

Cano J, Guarro J, Gené J (2004) Molecular and morphological identification of *Colletotrichum* species of clinical interest. J Clin Microbiol 42:2450–2454

Chithra S, Jasim B, Sachidanandan P, Jyothis M, Radhakrishnan EK (2014) Piperine production by endophytic fungus Colletotrichum gloeosporioides isolated from Piper nigrum. Phytomedicine 21:534–540

Christian N, Whitaker BK, Clay K (2015) Microbiomes: unifying animal and plant systems through the lens of community ecology theory. Front Microbiol 6:869

Clay K, Cheplick GP (1989) Effect of ergot alkaloids from fungal endophyte-infected grasses on fall armyworm (*Spodoptera frugiperda*). J Chem Ecol 15:169–182

Cui Y, Yi D, Bai X, Sun B, Zhao Y, Zhang Y (2012) Ginkgolide B produced endophytic fungus (*Fusarium oxysporum*) isolated from *Ginkgo biloba*. Fitoterapia 83:913–920

Dandekar AA, Chugani S, Greenberg EP (2012) Bacterial quorum sensing and metabolic incentives to cooperate. Science 338:264–266

Deng Y, Wu J, Tao F, Zhang LH (2011) Listening to a new language: DSF-based quorum sensing in gram-negative bacteria. Chem Rev 111:160–173

Dong L-H, Fan S-W, Ling Q-Z, Huang B-B, Wei Z-J (2014) Indentification of huperzine A-producing endophytic fungi isolated from *Huperzia serrata*. World J Microbiol Biotechnol 30:1011–1017

Eaton CJ, Cox MP, Scott B (2011) What triggers grass endophytes to switch from mutualism to pathogenism? Plant Sci 180:190–195

Eldar A (2011) Social conflict drives the evolutionary divergence of quorum sensing. Proc Natl Acad Sci 108:13635–13640

El-Elimat T, Raja HA, Graf TN, Faeth SH, Cech NB, Oberlies NH (2014) Flavonolignans from *Aspergillus iizukae*, a fungal endophyte of milk thistle (*Silybum marianum*). J Nat Prod 77:193–199

Esquenazi E, Yang YL, Watrous J, Gerwick WH, Dorrestein PC (2009) Imaging mass spectrometry of natural products. Nat Prod Rep 26:1521–1534

Eyberger AL, Dondapati R, Porter JR (2006) Endophyte fungal isolates from *Podophyllum peltatum* produce podophyllotoxin. J Nat Prod 69:1121–1124

Feldman M, Al-Quntar A, Polacheck I, Friedman M, Steinberg D (2014) Therapeutic potential of thiazolidinedione-8 as an antibiofilm agent against *Candida albicans*. PLoS ONE 9:e93225

Flores-Bustamante ZR, Rivera-Orduña FN, Martínez-Cárdenas A, Flores-Cotera LB (2010) Microbial paclitaxel: Advances and perspectives. J Antibiot (Tokyo) 63:460–467

Freeman S, Rodriguez RJ (1993) Genetic conversion of a fungal plant pathogen to a nonpathogenic, endophytic mutualist. Science 260:75–78

Goh CH, Vallejos DFV, Nicotra AB, Mathesius U (2013) The impact of beneficial plant-associated microbes on plant phenotypic plasticity. J Chem Ecol 39:826–839

Grice CM, Bertuzzi M, Bignell EM (2013) Receptor-mediated signaling in *Aspergillus fumigatus*. Front Microbiol 4:26

Hao X, Pan J, Zhu X (2013) Taxol producing fungi. Natural products. Springer, Berlin Heidelberg, pp 2797–2812

Howat S, Park B, Oh IS, Jin Y-W, Lee E-K, Loake GJ (2014) Paclitaxel: biosynthesis, production and future prospects. N Biotechnol 31:242–245

Huang Q, Roessner CA, Croteau R, Scott AI (2001) Engineering *Escherichia coli* for the synthesis of taxadiene, a key intermediate in the biosynthesis of taxol. Bioorganic Med Chem 9:2237–2242

Huxtable RJ (1992) The pharmacology of extinction. J Ethnopharmacol 37:1–11

Jennewein S, Park H, DeJong JM, Long RM, Bollon AP, Croteau RB (2005) Coexpression in yeast of taxus cytochrome P450 reductase with cytochrome p450 oxygenases involved in taxol biosynthesis. Biotechnol Bioeng 89:588–598

Kharwar RN, Mishra A, Gond SK, Stierle A, Stierle D (2011) Anticancer compounds derived from fungal endophytes: their importance and future challenges. Nat Prod Rep 28:1208–1228

Kim G, LeBlanc ML, Wafula EK, DePamphilis CW, Westwood JH (2014) Genomic-scale exchange of mRNA between a parasitic plant and its hosts. Science 345:808–811

Kogel KH, Franken P, Hückelhoven R (2006) Endophyte or parasite—what decides? Curr Opin Plant Biol 9:358–363

Kumar A, Patil D, Rajamohanan PR, Ahmad A (2013) Isolation, purification and characterization of vinblastine and vincristine from endophytic fungus *Fusarium oxysporum* isolated from *Catharanthus roseus*. PLoS ONE 8:e71805

Kusari S, Spiteller M (2011) Are we ready for industrial production of bioactive plant secondary metabolites utilizing endophytes? Nat Prod Rep 28:1203–1207

Kusari S, Lamshöft M, Zühlke S, Spiteller M (2008) An endophytic fungus from *Hypericum perforatum* that produces hypericin. J Nat Prod 71:159–162

Kusari S, Zühlke S, Košuth J, Čellárová E, Spiteller M (2009a) Light-independent metabolomics of endophytic *Thielavia subthermophila* provides insight into microbial hypericin biosynthesis. J Nat Prod 72:1825–1835

Kusari S, Zühlke S, Spiteller M (2009b) An endophytic fungus from *Camptotheca acuminata* that produces camptothecin and analogues. J Nat Prod 72:2–7

Kusari S, Zühlke S, Spiteller M (2011) Effect of artificial reconstitution of the interaction between the plant *Camptotheca acuminata* and the fungal endophyte *Fusarium solani* on camptothecin biosynthesis. J Nat Prod 74:764–775

Kusari S, Hertweck C, Spiteller M (2012a) Chemical ecology of endophytic fungi: origins of secondary metabolites. Chem Biol 19:792–798

Kusari S, Verma VC, Lamshöft M, Spiteller M (2012b) An endophytic fungus from *Azadirachta indica* A. Juss. that produces azadirachtin. World J Microbiol Biotechnol 28:1287–1294

Kusari S, Lamshöft M, Kusari P, Gottfried S, Zühlke S, Louven K et al (2014a) Endophytes are hidden producers of maytansine in *Putterlickia* roots. J Nat Prod 77:2577–2584

Kusari S, Singh S, Jayabaskaran C (2014b) Biotechnological potential of plant-associated endophytic fungi: hope versus hype. Trends Biotechnol 32:297–303

Kusari S, Singh S, Jayabaskaran C (2014c) Rethinking production of Taxol® (paclitaxel) using endophyte biotechnology. Trends Biotechnol 32:304–311

Kusari P, Kusari S, Spiteller M, Kayser O (2015) Implications of endophyte-plant crosstalk in light of quorum responses for plant biotechnology. Appl Microbiol Biotechnol 99:5383–5390

Kusari P, Kusari S, Eckelmann D, Zühlke S, Kayser O, Spiteller M (2016) Cross-species biosynthesis of maytansine in *Maytenus serrata*. RSC Adv 6:10011–10016

Lahrmann U, Zuccaro A (2012) Opprimo ergo sum–evasion and suppression in the root endophytic fungus *Piriformospora indica*. Mol Plant Microbe Interact 25:727–737

Li SM (2010) Prenylated indole derivatives from fungi: structure diversity, biological activities, biosynthesis and chemoenzymatic synthesis. Nat Prod Rep 27:57–78

Li YF, Tsai KJS, Harvey CJB, Li JJ, Ary BE, Berlew EE et al (2016) Comprehensive curation and analysis of fungal biosynthetic gene clusters of published natural products. Fungal Genet Biol. doi:10.1016/j.fgb.2016.01.012

Ling LL, Schneider T, Peoples AJ, Spoering AL, Engels I, Conlon BP et al (2015) A new antibiotic kills pathogens without detectable resistance. Nature 517:455–459

Luo Y, Li BZ, Liu D, Zhang L, Chen Y, Jia B et al (2015) Engineered biosynthesis of natural products in heterologous hosts. Chem Soc Rev 44:5265–5290

Lyons PC, Plattner RD, Bacon CW (1986) Occurrence of peptide and clavine ergot alkaloids in tall fescue grass. Science 232:487–489

Mallick EM, Bennett RJ (2013) Sensing of the microbial neighborhood by *Candida albicans*. PLoS Pathog 9:e1003661

Newman DJ, Cragg GM (2016) Natural products as sources of new drugs from 1981 to 2014. J Nat Prod 79:629–661

Nickerson KW, Atkin AL, Hornby JM (2006) Quorum sensing in dimorphic fungi: farnesol and beyond. Appl Environ Microbiol 72:3805–3813

Nisa H, Kamili AN, Nawchoo IA, Shafi S, Shameem N, Bandh SA (2015) Fungal endophytes as prolific source of phytochemicals and other bioactive natural products: a review. Microb Pathog 82:50–59

Nongkhlaw FMW, Joshi SR (2016) Horizontal gene transfer of the non-ribosomal peptide synthetase gene among endophytic and epiphytic bacteria associated with ethnomedicinal plants. Curr Microbiol 72:1–11

Oh DC, Kauffman CA, Jensen PR, Fenical W (2007) Induced production of emericellamides A and B from the marine-derived fungus *Emericella sp.* in competing co-culture. J Nat Prod 70:515–520

Paddon CJ, Westfall PJ, Pitera DJ, Benjamin K, Fisher K, McPhee D et al (2013) High-level semi-synthetic production of the potent antimalarial artemisinin. Nature 496:528–532

Parthasarathy R, Sathiyabama M (2014) Gymnemagenin-producing endophytic fungus isolated from a medicinal plant *Gymnema sylvestre* R.Br. Appl Biochem Biotechnol 172:3141–3152

Partida-Martínez LP, Heil M (2011) The microbe-free plant: fact or artifact? Front Plant Sci 2:100

Photita W, Taylor PWJ, Ford R, Hyde KD, Lumyong S (2005) Morphological and molecular characterization of *Colletotrichum* species from herbaceous plants in Thailand. Fungal Divers 18:117–133

Pu X, Qu X, Chen F, Bao J, Zhang G, Luo Y (2013) Camptothecin-producing endophytic fungus *Trichoderma atroviride* LY357: Isolation, identification, and fermentation conditions optimization for camptothecin production. Appl Microbiol Biotechnol 97:9365–9375

Puri SC, Nazir A, Chawla R, Arora R, Riyaz-ul-Hasan S, Amna T et al (2006) The endophytic fungus *Trametes hirsuta* as a novel alternative source of podophyllotoxin and related aryl tetralin lignans. J Biotechnol 122:494–510

Puri SG, Verma V, Amna T, Qazi GN, Spiteller M (2005) An endophytic fungus from *Nothapodytes foetida* that produces camptothecin. J Nat Prod 68:1717–1719

Redman RS, Ranson JC, Rodriguez RJ (1999) Conversion of the pathogenic fungus *Colletotrichum magna* to a nonpathogenic, endophytic mutualist by gene disruption. Mol Plant-Microbe Interact 12:969–975

Rehman S, Shawl AS, Kour A, Andrabi R, Sudan P, Sultan P et al (2008) An endophytic *Neurospora* sp. from *Nothapodytes foetida* producing camptothecin. Appl Biochem Micro 44:203–209

Saikkonen K, Gundel PE, Helander M (2013) Chemical ecology mediated by fungal endophytes in grasses. J Chem Ecol 39:962–968

Santhanam R, Luu VT, Weinhold A, Goldberg J, Oh Y, Baldwin IT (2015) Native root-associated bacteria rescue a plant from a sudden-wilt disease that emerged during continuous cropping. Proc Natl Acad Sci 112:E5013–E5020

Schaefer AL, Greenberg EP, Oliver CM, Oda Y, Huang JJ, Bittan-Banin G et al (2008) A new class of homoserine lactone quorum-sensing signals. Nature 454:595–599

Scherlach K, Hertweck C (2009) Triggering cryptic natural product biosynthesis in microorganisms. Org Biomol Chem 7:1753–1760

Shu S, Zhao X, Wang W, Zhang G, Cosoveanu A, Ahn Y et al (2014) Identification of a novel endophytic fungus from *Huperzia serrata* which produces huperzine A. World J Microbiol Biotechnol 30:3101–3109

Shweta S, Gurumurthy BR, Ravikanth G, Ramanan US, Shivanna MB (2013) Endophytic fungi from *Miquelia dentata* Bedd., produce the anti-cancer alkaloid, camptothecine. Phytomedicine 20:337–342

Soen Y (2014) Environmental disruption of host-microbe co-adaptation as a potential driving force in evolution. Front Genet 5:168

Stierle A, Strobel G, Stierle D (1993) Taxol and taxane production by *Taxomyces andreanae*, an endophytic fungus of pacific yew. Science 260:214–216

Su J, Yang M (2015) Huperzine A production by *Paecilomyces tenuis* YS-13, an endophytic fungus isolated from *Huperzia serrata*. Nat Prod Res 29:1035–1041

Suryanarayanan TS (2013) Endophyte research: going beyond isolation and metabolite documentation. Fungal Ecol 6:561–568

Suryanarayanan TS, Thirunavukkarasu N, Govindarajulu MB, Sasse F, Jansen R, Murali TS (2009) Fungal endophytes and bioprospecting. Fungal Biol Rev 23:9–19

Tan RX, Zou WX (2001) Endophytes: a rich source of functional metabolites. Nat Prod Rep 18:448–459

Tanaka A, Takemoto D, Chujo T, Scott B (2012) Fungal endophytes of grasses. Curr Opin Plant Biol 15:462–468

von Maltzahn G, Flavell RB, Toledo GV, Leff JW, Samayoa P, Marquez LM et al. (2016) Endophytes, associated compositions, and methods of use thereof. United States Patent Application 20160021891

Wada-Katsumata A, Zurek L, Nalyanya G, Roelofs WL, Zhang A, Schal C (2015) Gut bacteria mediate aggregation in the German cockroach. Proc Natl Acad Sci 112:15678–15683

Wang Y, Zeng QG, Zhang ZB, Yan RM, Wang LY, Zhu D (2011) Isolation and characterization of endophytic huperzine A-producing fungi from *Huperzia serrata*. J Ind Microbiol Biotechnol 38:1267–1278

Wang XJ, Min CL, Ge M, Zuo RH (2014) An endophytic sanguinarine-producing fungus from *Macleaya cordata, Fusarium proliferatum* BLH51. Curr Microbiol 68:336–341

Wang WX, Kusari S, Sezgin S, Lamshöft M, Kusari P, Kayser O et al (2015) Hexacyclopeptides secreted by an endophytic fungus *Fusarium solani* N06 act as crosstalk molecules in *Narcissus tazetta*. Appl Microbiol Biotechnol 99:7651–7662

Wang WX, Kusari S, Laatsch H, Golz C, Kusari P, Strohmann C et al (2016) Antibacterial azaphilones from an endophytic fungus, *Colletotrichum* sp. BS4. J Nat Prod 79:704–710

Wever WJ, Bogart JW, Baccile JA, Chan AN, Schroeder FC, Bowers AA (2015) Chemoenzymatic synthesis of thiazolyl peptide natural products featuring an enzyme-catalyzed formal [4 + 2] cycloaddition. J Am Chem Soc 137:3494–3497

Wilson D (1995) Endophyte: the evolution of a term, and clarification of its use and definition. Oikos 73:274–276

Xiong ZQ, Yang YY, Zhao N, Wang Y (2013) Diversity of endophytic fungi and screening of fungal paclitaxel producer from Anglojap yew. Taxus x media. BMC Microbiol 13:71

Xu Y, Masuko S, Takieddin M, Xu H, Liu R, Jing J et al (2011) Chemoenzymatic synthesis of homogeneous ultralow molecular weight heparins. Science 334:498–501

Yang YL, Xu Y, Straight P, Dorrestein PC (2009) Translating metabolic exchange with imaging mass spectrometry. Nat Chem Biol 5:885–887

Youk H, Lim WA (2014) Secreting and sensing the same molecule allows cells to achieve versatile social behaviors. Science 343:1242782

Young CA, Felitti S, Shields K, Spangenberg G, Johnson RD, Bryan GT et al (2006) A complex gene cluster for indole-diterpene biosynthesis in the grass endophyte *Neotyphodium lolii*. Fungal Genet Biol 43:679–693

Zhang Q, Wei X, Wang J (2012) Phillyrin produced by *Colletotrichum gloeosporioides*, an endophytic fungus isolated from *Forsythia suspensa*. Fitoterapia 83:1500–1505

Zhang G, Wang W, Zhang X, Xia Q, Zhao X, Ahn Y et al (2015) *De Novo* RNA sequencing and transcriptome analysis of *Colletotrichum gloeosporioides* ES026 reveal genes related to biosynthesis of huperzine A. PLoS ONE 10:e0120809

Zhao XM, Wang ZQ, Shu SH, Wang WJ, Xu HJ, Ahn YJ et al (2013) Ethanol and methanol can improve huperzine A production from endophytic *Colletotrichum gloeosporioides* ES026. PLoS ONE 8:e61777

Zuck KM, Shipley S, Newman DJ (2011) Induced production of N-formyl alkaloids from *Aspergillus fumigatus* by co-culture with *Streptomyces peucetius*. J Nat Prod 74:1653–1657

Chapter 15
Geomycology

Geoffrey Michael Gadd

15.1 Introduction

Geomycology can be simply defined as 'the scientific study of the roles of fungi in processes of fundamental importance to geology' (Gadd 2007a, 2011; Gadd et al. 2012). This includes such topics as the alteration and weathering of rocks and minerals, soil formation, the transformation and accumulation of metals, decomposition and nutrient cycling. The decomposition of organic substances can be included since this results in major geochemical cycling of elements in the biosphere, with the metabolism of organic compounds underpinning all fungal activities and interactions with environmental components. A variety of inorganic or organic products of fungal metabolism can serve as chemical reagents in reactions such as metal precipitation, metal solubilization, mineral weathering and dissolution while biomechanical effects on the substratum may result from the fungal branching filamentous growth form (Burford et al. 2003a; Gadd 2007a). However, appreciation of fungi as agents of biogeochemical change is limited and they are frequently neglected within broader geomicrobiological contexts (Gadd 2008b). While geochemical activities of bacteria and archaea receive considerable attention, especially in carbon-limited and/or anaerobic environments, in aerobic terrestrial environments fungi are of great importance, especially when considering rock surfaces, soil and the plant root–soil interface. For example, mycorrhizal fungi are associated with most plant species and are involved in major redistributions of inorganic nutrients (Finlay et al. 2009). Free-living fungi have major roles in the decomposition of plant and other organic materials, including xenobiotics, and therefore in the biogeochemical cycling of all the elements comprising such substances (e.g. C, N, P, S, metals) (Gadd 2004a, 2007a, 2008a). Lichens commonly inhabit exposed

G.M. Gadd (✉)
Geomicrobiology Group, School of Life Sciences, University of Dundee, Dundee, Scotland DD1 5EH, UK
e-mail: g.m.gadd@dundee.ac.uk

D. Purchase (ed.), *Fungal Applications in Sustainable Environmental Biotechnology*, Fungal Biology, DOI 10.1007/978-3-319-42852-9_15

rock (and other) substrates, and play fundamental roles in early stages of rock colonization and mineral soil formation (Haas and Purvis 2006). Fungi are also major biodeteriorative agents of stone, wood, plaster, cement and other building materials, and important components of rock-inhabiting microbial communities, participating in mineral dissolution and secondary mineral formation (Burford et al. 2003a, b, 2006; Gadd et al. 2005, 2007, 2014; Fomina et al. 2005a, b; Gadd 2007a). Fungi also have a role in the maintenance of soil structure due to their predominantly filamentous branching growth habit and exopolymer production (Ritz and Young 2004). In the aquatic environment, fungi are important decomposers, with evidence of their occurrence and activity even within deep marine sediments (Reitner et al. 2006; Edgcomb et al. 2011). This chapter outlines some important fungal roles and functions in rock, mineral, metal and soil transformations, and will emphasize the importance of fungi as agents of geochemical change, and the applied significance of these processes in environmental biotechnology.

15.2 Mineral Transformations

Fungi are involved in the formation and deterioration of minerals, the majority of such interactions being accompanied by changes in metal speciation and mobility, especially when metals are a component of the interacting mineral, or are present in the cellular microenvironment.

15.2.1 Mineral Formation

Biomineralization refers to the processes by which organisms form minerals and is usually categorized into biologically-induced mineralization (BIM) and biologically-controlled mineralization (BCM). Biologically-induced mineralization is where an organism modifies the local microenvironment creating conditions that favour extracellular chemical precipitation of mineral phases. The organism does not appear to control the biomineralization process in BIM while a great degree of control over biomineralization is exerted in BCM, e.g. complex cellular biomineral structures in certain eukaryotes (Gadd and Raven 2010). Fungal biomineralization examples therefore usually refer to biologically-induced mineralization. This can result from oxidation or reduction of a metal species, and metabolite excretion, e.g. CO_2, and oxalate (Gadd et al. 2012, 2014). It can also result from organic matter decomposition where released substances re-precipitate with metals in the microenvironment, and vice versa, with fungal surfaces providing reactive sites for sorption (≡biosorption) which can also lead to formation of mineral precipitates (Lloyd et al. 2008; Gadd 2009a, 2010). Apart from the more detailed biomineral examples that follow, many fungal-associated minerals have been recorded, e.g. birnessite, ferrihydrite, iron gluconate, calcium formate, forsterite, goethite, halloysite, hydrocerussite, todorokite, moolooite, montmorillonite, pyromorphite, anglesite and uranium phosphates (Burford et al. 2003a, b; Gadd 2007a, 2010; Fomina et al. 2007a, b, 2008; Gadd and Raven 2010; Rhee et al. 2012; Liang et al. 2015, 2016a, b).

15.2.2 Mineral Biodeterioration

Direct and indirect biomehanical and biochemical mechanisms are involved in mineral biodeterioration (Sand 1997; Edwards et al. 2005; Lian et al. 2008; McMaster 2012; Bonneville et al. 2009, 2011). Biomechanical deterioration of rocks and minerals can occur through penetration, boring and burrowing into porous or decaying material and along crystal planes in, e.g. calcitic and dolomitic rocks (Sterflinger 2000; Golubic et al. 2005; Smits 2006; Gadd 2007a; Cockell and Herrera 2008). Biochemical weathering of rocks and minerals can occur through excretion of, e.g. H^+, CO_2, organic acids, siderophores and other metabolites, and this is thought to be more important than mechanical degradation. This can result in pitting and etching to complete dissolution (Drever and Stillings 1997; Ehrlich 1998; Gharieb et al. 1998; Kumar and Kumar 1999; Adamo and Violante 2000; Adeyemi and Gadd 2005; Edwards et al. 2005; Wei et al. 2012b). Oxalate was particularly important in the biodeterioration of uranium oxides and depleted uranium (Fomina et al. 2007a, b, 2008; Gadd and Fomina 2011). Mineral dissolution may result in release of toxic (Sayer et al. 1999) or essential metals like K (Lian et al. 2008). Fungi can acidify their microenvironment via a number of mechanisms, which include the excretion of protons and organic acids, while respiratory CO_2 can result in carbonic acid formation. In addition, fungi excrete a variety of other metal-complexing metabolites (e.g. siderophores, carboxylic acids, amino acids and phenolic compounds) (Burgstaller and Schinner 1993). Fungal tunnels within soil minerals have been explained as a result of dissolution and 'burrowing' within the mineral matrix (Jongmans et al. 1997; Landeweert et al. 2001; Golubic et al. 2005; Cockell and Herrera 2008). Fungi may also explore pre-existing cracks, fissures, pores and weak points in weatherable minerals and build a matrix of secondary minerals of the same or different chemical composition as the substrate, e.g. secondary $CaCO_3$ precipitation in calcareous soil and rock (Verrecchia 2000) or oxalate formation (Fomina et al. 2010; Gadd et al. 2014). As a result, fissures and cracks become secondarily cemented with mycogenic minerals, and after death and degradation of fungal hyphae, tunnels may be left within the minerals (Fomina et al. 2010).

15.3 Common Mineral and Biomineral Transformations by Fungi

Fungi can potentially be involved in many non-specific mineral transformations in the environment at differing scales (Hutchens 2009; Rosling et al. 2009; Smits 2009), especially when considering their ubiquity and capacity for production of mineral-transforming metabolites, their symbiotic associations, and the consequences of their significant environmental properties such as organic matter decomposition and resultant element cycling. Only a few examples will be given here.

15.3.1 Carbonates

A significant proportion of insoluble carbonate at the Earth's surface is of biogenic origin. Certain fungi can deposit calcium carbonate extracellularly (Verrecchia et al. 1990; Burford et al. 2006; Li et al. 2014, 2015). A mixture of calcite ($CaCO_3$) and calcium oxalate monohydrate (whewellite; $CaC_2O_4{\cdot}H_2O$) was precipitated on hyphae of *Serpula himantioides* when grown in simulated limestone microcosms (Burford et al. 2006). Urease-positive fungi can be used for the precipitation of metal-containing carbonates, which provides a means of metal biorecovery and purification (Li et al. 2014). Incubation of urease-positive *Neurospora crassa* in urea-containing media provided a means for the formation of calcite, as well as carbonates containing other metals. When the carbonate-laden culture supernatant was mixed with $CdCl_2$, the Cd was precipitated in the form of otavite ($CdCO_3$) thus immobilizing the cadmium. The otavite was found to be of high purity, and a small proportion exhibited nanoscale dimensions, which may provide further advantages for industrial application than larger size biominerals (Li et al. 2014). After incubation in media amended with urea and $CaCl_2$ and/or $SrCl_2$, *Pestalotiopsis* sp. and *Myrothecium gramineum*, isolated from calcareous soil, could precipitate calcite ($CaCO_3$), strontianite ($SrCO_3$), vaterite in different forms [$CaCO_3$, $(Ca_xSr_{1-x})CO_3$] and olekminskite [$Sr(Sr,Ca)(CO_3)_2$] again suggesting that urease-positive fungi could play an important role in the environmental fate, bioremediation or biorecovery of Sr or other metals and radionuclides that form insoluble carbonates (Li et al. 2015). *Paecilomyces javanicus* was found to mediate formation of an unknown lead mineral phase after incubation in liquid media with lead shot. After 2 weeks incubation, precipitated mineral phase particles were found to contain plumbonacrite [$Pb_{10}(CO_3)_6O(OH)_6$]. However, after 4 weeks incubation, the lead particles that accumulated inside the fungal pellets were transformed into a white lead-containing secondary mineral phase composed of lead oxalate (PbC_2O_4), hydrocerussite [$Pb_3(CO_3)_2(OH)_2$] and a new species of lead hydroxycarbonate, thus revealing novel steps in lead carbonation by fungi (Rhee et al. 2016).

Insoluble carbonates may be broken down by fungal attack, usually the result of acid formation but may also involve physical processes (Lian et al. 2008), and various fungi and lichens have this property (Adamo and Violante 2000; Cockell and Herrera 2008; Lian et al. 2008). Such activity is particularly evident on limestones and marble used in building construction, but may also occur in natural limestone (Golubic et al. 2005; Cockell and Herrera 2008). Fungal attack on carbonate substrates (dolomites and limestones) can result in diagenesis of these substrates to dolomite [$CaMg(CO_3)_2$], glushinskite ($MgC_2O_4{\cdot}2H_2O$), weddellite ($CaC_2O_4{\cdot}2H_2O$), whewellite ($CaC_2O_4{\cdot}H_2O$), and possibly struvite ($NH_4MgPO_4{\cdot}6H_2O$)· Advanced stages of diagenesis may be characterized by dissolution and replacement of original minerals by the new substrates produced by fungal biomineralization (Kolo et al. 2007).

15.3.2 Oxalates

Calcium oxalate is the most common environmental form of oxalate, occurring as the dihydrate ($CaC_2O_4{\cdot}3H_2O$, weddellite) or the more stable monohydrate ($CaC_2O_4{\cdot}H_2O$, whewellite) (Gadd 1999; Gadd et al. 2014). The initial precipitation phase is the trihydrate ($CaC_2O_4{\cdot}3H_2O$), which loses water of crystallization to form either the dihydrate or monohydrate, depending on conditions. Calcium oxalate can be associated with free-living, pathogenic and plant symbiotic fungi, and lichens, and is formed by precipitation of soluble calcium as the oxalate (Gharieb et al. 1998; Gadd 1999; Adamo and Violante 2000; Adamo et al. 2002). Fungal calcium oxalate can exhibit a variety of crystalline forms (tetragonal, bipyramidal, plate-like, rhombohedral or needles) (Arnott 1995). Calcium oxalate precipitation has an important influence on biogeochemical processes in soils, acting as a calcium reservoir, and also influences phosphate availability. The natural dihydrate form of calcium sulphate, ($CaSO_4{\cdot}2H_2O$) (gypsum), found in gypsiferous soils and certain building construction materials, was solubilized by *Aspergillus niger* and *Serpula himantioides* with the production of oxalic acid resulting in precipitation of calcium oxalate (Gharieb et al. 1998). Fungi can produce many metal oxalates on interacting with a variety of different metals and metal-bearing minerals, e.g. Ca, Cd, Co, Cu, Mg, Mn, Sr, Zn, Ni and Pb (Sayer and Gadd 1997; Gadd 1999, 2007a; Sayer et al. 1999; Gadd et al. 2014). *A. niger* and *S. himantioides* were capable of transforming insoluble manganese oxide minerals, including those produced biogenically, into manganese oxalates. In some cases manganese oxalate trihydrate resulted, followed by conversion to manganese oxalate dihydrate (Wei et al. 2012a). The formation of toxic metal oxalates may contribute to fungal metal tolerance (Gadd 1993a; Jarosz-Wilkolazka and Gadd 2003). In many arid and semi-arid regions, calcareous soils and near surface limestones (calcretes) are secondarily cemented with calcite ($CaCO_3$) and whewellite (calcium oxalate monohydrate, $CaC_2O_4{\cdot}H_2O$) and the presence of fungal filaments mineralized with these substances has been reported (Verrecchia 2000). Calcium oxalate can also be degraded to calcium carbonate, and this may again cement pre-existing limestones (Verrecchia et al. 2006). Other experimental work has demonstrated fungal precipitation of secondary calcite, whewellite and glushkinskite ($MgC_2O_4{\cdot}2H_2O$) (Burford et al. 2003a, b, 2006; Gadd 2007a). Fungal attack on dolomitic and seawater substrates resulted in the formation of Ca-oxalates (weddellite, $CaC_2O_4{\cdot}2H_2O$; whewellite, $CaC_2O_4{\cdot}H_2O$) and glushinskite ($MgC_2O_4{\cdot}2H_2O$) (Kolo and Claeys 2005).

15.3.3 Oxides

Several fungi can promote Mn(II) oxidation to Mn(IV)O_2 including *Acremonium* spp. (Miyata et al. 2004, 2007; Saratovsky et al. 2009). In many cases, fungal oxidation is probably non-enzymatic and due to interaction with a metabolic product (e.g. a hydroxy acid) or a cellular component (Ehrlich and Newman 2009) although the involvement of laccase and/or multicopper oxidases have been shown in ascomycetes (Miyata et al. 2004, 2007). The MnO_x material produced by *Acremonium* KR21-2 is manifest as small crystalline particles which adopt a todorokite-like tunnel structure, which is in contrast to previously reported microbial MnO_x materials which adopt layered birnessite-type structures (Saratovsky et al. 2009). Non-enzymatic microbial Mn^{2+} oxidation may be effected through production of metabolites, e.g. hydroxycarboxylic acids such as citrate, lactate, malate, gluconate or tartrate. Some fungi can oxidize Mn(II) and Fe(II) in metal-bearing minerals such as siderite ($FeCO_3$) and rhodochrosite ($MnCO_3$) and precipitate them as oxides (Grote and Krumbein 1992). Manganese and iron oxides are major components (20–30 %) along with clay (~60 %) and various trace elements in the brown-to-black veneers known as desert varnish or rock varnish (Grote and Krumbein 1992; Gorbushina 2007). Conversely, manganese-reducing microbes may mobilize oxidized or fixed manganese, releasing it into the aqueous phase. Most of those fungi that reduce Mn(IV) oxides such as MnO_2 reduce them indirectly (non-enzymatically) with the likely mechanism being the production of metabolic products that can act as reductants for Mn(IV) oxides such as oxalate (Ehrlich and Newman 2009; Wei et al. 2012a).

15.3.4 Phosphates

Phosphorus occurs primarily as organic phosphate esters and inorganic forms, e.g. calcium, aluminium and iron phosphates. Organic phosphates are hydrolyzed by phosphatases which liberate orthophosphate during microbial decomposition of organic material. Fungi also liberate free orthophosphate from insoluble inorganic phosphates by producing acids or chelators, e.g. gluconate, citrate, oxalate and lactate, which complex the metal resulting in dissociation. Phosphate-solubilizing activity is very important in the plant mycorrhizosphere (Whitelaw et al. 1999). Microbes can also play a role in the formation of phosphate minerals such as vivianite [$Fe_3(PO_4)_2{\cdot}8H_2O$], strengite ($FePO_4{\cdot}2H_2O$) and variscite ($AlPO_4{\cdot}2H_2O$). Here, the orthophosphate may arise from organic phosphate degradation while Fe or Al may arise from microbial solubilization of other minerals. Such formation of phosphate minerals is probably most common in soil (Ehrlich and Newman 2009). Fungal biodeterioration of metallic lead can result in pyromorphite ($Pb_5[PO_4]_3X$ [X=F, Cl or OH]) formation (Rhee et al. 2012, 2014a, b). Previous research has demonstrated that many fungi exhibit uranium tolerance and can solubilize uranium oxides and depleted uranium and re-precipitate secondary uranium phosphate minerals of the meta-autunite group, uramphite and/or chernikovite, which can encrust fungal hyphae to high accumulation values of 300–400 mg U g dry wt^{-1} (Fomina et al. 2007a, b, 2008). Such minerals appear capable of long-term U retention (Fomina et al. 2007a, b, 2008; Gadd and Fomina 2011). *Aspergillus niger* and *Paecilomyces javanicus* could precipitate U-containing phosphate biominerals when grown with an organic source of P, with the hyphal matrix serving to localize the resultant uranium minerals. Several uranyl phosphate species were identified including potassium uranyl phosphate hydrate ($KPUO_6{\cdot}3H_2O$), meta-ankoleite [$(K_{1.7}Ba_{0.2})(UO_2)_2(PO_4)_2{\cdot}6H_2O$], uranyl phosphate hydrate [$(UO_2)_3(PO_4)_2{\cdot}4H_2O$], meta-ankoleite [$K(UO_2)(PO_4){\cdot}3H_2O$], uramphite ($NH_4UO_2PO_4{\cdot}3H_2O$) and chernikovite [$(H_3O)_2(UO_2)_2(PO_4)_2{\cdot}6H_2O$] (Liang et al. 2015). These organisms could also mediate lead bioprecipitation during growth on organic phosphate substrates, which resulted in almost complete removal of Pb from solution and extensive precipitation of lead-containing minerals around the biomass, confirming the importance of the mycelium as a reactive network for biomineralization (Liang et al. 2016a). The minerals were identified as pyromorphite [$Pb_5(PO_4)_3Cl$], only produced by *P. javanicus*, and lead oxalate (PbC_2O_4), produced by *A. niger* and *P. javanicus*. Two main lead biomineralization mechanisms were therefore distinguished: pyromorphite formation depending on organic phosphate hydrolysis and lead oxalate formation depending on oxalate excretion. This also indicated some species specificity in biomineralization depending on nutrition and physiology (Liang et al. 2016a). Several yeast species could also exhibit lead bioprecipitation when utilizing an organic phosphorus-containing substrate (glycerol 2-phosphate, phytic acid) as the sole phosphorus source. The minerals precipitated through phosphatase activity on the substrates were lead phosphate [$Pb_3(PO_4)_2$], pyromorphite [$Pb_5(PO_4)_3Cl$], anglesite ($PbSO_4$) and the lead oxides massicot and litharge (PbO), with variations in the mineral types produced between the different species. All test yeasts produced pyromorphite, and most produced anglesite (Liang et al. 2016b). Such processes may be relevant to metal immobilization biotechnologies for bioremediation, metal and P biorecovery, as well as the utilization of waste organic phosphates.

15.3.5 Silicates

Silicates are the largest class of minerals comprising 30 % of all minerals and making up 90 % of the Earth's crust (Ehrlich 1998; Brehm et al. 2005; Ehrlich and Newman 2009). Many kinds of fungi and lichens play an important role in the dissolution of silicates and therefore in the genesis of clay minerals, and in soil and sediment formation (Barker and Banfield 1996, 1998; Banfield et al. 1999; Arocena et al. 1999; Adamo and Violante 2000; Arocena et al. 2003; Tazaki 2006; Theng and Yuan 2008). The presence of clay minerals can be a typical symptom of biogeochemically weathered rocks, and this has been observed for lichens and ectomycorrhizas (Barker and Banfield 1998; Arocena et al. 1999). Bioweathering action is mainly indirect, either through the production of chelates or the production of acids (mineral or organic), or other metabolites, together with biomechanical effects (Cromack et al. 1979; De la Torre et al. 1993; Mandal et al. 2002). Metabolic agents may be excreted into the bulk phase but may also involve adhering organisms on surfaces of silica or silicates resulting in etching (Bennett et al. 2001; Wei et al. 2012b). After colonization of muscovite, a phyllosilicate mineral, in the form of a mineral sheet model system by *Aspergillus niger*, a network of hyphae covered the surface of the muscovite and mineral dissolution or degradation was clearly evidenced by a network of fungal 'footprints' that reflected coverage by the mycelium (Wei et al. 2012b). New biominerals resulted from fungal interactions with both zinc silicate and zinc sulphide, largely resulting from organic acid excretion. Zinc oxalate dihydrate was formed and the mineral surfaces showed varying patterns of bioweathering and biomineral formation. In addition, calcium oxalate was formed from the calcium present in the mineral ore fractions, as well as calcite (Wei et al. 2013). Such mechanisms of silicate dissolution may release limiting nutrients like bound P and Fe. In lichen weathering of silicates, calcium, potassium, iron, clay minerals and nanocrystalline aluminous iron oxyhydroxides become mixed with fungal organic polymers such as extracellular polysaccharides (Barker and Banfield 1998), while biotite [$K(Mg,Fe(II))_3AlSi_3O_{10}(OH,O,F)_2$] was interpenetrated by fungal hyphae along cleavages, partially converting it to vermiculite [$(Mg,Fe(II),Al)_3(Al,Si)_4O_{10}(OH)_2{\cdot}4H_2O)$] (Barker and Banfield 1996). The fungal partner has also been reported to be involved in formation of secondary silicates, such as opal ($SiO_2{\cdot}nH_2O$) and forsterite (Mg_2SiO_4), in lichen thalli (Gorbushina et al. 2001). The transformation rate of mica (the general formula for minerals of the mica group is $XY_{2-3}Z_4O_{10}(OH, F)_2$ with X=K, Na, Ba, Ca, Cs, (H_3O), (NH_4); Y=Al, Mg, Fe^{2+}, Li, Cr, Mn, V, Zn; and Z=Si, Al, Fe^{3+}, Be, Ti) and chlorite [$(Mg,Fe,Li)_6AlSi_3O_{10}(OH)_8)$] to 2:1 expandable clays was pronounced in ectomycorrhizosphere soil and probably a result of the high production of organic acids and direct extraction of K^+ and Mg^{2+} by fungal hyphae (Arocena et al. 1999). Fungal–clay mineral interactions also play an important role in soil development, aggregation and stabilization (Burford et al. 2003a). Fungi entangle soil particles in their hyphae forming microaggregates and also effect exopolysaccharide-mediated aggregation (Ritz and Young 2004). Such interactions between clay minerals and fungi alter the sorptive properties of both the clay minerals and fungal hyphae (Morley and Gadd 1995; Fomina and Gadd 2002a). Clay minerals (e.g. bentonite, palygorskite and kaolinite) can also influence the size, shape and structure of fungal mycelial pellets (Fomina and Gadd 2002b).

15.3.6 Reduction or Oxidation of Metals and Metalloids

Many fungi precipitate reduced forms of metals and metalloids, e.g. Ag(I) reduction to elemental silver Ag(0); selenate [Se(VI)] and selenite [Se(IV)] to elemental selenium [Se(0)]; tellurite [Te(IV)] to elemental tellurium [Te(0)] (Kierans et al. 1991; Gharieb et al. 1995, 1999). Reduction of Hg(II) to volatile Hg(0) can also be mediated by fungi (Gadd 1993b, 2000a). An *Aspergillus* sp. was able to grow at arsenate concentrations of 0.2 M and it was suggested that increased arsenate reduction contributed to tolerance (Canovas et al. 2003a, b). Mn oxidation/reduction has been mentioned previously.

15.3.7 Other Mycogenic Minerals

A range of other minerals have been found in association with fungi. Apart from those detailed above, mycogenic secondary minerals associated with fungal hyphae and lichen thalli include desert varnish (MnO and FeO), ferrihydrite ($5Fe_2O_3{\cdot}9H_2O$), iron gluconate, calcium formate, forsterite, goethite (α-$Fe^{3+}O(OH)$), halloysite [$Al_2Si_2O_5(OH)_4$] and hydroserussite [$Pb_3(CO_3)_2(OH)_2$] (Grote and Krumbein 1992; Hirsch et al. 1995; Verrecchia 2000; Gorbushina et al. 2001; Arocena et al. 2003; Burford et al. 2003b). *Lichenothelia* spp. can oxidize manganese and iron in metal-bearing minerals, such as siderite ($FeCO_3$) and rhodochrosite ($MnCO_3$), and precipitate them as oxides (Grote and Krumbein 1992). Similar oxidation of Fe(II) and Mn(II) by fungi can lead to the formation of dark patinas on glass surfaces (Eckhardt 1985). Another biogenic mineral (tepius) has been identified in association with a lichen carpet that covers high mountain ranges in Venezuela (Gorbushina et al. 2001).

15.4 Metal–Fungal Interactions and Transformations

Metals, metalloids, metal radionuclides, organometals and organometalloids, and their compounds interact with fungi in various ways depending on chemical speciation, organism and environment, with fungi also being capable of influencing metal speciation and mobility (Gadd and Griffiths 1978, 1980; Gadd 1984, 1992,

1993a, 2004a, b, 2007a, b, 2008c, 2009b; Gadd et al. 2005; Newby and Gadd 1987; Dutton and Evans 1996; Ramsay et al. 1999; Gadd et al. 2001, 2012; Fomina et al. 2003). Many metals are essential for life, e.g. Na, K, Cu, Zn, Co, Ca, Mg, Mn and Fe, but all can exert toxicity when present above certain threshold concentrations. Other metals, e.g. Cs, Al, Cd, Hg and Pb, have no known metabolic function in fungi but all can be accumulated. Metal toxicity is affected by environmental conditions and the chemical behaviour of the particular metal species. Metals exert toxic effects in many ways and can inhibit enzymes, displace or substitute for essential metal ions, cause disruption of membranes, interact with systems which normally protect against the harmful effects of free radicals, inhibit growth and spore germination of fungi, affect reproduction and metabolism and reduce the ability of mycorrhizal fungi to colonize host plant roots (Gadd 1993a; Howlett and Avery 1997; Fomina et al. 2005c). However, many fungi are found in metal-polluted locations and a variety of mechanisms, both active and incidental, contribute to tolerance. Mechanisms of toxic metal tolerance in fungi include reduction of metal uptake and/or increased efflux, metal immobilization (e.g. cell wall adsorption, extracellular precipitation of secondary minerals, extracellular binding by polysaccharides and extracellular metabolites and intracellular sequestration by, e.g. metallothioneins and phytochelatins, and localization in vacuoles (Gadd et al. 1987; Gadd 1993a; Joho et al. 1995; Blaudez et al. 2000; Perotto and Martino 2001; Baldrian 2003; Meharg 2003). Fungi therefore have many properties which influence metal mobility and toxicity. All the mechanisms by which fungi (and other microorganisms) effect changes in metal speciation and mobility are survival determinants but also components of biogeochemical cycles for metals, and many other associated elements including carbon, nitrogen, sulphur and phosphorus (Gadd 2004a, 2006, 2007a, b, 2008a). These may be considered in terms of metal mobilization or immobilization mechanisms.

15.4.1 Metal Mobilization

Metal mobilization from rocks, minerals, soil and other substrates can be achieved by protonolysis, carbonic acid formation from respiratory CO_2, complexation by Fe (III)-binding siderophores and other excreted metabolites, e.g. amino acids, phenolic compounds and organic acids, and methylation which can result in volatilization. Fungal-derived carboxylic acids can attack mineral surfaces (see previously) and these provide protons as well as a metal-chelating anion (Burgstaller and Schinner 1993). Oxalic acid can act as a leaching agent for metals that form soluble oxalate complexes, including Al and Fe (Strasser et al. 1994). Solubilization mechanisms also have consequences for mobilization of metals from toxic metal-containing minerals, e.g. pyromorphite [$Pb_5(PO_4)_3Cl$], contaminated soil and other solid wastes (Sayer et al. 1999; Fomina et al. 2004, 2005b, c). Fungi may also mobilize metals and attack mineral surfaces by redox processes: Fe(III) and Mn(IV) solubility is increased by reduction to Fe(II) and Mn(II), respectively. Reduction of Hg(II) to volatile elemental Hg(0) has also been recorded in fungi (Gadd 1993b).

15.4.2 *Metal Immobilization*

Fungal biomass provides a metal sink, either by biosorption to biomass (cell walls, pigments and extracellular polysaccharides), intracellular accumulation and sequestration, or precipitation of metal compounds onto and/or around hyphae (Gadd 1993a, 2000a, b; Gadd 2001a, b, c, 2007a, 2009a; Baldrian 2003; Fomina et al. 2007b, c; Fomina and Gadd 2014). Fungi are effective biosorbents for a variety of metals including Ni, Zn, Ag, Cu, Cd and Pb and this can occur in both living and dead biomass (Gadd 1990, 1993a; Sterflinger 2000; Fomina and Gadd 2014). The presence of chitin, and pigments like melanin, strongly influences the ability of fungi to act as sorbents. A variety of functional groups can be involved in biosorption and also metal association with other cellular components and macromolecules. Fungal biomineralization processes resulting from metabolite excretion or redox transformations also lead to metal immobilization as biominerals or elemental forms, as described previously (Gadd 2007a).

15.5 Fungal Symbioses in Metal and Mineral Transformations

Many fungi form remarkable partnerships with plants (mycorrhizas) and algae or cyanobacteria (lichens) that are of great significance in the terrestrial environment. In general terms, the mycobiont is provided with carbon by the photobionts, while the mycobiont may protect the symbiosis from harsh environmental conditions (e.g. desiccation, metal toxicity), and provide increased access to inorganic nutrients.

15.5.1 *Lichens*

Lichens are fungi that exist in facultative or obligate symbioses with one or more photosynthesizing partners and found in almost all surface terrestrial environments: an estimated 6–8 % of the Earth's land surface is dominated by lichen covering (Haas and Purvis 2006). Lichens play an important role in many biogeochemical processes, and are pioneer colonizers of fresh rock outcrops. Globally, lichens play important roles in retention and distribution of nutrient (e.g. C, N) and trace elements, in soil formation, and rock weathering. Alteration of bedrock minerals and biomineralization processes in the proximity of lichens results in differing chemical microenvironments and underpins their participation in mineral transformations and element cycling (Banfield et al. 1999; Adamo and Violante 2000; Chen et al. 2000). Lichens accumulate metals such as lead (Pb) and copper (Cu), and many other elements, including radionuclides, to high levels (Purvis and Pawlik-Skowronska 2008). They also form a variety of metal-organic biominerals, e.g. oxalates, especially during growth on metal-rich substrates (Chen et al. 2000; Adamo et al. 2002). On copper-sulphide bearing rocks, precipitation of copper oxalate (moolooite) can occur within lichen thalli (Purvis 1996; Purvis and Halls 1996).

15.5.2 *Mycorrhizas*

Almost all land plants depend on symbiotic mycorrhizal fungi (Smith and Read 1997; Wang and Qui 2006). Two main types include endomycorrhizas where the fungus colonizes the interior of host plant root cells (e.g. ericoid and arbuscular mycorrhizas) and ectomycorrhizas where the fungus is located outside plant root cells. Mycorrhizal fungi are involved in phosphate solubilization, proton-promoted and ligand-promoted metal mobilization from mineral sources, metal immobilization within biomass and extracellular precipitation of mycogenic metal oxalates (Lapeyrie et al. 1991; Blaudez et al. 2000; Christie et al. 2004; Fomina et al. 2004, 2005b, 2006; Bellion et al. 2006; Finlay et al. 2009; McMaster 2012; Smits et al. 2012). Biogeochemical activities of mycorrhizal fungi lead to changes in the physico-chemical characteristics of the root environment and enhanced weathering of soil minerals (McMaster 2012; Bonneville et al. 2009, 2011). Furthermore, ectomycorrhizal mycelia may respond to the presence of different soil silicate and phosphate minerals (apatite, quartz, potassium feldspar) by regulating growth and activity, e.g. colonization, carbon allocation and substrate acidification (Rosling et al. 2004a, b).

During growth, mycorrhizal fungi often excrete low molecular weight carboxylic acids and siderophores (Martino et al. 2003; Fomina et al. 2004). The weathering of hornblendes, feldspars and granitic bedrock in certain soils has been attributed to oxalic, citric, succinic, formic and malic acid excretion by ectomycorrhizal hyphae which can produce micro- to millimolar concentrations of these organic acids in their microenvironments. Ectomycorrhizal fungi can form narrow cylindrical pores in weatherable minerals in podzol E horizons, probably by exuding low molecular weight organic acids and/or siderophores at their hyphal tips, causing local dissolution of Al silicates (Jongmans et al. 1997; Van Breemen et al. 2000). Ectomycorrhizal fungi (*Suillus granulatus* and *Paxillus involutus*) can also release elements from apatite and wood ash (K, Ca, Ti, Mn, Pb) (Wallander et al. 2003). Ericoid mycorrhizal and ectomycorrhizal fungi can dissolve a variety of cadmium, copper, zinc and lead-bearing minerals including metal phosphates (Leyval and Joner 2001; Martino et al. 2003; Fomina et al. 2004, 2005b). Mobilization of phosphorus from inorganic and organic phosphorus sources is generally regarded as one of the most important functions of mycorrhizal fungi, and this will also result in redistribution of associated metals, and also the formation of other secondary minerals including other metal phosphates. The association of arbuscular mycorrhizal fungi (AMF) with the roots of *Lindenbergia philippensis*, sampled from a Zn-contaminated settling pond at a zinc smelter, significantly enhanced Zn accumulation in Zn-loaded rhizosphere sediment compared to treatments that suppressed AMF colonization. A significant proportion of Zn was present as crystalline and other solid materials that were associated with the root mucilaginous sheath (Kangwankraiphaisan et al. 2013). Such results may indicate a role for AMF in enhancing Zn immobilization in the rhizosphere of plants that successfully colonize Zn mining and smelting disposal sites (Christie et al. 2004; Turnau et al. 2012; Kangwankraiphaisan et al. 2013).

15.6 Environmental and Applied Significance of Geomycology

15.6.1 Biocorrosion of Metals

Microbial biodeterioration of metal due to microbial activity is termed biocorrosion or microbially-influenced corrosion (MIC) (Beech and Sunner 2004). While several groups of bacteria are more commonly associated with biocorrosion, e.g. sulphate-reducing bacteria (SRB), various fungi may colonize and/or be present within complex biofilm communities on metal surfaces (Beech and Sunner 2004; Gu 2009).

15.6.2 Bioleaching of Metals from Ores

Solubilization mechanisms provide a means for removal of metals from industrial wastes and by-products, low-grade ores and metal-bearing minerals, which is relevant to bioremediation of soil matrices and solid wastes, metal recovery and recycling (Burgstaller and Schinner 1993; Gadd 2000a; Gadd and Sayer 2000; Brandl 2001; Kartal et al. 2006). Extracellular ligands excreted by fungi, especially low molecular weight organic acids, have been used to leach metals such as Zn, Cu, Ni and Co from a variety of materials, including low grade ores (Brandl 2001; Santhiya and Ting 2005).

15.6.3 Bioweathering of Rocks and Minerals: Soil Formation

Weathering is a process in which rock is eroded or broken down into smaller particles and finally to constituent minerals, ultimately leading to mineral soil formation (Tazaki 2006; Ehrlich and Newman 2009). Physical, chemical and biological processes are involved: bioweathering can be defined as the erosion and decay of rocks and minerals mediated by living organisms. Many fungi are effective biological weathering agents as described earlier (Gorbushina et al. 1993; Sterflinger 2000; Verrecchia 2000; Burford et al. 2003a, b; Gadd 2007a; Sverdrup 2009; Gorbushina and Broughton 2009). Fungi are probably associated with all rocks and minerals, building stone and concrete (Burford et al. 2003a, b; Gleeson et al. 2005, 2006, 2007, 2010; Gorbushina 2007; Gorbushina and Broughton 2009). As mentioned previously, lichens are highly significant bioweathering agents (Adamo and Violante 2000; Adamo et al. 2002). Lithobiotic biofilm communities can interact with mineral substrates and as well as deterioration this can also result in the formation of patinas, films, varnishes, crusts and stromatolites (Gadd 2007a; Gorbushina 2007; Fomina et al. 2010). Mycorrhizal fungi are also very important in mineral weathering and dissolution of insoluble metal compounds in the soil. Acidification is an important fungal bioweathering mechanism with low molecular weight organic acid anions being especially significant (Gadd 1999; Hoffland et al. 2004). Because production of these substances has a carbon cost, symbiotic mycorrhizal fungi that are provided with organic carbon compounds by the plant host may have an advantage over free-living saprotrophs (Hoffland et al. 2004). It should be stressed that the activities of all groups of microbes and interactions between them should be considered in bioweathering. Fungal–bacterial interactions are likely to be significant in mineral weathering in the root environment (Balogh-Brunstad et al. 2008; Koele et al. 2009).

15.6.4 Bioweathering of Rocks and Minerals: Structural Decay of Stone and Mineral Artefacts

Deterioration of stone monuments represents a permanent loss of cultural heritage (Scheerer et al. 2009; Cutler and Viles 2010). The most common stone types affected are marble, limestone, sandstone and granite, with many overlaps between calcareous and siliceous rocks. Materials used to stabilize building blocks (mortar) and to coat surfaces prior to painting (plaster or stucco) are also susceptible to degradation. External stone surfaces are a complex ecosystem, including cyanobacteria, bacteria, fungi, protists and also 'higher' organisms such as small animals and plants (Scheerer et al. 2009). Microbial colonization generally initiates with phototrophic cyanobacteria and algae, usually in a biofilm, probably followed by lichens, and then general heterotrophs although establishment of heterotrophic rock communities is possible without initial phototroph involvement (Roeselers et al. 2007). Highly deteriorated stone surfaces provide appropriate conditions (a "proto-soil") for colonization by mosses, ferns and higher plants.

Mechanisms of stone deterioration are complex and include most direct and indirect mechanisms previously discussed for mineral dissolution and decay (Sand 1997; Scheerer et al. 2009). The formation of biofilms is significant, aiding colonization and survival, with the EPS also capable of metal complexation and weakening of the mineral lattice through wetting and drying cycles. The production of efflorescences ('salting') involves secondary minerals produced through reaction of anions from excreted acids with cations from the stone. Such secondary minerals can cause physical damage leading to blistering, flaking, scaling and granular disintegration, which may often be the main mechanism of stone decay (Wright 2002). Physical damage may be caused by penetration of fungal hyphae (Hirsch et al. 1995; Cockell and Herrera 2008). Weakened areas of the stone will be affected first. Lichens cause mechanical damage due to penetration of their rhizines, composed of fungal filaments, and the expansion/contraction of the thallus on wetting/drying, which can lift grains of stone from the surface (De los Rios et al. 2004; Gaylarde and Morton 2002). 'Lichen acids', principally oxalic acid cause damage at the stone/lichen interface, and lichen thalli may accumulate 1–50 % calcium oxalate, depending on the substrate (Seaward 2003; Lisci et al. 2003). In addition, carbonic acid formed in the lichen thallus can solubilize calcium and magnesium carbonates in calcareous stone (Tiano 2002). Fungal biodeterioration of ancient ivory (natural apatite; walrus tusk) was accompanied by widespread tunnelling by fungal hyphae as well as 'fungal footprints' where surfaces were etched as a consequence of mycelial colonization. Similar phenomena were observed with boar tusk ivory, while production of organic acid metabolites could lead to complete dissolution. Colonization of ivory and/or exposure to fungal activity lead to extensive secondary biomineral formation, and this was mainly identified as calcium oxalate monohydrate, whewellite (Pinzari et al. 2013).

15.6.5 Concrete Biodeterioration

All types of building and ceramic materials, concrete and cement can be biodeteriorated and in some environments, fungi dominate the concrete-deteriorating microbiota (Gu et al. 1998; Nica et al. 2000; Gu 2009; Scheerer et al. 2009; Cutler and Viles 2010). Apart from structural uses, cement and concrete are used as barriers in all kinds of nuclear waste repositories. Despite the theoretical service life of such concrete materials reaching up to one million years, biocorrosion is an important factor to take into account over such time periods. Microbial attack on concrete is mediated by protons, inorganic and organic acids and the production of hydrophilic slimes leading to biochemical and biophysical/biomechanical deterioration (Sand 1997; Fomina et al. 2007c; Scheerer et al. 2009). Fungal degradation may proceed more rapidly than bacterial degradation with complexation suggested as the main mechanism of calcium mobilization. Microfungi from the genera *Aspergillus, Alternaria* and *Cladosporium* were able to colonize samples of the concrete used as radioactive waste barrier in the Chernobyl reactor and leached iron, aluminium, silicon and calcium, and re-precipitated silicon and calcium oxalate in their microenvironment (Fomina et al. 2007c). Fungi are also important members of the microbial communities (including lichens) that colonize and deteriorate 'normal' concrete and cement used in buildings and other structures.

15.7 Bioremediation, Biotechnology and Bioprocessing

Several fungal metal and mineral transformations have potential for the treatment of environmental pollution (Gadd 2004a, 2005; Pumpel and Paknikar 2001). Fungi will be components of the microbiota in any metal-polluted sites where their activities may contribute to natural attenuation of the pollutants, and will also be involved in many standard soil and waste treatment processes, revegetation strategies, and effluent treatments. Fungi were clearly important in the bioremediation of selenium-contaminated soils (Thomson-Eagle and Frankenberger 1992). In addition, fungal mineral-solubilizing properties are important in plant nutrition and soil fertility especially regarding plant phosphate nutrition. In addition to bioremediation, metal- and mineral transformations have applications in other areas of biotechnology and bioprocessing, including biosensors, biocatalysis, electricity generation and nanotechnology.

15.7.1 Bioleaching

Fungal solubilization of metals from solid minerals, metal and mineral wastes, including contaminated soil, for metal recovery, recycling and bioremediation purposes have all been investigated, although fungal systems cannot compare with the efficiency of established bacterial bioleaching processes and may be more suited to specific bioreactor applications. Metals can be solubilized from fly ash (originating from municipal solid waste incineration), contaminated soil, electronic scrap and other waste materials by fungal activity (Brandl 2001; Brandl and Faramarzi 2006).

15.7.2 Biosorption,[1] Bioaccumulation and Biorecovery

Biosorption is a physico-chemical process, simply defined as 'the removal of substances from solution by biological material'. It is a property of both living and dead organisms (and their components), and has been proposed as a promising biotechnology for removal (and/or recovery) of metals, radionuclides, organic pollutants for many years because of its simplicity, analogous operation to conventional ion exchange technology and apparent efficiency (Gadd and Mowll 1985; Gadd 1986 1990, 2001a, b, 2009a; De Rome and Gadd 1987; Gadd and De Rome 1988; Gadd and White 1989, 1990, 1992, 1993; Volesky 1990, 2007; Garnham et al. 1992; White et al. 1995; Wang and Chen 2006, 2009; Fomina and Gadd 2014). However, there has been little or no exploitation in an industrial context, largely due to the greater selectivity and efficiency of commercial ion-exchange resins. Modification of biomass has been attempted to improve efficiency or selectivity of microbial biosorbents. Fungal–clay biomineral sorbents combined the sorptive advantages of the individual counterparts, i.e. the high density of metal-binding sites per unit area and high sorption capacity of fungal biomass, high sorption affinity and the high surface area per unit weight mechanical strength and efficient sorption at high metal concentrations of the clay minerals (Fomina and Gadd 2002a, 2014).

Urease-positive fungi can be used for the precipitation of metal-containing carbonates, which provides a means of metal biorecovery and purification (Li et al. 2014, 2015). Similarly, the formation of other insoluble metal compounds and minerals by fungi or their metabolites could also be considered as a means to biorecover metals, metalloids and radionuclides, e.g. oxalates, oxides, oxalates and phosphates, as well as the production of elemental metal or metalloid forms (Gadd 2010; Gadd et al. 2012).

15.7.3 Metalloid Bioremediation[2]

The ability of fungi, along with bacteria, to transform metalloids has been utilized successfully in the bioremediation of contaminated land and water. Selenium methylation results in volatilization, a process which has been used to remove selenium from the San Joaquin Valley and Kesterson Reservoir, California using evaporation pond management and primary pond operation (Thomson-Eagle and Frankenberger 1992).

---

[1]Additional information on biosorption in fungi is presented in Chap. 4—*Application of biosorption function of fungi in wastewater and sludge treatment.*

[2]Further information on metalloid bioremediation can be found in Chap. 10—*Mycoremedation of heavy metal/metaaloid-contaminated soil: current understanding and future prospects.*

15.7.4 Mycoremediation and the Mycorrhizosphere

Mycorrhizal associations may have application in the general area of phytoremediation (Rosen et al. 2005; Gohre and Paszkowski 2006): phytoremediation is the use of plants to remove or detoxify environmental pollutants (Salt et al. 1998). Mycorrhizas may enhance phytoextraction by increasing plant biomass, and some studies have shown increased plant accumulation of metals, especially when inoculated with mycorrhizal fungi isolated from metalliferous environments. However, the potential impact of mycorrhizal fungi on bioremediation may be dependent on many factors including their metal tolerance, and the nutritional status of contaminated soils (Meharg 2003). In addition, some studies have shown that mycorrhizas can reduce plant metal uptake (Tullio et al. 2003; Christie et al. 2004). A protective metal-binding effect of ectomycorrhizal fungi (EcM) has been postulated (e.g. Leyval et al. 1997). A Cu-adapted *Suillus luteus* isolate provided protection against Cu toxicity in pine seedlings exposed to elevated Cu. Such a metal-adapted *Suillus-Pinus* combination might be suitable for large-scale land reclamation at phytotoxic metalliferous and industrial sites (Adriaensen et al. 2005). Ectomycorrhizal fungi fixed Cd(II) and Pb(II), and formed a biological barrier that reduced movement of these metals in birch tissues (Krupa and Kozdroj 2004). Naturally occurring soil organic compounds can stabilize potentially toxic metals like Cu, Cd, Pb and Mn. The insoluble glycoprotein, glomalin, produced in copious amounts on hyphae of arbuscular mycorrhizal fungi can sequester such metals, and could be considered a useful stabilization phenomenon (Gonzalez-Chavez et al. 2004). Phytostabilization strategies may reduce the dispersion of uranium (U) and the environmental risks of U-contaminated soils. *Glomus intraradices* increased root U concentration and content, but decreased shoot U concentrations. AM fungi and root hairs improved not only P acquisition but also root uptake of U, and the mycorrhiza generally decreased U translocation from plant root to shoot (Rufyikiri et al. 2004; Chen et al. 2005a, b). For ericaceous mycorrhizas, clear host protection has been observed in, e.g. *Calluna*, *Erica*, and *Vaccinium* spp. growing on Cu- and Zn-polluted and/or naturally metalliferous soils, the fungus preventing metal translocation to plant shoots (Bradley et al. 1981, 1982; Smith and Read 1997). The development of stress-tolerant plant-mycorrhizal associations may therefore be a promising strategy for phytoremediation and soil amelioration (Schutzendubel and Polle 2002; Perotto et al. 2002; Martino et al. 2003; Cairney and Meharg 2003).

The importance of mycorrhizas in plant phosphorus nutrition has been appreciated for a long time because of their ability to dissolve and transform calcium-containing insoluble compounds and minerals (calcium phosphates, carbonate and sulphate) (Callot et al. 1985a, b; Lapeyrie et al. 1990, 1991; Gharieb and Gadd 1999). However, toxic metal mineral solubilization has received little attention, though this should be considered in any revegetation, natural attenuation or phytoremediation strategies. The ectomycorrhizal fungi *Suillus granulatus* and *Pisolithus tinctorius* can promote release of cadmium and phosphorus from rock phosphate (Leyval and Joner 2001) while the ericoid mycorrhizal fungus

Oidiodendron maius can solubilize zinc oxide and phosphate (Martino et al. 2003). Many ericoid mycorrhizal and ectomycorrhizal fungi are able to solubilize zinc, cadmium, copper phosphates and lead chlorophosphate (pyromorphite) releasing phosphate and metals (Fomina et al. 2004). Both non-mycorrhizal *Pinus sylvestris* and pines infected with the ectomycorrhizal *Paxillus involutus* were able to enhance zinc phosphate dissolution, withstand metal toxicity, and acquire the mobilized phosphorus, increasing the phosphorus amount in shoots when zinc phosphate was present in the growth matrix (Fomina et al. 2006).

15.7.5 Nanoparticle Formation[3] and Nano-Biotechnology

Metal-containing micro-/nanoparticles have applications as new ceramic–metal (cermet) or organic–metal (orgmet) composites or structured materials for a variety of applications (Hennebel et al. 2009). The use of metal-accumulating microbes for the production of nanoparticles, and their assembly, may allow control over size, morphology, composition and crystallographic orientation. The production of such biomimetic materials is relevant to the production of new advanced materials, with applications in metal and radionuclide bioremediation, antimicrobial treatments (e.g. nano-silver), solar energy and electrical battery applications, and microelectronics (Dameron et al. 1989; Klaus-Joerger et al. 2001). Because of their high specific surface area and high catalytic properties, biogenic metal products also offer potential for sorption and degradation of organic contaminants, as well as a variety of other applications, e.g. electricity generation in fuel cells, novel catalysts and sensors. Biogenic Mn oxides can sequester metals like Pb, Zn, Co, Ni, As and Cr and also oxidize certain organic pollutants (Hennebel et al. 2009). In contrast to bacteria, rather less attention has been given to fungal systems in this context although fungal reductive transformations of metalloids, Ag and Au species to nano- or colloidal forms are well known, as well as metal-containing reactive crystallites (Dameron et al. 1989) and Mn oxides (Miyata et al. 2004, 2007). Urease-positive fungi can be used for the precipitation of metal-containing carbonates, some in nanoscale dimensions, which also provides a means of metal biorecovery and purification (Li et al. 2014, 2015).

15.7.6 Soil Treatment Processes and Microbial Influence

Application to soils of certain amendments that immobilize metals, e.g. lime or phosphate treatment, have demonstrated enhanced natural remediation resulting in improved vegetation growth, increased microbial activity and diversity and reduced

[3]For additional information on the application of fungi in nanoparticle formation, please refer to Chap. 13—*Fungal biosynthesis of nanoparticles, a cleaner alternative.*

offsite metal transport. However, while long-term stability of certain metal complexes and compounds has been shown in model systems (Adriano et al. 2004), the influence of plant roots, microbes and mycorrhizal associations on such stability has often been neglected. In nature, apatite $Ca_5(PO_4)_3(F,Cl,OH)$, pyromorphite $Pb_5(PO_4)_3Cl$, mimetite $Pb_5(AsO_4)_3Cl$ and vanadinite $Pb_5(VO_4)_3Cl$ are the most common prototypes of the apatite family of minerals. These minerals, as well as others belonging to the group, hold promise for stabilization and recycling of industrial and nuclear waste and have been explored for treatment of lead-contaminated soils and waters (Ruby et al. 1994; Cotter-Howells 1996; Cotter-Howells and Caporn 1996; Ioannidis and Zouboulis 2003; Manning 2008; Oelkers and Montel 2008; Oelkers and Valsami-Jones 2008). Hence, the stability of these minerals is of interest in any remediation strategy to reduce the effects of potentially toxic elements, like Pb, V and As in soil. In particular, pyromorphite, the most stable lead phosphate mineral under a wide range of geochemical conditions, has often been suggested as a means to reduce the bioavailability of lead. However, solubilization of pyromorphite, and formation of lead oxalate, by several fungi demonstrates that pyromorphite may not be as effective at immobilizing lead as some previous studies have suggested (Sayer et al. 1999; Fomina et al. 2004). Similarly, despite the insolubility of vanadinite, fungi exerted both biochemical and biophysical effects on the mineral including etching, penetration and formation of new biominerals (Ceci et al. 2015a). Lead oxalate was precipitated by *Aspergillus niger* during the bioleaching of natural and synthetic vanadinite, as well as mimetite, and this suggested a general fungal mechanism for the transformation of lead-containing apatite group minerals (e.g. vanadinite, pyromorphite, mimetite) (Ceci et al. 2015a, b). This pattern of fungal bioweathering of lead apatites can be extended to other metal apatites, such as calcium apatite $[Ca_5(PO_4)_3(OH,F,Cl)]$. Here, the formation of monohydrated (whewellite) and dihydrated (weddellite) calcium oxalate can be mediated by many different fungal species (Burford et al. 2006; Guggiari et al. 2011; Pinzari et al. 2010, 2013; Gadd et al. 2014). The ability of free-living and mycorrhizal fungi to transform toxic metal-containing minerals should therefore be taken into account in risk assessments of the long-term environmental consequences of in situ chemical remediation techniques, revegetation strategies or natural attenuation of contaminated sites. The bioweathering potential of fungi has been envisaged as a possible means for the bioremediation of asbestos rich soils. Several fungi could extract iron from asbestos mineral fibres (e.g. 7.3 % from crocidolite and 33.6 % from chrysotile by a *Verticillium* sp.), thereby removing the reactive iron ions responsible for DNA damage (Daghino et al. 2006).

15.8 Conclusions

Geomycological roles of fungi have often been neglected in wider geomicrobiological contexts but they are of significant importance of in several key areas. These include organic and inorganic transformations, nutrient and element cycling, rock and mineral transformations, bioweathering, mycogenic biomineral formation, fungal-clay interactions and metal-fungal interactions. It is probably within the terrestrial environment where fungi have the greatest geochemical influence especially when considering soil, rock and mineral surfaces, and the plant root–soil interface. However, they are also important in aquatic habitats and since they are now recognized as significant components of aquatic sediments, their importance here may be underestimated. Mutualistic relationships of fungi with phototrophic organisms, lichens (algae, cyanobacteria) and mycorrhizas (plants), are of special significance as geoactive agents. Transformations of metals and minerals are central to many geomycological processes, and fungi possess many properties that can effect changes in metal speciation, toxicity and mobility, as well as mineral formation or mineral dissolution. Such mechanisms are important in natural biogeochemical cycles for metals as well as associated elements in biomass, soil, rocks and minerals, e.g. sulphur and phosphorus, and metalloids, actinides and metal radionuclides. Apart from being important in natural biosphere processes, metal and mineral transformations can have beneficial or detrimental consequences in a human context. Some fungal transformations have beneficial applications in environmental biotechnology, e.g. in metal and radionuclide leaching, biorecovery, detoxification, and bioremediation, and in the production or deposition of biominerals or metallic elements with catalytic or other properties in nanoparticle, crystalline or colloidal forms. Such substances may be relevant to the development of novel biomaterials for technological and antimicrobial purposes. Metal and mineral transformations may also result in adverse effects when these processes result in spoilage and destruction of natural and synthetic materials, rock and mineral-based building materials (e.g. concrete), acid mine drainage and associated metal pollution, biocorrosion of metals, alloys and related substances, and adverse effects on radionuclide speciation, mobility and containment. The ubiquity and importance of fungi in biosphere processes underline the importance of geomycology as an interdisciplinary subject area within microbiology and mycology and areas of Earth Sciences and mineralogy concerned with abiotic and biotic interfaces.

Acknowledgments The author gratefully acknowledges research support from the Natural Environment Research Council (NE/M010910/1 (TeaSe); NE/M011275/1 (COG3)), the National Natural Science Foundation of China (U1503281), the Biotechnology and Biological Sciences Research Council, the Royal Societies of London and Edinburgh, CCLRC Daresbury SRS, British Nuclear Fuels plc, the National Nuclear Laboratory and Nuclear Decommissioning Agency, and an award under the 1000 Talents Plan with the Xinjiang Institute of Geography and Ecology, Chinese Academy of Sciences, Urumqi, China.

References

Adamo P, Violante P (2000) Weathering of rocks and neogenesis of minerals associated with lichen activity. Appl Clay Sci 16:229–256

Adamo P, Vingiani S, Violante P (2002) Lichen-rock interactions and bioformation of minerals. Dev Soil Sci 28B:377–391

Adeyemi AO, Gadd GM (2005) Fungal degradation of calcium-, lead- and silicon-bearing minerals. Biometals 18:269–281

Adriaensen K, Vralstad T, Noben JP, Vangronsveld J, Colpaert JV (2005) Copper-adapted *Suillus luteus*, a symbiotic solution for pines colonizing Cu mine spoils. Appl Environ Microbiol 71:7279–7284

Adriano DC, Wenzel WW, Vangronsveld J, Bolan NS (2004) Role of assisted natural remediation in environmental cleanup. Geoderma 122:121–142

Arnott HJ (1995) Calcium oxalate in fungi. In Khan SR (ed) Calcium oxalate in biological systems, pp 73–111. CRC Press, Boca Raton, FL

Arocena JM, Glowa KR, Massicotte HB, Lavkulich L (1999) Chemical and mineral composition of ectomycorrhizosphere soils of subalpine fir (*Abies lasiocarpa* (Hook.) Nutt.) in the AE horizon of a Luvisol. Can J Soil Sci 79:25–35

Arocena JM, Zhu LP, Hall K (2003) Mineral accumulations induced by biological activity on granitic rocks in Qinghai Plateau, China. Earth Surf Proc Landforms 28:1429–1437

Baldrian P (2003) Interaction of heavy metals with white-rot fungi. Enzyme Microb Technol 32:78–91

Balogh-Brunstad Z, Keller CK, Gill RA, Bormann BT, Li CY (2008) The effect of bacteria and fungi on chemical weathering and chemical denudation fluxes in pine growth experiments. Biogeochem 88:153–167

Banfield JP, Barker WW, Welch SA, Taunton A (1999) Biological impact on mineral dissolution: application of the lichen model to understanding mineral weathering in the rhizosphere. Proc Nat Acad Sci USA 96:3404–3411

Barker WW, Banfield JF (1996) Biologically versus inorganically mediated weathering reactions: relationships between minerals and extracellular microbial polymers in lithobiotic communities. Chem Geol 132:55–69

Barker WW, Banfield JF (1998) Zones of chemical and physical interaction at interfaces between microbial communities and minerals: a model. Geomicrobiol J 15:223–244

Beech IB, Sunner J (2004) Biocorrosion: towards understanding interactions between biofilms and metals. Curr Opinion Biotechnol 15:181–186

Bellion M, Courbot M, Jacob C, Blaudez D, Chalot M (2006) Extracellular and cellular mechanisms sustaining metal tolerance in ectomycorrhizal fungi. FEMS Microbiol Lett 254:173–181

Bennett PC, Rogers JA, Hiebert FK, Choi WJ (2001) Silicates, silicate weathering, and microbial ecology. Geomicrobiol J 18:3–19

Blaudez D, Botton B, Chalot M (2000) Cadmium uptake and subcellular compartmentation in the ectomycorrhizal fungus *Paxillus involutus*. Microbiol 146:1109–1117

Bonneville S, Smits MM, Brown A, Harrington J, Leake JR, Brydson R et al (2009) Plant-driven fungal weathering: early stages of mineral alteration at the nanometer scale. Geology 37: 615–618

Bonneville S, Morgan DJ, Schmalenberger A, Bray A, Brown A, Banwart SA et al (2011) Tree-mycorrhiza symbiosis accelerate mineral weathering: evidences from nanometer-scale elemental fluxes at the hypha-mineral interface. Geochim Cosmochim Acta 75:6988–7005

Bradley R, Burt AJ, Read DJ (1981) Mycorrhizal infection and resistance to heavy metals. Nature 292:335–337

Bradley B, Burt AJ, Read DJ (1982) The biology of mycorrhiza in the Ericaceae. VIII. The role of mycorrhizal infection in heavy metal resistance. New Phytol 91:197–209

Brandl H (2001) Heterotrophic leaching. In Gadd GM (ed) Fungi in bioremediation, pp 383–423. Cambridge University Press, Cambridge

Brandl H, Faramarzi MA (2006) Microbe-metal-interactions for the biotechnological treatment of metal-containing solid waste. China Partic 4:93–97

Brehm U, Gorbushina A, Mottershead D (2005) The role of microorganisms and biofilms in the breakdown and dissolution of quartz and glass. Palaeogeo Palaeoclim Palaeoecol 219:117–129

Burford EP, Fomina M, Gadd GM (2003a) Fungal involvement in bioweathering and biotransformation of rocks and minerals. Mineral Mag 67:1127–1155

Burford EP, Kierans M, Gadd GM (2003b) Geomycology: fungal growth in mineral substrata. Mycologist 17:98–107

Burford EP, Hillier S, Gadd GM (2006) Biomineralization of fungal hyphae with calcite ($CaCO_3$) and calcium oxalate mono- and dihydrate in carboniferous limestone microcosms. Geomicrobiol J 23:599–611

Burgstaller W, Schinner F (1993) Leaching of metals with fungi. J Biotechnol 27:91–116

Cairney JWG, Meharg AA (2003) Ericoid mycorrhiza: a partnership that exploits harsh edaphic conditions. Eur J Soil Sci 54:735–740

Callot G, Guyon A, Mousain D (1985a) Inter-relation entre aiguilles de calcite et hyphes mycéliens. Agronomie 5:209–216

Callot G, Mousain D, Plassard C (1985b) Concentrations de carbonate de calcium sur les parois des hyphes mycéliens. Agronomie 5:143–150

Canovas D, Duran C, Rodriguez N, Amils R, de Lorenzo V (2003a) Testing the limits of biological tolerance to arsenic in a fungus isolated from the River Tinto. Environ Microbiol 5:133–138

Canovas D, Mukhopadhyay R, Rosen BP, de Lorenzo V (2003b) Arsenate transport and reduction in the hyper-tolerant fungus *Aspergillus* sp P37. Environ Microbiol 5:1087–1093

Ceci A, Rhee YJ, Kierans M, Hillier S, Pendlowski H, Gray N et al (2015a) Transformation of vanadinite ($Pb_5(VO_4)_3Cl$) by fungi. Environ Microbiol 17:2018–2034

Ceci A, Kierans M, Hillier S, Persiani AM, Gadd GM (2015b) Fungal bioweathering of mimetite and a general geomycological model for lead apatite mineral biotransformations. Appl Environ Microbiol 81:4955–4964

Chen BD, Jakobsen I, Roos P, Zhu YG (2005a) Effects of the mycorrhizal fungus *Glomus intraradices* on uranium uptake and accumulation by *Medicago truncatula* L. from uranium-contaminated soil. Plant Soil 275:349–359

Chen BD, Zhu YG, Zhang XH, Jakobsen I (2005b) The influence of mycorrhiza on uranium and phosphorus uptake by barley plants from a field-contaminated soil. Environ Sci Poll Res 12:325–331

Chen J, Blume H-P, Beyer L (2000) Weathering of rocks induced by lichen colonization—a review. Catena 39:121–146

Christie P, Li XL, Chen BD (2004) Arbuscular mycorrhiza can depress translocation of zinc to shoots of host plants in soils moderately polluted with zinc. Plant Soil 261:209–217

Cockell CS, Herrera A (2008) Why are some microorganisms boring? Trends Microbiol 16: 101–106

Cotter-Howells J (1996) Lead phosphate formation in soils. Environ Pollut 93:9–16

Cotter-Howells J, Caporn S (1996) Remediation of contaminated land by formation of heavy metal phosphates. Appl Geochem 11:335–342

Cromack K Jr, Solkins P, Grausten WC, Speidel K, Todd AW, Spycher G et al (1979) Calcium oxalate accumulation and soil weathering in mats of the hypogeous fungus *Hysterangium crassum*. Soil Biol Biochem 11:463–468

Cutler N, Viles H (2010) Eukaryotic microorganisms and stone biodeterioration. Geomicrobiol J 27:630–646

Daghino S, Turci F, Tomatis M, Favier A, Perotto S, Douki T et al (2006) Soil fungi reduce the iron content and the DNA damaging effects of asbestos fibers. Environ Sci Technol 40:5793–5798

Dameron CT, Reese RN, Mehra RK, Kortan AR, Carroll PJ, Steigerwald ML et al (1989) Biosynthesis of cadmium sulphide quantum semiconductor crystallites. Nature 338:596–597

De la Torre MA, Gomez-Alarcon G, Vizcaino C, Garcia MT (1993) Biochemical mechanisms of stone alteration carried out by filamentous fungi living on monuments. Biogeochemistry 19:129–147

De los Rios A, Galvan V, Ascaso C (2004) *In situ* microscopical diagnosis of biodeterioration processes at the convent of Santa Cruz la Real, Segovia, Spain. Int Biodeterior Biodegrad 51:113–120

De Rome L, Gadd GM (1987) Copper adsorption by *Rhizopus arrhizus*, *Cladosporium resinae* and *Penicillium italicum*. Appl Microbiol Biotechnol 26:84–90
Drever JI, Stillings LL (1997) The role of organic acids in mineral weathering. Coll Surf 120: 167–181
Dutton MV, Evans CS (1996) Oxalate production by fungi: its role in pathogenicity and ecology in the soil environment. Can J Microbiol 42:881–895
Edgcomb VP, Beaudoin D, Gast R, Biddle JF, Teske A (2011) Marine subsurface eukaryotes: the fungal majority. Environ Microbiol 13:172–183
Edwards KJ, Bach W, McCollom TM (2005) Geomicrobiology in oceanography: microbe-mineral interactions at and below the seafloor. Trends Microbiol 13:449–456
Ehrlich HL (1998) Geomicrobiology: its significance for geology. Earth Sci Rev 45:45–60
Ehrlich HL, Newman DK (2009) Geomicrobiology, 5th edn. CRC Press/Taylor and Francis Group, Boca Raton, FL
Eckhardt FEW (1985) Solubilisation, transport, and deposition of mineral cations by microorganisms-efficient rock-weathering agents. In: Drever J (ed) The chemistry of weathering, pp 161–173. D. Reidel Publishing Company, Dordrecht
Finlay R, Wallander H, Smits M, Holmstrom S, Van Hees P, Lian B et al (2009) The role of fungi in biogenic weathering in boreal forest soils. Fungal Biol Rev 23:101–106
Fomina M, Gadd GM (2002a) Metal sorption by biomass of melanin-producing fungi grown in clay-containing medium. J Chem Technol Biotechnol 78:23–34
Fomina M, Gadd GM (2002b) Influence of clay minerals on the morphology of fungal pellets. Mycol Res 106:107–117
Fomina M, Gadd GM (2014) Biosorption: current perspectives on concept, definition and application. Biores Technol 160:3–14
Fomina M, Ritz K, Gadd GM (2003) Nutritional influence on the ability of fungal mycelia to penetrate toxic metal-containing domains. Mycol Res 107:861–871
Fomina MA, Alexander IJ, Hillier S, Gadd GM (2004) Zinc phosphate and pyromorphite solubilization by soil plant-symbiotic fungi. Geomicrobiol J 21:351–366
Fomina M, Hillier S, Charnock JM, Melville K, Alexander IJ, Gadd GM (2005a) Role of oxalic acid over-excretion in toxic metal mineral transformations by *Beauveria caledonica*. Appl Environ Microbiol 71:371–381
Fomina MA, Alexander IJ, Colpaert JV, Gadd GM (2005b) Solubilization of toxic metal minerals and metal tolerance of mycorrhizal fungi. Soil Biol Biochem 37:851–866
Fomina M, Burford EP, Gadd GM (2005c) Toxic metals and fungal communities. In: Dighton J, White JF, Oudemans P (eds) The fungal community. Its organization and role in the ecosystem, pp 733–758. CRC Press, Boca Raton, FL
Fomina M, Charnock JM, Hillier S, Alexander IJ, Gadd GM (2006) Zinc phosphate transformations by the *Paxillus involutus*/pine ectomycorrhizal association. Microbial Ecol 52:322–333
Fomina M, Charnock JM, Hillier S, Alvarez R, Gadd GM (2007a) Fungal transformations of uranium oxides. Environ Microbiol 9:1696–1710
Fomina M, Charnock J, Bowen AD, Gadd GM (2007b) X-ray absorption spectroscopy (XAS) of toxic metal mineral transformations by fungi. Environ Microbiol 9:308–321
Fomina M, Podgorsky VS, Olishevska SV, Kadoshnikov VM, Pisanska IR, Hillier S et al (2007c) Fungal deterioration of barrier concrete used in nuclear waste disposal. Geomicrobiol J 24:643–653
Fomina M, Charnock JM, Hillier S, Alvarez R, Livens F, Gadd GM (2008) Role of fungi in the biogeochemical fate of depleted uranium. Curr Biol 18:375–377
Fomina M, Burford EP, Hillier S, Kierans M, Gadd GM (2010) Rock-building fungi. Geomicrobiol J 27:624–629
Gadd GM (1984) Effect of copper on *Aureobasidium pullulans* in solid medium: adaptation not necessary for tolerant behaviour. Trans Brit Mycol Soc 82:546–549

Gadd GM (1986) The uptake of heavy metals by fungi and yeasts: the chemistry and physiology of the process and applications for biotechnology. In: Eccles H, Hunt S (eds) Immobilisation of ions by bio–sorption. Ellis Horwood Ltd, Chichester, pp 135–147

Gadd GM (1990) Fungi and yeasts for metal binding. In: Ehrlich H, Brierley CL (eds) Microbial mineral recovery. McGraw-Hill, New York, pp 249–275

Gadd GM (1992) Metals and microorganisms: a problem of definition. FEMS Microbiol Lett 100:197–204

Gadd GM (1993a) Interactions of fungi with toxic metals. New Phytol 124:25–60

Gadd GM (1993b) Microbial formation and transformation of organometallic and organometalloid compounds. FEMS Microbiol Rev 11:297–316

Gadd GM (1999) Fungal production of citric and oxalic acid: importance in metal speciation, physiology and biogeochemical processes. Adv Microb Physiol 41:47–92

Gadd GM (2000a) Bioremedial potential of microbial mechanisms of metal mobilization and immobilization. Curr Opinion Biotechnol 11:271–279

Gadd GM (2000b) Microbial interactions with tributyltin compounds: detoxification, accumulation, and environmental fate. Sci Total Environ 258:119–127

Gadd GM (ed) (2001a) Fungi in bioremediation. Cambridge University Press, Cambridge

Gadd GM (2001b) Accumulation and transformation of metals by microorganisms. In: Rehm H-J, Reed G, Puhler A, Stadler P (eds) Biotechnology, a multi-volume comprehensive treatise, volume 10: special processes. Wiley-VCH Verlag GmbH, Weinheim, Germany, pp 225–264

Gadd GM (2001c) Metal transformations. In: Gadd GM (ed) Fungi in bioremediation. Cambridge University Press, Cambridge, pp 359–382

Gadd GM (2004a) Mycotransformation of organic and inorganic substrates. Mycologist 18:60–70

Gadd GM (2004b) Microbial influence on metal mobility and application for bioremediation. Geoderma 122:109–119

Gadd GM (2005) Microorganisms in toxic metal polluted soils. In: Buscot F, Varma A (eds) Microorganisms in soils: roles in genesis and functions. Springer-Verlag, Berlin, pp 325–356

Gadd GM (ed) (2006) Fungi in biogeochemical cycles. Cambridge University Press, Cambridge

Gadd GM (2007a) Geomycology: biogeochemical transformations of rocks, minerals, metals and radionuclides by fungi, bioweathering and bioremediation. Mycol Res 111:3–49

Gadd GM (2007b) Fungi and industrial pollutants. In: Kubicek CP, Druzhinina IS (eds) The mycota, volume IV: environmental and microbial relationships. Springer-Verlag, Berlin, pp 68–84

Gadd GM (2008a) Fungi and their role in the biosphere. In: Jorgensen SE, Fath B (eds) Encyclopedia of ecology. Elsevier, Amsterdam, pp 1709–1717

Gadd GM (2008b) Bacterial and fungal geomicrobiology: a problem with communities? Geobiol 6:278–284

Gadd GM (2008c) Transformation and mobilization of metals by microorganisms. In: Violante A, Huang PM, Gadd GM (eds) Biophysico-chemical processes of heavy metals and metalloids in soil environments. Wiley, Chichester, pp 53–96

Gadd GM (2009a) Biosorption: critical review of scientific rationale, environmental importance and significance for pollution treatment. J Chem Technol Biotechnol 84:13–28

Gadd GM (2009b) Heavy metal pollutants: environmental and biotechnological aspects. In: Schaechter M (ed) Encyclopedia of microbiology. Elsevier, Oxford, pp 321–334

Gadd GM (2010) Metals, minerals and microbes: geomicrobiology and bioremediation. Microbiol 156:609–643

Gadd GM (2011) Geomycology. In: Reitner J, Thiel V (eds) Encyclopedia of geobiology, part 7. Springer, Heidelberg, pp 416–432

Gadd GM, De Rome L (1988) Biosorption of copper by fungal melanin. Appl Microbiol Biotechnol 29:610–617

Gadd GM, Fomina M (2011) Uranium and fungi. Geomicrobiol J 28:471–482

Gadd GM, Griffiths AJ (1978) Microorganisms and heavy metal toxicity. Microb Ecol 4:303–317

Gadd GM, Griffiths AJ (1980) Effect of copper on morphology of *Aureobasidium pullulans*. Trans Brit Mycol Soc 74:387–392

Gadd GM, Mowll JL (1985) Copper uptake by yeast-like cells, hyphae and chlamydospores of *Aureobasidium pullulans*. Exp Mycol 9:230–240

Gadd GM, Raven JA (2010) Geomicrobiology of eukaryotic microorganisms. Geomicrobiol J 27:491–519

Gadd GM, Sayer JA (2000) Fungal transformations of metals and metalloids. In: Lovley DR (ed) Environmental microbe-metal interactions, pp 237–256. American Society for Microbiology, Washington

Gadd GM, White C (1989) The removal of thorium from simulated acid process streams by fungal biomass. Biotechnol Bioeng 33:592–597

Gadd GM, White C (1990) Biosorption of radionuclides by yeast and fungal biomass. J Chem Technol Biotechnol 49:331–343

Gadd GM, White C (1992) Removal of thorium from simulated acid process streams by fungal biomass: potential for thorium desorption and reuse of biomass and desorbent. J Chem Technol Biotechnol 55:39–44

Gadd GM, White C (1993) Microbial treatment of metal pollution—a working biotechnology? Trends Biotechnol 11:353–359

Gadd GM, White C, Mowll JL (1987) Heavy metal uptake by intact cells and protoplasts of *Aureobasidium pullulans*. FEMS Microbiol Ecol 45:261–267

Gadd GM, Ramsay L, Crawford JW, Ritz K (2001) Nutritional influence on fungal colony growth and biomass distribution in response to toxic metals. FEMS Microbiol Lett 204:311–316

Gadd GM, Fomina M, Burford EP (2005) Fungal roles and function in rock, mineral and soil transformations. In: Gadd GM, Semple KT, Lappin-Scott HM (eds) Microorganisms in Earth systems—advances in geomicrobiology. Cambridge University Press, Cambridge, pp 201–231

Gadd GM, Burford EP, Fomina M, Melville K (2007) Mineral transformations and biogeochemical cycles: a geomycological perspective. In: Gadd GM, Dyer P, Watkinson S (eds) Fungi in the environment. Cambridge University Press, Cambridge, pp 78–111

Gadd GM, Rhee YJ, Stephenson K, Wei Z (2012) Geomycology: metals, actinides and biominerals. Environ Microbiol Reports 4:270–296

Gadd GM, Bahri-Esfahani J, Li Q, Rhee YJ, Wei Z, Fomina M et al (2014) Oxalate production by fungi: significance in geomycology, biodeterioration and bioremediation. Fungal Biol Rev 28:36–55

Garnham GW, Codd GA, Gadd GM (1992) Accumulation of cobalt, zinc and manganese by the estuarine green microalga *Chlorella salina* immobilized in alginate microbeads. Environ Sci Technol 26:1764–1770

Gaylarde C, Morton G (2002) Biodeterioration of mineral materials. In: Bitton G (ed) Environmental microbiology, vol 1, pp 516–528. Wiley, New York

Gharieb MM, Gadd GM (1999) Influence of nitrogen source on the solubilization of natural gypsum ($CaSO_4.2H_2O$) and the formation of calcium oxalate by different oxalic and citric acid-producing fungi. Mycol Res 103:473–481

Gharieb MM, Wilkinson SC, Gadd GM (1995) Reduction of selenium oxyanions by unicellular, polymorphic and filamentous fungi: cellular location of reduced selenium and implications for tolerance. J Industrial Microbiol 14:300–301

Gharieb MM, Kierans M, Gadd GM (1999) Transformation and tolerance of tellurite by filamentous fungi: accumulation, reduction and volatilization. Mycol Res 103:299–305

Gharieb MM, Sayer JA, Gadd GM (1998) Solubilization of natural gypsum ($CaSO_4.2H_2O$) and the formation of calcium oxalate by *Aspergillus niger* and *Serpula himantioides*. Mycol Res 102:825–830

Gleeson DB, Clipson NJW, Melville K, Gadd GM, McDermott FP (2005) Mineralogical control of fungal community structure in a weathered pegmatitic granite. Microbial Ecol 50:360–368

Gleeson DB, Kennedy NM, Clipson NJW, Melville K, Gadd GM, McDermott FP (2006) Mineralogical influences on bacterial community structure on a weathered pegmatitic granite. Microbial Ecol 51:526–534

Gleeson D, McDermott F, Clipson N (2007) Understanding microbially active biogeochemical environments. Adv Appl Microbiol 62:81–104

Gleeson DB, Melville K, McDermott FP, Clipson NJW, Gadd GM (2010) Molecular characterization of fungal communities in sandstone. Geomicrobiol J 27:559–571

Gohre V, Paszkowski U (2006) Contribution of the arbuscular mycorrhizal symbiosis to heavy metal phytoremediation. Planta 223:1115–1122

Golubic S, Radtke G, Le Campion-Alsumard T (2005) Endolithic fungi in marine ecosystems. Trends Microbiol 13:229–235

Gonzalez-Chavez MC, Carrillo-Gonzalez R, Wright SF, Nichols KA (2004) The role of glomalin, a protein produced by arbuscular mycorrhizal fungi, in sequestering potentially toxic elements. Environ Poll 130:317–323

Gorbushina AA (2007) Life on the rocks. Environ Microbiol 9:1613–1631

Gorbushina AA, Broughton WJ (2009) Microbiology of the atmosphere-rock interface: how biological interactions and physical stresses modulate a sophisticated microbial ecosystem. Ann Rev Microbiol 63:431–450

Gorbushina AA, Boettcher M, Brumsack HJ, Krumbein WE, Vendrell-Saz M (2001) Biogenic forsterite and opal as a product of biodeterioration and lichen stromatolite formation in table mountain systems (tepuis) of Venezuela. Geomicrobiol J 18:117–132

Gorbushina AA, Krumbein WE, Hamann R, Panina L, Soucharjevsky S, Wollenzien U (1993) On the role of black fungi in colour change and biodeterioration of antique marbles. Geomicrobiol J 11:205–221

Grote G, Krumbein WE (1992) Microbial precipitation of manganese by bacteria and fungi from desert rock and rock varnish. Geomicrobiol J 10:49–57

Gu JD (2009) Corrosion, microbial. In: Schaechter M (ed) Encyclopedia of microbiology, 3rd edn, pp 259–269. Elsevier, Amsterdam

Gu JD, Ford TE, Berke NS, Mitchell R (1998) Biodeterioration of concrete by the fungus *Fusarium*. Int Biodet Biodegrad 41:101–109

Guggiari M, Bloque R, Aragno M, Verrecchia E, Job D, Junier P (2011) Experimental calcium-oxalate crystal production and dissolution by selected wood-rot fungi. Int Biodet Biodegrad 65:803–809

Haas JR, Purvis OW (2006) Lichen biogeochemistry. In: Gadd GM (ed) Fungi in biogeochemical cycles, pp 344–276. Cambridge University Press, Cambridge

Hennebel T, Gusseme BD, Verstraete W (2009) Biogenic metals in advanced water treatment. Trends Biotechnol 27:90–98

Hirsch P, Eckhardt FEW, Palmer RJ Jr (1995) Methods for the study of rock inhabiting microorganisms—a mini review. J Microbiol Meth 23:143–167

Hoffland E, Kuyper TW, Wallander H, Plassard C, Gorbushina AA, Haselwandter K, Holmstrom S, Landeweert R, Lundstrom US, Rosling A, Sen R, Smits MM, Van Hees PAW, Van Breemen N (2004) The role of fungi in weathering. Front Ecol Environ 2:258–264

Howlett NG, Avery SV (1997) Relationship between cadmium sensitivity and degree of plasma membrane fatty acid unsaturation in *Saccharomyces cerevisiae*. Appl Microbiol Biotechnol 48:539–545

Hutchens E (2009) Microbial selectivity on mineral surfaces: possible implications for weathering processes. Fungal Biol Rev 23:115–121

Ioannidis TA, Zouboulis AI (2003) Detoxification of a highly toxic lead-loaded industrial solid waste by stabilization using apatites. J Hazard Mater 97:173–191

Jarosz-Wilkołazka A, Gadd GM (2003) Oxalate production by wood-rotting fungi growing in toxic metal-amended medium. Chemosphere 52:541–547

Joho M, Inouhe M, Tohoyama H, Murayama T (1995) Nickel resistance mechanisms in yeasts and other fungi. J Industrial Microbiol 14:164–168

Jongmans AG, van Breemen N, Lundstrom US, van Hees PAW, Finlay RD, Srinivasan M et al (1997) Rock-eating fungi. Nature 1997(389):682–683

Kangwankraiphaisan T, Suntornvongsagul K, Sihanonth P, Klysubun W, Gadd GM (2013) Influence of arbuscular mycorrhizal fungi (AMF) on zinc biogeochemistry in the rhizosphere of *Lindenbergia philippensis* growing in zinc-contaminated sediment. Biometals 26:489–505

Kartal SN, Katsumata N, Imamura Y (2006) Removal of copper, chromium, and arsenic from CCA-treated wood by organic acids released by mold and staining fungi. Forest Products J 56:33–37

Kierans M, Staines AM, Bennett H, Gadd GM (1991) Silver tolerance and accumulation in yeasts. Biol Metals 4:100–106

Klaus-Joerger T, Joerger R, Olsson E, Granquist C-G (2001) Bacteria as workers in the living factory: metal-accumulating bacteria and their potential for materials sciences. Trends Biotechnol 19:15–20

Koele N, Turpault M-P, Hildebrand EE, Uroz S, Frey-Klett P (2009) Interactions between mycorrhizal fungi and mycorrhizosphere bacteria during mineral weathering: budget analysis and bacterial quantification. Soil Biol Biochem 41:1935–1942

Kolo K, Claeys P (2005) *In vitro* formation of Ca-oxalates and the mineral glushinskite by fungal interaction with carbonate substrates and seawater. Biogeosciences 2:277–293

Kolo K, Keppens E, Preat A, Claeys P (2007) Experimental observations on fungal diagenesis of carbonate substrates. J Geophys Res 112:1–20

Krupa P, Kozdroj J (2004) Accumulation of heavy metals by ectomycorrhizal fungi colonizing birch trees growing in an industrial desert soil. World J Microbiol Biotechnol 20:427–430

Kumar R, Kumar AV (1999) Biodeterioration of stone in tropical environments: an overview. The J. Paul Getty Trust, Madison, USA

Landeweert R, Hoffland E, Finlay RD, Kuyper TW, Van Breemen N (2001) Linking plants to rocks: ectomycorrhizal fungi mobilize nutrients from minerals. Trends Ecol Evolution 16: 248–254

Lapeyrie F, Picatto C, Gerard J, Dexheimer J (1990) TEM Study of intracellular and extracellular calcium oxalate accumulation by ectomycorrhizal fungi in pure culture or in association with *Eucalyptus* seedlings. Symbiosis 9:163–166

Lapeyrie F, Ranger J, Vairelles D (1991) Phosphate-solubilizing activity of ectomycorrhizal fungi *in vitro*. Can J Bot 69:342–346

Leyval C, Joner EJ (2001) Bioavailability of heavy metals in the mycorrhizosphere. In: Gobran GR, Wenzel WW, Lombi E (eds) Trace elements in the rhizosphere, pp. 165–185. CRC Press, Boca Raton, FL

Leyval C, Turnau K, Haselwandter K (1997) Effect of heavy metal pollution on mycorrhizal colonization and function: physiological, ecological and applied aspects. Mycorrhiza 7:139–153

Li Q, Csetenyi L, Gadd GM (2014) Biomineralization of metal carbonates by *Neurospora crassa*. Environ Sci Technol 48:14409–14416

Li Q, Csetenyi L, Paton GI, Gadd GM (2015) $CaCO_3$ and $SrCO_3$ bioprecipitation by fungi isolated from calcareous soil. Environ Microbiol 17:3082–3097

Lian B, Wang B, Pan M, Liu C, Teng HH (2008) Microbial release of potassium from K-bearing minerals by thermophilic fungus *Aspergillus fumigatus*. Geochim Cosmochim Acta 72:87–98

Liang X, Hillier S, Pendlowski H, Gray N, Ceci A, Gadd GM (2015) Uranium phosphate biomineralization by fungi. Environ Microbiol 17:2064–2075

Liang X, Kierans M, Ceci A, Hillier S, Gadd GM (2016a) Phosphatase-mediated bioprecipitation of lead by soil fungi. Environ Microbiol 18:219–231

Liang X, Csetenyi L, Gadd GM (2016b) Lead bioprecipitation by yeasts utilizing organic phosphorus substrates. Geomicrobiol J. 33:294–307

Lisci L, Monte M, Pacini E (2003) Lichens and higher plants on stone: a review. Int Biodet Biodegrad 51:1–17

Lloyd JR, Pearce CI, Coker VS, Pattrick RADP, van der Laan G, Cutting R et al (2008) Biomineralization: linking the fossil record to the production of high value functional materials. Geobiology 6:285–297

Mandal SK, Roy A, Banerjee PC (2002) Iron leaching from china clay by fungal strains. Trans Indian Inst Metals 55:1–7

Manning DAC (2008) Phosphate minerals, environmental pollution and sustainable agriculture. Elements 4:105–108

Martino E, Perotto S, Parsons R, Gadd GM (2003) Solubilization of insoluble inorganic zinc compounds by ericoid mycorrhizal fungi derived from heavy metal polluted sites. Soil Biol Biochem 35:133–141

McMaster TJ (2012) Atomic force microscopy of the fungi-mineral interface: applications in mineral dissolution, weathering and biogeochemistry. Curr Opinion Biotechnol 23:562–569

Meharg AA (2003) The mechanistic basis of interactions between mycorrhizal associations and toxic metal cations. Mycol Res 107:1253–1265

Miyata N, Tani Y, Iwahori K, Soma M (2004) Enzymatic formation of manganese oxides by an *Acremonium*-like hyphomycete fungus, strain KR21-2. FEMS Microbiol Ecol 47:101–109

Miyata M, Tani Y, Sakata M, Iwahori K (2007) Microbial manganese oxide formation and interaction with toxic metal ions. J Biosci Bioeng 104:1–8

Morley GF, Gadd GM (1995) Sorption of toxic metals by fungi and clay minerals. Mycol Res 99:1429–1438

Newby PJ, Gadd GM (1987) Synnema induction in *Penicillium funiculosum* by tributyltin compounds. Trans Brit Mycol Soc 89:381–384

Nica D, Davis JL, Kirby L, Zuo G, Roberts DJ (2000) Isolation and characterization of microorganisms involved in the biodeterioration of concrete in sewers. Int Biodet Biodegrad 46:61–68

Oelkers EH, Valsami-Jones E (2008) Phosphate mineral reactivity and global sustainability. Elements 4:83–87

Oelkers EH, Montel J-M (2008) Phosphates and nuclear waste storage. Elements 4:113–116

Ruby MV, Davis A, Nicholson A (1994) In situ formation of lead phosphates in soils as a method to immobilize lead. Environ Sci Technol 28:646–654

Perotto S, Martino E (2001) Molecular and cellular mechanisms of heavy metal tolerance in mycorrhizal fungi: what perspectives for bioremediation? Minerva Biotechnol 13:55–63

Perotto S, Girlanda M, Martino E (2002) Ericoid mycorrhizal fungi: some new perspectives on old acquaintances. Plant Soil 244:41–53

Pinzari F, Zotti M, De Mico A, Calvini P (2010) Biodegradation of inorganic components in paper documents: formation of calcium oxalate crystals as a consequence of *Aspergillus terreus* Thom growth. Int Biodet Biodegrad 64:499–505

Pinzari F, Tate J, Bicchieri M, Rhee YJ, Gadd GM (2013) Biodegradation of ivory (natural apatite): possible involvement of fungal activity in biodeterioration of the Lewis Chessmen. Environ Microbiol 15:1050–1062

Pumpel T, Paknikar KM (2001) Bioremediation technologies for metal-containing wastewaters using metabolically active microorganisms. Adv Appl Microbiol 48:135–169

Purvis OW (1996) Interactions of lichens with metals. Science Prog 79:283–309

Purvis OW, Halls C (1996) A review of lichens in metal-enriched environments. Lichenologist 28:571–601

Purvis OW, Pawlik-Skowronska B (2008) Lichens and metals. In Stress in yeasts and filamentous fungi, pp. 175–200. Elsevier, Amsterdam

Ramsay LM, Sayer JA, Gadd GM (1999) Stress responses of fungal colonies towards metals. In: Gow NAR, Robson GD, Gadd GM (eds) The fungal colony. Cambridge University Press, Cambridge, pp 178–200

Reitner J, Schumann G, Pedersen K (2006) Fungi in subterranean environments. In: Gadd GM (ed) Fungi in biogeochemical cycles. Cambridge University Press, Cambridge, pp 377–403

Rhee YJ, Hillier S, Gadd GM (2012) Lead transformation to pyromorphite by fungi. Curr Biol 22:237–241

Rhee YJ, Hillier S, Pendlowski H, Gadd GM (2014a) Pyromorphite formation in a fungal biofilm community growing on lead metal. Environ Microbiol 16:1441–1451

Rhee YJ, Hillier S, Pendlowski H, Gadd GM (2014b) Fungal transformation of metallic lead to pyromorphite in liquid medium. Chemosphere 113:17–21
Rhee YJ, Hillier S, Gadd GM (2016) A new lead hydroxycarbonate produced during transformation of lead metal by the soil fungus *Paecilomyces javanicus*. Geomicrobiol J 33:250–260
Ritz K, Young IM (2004) Interaction between soil structure and fungi. Mycologist 18:52–59
Roeselers G, van Loosdrecht MCM, Muyzer G (2007) Heterotrophic pioneers facilitate phototrophic biofilm development. Microbial Ecol 54:578–585
Rosen K, Zhong WL, Martensson A (2005) Arbuscular mycorrhizal fungi mediated uptake of Cs-137 in leek and ryegrass. Sci Total Environ 338:283–290
Rosling A, Lindahl BD, Taylor AFS, Finlay RD (2004a) Mycelial growth and substrate acidification of ectomycorrhizal fungi in response to different minerals. FEMS Microbiol Ecol 47:31–37
Rosling A, Lindahl BD, Finlay RD (2004b) Carbon allocation to ectomycorrhizal roots and mycelium colonising different mineral substrates. New Phytol 162:795–802
Rosling A, Roose T, Herrmann AM, Davidson FA, Finlay RD, Gadd GM (2009) Approaches to modelling mineral weathering by fungi. Fungal Biol Rev 23:1–7
Rufyikiri G, Huysmans L, Wannijn J, Van Hees M, Leyval C, Jakobsen I (2004) Arbuscular mycorrhizal fungi can decrease the uptake of uranium by subterranean clover grown at high levels of uranium in soil. Environ Poll. 130:427–436
Salt DE, Smith RD, Raskin I (1998) Phytoremediation. Ann Rev Plant Physiol Plant Mol Biol 49:643–668
Sand W (1997) Microbial mechanisms of deterioration of inorganic substrates: a general mechanistic overview. Int Biodeter Biodeg 40:183–190
Santhiya D, Ting YP (2005) Bioleaching of spent refinery processing catalyst using *Aspergillus niger* with high-yield oxalic acid. J Biotechnol 116:171–184
Saratovsky I, Gurr SJ, Hayward MA (2009) The structure of manganese oxide formed by the fungus *Acremonium* sp. strain KR21-2. Geochim Cosmochim Acta 73:3291–3300
Sayer JA, Gadd GM (1997) Solubilization and transformation of insoluble metal compounds to insoluble metal oxalates by *Aspergillus niger*. Mycol Res 101:653–661
Sayer JA, Cotter-Howells JD, Watson C, Hillier S, Gadd GM (1999) Lead mineral transformation by fungi. Curr Biol 9:691–694
Scheerer S, Ortega-Morales O, Gaylarde C (2009) Microbial deterioration of stone monuments: an updated overview. Adv Appl Microbiol 66:97–139
Schutzendubel A, Polle A (2002) Plant responses to abiotic stresses: heavy metal-induced oxidative stress and protection by mycorrhization. J Exp Bot 53:1351–1365
Seaward MRD (2003) Lichens, agents of monumental destruction. Microbiol Today 30:110–112
Smith SE, Read DJ (1997) Mycorrhizal symbiosis, 2nd edn. Academic Press, San Diego
Smits M (2006) Mineral tunnelling by fungi. In: Gadd GM (ed) Fungi in biogeochemical cycles. Cambridge University Press, Cambridge, pp 311–327
Smits MM (2009) Scale matters? Exploring the effect of scale on fungal-mineral interactions. Fungal Biol Rev 23:132–137
Smits MM, Bonneville S, Benning LG, Banwart SA, Leake JR (2012) Plant-driven weathering of apatite—the role of an ectomycorrhizal fungus. Geobiology 10:445–456
Sterflinger K (2000) Fungi as geologic agents. Geomicrobiol J 17:97–124
Strasser H, Burgstaller W, Schinner F (1994) High yield production of oxalic acid for metal leaching purposes by *Aspergillus niger*. FEMS Microbiol Lett 119:365–370
Sverdrup H (2009) Chemical weathering of soil minerals and the role of biological processes. Fungal Biol Rev 23:94–100
Tazaki K (2006) Clays, microorganisms, and biomineralization. In: Bergaya F, Theng BKG, Lagaly G (eds) Handbook of clay science, developments in clay science, vol 1, pp 477–497. Elsevier, Amsterdam
Theng BKG, Yuan G (2008) Nanoparticles in the soil environment. Elements 4:395–399

Thomson-Eagle ET, Frankenberger WT (1992) Bioremediation of soils contaminated with selenium. In: Lal R, Stewart BA (eds) Advances in soil science. pp 261–309. Springer, New York

Tiano P (2002) Biodegradation of cultural heritage: decay, mechanisms and control methods. Seminar article, New University of Lisbon, Department of Conservation and Restoration, 7–12 Jan 2002 (http://www.arcchip.cz/w09/w09_tiano.pdf)

Tullio M, Pierandrei F, Salerno A, Rea E (2003) Tolerance to cadmium of vesicular arbuscular mycorrhizae spores isolated from a cadmium-polluted and unpolluted soil. Biol Fertil Soils 37:211–214

Turnau K, Gawroński S, Ryszka P, Zook D (2012) Mycorrhizal-based phytostabilization of Zn–Pb tailings: lessons from the Trzebionka mining works (Southern Poland). In: Kothe E, Varma A (eds) Bio-Geo interactions in metal-contaminated soils. Springer-Verlag, Berlin, pp 327–348

Van Breemen N, Lundstrom US, Jongmans AG (2000) Do plants drive podzolization via rock-eating mycorrhizal fungi? Geoderma 94:163–171

Verrecchia EP (2000) Fungi and sediments. In: Riding RE, Awramik SM (eds) Microbial sediments, pp. 69–75. Springer, Berlin

Verrecchia EP, Dumont JL, Rolko KE (1990) Do fungi building limestones exist in semi-arid regions? Naturwissenschaften 77:584–586

Verrecchia EP, Braissant O, Cailleau G (2006) The oxalate-carbonate pathway in soil carbon storage: the role of fungi and oxalotrophic bacteria. In Gadd GM (ed) Fungi in biogeochemical cycles, pp. 289–310. Cambridge University Press, Cambridge

Volesky B (1990) Biosorption of heavy metals. CRC Press, Boca Raton, FL

Volesky B (2007) Biosorption and me. Wat Res 41:4017–4029

Wallander H, Mahmood S, Hagerberg D, Johansson L, Pallon J (2003) Elemental composition of ectomycorrhizal mycelia identified by PCR-RFLP analysis and grown in contact with apatite or wood ash in forest soil. FEMS Microbiol Ecol 44:57–65

Wang B, Qiu Y-L (2006) Phylogenetic distribution and evolution of mycorrhizas in land plants. Mycorrhiza 16:299–363

Wang HL, Chen C (2006) Biosorption of heavy metals by *Saccharomyces cerevisiae*: a review. Biotechnol Adv 24:427–451

Wang J, Chen C (2009) Biosorbents for heavy metals removal and their future. Biotechnol Adv 27

Wei Z, Hillier S, Gadd GM (2012a) Biotransformation of manganese oxides by fungi: solubilization and production of manganese oxalate biominerals. Environ Microbiol 14:1744–1753

Wei Z, Kierans M, Gadd GM (2012b) A model sheet mineral system to study fungal bioweathering of mica. Geomicrobiol J 29:323–331

Wei Z, Liang X, Pendlowski H, Hillier S, Suntornvongsagul K, Sihanonth P et al (2013) Fungal biotransformation of zinc silicate and sulfide mineral ores. Environ Microbiol 15:2173–2186

White C, Wilkinson SC, Gadd GM (1995) The role of microorganisms in biosorption of toxic metals and radionuclides. Int Biodeter Biodegrad 35:17–40

Whitelaw MA, Harden TJ, Helyar KR (1999) Phosphate solubilization in solution culture by the soil fungus *Penicillium radicum*. Soil Biol Biochem 31:655–665

Wright JS (2002) Geomorphology and stone conservation: sandstone decay in Stoke-on-Trent. Struct Surv 20:50–61

Index

D. Purchase (ed.), *Fungal Applications in Sustainable Environmental Biotechnology*, Fungal Biology, DOI 10.1007/978-3-319-42852-9

MIX
Papier aus verantwortungsvollen Quellen
Paper from responsible sources
FSC® C105338

If you have any concerns about our products,
you can contact us on
ProductSafety@springernature.com

In case Publisher is established outside the EU,
the EU authorized representative is:
Springer Nature Customer Service Center GmbH
Europaplatz 3, 69115 Heidelberg, Germany

Printed by Libri Plureos GmbH
in Hamburg, Germany